Microbiology and Microbial Physiology

Microbiology and Microbial Physiology

Edited by Dean Watson

Syrawood Publishing House
New York

Published by Syrawood Publishing House,
750 Third Avenue, 9th Floor,
New York, NY 10017, USA
www.syrawoodpublishinghouse.com

Microbiology and Microbial Physiology
Edited by Dean Watson

International Standard Book Number: 978-1-68286-410-4 (Hardback)

Cataloging-in-publication Data

Microbiology and microbial physiology / edited by Dean Watson.
p. cm.
Includes bibliographical references and index.
ISBN 978-1-68286-410-4
1. Microbiology. 2. Microorganisms--Physiology. 3. Microbial biotechnology. I. Watson, Dean.
QR41.2 .M53 2017
579--dc23

Printed in China.

TABLE OF CONTENTS

PREFACE

Microbiology refers to the study of unicellular, acellular and multicellular organisms. The main branches of microbiology include mycology, virology, bacteriology, parasitology, etc. It often overlaps with microbial physiology, applied microbiology, etc. This book gives a detailed explanation of the various concepts and applications of the subject. It elucidates new techniques and their applications in a multidisciplinary approach. Some of the diverse topics covered in the book address the varied branches that fall under this category. As this field is emerging at a rapid pace, the contents of this text will help the readers understand the modern concepts and applications of microbiology and microbial physiology. This book is an essential guide for both academicians and those who wish to pursue this discipline further.

The purpose of the book is to provide a glimpse into the dynamics and to present opinions and studies of some of the scientists engaged in the development of new ideas in the field from very different standpoints. This book will prove useful to students and researchers owing to its high content quality.

At the end, I would like to appreciate all the efforts made by the authors in completing their chapters professionally. I express my deepest gratitude to all of them for contributing to this book by sharing their valuable works. A special thanks to my family and friends for their constant support in this journey.

Editor

1

A Conservative Amino Acid Mutation in the Master Regulator FleQ Renders *Pseudomonas aeruginosa* Aflagellate

Ruchi Jain[1], Barbara I. Kazmierczak[1,2]*

1 Department of Internal Medicine (Infectious Diseases), Yale University School of Medicine, New Haven, Connecticut, United States of America, **2** Department of Microbial Pathogenesis, Yale University School of Medicine, New Haven, Connecticut, United States of America

Abstract

Flagellar-based motility plays a critical role in *Pseudomonas aeruginosa* pathogenesis, influencing both the establishment of bacterial infection and the host's response to the pathogen. Nonetheless, aflagellate clinical strains are often isolated from acutely and chronically infected patients and include the virulent laboratory strain PA103. We determined that PA103's aflagellate phenotype is the result of a single amino acid change (G240V) in the master flagellar regulator, FleQ. This mutation, which lies just outside the Walker B box of FleQ, abrogates the ability of FleQ to positively regulate flagellar gene expression. Reversal of this seemingly conservative amino acid substitution is sufficient to restore swimming motility to PA103, despite the presence of mutations in other flagellar genes of PA103. We also investigated the consequences of restoring flagellar assembly on PA103 virulence. Although a negative correlation between flagellar assembly and Type 3 secretion system (T3SS) expression has been reported previously, we did not observe downregulation of T3SS expression or function in Fla+ PA103. Restoration of flagellar assembly did, however, amplify IL-1 signals measured during murine pulmonary infection and was associated with increased bacterial clearance. These experiments suggest that loss of flagellar motility may primarily benefit PA103 by attenuating pathogen recognition and clearance during acute infection.

Editor: Gunnar F. Kaufmann, The Scripps Research Institute and Sorrento Therapeutics, Inc., United States of America

Funding: This work was supported by a Brown-Coxe Fellowship (R.J.), the National Institute of Allergy and Infectious Diseases (R01 AI075051, R01 AI081825 (B.I.K.)) and a Burroughs Wellcome Fund Investigator in Pathogenesis of Infectious Diseases Award (B.I.K.). The content is solely the responsibility of the authors and does not necessarily represent the official views of the National Institute of Allergy and Infectious Diseases or the National Institutes of Health. The funders had no role in study design, data collection and analysis, decision to publish, or preparation of the manuscript.

Competing Interests: The authors have declared that no competing interests exist.

* E-mail: barbara.kazmierczak@yale.edu

Introduction

Pseudomonas aeruginosa is a Gram-negative human pathogen that causes acute and chronic infections in individuals with compromised local barrier defenses (e.g. burns, corneal ulcers, Cystic Fibrosis (CF)), or systemic immunocompromise [1]. Bacterial motility is crucial for *P. aeruginosa* attachment to and colonization of environmental reservoirs, as well as of host tissues during infection. *P. aeruginosa* exhibits three distinct types of motility, which are powered by flagella (swimming) [2], Type IV pili (twitching) [3] or both motility organelles (swarming) [4]. Flagella and pili also contribute to the formation and maturation of biofilms, which are relatively resistant to clearance by antibiotics and phagocytic cells *in vitro* and *in vivo* [5,6].

The innate immune system recognizes bacterial flagellin via both extra- and intracellular pathogen recognition receptors, leading to the activation of host defense mechanisms that favor bacterial clearance [7–10]. Thus it is not surprising that many bacteria, including *P. aeruginosa*, downregulate flagellar expression in the host environment [11,12]. Downregulation of flagellar gene expression results from neutrophil elastase-mediated degradation of the hook-basal-body FlgE protein and subsequent intracellular accumulation of the anti-sigma factor FlgM [13].

Flagellar gene expression in *P. aeruginosa* is under the control of a master transcriptional regulator, FleQ [14]. FleQ is an enhancer binding protein (EBP) that belongs to the NtrC family of bacterial transcriptional factors. FleQ contains an N-terminal FleQ domain, a central AAA+/ATPase σ^{54}-interaction domain and a C-terminal helix-turn-helix DNA-binding domain [15]. FleQ activates transcription of genes involved in flagellar export (*flhA*, *fliLMNOPQ*) and localization (*flhF*, *fleN*), as well as structural components of the flagellar basal body (*fliEFG*) and the FleSR two-component system in a σ^{54}-dependent manner [16]. FleQ also binds the *pel* and *psl* operon promoters, where it activates or represses the expression of these exopolysaccharide biosynthetic operon in the presence or absence of cyclic-di-GMP, respectively, in σ^{54}-independent manner [17]. Thus FleQ activity is implicated in the control of loci important for both motile and sessile phenotypes.

FleQ exhibits ATPase activity [18], which is thought to provide energy for EBP remodeling of the σ^{54}-RNA polymerase closed complex and subsequent loading of DNA template into the active site of the RNA polymerase [15]. Recent work shows that cyclic-di-GMP inhibits FleQ ATPase activity, most likely by competing with ATP for binding to the Walker A box [18]. Inhibition is enhanced when FleQ is in complex with another flagellar regulator, FleN [18], and this likely accounts for earlier observations that FleN functions as a FleQ anti-activator *in vivo* [19,20].

Most environmental and clinical isolates exhibit flagellar motility, but inactivating mutations in flagellar genes, among them FleQ, are observed in sequenced isolates from chronically infected CF patients [21–23]. Non-flagellated strains isolated from acutely infected non-CF patients are also described and include PA103, a well-studied serogroup O11 strain initially isolated from a human pneumonia patient [24]. Loss of virulence factor expression by clinical isolates is often due to mutations in master regulators. This was described for Type 3 secretion system (T3SS)-negative *P. aeruginosa* strains isolated from CF patient airways [23]. Over-expression of the T3SS transcriptional activator, ExsA, was able to restore a functional T3SS to these isolates, demonstrating that the remainder of this complex protein secretion system was intact [25]. We tested whether an analogous loss of transcriptional activator function might account for the aflagellate phenotype of PA103.

Results

PA103 expresses a mutant FleQ protein that does not support flagellar assembly

PA103, a well-studied *P. aeruginosa* strain originally isolated from a patient with acute pneumonia, does not express flagella [26], but the basis for its aflagellate phenotype is not known. We hypothesized that PA103 lacked FleQ expression or function, and tested this by amplifying and cloning the *fleQ* gene from the flagellated reference strain PAO1 and expressing it in PA103. PA103 bacteria expressing $FleQ_{PAO1}$ swam, while PA103 bacteria transformed with empty vector (VC) did not (Figure 1). This suggested that the flagellar motility genes controlled by FleQ were present and encoded functional proteins in PA103.

Since PA103 swimming motility could be rescued by expression of $FleQ_{PAO1}$, we tested whether PA103 was defective in expression of endogenous FleQ. *fleQ* mRNA levels were assayed in exponential phase PA103 and PAO1 by quantitative RT-PCR. The *fleQ* gene was expressed by both strains, although *fleQ* mRNA levels were 6±2-fold higher in PAO1 as compared to PA103. FleQ protein was detected at similar steady state levels in whole cell lysates prepared from exponential phase PA103 and PAO1 by Western blotting with anti-FleQ antiserum (Figure 2).

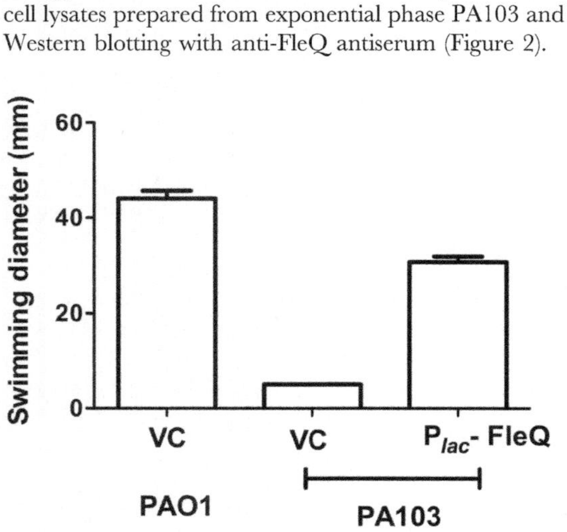

Figure 1. Expression of $FleQ_{PAO1}$ is sufficient for PA103 to swim. Swimming zone diameters were measured 16 h after inoculating 0.3% LB agar plates. Bars represent mean ± S.D. of three replicates. VC: vector control.

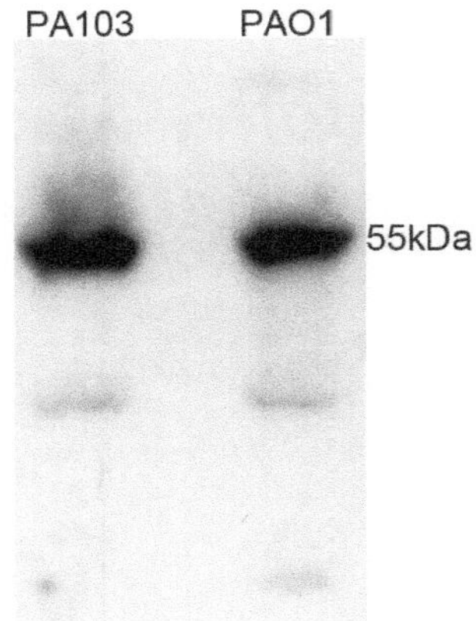

Figure 2. FleQ is expressed in both PAO1 and PA103. Whole cell lysates of PAO1 and PA103 strains (normalized to total protein) were separated by SDS-PAGE and immunoblotted with anti-FleQ antiserum.

We sequenced the *fleQ* gene from PAO1 and PA103 and identified only one missense mutation in $fleQ_{PA103}$ (G718T) leading to an amino acid substitution of Gly240Val, in addition to multiple silent mutations (T194G, A215G, T407C, T642C, T662C, T809C, C833A, T1010C, A1046G, C1379T, C1400T, A1415G and T1448C). $Glycine_{240}$ lies just before the putative Mg^{++} binding sequence (hhhhDE) of the Walker B motif of FleQ (Figure S1A). We aligned FleQ sequences from different *Pseudomonas* spp. (Figure S1B) as well as homologs from other flagellated bacteria (Figure S1C), and found that $Glycine_{240}$ was conserved in all sequences except that of $FleQ_{PA103}$.

To complete our analysis of the flagellar operons in PA103, we sequenced a PA103 genomic DNA library (2×250 bp paired-end) on a MiSeq, generating 696,082 raw reads (348,041,000 bp). De novo assembly with Velvet generated 536 contigs (n_{50} = 214,361 bp; max 595,194 bp) with median 18-fold coverage. Flagellar genes were contained within two contigs of length 179,321 bp (*flgB-fliJ*) and 40,082 bp (*fliK-motD*), and followed the same gene organization as PAO1, with the exception of the region between *flgL* and *fliC*. This region contains nine open reading frames predicted to encode glycosylation genes (Table 1); glycosylation islands of variable size appear in this position in many *P. aeruginosa* strains including the commonly studied reference strain PAK [27], which is included in this analysis (Genbank ASWU01000018). We again observed the FleQ(G240V) mutation, as well as missense mutations in other flagellar genes relative to PAO1. These are tabulated in Table 1, and clearly did not prevent flagellar assembly or function. We note that PA103 encodes an a-type flagellin, while PAO1 encodes a b-type flagellin [28]. This difference accounts for the low % identity between the *fliC* and *fliD* genes of PA103 and PAO1.

Mutation of Valine240 to Glycine in $FleQ_{PA103}$ is sufficient to restore flagellar expression and swimming to PA103

To test whether the conservative G240V amino acid change is responsible for the apparent loss of function of $FleQ_{PA103}$, site-directed mutagenesis was used to revert $Valine_{240}$ to Glycine in $FleQ_{PA103}$. The presence of the V240G substitution was confirmed by Sanger sequencing, and the resulting protein was named $FleQ(V240G)_{PA103}$. PA103 expressing the "corrected" $FleQ(V240G)_{PA103}$ allele swam as well as PA103 expressing $FleQ_{PAO1}$, while bacteria expressing $FleQ_{PA103}$ did not swim (Figure 3A). These results confirm that the single conservative

Table 1. Details of flagellar genes in PA103.

Gene name	PA locus ID (PAK locus ID*)	% identity, vis a vis PAO1 (PAK*)	Protein function
flgB	PA1077 (PAK_04305)	99% (99%)	flagellar basal-body rod protein
flgC	PA1078 (PAK_04304)	100% (100%)	flagellar basal-body rod protein
flgD	PA1079 (PAK_04303)	100% (100%)	flagellar basal-body rod modification protein
flgE	PA1080 (PAK_04302)	98% (98%)	flagellar hook protein
flgF	PA1081 (PAK_04301)	100% (100%)	flagellar basal-body rod protein
flgG	PA1082 (PAK_04300)	100% (100%)	flagellar basal-body rod protein
flgH	PA1083 (PAK_04299)	99% (99%)	flagellar L-ring protein
flgI	PA1084 (PAK_04298)	99% (99%)	flagellar P-ring protein
flgJ	PA1085 (PAK_04297)	99% (99%)	Peptidoglycan hydrolase FlgJ
flgK	PA1086 (PAK_04296)	88% (99%)	flagellar hook-associated protein FlgK
flgL	PA1087 (PAK_04295)	68% (100%)	flagellar hook-associated protein 3
orfA	(PAK_04294)	(100%)	Aminotransferase
orfB	(PAK_04293)	(100%)	Acyl carrier protein
orfC	(PAK_04292)	(99%)	3-oxoacyl-ACP
orfD	(PAK_04291)	(86%)	3-oxoacyl-ACP reductase
orfE	(PAK_04290)	(99%)	Transferase
orfF	(PAK_04289)	(99%)	2Fe-2S ferredoxin
orfG	(PAK_04288)	(98%)	Acetyltransferase
orfH	(PAK_04287)	(78%)	Methyltransferase
orfI	(PAK_04281)	(98%)	O-antigen biosynthesis protein
fliC	PA1092 (PAK_04280)	77% (94%)	A-type flagellin
flaG	PA1093 (PAK_04279)	41% (100%)	
fliD	PA1094 (PAK_04278)	43% (99%)	Flagellar cap protein
fliS	(PAK_04277)	(100%)	A-type flagellar protein FliS
fliS	PA1095 (PAK_04276)	66% (100%)	Flagellar protein FliS
fliT	PA1096 (PAK_04275)	47% (100%)	
fleQ	PA1097 (PAK_04274)	99% (99%)	Transcriptional regulator
fleS	PA1098 (PAK_04273)	100% (99%)	Two component sensor
fleR	PA1099 (PAK_04272)	100% (99%)	Two component response regulator
fliE	PA1100 (PAK_04271)	100% (98%)	Hook basal body complex protein
fliF	PA1101 (PAK_04270)	100% (99%)	M-ring outer membrane protein precursor
fliG	PA1102 (PAK_04269)	100% (100%)	Motor switch protein
fliH	PA1103 (PAK_04268)	99% (99%)	Probable assembly protein
fliI	PA1104 (PAK_04267)	100% (100%)	flagellum-specific ATP synthase
fliJ	PA1105 (PAK_04266)	99% (99%)	flagellar protein
fliK	PA1441 (PAK_03919)	97% (97%)	putative flagellar hook-length control protein
fliL	PA1442 (PAK_03918)	100% (100%)	Flagella assembly
fliM	PA1443 (PAK_03917)	100% (100%)	flagellar motor switch protein
fliN	PA1444 (PAK_03916)	100% (99%)	flagellar motor switch protein
fliO	PA1445 (PAK_03915)	99% (99%)	flagellar protein
fliP	PA1446 (PAK_03914)	100% (100%)	flagellar biosynthetic protein
fliQ	PA1447 (PAK_03913)	99% (99%)	flagellar biosynthetic protein
fliR	PA1448 (PAK_03912)	99% (99%)	flagellar biosynthetic protein
flhB	PA1449 (PAK_03911)	100% (100%)	flagellar biosynthetic protein
PA1450	PA1450 (PAK_03910)	99% (99%)	Hypothetical protein
PA1451	PA1451 (PAK_03909)	99% (99%)	Hypothetical protein
flhA	PA1452 (PAK_03908)	99% (99%)	flagellar biosynthesis protein
flhF	PA1453 (PAK_03907)	98% (99%)	flagellar biosynthesis protein
fleN	PA1454 (PAK_03906)	100% (100%)	flagellar synthesis regulator
fliA	PA1455 (PAK_03905)	100% (100%)	Sigma factor

Table 1. Cont.

Gene name	PA locus ID (PAK locus ID*)	% identity, vis a vis PAO1 (PAK*)	Protein function
cheY	PA1456 (PAK_03904)	100% (100%)	two-component response regulator
cheZ	PA1457 (PAK_03903)	100% (100%)	Chemotaxis protein
PA1458	PA1458 (PAK_03902)	99% (99%)	Probable two component sensor (CheA)
PA1459	PA1459 (PAK_03901)	99% (99%)	Chemotaxis-specific methyltransferase (CheB)
motC	PA1460 (PAK_03900)	100% (99%)	Motor protein
motD	PA1461 (PAK_03899)	99% (99%)	Motor protein

*PAK sequences from Genbank accession ASWU01000018.

amino acid change of Glycine240 to Valine abrogates FleQ function. The same series of plasmids was also introduced into a PA14 *fleQ*::Tn mutant. Again, the $FleQ_{PA103}$ allele did not complement the mutant for swimming, but either $FleQ_{PAO1}$ or $FleQ(V240G)_{PA103}$ could do so (Figure 3B).

Restoration of swimming motility was accompanied by expression of surface associated and whole cell flagellin in PA103 expressing $FleQ_{PAO1}$ and $FleQ(V240G)_{PA103}$; in contrast, no surface flagellin was detected in PA103 carrying an episomal copy of $FleQ_{PA103}$ or the vector alone (Figure 4). We also confirmed that the naturally-occurring FleQ mutation in PA103 was associated with decreased expression of FleQ-dependent transcripts. Expression of *fliL* and *flhA* in PAO1 and PA103 was measured by qRT-PCR. In line with our observations for flagellin expression, mRNA levels of *fliL* and *flhA* in PA103 were 9.5±4% and 1.6±0.7%, respectively, of those measured in PAO1. Expression of the "corrected" $FleQ(V240G)_{PA103}$ in PA103 increased mRNA levels of *flhA* 22±5.6 fold as compared to PA103 carrying the empty vector.

Although $FleQ_{PA103}$ did not support flagellar gene expression or swimming motility, over-expression of this non-functional allele in PAO1 did not alter swimming zone size of PAO1, suggesting that it has no dominant negative effect.

Restoring flagellar expression in PA103 has no measurable effect on T3SS

Soscia *et al.* previously observed an inverse relationship between flagellar assembly and Type 3 secretion system (T3SS) expression in *P. aeruginosa* [29]. We therefore tested whether restoration of flagellar assembly in PA103 altered the expression of genes encoding the T3SS effectors ExoU, ExoT or ExoS, using *lux* transcriptional reporters constructed for each of these promoters [30]. Contrary to expectations, we did not observe decreased transcriptional reporter activity when flagellar assembly was induced by the expression of $FleQ(V240G)_{PA103}$ (Figure 5A). We also examined secretion of T3SS effectors by Western blot analysis, and did not observe any consistent changes in T3SS effector secretion between PA103 strains expressing functional ($FleQ_{PAO1}$, $FleQ(V240G)_{PA103}$) or non-functional ($FleQ_{PA103}$, vector control) alleles of FleQ (Figure 5B). Lastly, T3SS-dependent cytotoxicity of bacteria expressing either $FleQ_{PAO1}$ or $FleQ_{PA103}$ toward HeLa cells was assayed by LDH release; bacteria were centrifuged onto cells after inoculation to minimize any effect of motility on efficiency of infection. Again, flagellated PA103 showed greater cytotoxicity in this assay than either aflagellate construct (Figure 5C). Thus, restoration of flagellar assembly to PA103 did not appreciably diminish T3SS expression or T3SS-dependent phenotypes *in vitro*.

Increased clearance of Fla+ PA103 *in vivo*

Bacterial flagellin is a potent activator of the innate immune system, as both surface receptors (TLR5) and intracellular receptors (mNAIP5) respond to its presence. We tested whether flagellated PA103 would elicit an altered immune response during the course of infection or show differences in bacterial clearance, using a murine acute pneumonia model that is well-characterized for PA103 [31,32]. For these experiments, the corrected *fleQ(V240G)*$_{PA103}$ gene was recombined onto the PA103 chromosome in place of the endogenous *fleQ* gene by allelic exchange, generating Fla$^+$ PA103. Restoration of surface associated flagella (Figure S2A) and swimming motility were confirmed for this strain (Figure S2B). The specific growth rate (μ_{max}) for Fla$^+$ PA103 in LB was determined and found to be 1.5±0.2 h^{-1}; this did not differ from that measured for wild-type PA103 in LB (1.4±0.2 h^{-1}).

Mice were intranasally infected with 5×10^5 cfu of PA103 or Fla$^+$ PA103. At 6h post-infection both groups of mice showed the typical recruitment of neutrophils to the airways, a response that can be seen as early as 2h post-infection and which peaks ca. 12h post-infection [31,32]. Although the number of neutrophils and macrophages recruited to the airway did not differ between the two groups of animals, Fla$^+$ PA103 was recovered in lower numbers from mouse lungs at this time point (Figure 6). We also measured IL-1α (Figure 7A) and IL-1β(Figure 7B) concentrations in BAL fluid, and found higher levels of both cytokines in mice infected with the Fla$^+$ PA103 strain.

Discussion

The basis for PA103's inability to swim was first examined some 30 years ago[26], and has remained unresolved. We describe a naturally occurring mutation in PA103 *fleQ*, the gene encoding the master regulator of flagellar biosynthesis, that abrogates function of this transcriptional regulator and accounts for the aflagellate phenotype of PA103, a virulent acute pneumonia isolate. Reversion of the single conservative amino acid change (G240V) resulting from this mutation is sufficient to restore flagellar assembly and swimming motility to PA103. The amino acid substitution in $FleQ_{PA103}$ does not appear to destabilize the protein, which is detected in comparable steady-state amounts in lysates of PA103 and PAO1. Rather, the mutation interferes with FleQ's function as a transcriptional activator. Amino acid 240 lies immediately upstream of the conserved hhhhDE motif within the Walker B box of FleQ. Mutational analysis of FleQ and homologous EBPs has demonstrated that this region of the protein is involved in hydrolyzing ATP, while Walker A box amino acids are responsible for nucleotide binding [15,18]. Disruption of FleQ ATPase activity, which is thought to power remodelling of the RNAP complex and template loading, significantly reduces its

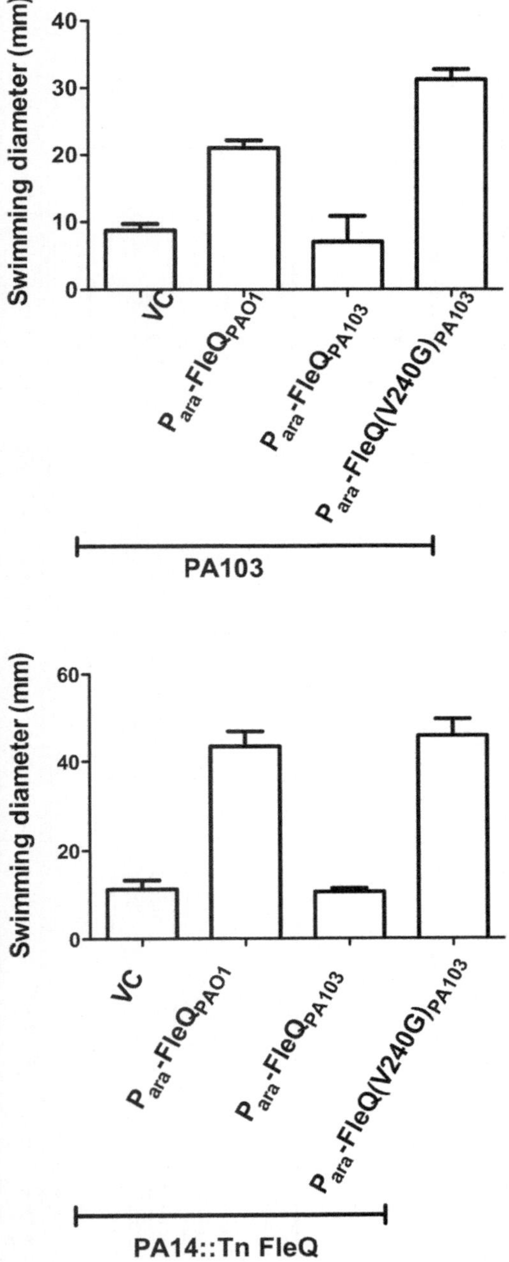

Figure 3. Reversion of the Valine240 mutation to Glycine in FleQ$_{PA103}$ restores FleQ function. (A) Swimming zones were measured for PA103 carrying pMQ72 (VC), or plasmids encoding FleQ$_{PAO1}$, FleQ$_{PA103}$, or the "corrected" allele FleQ(V240G)$_{PA103}$ 16 hr after inoculating 0.3% LB agar plates containing gentamicin and 0.2% arabinose. All genes are expressed from an arabinose inducible promoter (P$_{ara}$). Bars represent mean ± SD of three replicates. (**B**) The same set of plasmids was transformed into PA14 *fleQ*::Tn. Swimming was measured on 0.3% LB agar containing gentamicin and 0.2% arabinose. Bars represent mean ± SD of three replicates.

ability to activate transcription, as shown by the effects of two physiological inhibitors, c-di-GMP and FleN, on flagellar gene expression.

Swimming motility has been correlated with *P. aeruginosa* virulence in murine models and in human infection [33–35].

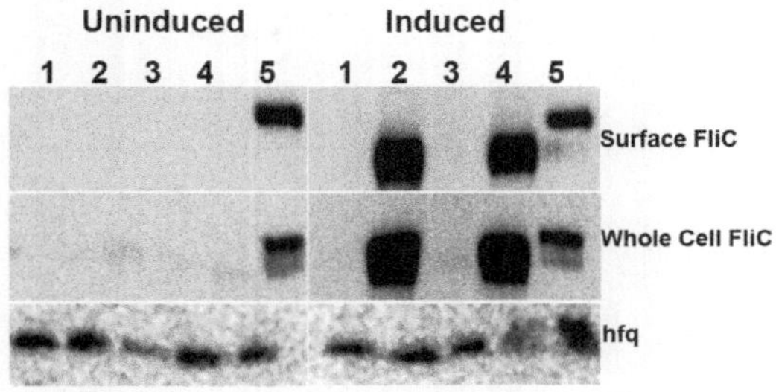

Figure 4. PA103 assembles flagella upon expression of corrected FleQ(V240G)$_{PA103}$. Western blots of surface-associated **(upper panel)** and total flagellin **(middle panel)** were prepared from bacteria grown in LB without (uninduced) or with (induced) 0.2% arabinose. Samples were normalized to total protein (as measured in the cell pellet) prior to SDS-PAGE and Western blotting. Whole cell pellets were also probed with anti-Hfq antiserum to confirm equal loading **(lower panel)**. Lanes 1-4 are PA103 carrying (1) empty pMQ72 plasmid, (2) FleQ$_{PAO1}$, (3) FleQ$_{PA103}$ and (4) FleQ(V240G)$_{PA103}$. Lane 5: PAO1 carrying an empty pMQ72 plasmid. Note difference in electrophoretic mobility between a-type flagellin of PA103 and b-type flagellin of PAO1.

Flagellar motility promotes bacterial dissemination and is necessary for initial surface colonization and biofilm formation; at the same time, flagellin is a potent immunogen and is recognized by several receptors of the innate immune system, e.g. TLR5 and NAIP5/NLRC4. *P. aeruginosa* bacteria exhibiting flagellar motility are taken up more efficiently by phagocytes than non-motile organisms, and elicit more robust innate immune responses during infection [36]. Thus, it is not surprising that bacteria can modulate the expression of flagellin in the host by a number of mechanisms that include both programmed switches into non-motile lifestyles [37] as well as regulated responses to immune system products, such as neutrophil elastase [38]. Although it is not possible to determine why a loss-of-function FleQ mutation was selected in a clinical isolate such as PA103, we did test whether restoring flagellar expression had any measurable effects on *in vitro* or *in vivo* virulence of this strain.

An earlier report by Soscia et al. suggested that T3SS activity and flagellar assembly are inversely related, as a PAO1*fliC* mutant showed increased Type 3 effector secretion relative to its isogenic wild-type parent [29]. Since PA103 secretes more T3SS effectors than 231 of 237 unique clinical isolates tested by quantitative ELISA [34], we investigated whether this was related to its lack of flagellar assembly. However, the restoration of flagellar assembly that accompanied FleQ(V240G)$_{PA103}$ expression had no measurable effect on T3SS effector transcription, secretion, or activity in cell-based assays. It is possible that the phenotype observed by Soscia and colleagues is specific to the loss of *fliC* expression in a wild-type FleQ background, and requires expression of other FleQ-dependent genes. Alternatively, T3SS activity in PA103 may be determined by regulatory inputs other than the absence or presence of flagella.

Many studies have examined the consequences of a loss of flagellar motility or expression on virulence; we took advantage of our findings to test how restoration of flagellar expression would affect PA103 fitness in an acute pneumonia model. We observed that Fla^{+} bacteria were cleared more efficiently than Fla^{-} PA103. Our *in vitro* data suggest that this is not likely the result of decreased T3SS expression by the Fla^{+} strain. Instead, our observations support the findings of Berwin and colleagues, who noted that motile *P. aeruginosa* (PA14) elicit higher levels of inflammasome activation *in vitro* and elevated IL-1β responses *in vivo* following

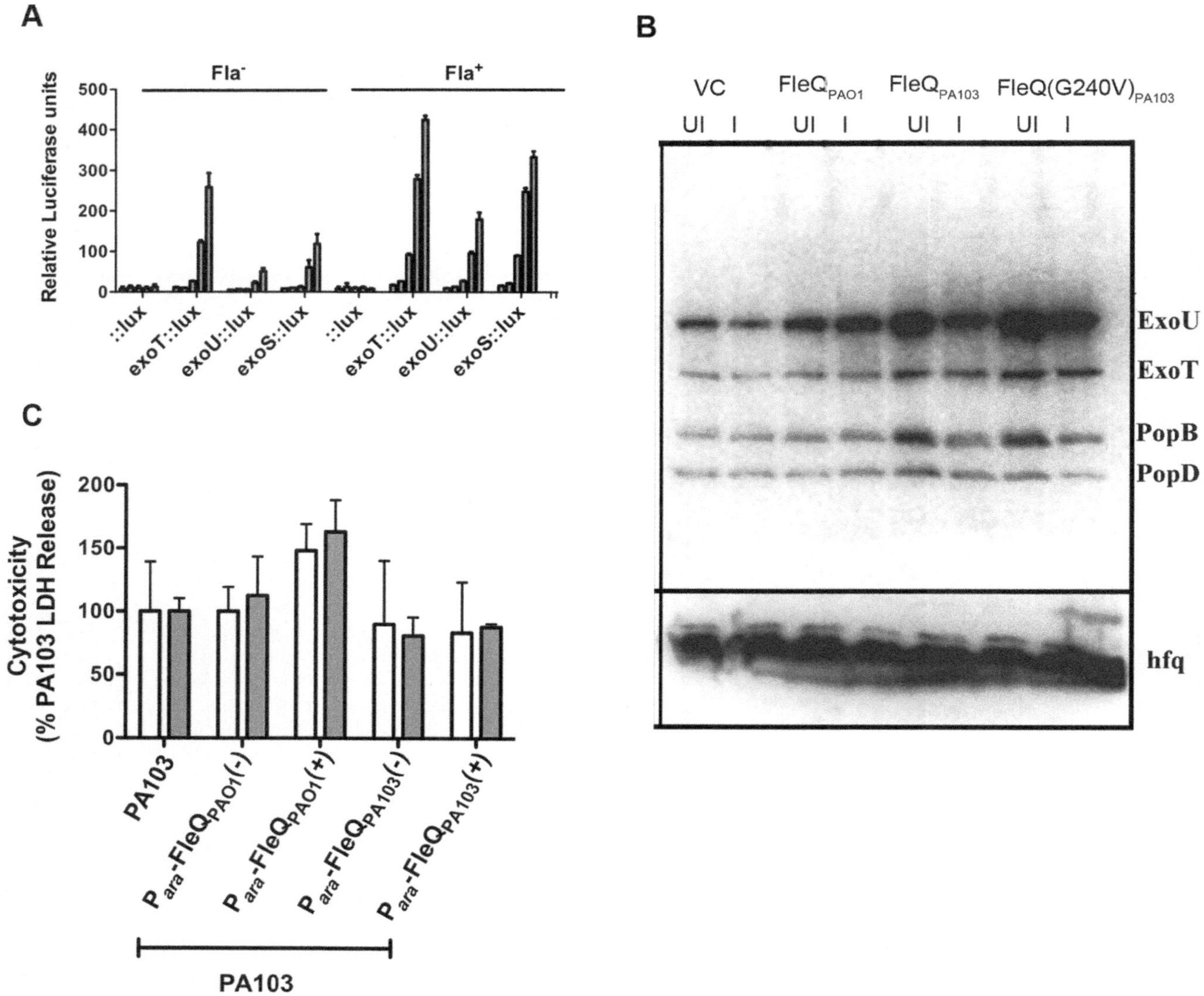

Figure 5. T3SS expression by PA103 is not altered when flagellar assembly is restored. (A) PA103 carrying *lux* transcriptional reporters for the *exoU*, *exoS* or *exoT* promoters, or a promoterless control, was transformed with P_{ara}-FleQ(V240G)$_{PA103}$ (Fla$^+$) or empty vector (Fla$^-$). RLU (luminescence/OD_{600}) was measured hourly (0–4 h) after diluting bacteria into T3SS inducing media (MinS+NTA). Bars represent the mean ± SD of triplicate samples, and the experiment was independently reproduced three times. (**B**) Western blot of secreted T3SS effectors. Culture supernatants of PA103 carrying pMQ72 (VC), FleQ$_{PAO1}$, FleQ$_{PA103}$ or FleQ(V240G)$_{PA103}$ were prepared from bacteria grown in MinS+NTA with (induced) or without (uninduced) 0.2% arabinose. Each lane was loaded with supernatant equivalent to 15 µg total protein, as measured in the cell pellet. ExoU, ExoT, PopB and PopD were detected by Western blotting with polyclonal antiserum. Cell pellets were immunoblotted with anti-Hfq antiserum to confirm equal loading. (**C**) Cytotoxicity toward HeLa cells as measured by LDH release. HeLa cells were infected with bacteria (MOI = 10) in triplicate and sampled at 1 h (white bars) and 2 h (grey bars) post-infection. Values are normalized to cytotoxicity caused by PA103 at each time point. The assay was repeated three times, and bars show the mean ± S.D. of an assay, which is representative of three independent experiments.

peritoneal infection as compared to non-motile mutants [36]. We likewise observed higher IL-1β and IL-1α levels in BAL fluid from lungs infected with Fla$^+$ PA103 than the Fla$^-$ strain. Both Fla$^+$ and Fla$^-$ infections elicited robust neutrophil recruitment to the airways, but Fla$^+$ bacteria were cleared more efficiently. As non-flagellated *P. aeruginosa* mutants are more resistant to phagocytosis than motile counterparts [39], aflagellate PA103 may benefit not only by eliciting a diminished or delayed pro-inflammatory IL-1 host response, but also by evading clearance by resident and recruited phagocytes.

Sequencing of *P. aeruginosa* clinical isolates from chronically infected patients has demonstrated that many of the virulence factors essential for the establishment of infection are mutationally inactivated in these highly adapted strains. It is not clear if these genes are lost in the absence of positive selection, or if their expression confers decreased fitness upon bacteria colonizing the CF airways. Environmental *P. aeruginosa* strains and those associated with acute infections are much more likely to express virulence factors such as motility organelles and protein secretion systems, and thus it is of interest that a hypervirulent strain such as PA103 is aflagellate. Restoring flagellar assembly and swimming motility to this strain did not have the anticipated negative effect on its expression and secretion of T3SS effectors, but did increase both host inflammatory responses to infection and bacterial

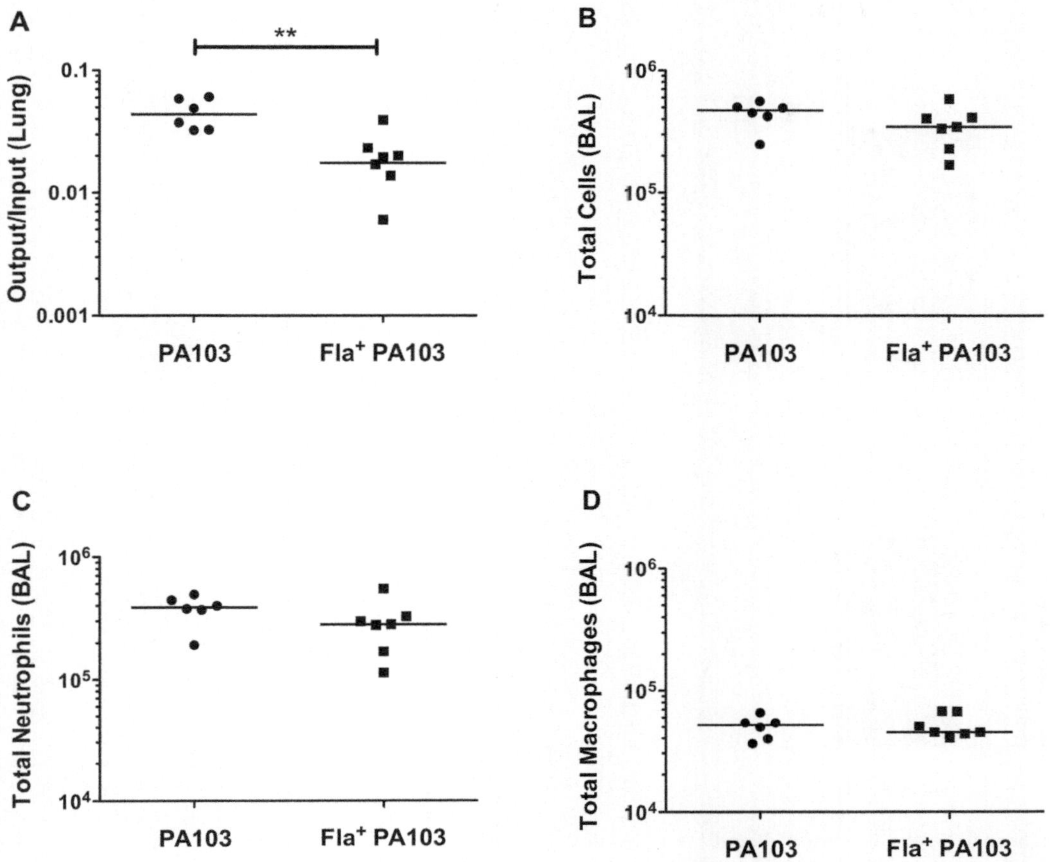

Figure 6. Flagellated bacteria are cleared more efficiently in mouse lung. C57BL/6 mice were intranasally infected with *P. aeruginosa* strain PA103 or *fla*$^{+}$ PA103, then sacrificed 6 hpi. (**A**) Lung bacterial burden expressed as number of bacteria recovered per lung (output) divided by measured inoculum administered (input). Each symbol represents an individual animal; the line indicates the geometric mean for each group. (**B**) Total cells, (**C**) neutrophils and (**D**) macrophages were counted in BAL. Each symbol represents an individual animal; lines indicate medians. Log-transformed data were analyzed using Mann-Whitney test (**, $p<0.01$).

clearance, revealing the "cost" of possessing flagella in the setting of acute infection.

Materials and Methods

Ethics statement

All animal work was conducted according to relevant national and international guidelines. Protocols for all animal studies were approved by the Yale Institutional Animal Care and Use Committee.

Bacterial strains and growth conditions

Strains and plasmids used in this study are listed in Table 2. *Escherichia coli* was cultured in Luria Broth (LB) and *Pseudomonas aeruginosa* was cultured in LB or Vogel–Bonner minimal (VBM) media [40] with antibiotics as required at the following concentrations: gentamicin, 15 μg ml^{-1} for *E. coli* or 100 μg ml^{-1} for *P. aeruginosa*; ampicillin, 100 μg ml^{-1} for *E. coli*; carbenicillin, 200 μg ml^{-1} for *P. aeruginosa*. To induce T3SS expression, *P. aeruginosa* was grown in MinS medium containing nitriloacetate (NTA) at 37°C with aeration [41]. Chemically competent *E. coli* DH5α was used for transformations (Invitrogen) and *P. aeruginosa* was transformed by electroporation [42]. All bacterial strains were stored at −80°C as 15% (v/v) glycerol stocks. Growth rates of *P. aeruginosa* cultures were determined as described [43].

Plasmid and Strain Construction

The *fleQ* gene was amplified from PAO1 and PA103 genomic DNA using primers FleQF and FleQR (Table 3). Amplicons were digested with EcoRI and HindIII, then cloned into pMQ72 and pUCP-SK for expression in *P. aeruginosa*; all constructs were confirmed by nucleotide sequencing (Keck Sequencing Facility, Yale). Arabinose was used at a concentration of 0.2% to induce expression of genes from the pMQ72 vector.

$FleQ_{PA103}$ was mutated to $FleQ(V240G)_{PA103}$ by changing Valine 240 to Glycine (nt: T718G) according to the QuikChange Site-Directed Mutagenesis kit protocol (Stratagene), using primers FQV240GF and FQV240GR (Table 3). $FleQ_{PA103}$ cloned in pUCP was used as a template for the mutagenesis reaction, and the introduction of only the desired mutation was confirmed by nucleotide sequencing. $FleQ(V240G)_{PA103}$ was then subcloned into pMQ72 and pEX18-Gm^{R} as a EcoR1-HindIII fragment. The V240G mutation was introduced into the endogenous chromosomal copy of *fleQ* in PA103 via allelic exchange [44]. Briefly, pEX-*fleQ*$(V240G)_{PA103}$ was transformed into *E. coli* S17-1 and was mobilized into *P. aeruginosa* by mating. Selection of double recombinants was carried out on 10% sucrose and mutant

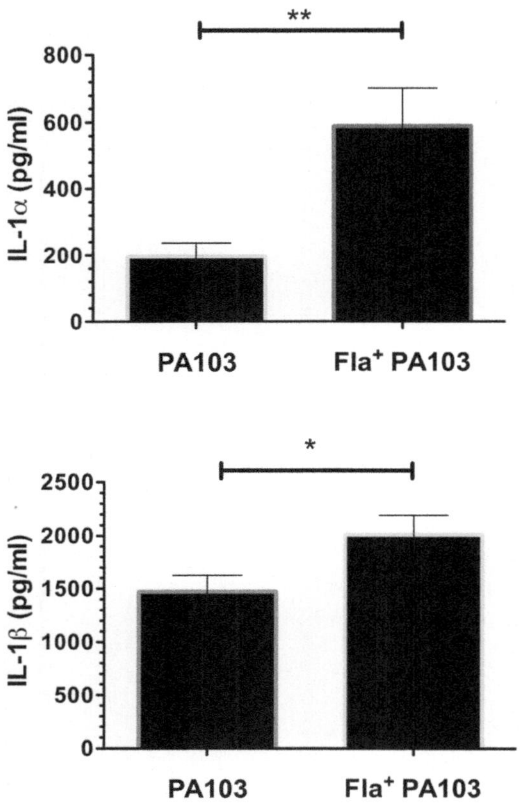

Figure 7. Fla⁺ PA103 elicits more IL-1 cytokines than aflagellate (wild-type) PA103. BAL fluid was collected 6 hpi, and IL-1α (**A**) and IL-1β (**B**) were measured by sandwich ELISA. Bars indicate mean ± S.E.M. for groups of mice (n = 6-7); these were compared by t-test (*, $p<0.05$; **, $p<0.01$).

constructs were confirmed by sequencing and phenotypic characterization.

RNA isolation and quantitative real time PCR

Bacterial RNA was isolated from early log-phase cells ($OD_{600} = 0.4$) as previously described [45]. cDNA was synthesized from 1 µg RNA using iScript cDNA Synthesis kit (BioRad). For qRT-PCR, 1 µl of cDNA was added to SYBRGreen qPCR mastermix (BioRad) containing 500 nM of forward (-RTF) and reverse (-RTR) primers (Table 3). Transcript levels of all genes tested were normalized to transcript levels of the ribosomal gene, *rplU*. Data presented are the mean ± S.D. of triplicate determinations from two independent experiments.

Whole Genome Sequencing and Sequence analysis

Bacterial genomic DNA was prepared for sequencing on the Illumina MiSeq by the Yale Center for Genomic Analysis using the TruSeq DNA LT Sample Prep Kit (Illumina, Inc. San Diego, CA USA). Steps were performed as described in the TruSeq DNA Sample Preparation Guide (#15026486 Rev.C, July 2012). Purified libraries were barcoded, pooled and sequenced on the MiSeq using a 2×250 paired-end protocol. Initial basecalls were converted to fastq files using MiSeq CASAVA software suite, demultiplexed and clipped for adaptors. Sequences were examined using FastQC (Galaxy). Reads were mapped to the *P. aeruginosa* PAO1 reference genome with the BWA (v. 0.6.2) software package using default parameters [46,47]. Bowtie (v. 2.1.0) was used to align reads [48], and SNPs were called using the Samtools mpileup [49]. We also carried out a de novo assembly with Velvet (v. 1.1.04) [50] using default parameters, and *k*-mer length of 35. Contigs containing flagellar genes were identified with nucleotide BLAST (blastn) against *Pseudomonas aeruginosa* genomes. Restriction enzyme site analyses of DNA sequences and multiple-sequence alignments (with CLUSTAL W) were performed using the BioEdit sequence alignment editor (version 7.0; Tom Hall) or MacVector (Version 11.1). Secondary structure analysis of the protein

Table 2. Strains and plasmids used in this study.

Strains	**Description/Relevant genotype**	**Source**
E. coli		
DH5α	*supE44 ΔlacU169*(Φ80d*lacZ*ΔM15) *hsdR17 thi-1 relA1 recA1*	Invitrogen
S17.1	*thi pro hsdR recA* RP4-2 (Tc::Mu) (Km::Tn7)	[55]
P. aeruginosa		
PA103		[24]
PAO1		[56]
PA14*fleQ*::Tn		G. O'Toole
Fla⁺-PA103	PA103 with "corrected" FleQ(V240G)$_{PA103}$ allele replacing endogeneous allele	This work
Plasmids		
pMQ72	Expression vector; Gm^r URA3 P$_{BAD}$; *araC*	[57]
pUCP-SK	Expression vector; Ap^r (Cb^r)	[58]
pEX18-Gm	Allelic exchange vector; Gm^r *sacB oriT*	[59]
pEX-*fleQ*(V240G)$_{PA103}$	"Corrected" FleQ(V240G)$_{PA103}$ cloned in pEX18 vector; Gm^r	This work
P$_{lac}$-FleQ$_{PAO1}$	FleQ$_{PAO1}$ cloned in pUCP-SK under *lac* promoter; Cb^r	This work
P$_{ara}$- FleQ$_{PAO1}$	FleQ$_{PAO1}$ cloned under pBAD promoter in pMQ72; Gm^r	This work
P$_{ara}$- FleQ$_{PA103}$	FleQ$_{PA103}$ cloned under pBAD promoter in pMQ72; Gm^r	This work
P$_{ara}$- FleQ(V240G)$_{PA103}$	FleQ(V240G)$_{PA103}$ cloned under pBAD promoter pMQ72; Gm^r	This work

Table 3. Primers used in this study.

Primer Name	Primer Sequence (5′-3′)
FQRTF1	GCGCGAAACCAAACTCTTGCTGAT
FQRTR1	TCGCCGAGGAAGTTGAGAATGACT
FliLRTF	GCTGTTCTCGAGTCAGAGTTT
FliFRTR	TTGGCCGGTTTCCTTCTT
FlhARTF	CGGACAACAAGCAAGTCACCATCG
FlhARTR	ATCGGCGGCGAAGAAGCGTTTGAC
FQV240GF	CGAGCTGGCCAACGGCGGCACCCTGTTCCTCGACG
FQV240GR	CGTCGAGGAACAGGGTGCCGCCGTTGGCCAGCTCG
FleQF	GACGAATTCGATGCAAGGCAGCTGATCAAAATGTGGCGCG
FleQR	GAAAATCAAAAGCTTGCGAAACGACCTGTCAATCATC

sequence was carried out using NCBI CD Search [51]. Nucleotide sequences for regions encoding the flagellar genes *flgB-fliJ* and *fliK-motD* have been deposited to Genbank (KJ411884 and KJ411885).

Swimming Assay

Swimming assays were performed by spotting 5 µl of an overnight bacterial culture on 0.3% LB agar plates supplemented with antibiotics (and 0.2% arabinose as appropriate). The swimming zone was measured after overnight incubation at 30°C. Plates were photographed with an Image Station 2000R (Kodak).

Surface expression of flagellin

Surface flagella were harvested as previously described [52]. Briefly, *P. aeruginosa* strains were grown at 37°C in LB (with antibiotics as required), harvested by centrifugation, and resuspended in 50 mM sodium phosphate (pH 7.0) containing 10 mM $MgCl_2$. Flagella were sheared by blending bacterial suspensions for 20 seconds in a Waring blender. Flagellar detachment was monitored by viewing cells under the microscope for loss of swimming. Samples were centrifuged at 12,000×*g* for 30 min at 4°C to pellet cell bodies. The supernatants containing sheared flagella were separated by SDS-PAGE and blotted to PVDF. Flagellin was detected using anti-FliC antiserum (1:4000) [53]. A goat antirabbit IgG horseradish peroxidase conjugate was used as a secondary antibody (1:4000, Bio-Rad). Detection was performed using ECL substrate (100 mM Tris-HCL pH 8.5, 250 mM luminol, 90 mM coumarate, 0.009% H_2O_2), and blots were visualized and anlysed using an Image Station 2000R running Carestream MI Software Version 5.0.2 (Kodak). Protein in cell pellets was determined by BCA assay (Pierce) and supernatant equivalent to 25 µg was loaded on the gel for analysis. Rabbit polyclonal antiserum generated against *P. aeruginosa* Hfq was used to confirm equal loading of protein samples.

T3SS secretion assay

Single bacterial colonies were used to inoculate 10 mL MinS supplemented with antibiotics and arabinose, as appropriate. Cultures were grown with shaking at 37°C for 12–14 h. Cells were pelleted, resuspended in buffer [20 mM Tris.Cl, pH 8.0, 137 mM NaCl, 10% glycerol, 1% Triton X-100, 2 mM EDTA and protease inhibitors] and lysed by sonication. Total protein in cell pellets was determined using BCA Assay (Pierce). Supernatants were normalized based on total protein of corresponding cell pellets, separated by SDS-PAGE and blotted to PVDF. Secreted T3SS proteins were detected using anti-T3SS polyclonal antiserum (1:10,000) [54], followed by incubation with secondary antibody and ECL substrate as described above.

Luciferase constructs and the luminescence reading

Bacterial strains were grown in LB with aeration at 37°C for 14–16 h. Cultures were centrifuged and bacterial pellets were resuspended and washed twice in MinS containing NTA. Samples were subcultured 1:100 into 10 ml of MinS and incubated at 37°C with shaking. Luminescence was measured at 1 h intervals starting at time zero using a luminometer (Turner Designs, Sunnyvale, CA). Relative Luciferase Units (RLU) was calculated by dividing luminescence by the optical density (OD_{600}) of each culture at time of sampling.

LDH release assay

HeLa cells (ATCC) were cultured in DMEM (Gibco) supplemented with 10% heat-inactivated fetal calf serum (Atlanta Biologicals) and 10 mM L-glutamate in a 5% CO_2 humidified incubator at 37°C. For LDH assays, 2×10^5 cells/well were seeded 24 h prior to infection into 24-well tissue culture plates. Overnight bacterial cultures in LB (with antibiotics and arabinose if required) were subcultured into fresh media, grown to early exponential phase, and diluted in DMEM plus 1% FCS. HeLa cells were washed with PBS prior to infection, which was initiated by centrifuging bacteria (MOI = 10) onto Hela cell monolayers (1000×*g* for 5 minutes at RT). LDH released into culture media at different time points was measured according to manufacturer's instructions (Takara Bio).

Mouse infections

C57Bl/6 mice were purchased from the National Cancer Institute (NIH) and housed in isolator cages under specific pathogen-free conditions. Groups of 8 wk-old female mice were intranasally infected with ca. 5×10^5 CFU of early log-phase bacteria as described previously [31]. Mice were euthanized 6 h post-infection. Bronchoalveolar lavage fluid was collected and processed for cytokine assays and differential counts as previously described [31]. Lungs were aseptically removed, homogenized in PBS +0.1% Triton-X100, and passed through a sterile screen to obtain single cell suspensions. Samples were serially diluted and plated on VBM agar. Colonies were counted after 36 h incubation at 37°C.

Measurement of cytokines

IL-1α and IL-1β in BAL fluid were measured by sandwich ELISA (R&D Systems) according to the manufacturer's protocol.

Statistical analysis

Data were analyzed in GraphPad Prism (v5.0). Mann-Whitney test was used for pair-wise comparisons. p values <0.05 were considered statistically significant.

Supporting Information

Figure S1 Sequence analysis of FleQ. (A) Domain structure of FleQ. Conserved domains of FleQ were identified using the NCBI CD search tool and include an amino-terminal FleQ domain, an ATP binding domain (AAA) containing Walker A and Walker B boxes, and a helix-turn-helix (HTH) domain. **(B) Multiple sequence alignment of FleQ from Pseudomonads.** Analysis includes *P. aeruginosa* strains LESB58, UCBPP-PA14, PAO1, PAK, PA103, and *P. fluorescens*, *P. putida*, *P. syringae* and *P. stutzeri*. (**C**) **Multiple Sequence alignment of $FleQ_{PAO1}$ and $FleQ_{PA103}$ with homologous transcriptional regulators from other organisms.** Accession numbers corresponding to the proteins used in this analysis are as follows: FleQ [*Legionella pneumophila*], CAD97470; FlbD [*Caulobacter crescentus*], AAA23039; NifA [*Klebsiella pneumoniae*], YP_002237549; FleQ [*Vibrio cholerae*], WP_001961246; FleQ [*Xanthomonas campestris*], YP_363745 and FleR [*Pseudomonas aeruginosa*], NP_249790. The residue corresponding to $FleQ_{PAO1}$ G240 is boxed.

Figure S2 Fla^{+} PA103 expresses functional surface flagella. (A) Western blot of surface flagellin preps and whole cells prepared from overnight cultures of PA103, PAO1 or Fla^{+} PA103; flagellin was detected with anti-FliC polyclonal rabbit serum. Whole cell lysates were also probed with anti-Hfq antiserum to confirm equal loading. (**B**) Swimming zones of *P. aeruginosa* strains as measured on 0.3% LB agar after incubation for 16 h at 30°C. Bars represent mean ± S.D. of three independent experiments.

Acknowledgments

We are grateful to G. A. O'Toole for providing plasmid pMQ72 and the PA14 *fleQ*::Tn mutant, and to Ruben Ramphal for providing FleQ antiserum. We thank Ming-Jie Wu for excellent technical assistance.

The content is solely the responsibility of the authors and does not necessarily represent the official views of the National Institute of Allergy and Infectious Diseases or the National Institutes of Health.

Author Contributions

Conceived and designed the experiments: RJ BIK. Performed the experiments: RJ BIK. Analyzed the data: RJ BIK. Contributed reagents/materials/analysis tools: RJ BIK. Wrote the paper: RJ BIK.

References

1. Lyczak JB, Cannon CL, Pier GB (2000) Establishment of *Pseudomonas aeruginosa* infection: lessons from a versatile opportunist. Microbes Infect 2: 1051–1060.
2. Bardy SL, Ng SY, Jarrell KF (2003) Prokaryotic motility structures. Microbiology 149: 295–304.
3. Mattick JS (2002) Type IV pili and twitching motility. Annu Rev Microbiol 56: 289–314.
4. Kohler T, Curty LK, Barja F, van Delden C, Pechere JC (2000) Swarming of *Pseudomonas aeruginosa* is dependent on cell-to-cell signaling and requires flagella and pili. J Bacteriol 182: 5990–5996.
5. O'Toole G, Kaplan HB, Kolter R (2000) Biofilm formation as microbial development. Annu Rev Microbiol 54: 49–79.
6. Parsek MR, Singh PK (2003) Bacterial biofilms: an emerging link to disease pathogenesis. Annu Rev Microbiol 57: 677–701.
7. Miao EA, Ernst RK, Dors M, Mao DP, Aderem A (2008) *Pseudomonas aeruginosa* activates caspase 1 through Ipaf. Proc Natl Acad Sci U S A 105: 2562–2567.
8. Ramos HC, Rumbo M, Sirard JC (2004) Bacterial flagellins: mediators of pathogenicity and host immune responses in mucosa. Trends Microbiol 12: 509–517.
9. Smith KD, Andersen-Nissen E, Hayashi F, Strobe K, Bergman MA, et al. (2003) Toll-like receptor 5 recognizes a conserved site on flagellin required for protofilament formation and bacterial motility. Nat Immunol 4: 1247–1253.
10. Sutterwala FS, Mijares LA, Li L, Ogura Y, Kazmierczak BI, et al. (2007) Immune recognition of *Pseudomonas aeruginosa* mediated by the IPAF/NLRC4 inflammasome. J Exp Med 204: 3235–3245.
11. Palmer KL, Mashburn LM, Singh PK, Whiteley M (2005) Cystic fibrosis sputum supports growth and cues key aspects of *Pseudomonas aeruginosa* physiology. J Bacteriol 187: 5267–5277.
12. Wolfgang MC, Jyot J, Goodman AL, Ramphal R, Lory S (2004) *Pseudomonas aeruginosa* regulates flagellin expression as part of a global response to airway fluid from cystic fibrosis patients. Proc Natl Acad Sci U S A 101: 6664–6668.
13. Jyot J, Sonawane A, Wu W, Ramphal R (2007) Genetic mechanisms involved in the repression of flagellar assembly by *Pseudomonas aeruginosa* in human mucus. Mol Microbiol 63: 1026–1038.
14. Arora SK, Ritchings BW, Almira EC, Lory S, Ramphal R (1997) A transcriptional activator, FleQ, regulates mucin adhesion and flagellar gene expression in *Pseudomonas aeruginosa* in a cascade manner. J Bacteriol 179: 5574–5581.
15. Bush M, Dixon R (2012) The role of bacterial enhancer binding proteins as specialized activators of sigma54-dependent transcription. Microbiol Mol Biol Rev 76: 497–529.
16. Dasgupta N, Wolfgang MC, Goodman AL, Arora SK, Jyot J, et al. (2003) A four-tiered transcriptional regulatory circuit controls flagellar biogenesis in *Pseudomonas aeruginosa*. Mol Microbiol 50: 809–824.
17. Baraquet C, Murakami K, Parsek MR, Harwood CS (2012) The FleQ protein from *Pseudomonas aeruginosa* functions as both a repressor and an activator to control gene expression from the pel operon promoter in response to c-di-GMP. Nucleic Acids Res 40: 7207–7218.
18. Baraquet C, Harwood CS (2013) Cyclic diguanosine monophosphate represses bacterial flagella synthesis by interacting with the Walker A motif of the enhancer-binding protein FleQ. Proc Natl Acad Sci U S A 110: 18478–18483.
19. Dasgupta N, Arora SK, Ramphal R (2000) *fleN*, a Gene That Regulates Flagellar Number in *Pseudomonas aeruginosa*. J Bacteriol 182: 357–364.
20. Dasgupta N, Ramphal R (2001) Interaction of the antiactivator FleN with the transcriptional activator FleQ regulates flagellar number in *Pseudomonas aeruginosa*. J Bacteriol 183: 6636–6644.
21. Luzar MA, Thomassen MJ, Montie TC (1985) Flagella and motility alterations in *Pseudomonas aeruginosa* strains from patients with cystic fibrosis: relationship to patient clinical condition. Infect Immun 50: 577–582.
22. Mahenthiralingam E, Campbell ME, Speert DP (1994) Nonmotility and phagocytic resistance of *Pseudomonas aeruginosa* isolates from chronically colonized patients with cystic fibrosis. Infect Immun 62: 596–605.
23. Smith EE, Buckley DG, Wu Z, Saenphimmachak C, Hoffman LR, et al. (2006) Genetic adaptation by *Pseudomonas aeruginosa* to the airways of cystic fibrosis patients. Proc Natl Acad Sci 103: 8487–8492.
24. Liu Y, Davin-Regli A, Bosi C, Charrel RN, Bollet C (1996) Epidemiological investigation of *Pseudomonas aeruginosa* nosocomial bacteraemia isolates by PCR-based DNA fingerprinting analysis. J Med Microbiol 45: 359–365.
25. Dacheux D, Attree I, Toussaint B (2001) Expression of ExsA in trans confers type III secretion system-dependent cytotoxicity on noncytotoxic *Pseudomonas aeruginosa* cystic fibrosis isolates. Infect Immun 69: 538–542.
26. Montie TC, Craven RC, Holder IA (1982) Flagellar preparations from *Pseudomonas aeruginosa*: isolation and characterization. Infect Immun 35: 281–288.
27. Arora SK, Bangera MG, Lory S, Ramphal R (2001) A genomic island in *Pseudomonas aeruginosa* carries the determinants of flagellin glycosylation. Proc Natl Acad Sci U S A 98: 9342–9347.
28. Spangenberg C, Heuer T, Burger C, Tummler B (1996) Genetic diversity of flagellins of *Pseudomonas aeruginosa*. FEBS Lett 396: 213–217.
29. Soscia C, Hachani A, Bernadac A, Filloux A, Bleves S (2007) Cross talk between type III secretion and flagellar assembly systems in *Pseudomonas aeruginosa*. J Bacteriol 189: 3124–3132.
30. Laskowski MA, Osborn E, Kazmierczak BI (2004) A novel sensor kinase-response regulator hybrid regulates type III secretion and is required for virulence in *Pseudomonas aeruginosa*. Mol Microbiol 54: 1090–1103.
31. Mijares LA, Wangdi T, Sokol C, Homer R, Medzhitov R, et al. (2011) Airway epithelial MyD88 restores control of *Pseudomonas aeruginosa* murine infection via an IL-1-dependent pathway. J Immunol 186: 7080–7088.
32. Wangdi T, Mijares LA, Kazmierczak BI (2010) *In vivo* discrimination of T3SS-positive and –negative *Pseudomonas aeruginosa* via a caspase-1-dependent pathway. Infect Immun 78: 4744–4753.
33. Feldman M, Bryan R, Rajan S, Scheffler L, Brunnert S, et al. (1998) Role of flagella in pathogenesis of *Pseudomonas aeruginosa* pulmonary infection. Infect Immun 66: 43–51.
34. Ledizet M, Murray TS, Puttagunta S, Slade MD, Quagliarello VJ, et al. (2012) The ability of virulence factor expression by *Pseudomonas aeruginosa* to predict clinical disease in hospitalized patients. PLoS ONE 7: e49578.
35. Montie TC, Doyle-Huntzinger D, Craven RC, Holder IA (1982) Loss of virulence associated with absence of flagellum in an isogenic mutant of *Pseudomonas aeruginosa* in the burned-mouse model. Infect Immun 38: 1296–1298.
36. Patankar YR, Lovewell RR, Poynter ME, Jyot J, Kazmierczak BI, et al. (2013) Flagellar motility is a key determinant of the magnitude of the inflammasome response to *Pseudomonas aeruginosa*. Infect Immun 81: 2043–2052.
37. Tart AH, Wolfgang MC, Wozniak DJ (2005) The alternative sigma factor AlgT represses *Pseudomonas aeruginosa* flagellum biosynthesis by inhibiting expression of fleQ. J Bacteriol 187: 7955–7962.
38. Jyot J, Sonawane A, Wu W, Ramphal R (2007) Genetic mechanisms involved in the repression of flagellar assembly by *Pseudomonas aeruginosa* in human mucus. Mol Microbiol 63: 1026–1038.
39. Amiel E, Lovewell RR, O'Toole GA, Hogan DA, Berwin B (2010) *Pseudomonas aeruginosa* evasion of phagocytosis is mediated by loss of swimming motility and is independent of flagellum expression. Infect Immun 78: 2937–2945.
40. Vogel HJ, Bonner DM (1956) Acetylornithinase of *Escherichia coli*: partial purification and some properties. J Biol Chem 218: 97–106.
41. Nicas T, Iglewski BH (1985) The contribution of exoproducts to virulence of *Pseudomonas aeruginosa*. Can J Microbiol 31: 387–392.
42. Choi KH, Kumar A, Schweizer HP (2006) A 10-min method for preparation of highly electrocompetent *Pseudomonas aeruginosa* cells: application for DNA

fragment transfer between chromosomes and plasmid transformation J Microbiol Methods 64: 391–397.
43. van Ditmarsch D, Xavier JB (2011) High-resolution time series of *Pseudomonas aeruginosa* gene expression and rhamnolipid secretion through growth curve synchronization. BMC Microbiol 11: 140.
44. Schweizer HP, Hoang TT (1995) An improved system for gene replacement and xylE fusion analysis in *Pseudomonas aeruginosa*. Gene 158: 15–22.
45. Schuster M, Lostroh CP, Ogi T, Greenberg EP (2003) Identification, timing, and signal specificity of *Pseudomonas aeruginosa* quorum-controlled genes: a transcriptome analysis. J Bacteriol 185: 2066–2079.
46. Li H, Durbin R (2009) Fast and accurate short read alignment with Burrows-Wheeler transform. Bioinformatics 25: 1754–1760.
47. Li H, Durbin R (2010) Fast and accurate long-read alignment with Burrows-Wheeler transform. Bioinformatics 26: 589–595.
48. Langmead B, Trapnell C, Pop M, Salzberg SL (2009) Ultrafast and memory-efficient alignment of short DNA sequences to the human genome. Genome Biol 10: R25.
49. Li H, Handsaker B, Wysoker A, Fennell T, Ruan J, et al. (2009) The Sequence Alignment/Map format and SAMtools. Bioinformatics 25: 2078–2079.
50. Zerbino DR, Birney E (2008) Velvet: algorithms for de novo short read assembly using de Bruijn graphs. Genome Res 18: 821–829.
51. Marchler-Bauer A, Zheng C, Chitsaz F, Derbyshire MK, Geer LY, et al. (2013) CDD: conserved domains and protein three-dimensional structure. Nucleic Acids Res 41: D348–352.
52. Murray TS, Kazmierczak BI (2006) FlhF is required for swimming and swarming in *Pseudomonas aeruginosa*. J Bacteriol 188: 6995–7004.
53. Schniederberend M, Abdurachim K, Murray TS, Kazmierczak BI (2013) The GTPase Activity of FlhF Is Dispensable for Flagellar Localization, but Not Motility, in *Pseudomonas aeruginosa*. J Bacteriol 195: 1051–1060.
54. Li L, Ledizet M, Kar K, Koski RA, Kazmierczak BI (2005) An indirect enzyme-linked immunosorbent assay for rapid and quantitative assessment of Type III virulence phenotypes of *Pseudomonas aeruginosa* isolates. Ann Clin Microbiol Antimicrob 4: 22.
55. Simon R, Priefer U, Puhler A (1983) A broad host range mobilization system for in vivo genetic engineering: transposon mutagenesis in Gram negative bacteria. Biotechnology 1: 784–791.
56. Stover CK, Pham XQ, Erwin AL, Mizoguchi SD, Warrener P, et al. (2000) Complete genome sequence of *Pseudomonas aeruginosa* PAO1, an opportunistic pathogen. Nature 406: 959–964.
57. Shanks RM, Caiazza NC, Hinsa SM, Toutain CM, O'Toole GA (2006) *Saccharomyces cerevisiae*-based molecular tool kit for manipulation of genes from gram-negative bacteria. Appl Environ Microbiol 72: 5027–5036.
58. Watson AA, Alm RA, Mattick JS (1996) Construction of improved vectors for protein production in *Pseudomonas aeruginosa*. Gene 172: 163–164.
59. Hoang TT, Karkhoff-Schweizer RR, Kutchma AJ, Schweizer HP (1998) A broad-host-range Flp-FRT recombination system for site-specific excision of chromosomally-located DNA sequences: application for isolation of unmarked *Pseudomonas aeruginosa* mutants. Gene 212: 77–86.

2

Macrophage Control of Phagocytosed Mycobacteria Is Increased by Factors Secreted by Alveolar Epithelial Cells through Nitric Oxide Independent Mechanisms

Dagbjort H. Petursdottir*°, Olga D. Chuquimia°, Raphaela Freidl, Carmen Fernández

Department of Molecular Biosciences, The Wenner-Gren Institute, Stockholm University, Stockholm, Sweden

Abstract

Tissue-resident macrophages are heterogeneous with tissue-specific and niche-specific functions. Thus, simplified models of macrophage activation do not explain the extent of heterogeneity seen *in vivo*. We focus here on the respiratory tract and ask whether factors secreted by alveolar epithelial cells (AEC) can influence the functionality of resident pulmonary macrophages (PuM). We have previously reported that factors secreted by AEC increase control of intracellular growth of BCG in macrophages. In the current study, we also aimed to investigate possible mechanisms by which AEC-derived factors increase intracellular control of BCG in both primary murine interstitial macrophages, and bone marrow-derived macrophages and characterize further the effect of these factors on macrophage differentiation. We show that; a) in contrast to other macrophage types, IFN-γ did not increase intracellular growth control of *Mycobacterium bovis*, Bacillus Calmette-Guérin (BCG) by interstitial pulmonary macrophages although the same macrophages could be activated by factors secreted by AEC; b) the lack of response of pulmonary macrophages to IFN-γ was apparently regulated by suppressor of cytokine signaling (SOCS)1; c) AEC-derived factors did not induce pro-inflammatory pathways induced by IFN-γ e.g. expression of inducible nitric oxide synthase (iNOS), secretion of nitric oxide (NO), or IL-12, d) in contrast to IFN-γ, intracellular bacterial destruction induced by AEC-derived factors was not dependent on iNOS transcription and NO production. Collectively, our data show that PuM were restricted in inflammatory responses mediated by IFN-γ through SOCS1 and that factors secreted by AEC- enhanced the microbicidal capacities of macrophages by iNOS independent mechanisms.

Editor: Joyoti Basu, Bose Institute, India

Funding: This work was supported by grant FP7 (HEALTH-F3-2008-200732) from the European Commission. The funders had no role in study design, data collection and analysis, decision to publish, or preparation of the manuscript.

Competing Interests: The authors have declared that no competing interests exist.

* Email: dagbjort.petursdottir@su.se

° These authors contributed equally to this work.

Introduction

Macrophages are essential elements in both innate and adaptive immunity playing a crucial role in host defense. In addition to being a first line of resistance against pathogens, macrophages contribute to remodeling and repair. Tissue-resident macrophages are a heterogeneous population of immune cells that fulfill tissue-specific and niche-specific functions. These range from dedicated homeostatic functions, such as clearance of cellular debris and iron processing, to central roles in tissue immune surveillance, response to infection and the resolution of inflammation. Understanding the mechanisms that dictate tissue macrophage heterogeneity should explain why simplified models of macrophage activation do not explain the extent of heterogeneity seen *in vivo*.

In the lung tissue, although a number of immune cells secrete many of these factors, alveolar epithelial cells (AEC) and other non-immune cells are also able to produce a broad variety of factors including cytokines and chemokines that affect inflammatory responses in macrophages [1–6]. We have reported that factors secreted by AEC can influence both the phenotype and other characteristics of resident pulmonary macrophages (PuM) [7], confirming that the micro-environment provided by cytokines and other factors contribute to macrophage polarization and hence to their antimicrobial properties.

Polarized macrophages have been broadly classified into two groups: M1 (classically activated macrophages) or M2 (alternatively activated macrophages) [8]. M1 macrophages respond to type 1 inflammatory cytokines and microbial products such as lipopolysaccharide (LPS) to active inducible nitric oxide synthase (iNOS) and the production of reactive oxygen species (ROS) [8–10]. Thus, M1 cells are thought to represent a "common host response" for microbicidal capacities [8–10] while M2 cells are immunoregulatory with poor microbicidal activity [8]. Since M2 macrophages display a very broad functional spectrum, a further subcategorization has been proposed based on the *in vitro* stimuli used for the generation of the different phenotypes, where culturing with IL-4 or IL-13 yields M2a; induced by immune complexes, TLR agonists or IL-1R yield M2b; IL-10 or glucocorticoids yield M2c [11].

Previously, we have shown that treatment of PuM with AEC-derived factors affect their functional properties seen e.g. in morphological alterations demonstrated by elongated cell shape as well as increased expression of the marker macrophage mannose receptor (MMR), both features indicating M2 polarization [7,12,13]. However, treatment with AEC-derived factors also increases the PuM phagocytic activity as well as their capacity to control mycobacteria intracellular growth upon infection with *Mycobacterium bovis*, Bacillus Calmette-Guérin (BCG), characteristics more related to M1 polarization [7].

In the present study we investigated possible mechanisms for the induction of intracellular bacterial growth control in macrophages mediated by AEC-derived factors and compared this with treatment with IFN-γ, a classical inducer of M1 activation. Our data suggested that AEC-derived factors induced microbicidal functions through iNOS-independent mechanisms and that suppressor of cytokine signaling 1 (SOCS1), a known inhibitor of IFN-γ signaling, [14]was involved in the inability of PuM to respond to IFN-γ.

Materials and Methods

Mice

All mouse experiments were approved and performed in accordance with the guidelines of the Animal Research Ethics Board at Stockholm University (ethical approval ID: N463/12). The studies were performed using 8–12-week-old female C57BL/6 mice purchased from NOVA-SCB, Sweden. All animals were kept at the Animal Department of the Arrhenius Laboratories, Stockholm University, Sweden and housed in pathogen-free conditions. Mice were supervised daily and sentinel mice were used to assess and ensure pathogen free conditions in the facility. IFN-$\gamma^{-/-}$SOCS1$^{-/-}$ mice were kindly provided by M. Rottenberg (MTC, Karolinska Institutet, Stockholm, Sweden) and experiments conducted in accordance with the guidelines of Karolinska Institutet.

Bacteria

We used BCG, transformed with a dual reporter plasmid containing the human codon-optimised and fluorescence-enhanced EGFP and the bacterial luciferase A and B (luxAB) genes from *Vibrio harveyi* [15] herein called GFP-BCG. Bacterial contents are expressed as relative luminescence units (RLU). The increase of OD and RLU was followed for all batches and experiments. All batches were also routinely controlled for CFU and RLU values. The correlation between CFU and RLU has been previously published as supplementary data [16,17]. Furthermore, a correlation between CFU and RLU has even been confirmed in the experimental setup used in this study, i.e. in an assay determining intracellular survival of bacteria within macrophages (unpublished data). GFP-BCG was grown in Middlebrook 7H9 broth (DIFCO, Sparks, MD, USA) supplemented with albumin-dextrose-catalase (ADC), 0.5% glycerol, 0.05% Tween 80 (vol/vol), and 50 µg/ml hygromycin for 10–15 days. Bacteria were collected at a log phase of growth (absorbance 1.0 measured at OD_{650}) and frozen in PBS with 10% glycerol and kept at −80°C. Before infection of cell cultures, a vial was thawed and placed in culture as above for 4–5 days, reaching an early log phase (OD_{650} ~0.3) corresponding to approximately 3×10^8 CFU per ml. To determine the RLU, Decanal (0.01% v/v in ethanol) (Sigma-Aldrich) was used as a specific substrate for the bacterial enzyme luxAB. The samples were mixed immediately and luminescence measured after 15 sec in a Modulus, Turner Bio Systems luminometer.

Isolation of alveolar epithelial cells (AEC) and interstitial pulmonary macrophages (PuM)

Total pulmonary cells were prepared using Corti's protocol [18] with previously described modifications [16,17]. PuM were isolated from either C57BL/6 or IFN-$\gamma^{-/-}$SOCS1$^{-/-}$ mice [19] by positively selecting CD45$^+$ cells from total pulmonary cells using MACS (Miltenyi Biotec, Bergish Gladbach, Germany) and a subsequent adhesion isolation. After isolation, cells were cultured for 48 h in RPMI (Gibco-Invitrogen, Paisley, UK) supplemented with 10% FCS, 2 mM L-glutamine, 100 U/ml penicillin, 100 µg/ml streptomycin, 0.02 M Hepes, and 0.05 M 2-mercaptoethanol (Sigma) (complete RPMI) at 37°C and 5% CO_2. Next, cells were washed to remove non-adherent cells and debris. In average, 98% of the cells were positive for the macrophage marker F4/80, as determined by flow cytometry. AEC were obtained by depleting CD45$^+$ and CD146$^+$ cells from lung cell preparations using MACS. In average, 92-95% of these CD45$^-$CD146$^-$ cells exhibited an AEC phenotype, where approximately 22% expressed podoplanin (type I AEC) and approximately 72% expressed CD74 (type II AEC) as determined by flow cytometry. Cell viability was determined by trypan blue exclusion to be over 90%.

For the preparation of AEC-derived media, AEC were plated (5×10^4 cells/well) in 96-well flat bottom plates (Costar, NY, USA) and cultured for 48 h. After washing the cells three times, cells were again kept unstimulated for 24 h. These cell culture supernatants were collected, pooled, filtered through a 0.2 µm filter and stored at −70°C until use. These cell culture supernatants were named AEC_{sup}.

Isolation of alveolar macrophages (AM) and peritoneal exudate macrophages (PEM)

AM were isolated from bronchoalveolar lavage (BAL) using a previously described protocol [20] with some modifications. Briefly, a bronchoalveolar lavage (BAL) was performed with 1–2 ml of pre-warmed (37°C) PBS-0.5 mM EDTA (Gibco) via the trachea 3-4 times. After lavage, BAL fluids were pooled and put on ice. PEM were isolated as previously described [20]. The mouse's skin was removed carefully leaving the peritoneal membrane intact. The peritoneal cavity was rinsed twice with RPMI (5 ml) and the collected peritoneal exudate cells were pooled and placed on ice. BAL and peritoneal exudate cells (PEC) were then washed twice in ice-cold RPMI and plated in 48-well flat bottom plates (Costar, NY, USA) at a cell concentration of 2×10^5 cells/well (BAL) and 1.5×10^6 cells/well (PEC) for 2 h in complete RPMI. Cells were washed 2–3 times to eliminate debris and non-adherent cells and cultured in complete RPMI overnight. After 24 h, ~98 ± 1% of BAL cells and 98 ± 1% of PEC were positive for the macrophage marker F4/80 and were considered as AM and PEM, respectively. A further characterization showed that, 83 ± 8% of AM were F4/80$^+$ CD11c$^+$, a typical AM phenotype, 15 ± 3% were F4/80$^+$CD11b$^+$, and 8±1% were F4/80$^+$CD11b$^+$CD11c$^+$ whereas approximately 99 ± 1% of PEM were F4/80$^+$ CD11b$^+$.

Generation of bone marrow derived macrophages (BMM)

BMM were harvested and frozen as previously described [21]. After thawing, the bone marrow cells were cultured in complete RPMI supplemented with 20% L929 cell conditioned medium (as a source of macrophage colony-stimulating factor) and cultured for 6–7 days, replacing medium every second day. Before use, growth factor-containing media were removed and the BMM cultured in complete RPMI for 24 h. BMM from caspase 1$^{-/-}$ mice were kindly provided by S. Musciol and B. Henriques-Normark (MTC, Karolinska Institutet, Solna, Sweden)

Infection and intracellular bacterial growth

BMM, PuM, AM and PEM were cultured in 48 or 12-well plates as described above 1×10^5 and 4×10^5 cells per well respectively. Cells were kept with complete RPMI without antibiotics for 24 h prior to infection with GFP-BCG at a MOI 10:1 (bacteria:cell) for 4 h. After infection, cells were washed three times and treated with gentamicin (100 µg/ml) for 30–60 min at 37°C. After further washing, cells were either kept untreated or treated with 20 ng/ml of IFN-γ (Mabtech, Sweden), or IL-1β (1 or 10 ng/ml) or AEC_{sup} (diluted 1:2 in cell culture media). For inhibition of iNOS 1 mM of N^G-monomethyl-L-arginine (NMMLA) (Sigma) was added together with either IFN-γ or AEC_{sup}. For reducing reactive oxygen species (ROS) in the cultures, cells were incubated with the ROS scavengers superoxide dismutase (SOD, 15 U/ml, Sigma-Aldrich) and catalase 500 µg/ml (Sigma-Aldrich). The bacterial loads were measured by determining RLU and data is displayed as RLU/10^6 cells.

Real- time quantitative PCR

Total RNA was harvested from BMM or PuM untreated or treated with either IFN-γ or AEC_{sup} at 4 or 24 h after the end of BCG infection using QIAzol Lysis Reagent (Qiagen). After isolation, the RNA was run on a 1% agarose gel to verify that the RNA was intact. RT-PCR was performed on 1 µg of RNA using SuperScript III Reverse Transcriptase (Invitrogen) with Oligo$(dT)_{20}$ Primers (Invitrogen). Transcripts were quantified on a RotorGene 6000 real-time PCR cycler (Corbett Life Sciences) using Kapa Sybrfast qPCR Kit (Kapa Biosystems). Primer sequences used were: iNOS sense: CAGCTCGGCTGTA-CAAACCTT, antisense: CATTGGAAGTGAAGCGTTTCG, IL-1β sense: TGGTGTGTGACGTTCCCATT antisense: CAG-CACGAGGCTTTTTTGTTG, Arg-1 sense: GCAGTGGCTT-TAACCTTGGC, antisense: TGGCGCATTCACAGTCACTT. HPRT sense CCC AGC GTC GTG ATT AGC, antisense GGA ATA AAC ACT TTT TCC AAA TCC was used as a control gene and used to calculate dCt values. To determine the effect of treatment with IFN-γ or AEC_{sup}, values were normalized to that of cells infected with BCG receiving no treatment and the fold induction (RQ) calculated.

Cytokine and nitric oxide (NO) determinations

Cytokines and nitrite (an indirect indicator of NO production) were measured in cell culture supernatants from BMM and PuM un-treated or treated with either IFN-γ or AEC_{sup} at 4, 24 and 48 h after the end of infection with BCG. For ELISA determinations, the commercially, TNF, IL-6, CXCL10/Interferon gamma-induced protein 10 kDa (IP-10), IL-1β (R&D Systems) and IL-12, (Mabtech) were used to determine the cytokine levels in the culture supernatants according to the manufacturer's recommendations. The enzyme-substrate reaction was developed using p-nitrophenyl phosphate (Sigma) for IL-12 and tetramethylbenzidine substrate (R&D Systems) for the rest of determinations. Depending on the substrate used, the optical density was measured in a multiscan ELISA reader (Anthos Labtech Instruments, Salzburg, Austria) at 405 or 450 nm. NO production was determined by measuring nitrite concentration using the Griess reaction according to the manufacturers protocol (Sigma).

Analysis of ROS production

BMM were prepared as described above and then pretreated with either IFN-γ (20 ng/ml) or AEC_{sup} (diluted 1:2 in cell culture medium) for 24 h. Cells were then infected with BCG (MOI 1:10) and 500 µM luminol (Sigma) added to the cultures. Plates were then placed in an EnSpire Multimode Plate Reader (PerkinElmer) and luminescence monitored at 37°C every 3 min for 15 h.

Statistical analysis

Data are presented as the mean ± SD. Differences between treatments in the groups were analyzed using one-way ANOVA followed by Bonferroni's post-test for multiple comparisons. For qPCR data differences between treatment at different time-points were analyzed using a non-parametric one-way ANOVA (Kruskal-Wallis) with Dunn's post-test. Differences were considered significant at $p<0.05$. All data were analyzed using the GraphPad InStat version 5.0 (GraphPad Software, San Diego, CA, USA).

Results

Impaired control of intracellular BCG growth by PuM upon IFN-γ treatment

Since IFN-γ is known to mediate the microbicidal activation of macrophages against mycobacteria [22] we evaluated the response of PuM and BMM to IFN-γ. BMM and PuM were infected with GFP-BCG as described in Materials and Methods, treated with different concentrations of IFN-γ (20, 40 and 80 ng/ml) or left untreated. Bacterial loads were monitored by measuring RLU. As expected, BMM reduced the intracellular bacterial content upon treatment with IFN-γ (Figure 1). In contrast, PuM were not able to reduce bacterial load even after treatment with doses of IFN-γ as high as 80 ng/ml (Figure 1). We further evaluated whether PuM needed to be activated prior to mycobacterial infection since it has been described that IFN-γ "activated" macrophages could better control mycobacteria growth [23,24]. For this, PuM were treated with IFN-γ prior infection and bacterial loads monitored. We found that PuM pre-treated with IFN-γ were still unable to control intracellular BCG growth (Fig S1).

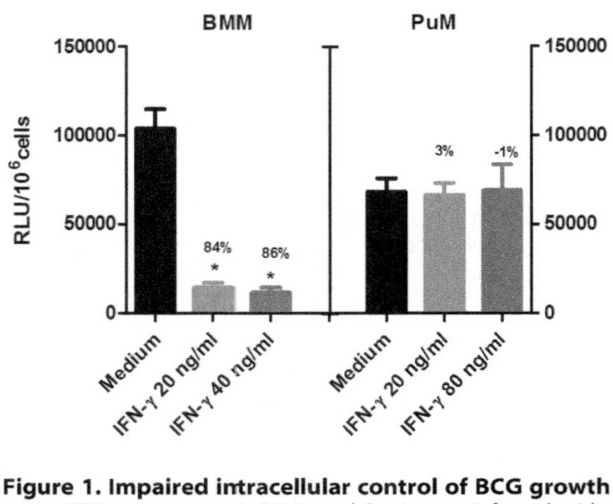

Figure 1. Impaired intracellular control of BCG growth by PuM upon IFN-γ treatment. BMM and PuM were infected with GFP-BCG. After infection, cells were left untreated or treated with IFN-γ (20, 40 and 80 ng/ml) for 48 h. Bacterial growth was evaluated by determining RLU. Data are shown as % reduction of phagocytosed bacteria evaluated as RLU. Values are means ± SD of the mean of a representative experiment from 3 independent experiments with 4 replicates each. The differences between groups of BMM and PuM were analyzed using a one-way ANOVA followed by Bonferroni's Multiple Comparison Test, * significantly different from medium control, $P<0.05$.

PuM are specifically unresponsive to IFN-γ in the control of intracellular bacterial growth

To understand whether the unresponsiveness to IFN-γ was a characteristic of pulmonary macrophages, we compared PuM with other tissue derived macrophages, namely AM and PEM obtained from lung and peritoneal exudates, respectively. Macrophages were treated with IFN-γ and with AEC-derived factors (AEC_{sup}) since we have described previously [7] that treatment with AEC-derived media decreased intracellular bacterial loads in PuM. AM, PuM and PEM were isolated and infected with GFP-BCG and later, either left untreated or treated with IFN-γ or AEC_{sup}. Bacterial loads were monitored by measuring RLU. In contrast to the effect on PuM, IFN-γ was able to induce a significant control of intracellular BCG growth in AM and PEM (49% and 45% reduction respectively) (Figure 2). Importantly, treatment with AEC_{sup} mediated reduction of intracellular BCG in the three types of primary macrophages, PuM, AM and PEM (Figure 2) indicating that PuM was specifically unable to respond to IFN-γ.

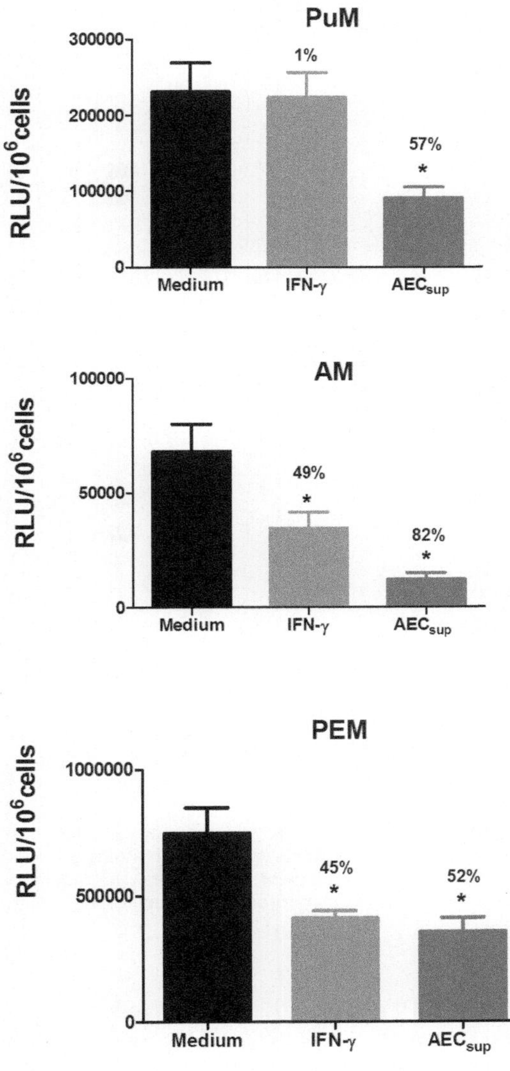

Figure 2. PuM are specifically unresponsive to IFN-γ in the control of intracellular bacteria. PuM, AM and PEM were infected with GFP-BCG. After infection, cells were treated with either IFN-γ or AEC_{sup} for 48 h or left untreated. Bacterial growth was evaluated by determining RLU. Data are shown as % reduction of phagocytosed bacteria evaluated as RLU. Values are means ± SD of the mean of 2 independent experiments with 4 replicates each. The differences between groups of PuM, AM and PEM were analyzed using a one-way ANOVA followed by Bonferroni's Multiple Comparison Test, * significantly different from medium control, $P<0.05$.

IFN-γ and AEC_{sup} induce different cytokine profiles in PuM and BMM

To further understand the different responses elicited in PuM and BMM by AEC_{sup} and IFN-γ, we stimulated the two cell types *in vitro* as described in Materials and Methods and collected supernatants 4 and 24 h later. We monitored the production of IL-12, IP-10 and IL-6 since these factors are associated with IFN-γ mediated signaling through the activation of receptor-associated JAKs. Two major observations are obvious from the data shown in Figure 3. Firstly, comparing the stimulation capacity of IFN-γ and AEC_{sup}, the secretory profile was different in both cell types. AEC_{sup} did not induce production of IL-12 and IP-10 statistically above that of unstimulated cells but induced IL-6 secretion significantly in BMM at 4 h. Similarly, AEC_{sup} also tended to increase IL-6 secretion in PuMSecondly, although the profile of stimulation was similar, the secretion of all factors measured was lower in PuM compared with BMM.

Impaired intracellular control of BCG by PuM upon IFN-γ treatment is regulated by SOCS1

To assess whether the reduced capacity of PuM to respond to IFN-γ was related to intracellular regulations of signaling, we tested this in PuM derived from IFN-$\gamma^{-/-}$SOCS$1^{-/-}$ mice. SOCS1 has been described to be a critical inhibitor of IFN-γ signaling [25] and also able to dampen early responses to BCG and *Mycobacterium tuberculosis* [19]. Thus, we evaluated the effects of IFN-γ and AEC_{sup} treatments on intracellular BCG control by PuM from IFN-$\gamma^{-/-}$SOCS$1^{-/-}$ mice. After IFN-γ treatment, PuM from IFN-$\gamma^{-/-}$SOCS$1^{-/-}$ mice controlled intracellular growth of BCG, to a similar extent as cells treated with AEC_{sup}, indicating that the lack of response to IFN-γ is likely not due to lack of surface IFN-γ receptor expression but rather that the response is under regulation of SOCS1 (Figure 4a). However, there were no differences seen in SOCS1 expression between BMM and PuM and both cell types upregulated SOCS1 upon stimulation with IFN-γ indicating that the selective lack of response to IFN-γ in PuM was not due to a lack of SOCS1 but rather an event downstream of SOCS1 (data not shown). Treating cells with both IFN-γ and AEC_{sup} tended to show an additive effect suggesting independent mechanisms of intracellular bacterial control (Figure 4a). Similarly to the results observed in the wild type mice, upon treatment with IFN-γ, the levels of IL-12 induced in PuM derived from IFN-$\gamma^{-/-}$SOCS$1^{-/-}$ mice were lower than in BMM (data not shown) Thus, reduced IL-12 secretion may not only be subjected to SOCS1 regulation but also to other mechanisms inherent to different tissue-derived macrophages.

AEC_{sup} increases intracellular growth control without an effect on iNOS

Up-regulation of iNOS and the subsequent synthesis of reactive nitrogen species is currently believed to be one of the primary mechanisms downstream of IFN-γ-induced killing of mycobacteria [22]. On the other hand, Arg-1 can compete with iNOS for their common substrate L-arginine, and consequently, the expression of iNOS and Arg-1 is often reciprocally regulated in macrophages [26]. To reveal possible mechanisms for the observed intracellular bacterial control, we followed the expression of iNOS and Arg-1

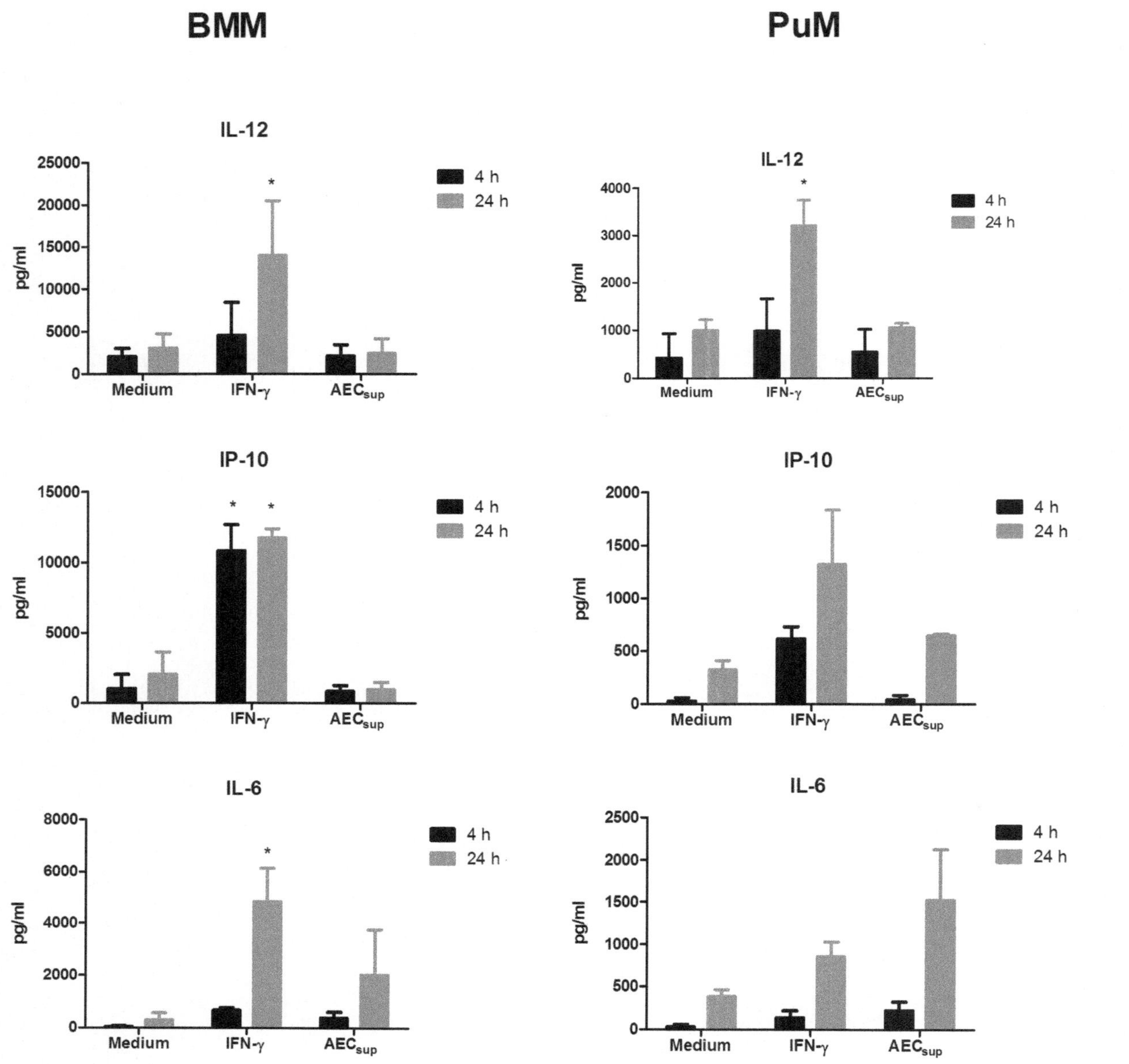

Figure 3. IFN-γ and AECsup induce different cytokine profiles in PuM and BMM. BMM and PuM were infected with GFP-BCG. After infection, cells were left untreated or treated with either IFN-γ or AEC_{sup}. Cell culture supernatants were collected at 4 and 24 h after treatments. IL-12, IP-10 and IL-6, levels were measured using ELISA. Values are expressed as means ± SD, from 3 independent experiments. The differences between groups of PuM and BMM were analyzed using a one-way ANOVA followed by Bonferroni's Multiple Comparison Test, * Significantly different from medium control, $p<0.05$.

induced in PuM and BMM upon treatment with AEC_{sup} or IFN-γ. We found that IFN-γ but not AEC_{sup} induced iNOS in both BMM and PuM (Figure 5a). In contrast, in both PuM and BMM, Arg-1 was induced only by AEC_{sup} and not by IFN-γ (Figure 5b). Notably, the levels of Arg-1 induced in BMM were higher than the levels induced in PuM. Addition of NMMLA, an inhibitor of iNOS, to cultures of BMM infected with BCG and treated with AEC_{sup} or IFN-γ, inhibited the bacterial load reduction mediated by IFN-γ but had no effect on AEC_{sup} activity (Figure 5c) confirming that AEC_{sup} increased intracellular killing using an iNOS independent pathway.

AEC_{sup} induces IL-1β transcription and secretion

IL-1β has been shown to be involved in the intracellular killing of mycobacteria through the induction of autophagy related mechanisms [27]. We therefore investigated whether AEC_{sup} affect IL-1β production. We observed a difference in the profile of induction of this interleukin by the two stimuli where only AEC_{sup} induced IL-1β transcription (Figure 6a) and secretion (Figure 6b) above that of cells only infected with BCG. We next tested whether this increase in IL-1β secretion in AEC_{sup}-treated BMM might mediate the increased intracellular killing of BCG by these cells. However, addition of exogenous IL-1β to the cultures did not affect intracellular bacterial killing (Figure 6c). Finally, we asked

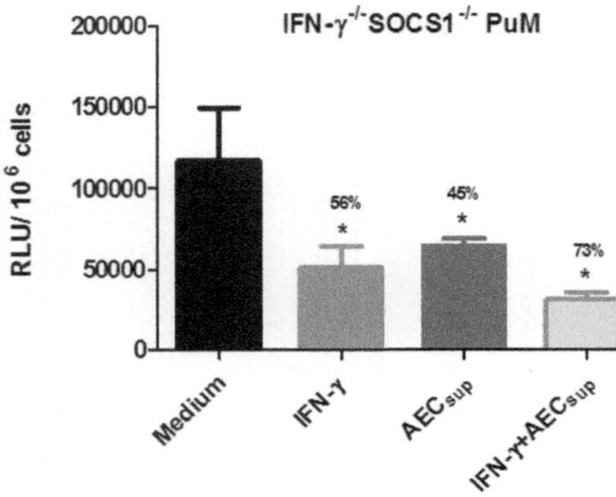

Figure 4. Impaired intracellular control of BCG growth by PuM upon IFN-γ treatment is regulated by SOCS1. PuM from IFN-$\gamma^{-/-}$ SOCS1$^{-/-}$ mice were infected with GFP-BCG. After infection, cells were left untreated or treated with IFN-γ, AEC_{sup} and IFN-γ + AEC_{sup} for 48 h. Bacterial growth was evaluated by determining RLU. Data are shown as % reduction of phagocytosed bacteria evaluated as RLU. Values are means ± SD of the mean of a representative experiment from 2 independent experiments with 4 replicates. The differences between groups of IFN-$\gamma^{-/-}$ SOCS1$^{-/-}$ PuM were analyzed using a one-way ANOVA followed by Bonferroni's Multiple Comparison Test. * significantly different from medium control, $P<0.05$.

the same question using BMM derived from caspase 1$^{-/-}$ mice that are unable to secrete the mature form of IL-1β. We observed that AEC_{sup} activated BMM derived from caspase 1$^{-/-}$ mice were able to control efficiently intracellular bacteria growth supporting the notion that IL-1β was not the key molecule in mediating the killing of BCG under these conditions (Figure 6d).

AECsup induces the production of reactive oxygen species (ROS)

Oxidative burst and the formation of ROS is another mechanism by which intracellular mycobacterial growth may be controlled [28]. Therefore, we next measured the production of ROS in our infection model. BMM treated with IFN-γ or AEC both displayed a significantly increased total production of ROS which peaked around 10 h after infection by BCG (Figure 7a). Using superoxide dismutase (SOD) and catalase; enzymes that convert superoxide to peroxide and peroxide to water and oxygen, respectively, did not affect the AEC_{sup}-induced killing making it unlikely that the increased killing in cells incubated with AEC_{sup} was due to an increased oxidative burst (Figure 7b). On the other hand, treatment with SOD and catalase partially decreased the IFN-γ-induced killing perhaps because of its effect on the formation of peroxynitrites, as these have been shown to mediate killing of BCG [29].

Choloroquine does not affect intracellular bacterial growth control promoted by AEC_{sup}

Since BCG reside in the phagosomal compartment we next addressed whether the increased intracellular killing might be because of increased processing through this compartment. For this, we treated BMM with chloroquine, an inhibitor of phagosomal acidification, after the infection but found that this treatment did not affect the intracellular bacterial control in macrophages receiving IFN-γ or AEC_{sup} (Figure 8). It is therefore unlikely that treatment with AEC_{sup} is acting by increasing phagosomal processing.

Discussion

Macrophages control intracellular pathogens by various anti-microbial mechanisms such as phagosome acidification, induction of reactive oxygen and nitrogen species, phagosome fusion with lysosomes or by autophagy [27,30,31]. The ability of macrophages to control intracellular bacterial growth is fundamental in the defense of the respiratory tract therefore; we studied here the modulation of macrophage activity in the lung tissue. It is broadly accepted that IFN-γ is a potent activator of macrophages. We have also reported previously that factors secreted by AEC increase control of intracellular bacterial growth by macrophages [7]. In the current study, we aimed to investigate possible mechanisms responsible for this activation. The main findings shown are that; a) in contrast to other macrophage types, IFN-γ did not activate interstitial pulmonary macrophages to control intracellular bacterial growth even if the same macrophages could be activated by factors secreted by surrounding cells namely AEC; b) the unresponsiveness of PuM to IFN-γ was apparently regulated by SOCS1; c) the profile of cytokine secretion induced on macrophages by IFN-γ and AEC-derived factors was different and; d) in contrast to IFN-γ, intracellular bacterial destruction induced by AEC factors was not dependent on iNOS.

The Th1 cytokine IFN-γ, is able to induce downstream iNOS and consequently a potent macrophage activator in the killing of intracellular mycobacteria [22].We first compared the microbicidal activity induced by IFN-γ on BMM and PuM infected with mycobacteria. BMM but not PuM were able to control intracellular growth of BCG after IFN-γ treatment. Pre-activation of PuM with IFN-γ before infection, did not revert the unresponsiveness of the PuM. Further, this was only observed in PuM and not in other cells of the respiratory tract such us alveolar macrophages or macrophages obtained from the peritoneal compartment. All this suggests that macrophages integrated in tissues are regulated differently than macrophages operating in secretions.

Interestingly, PuM were not completely unresponsive to IFN-γ since even to a lesser extent than for BMM, treatment with IFN-γ induced the secretion of NO and various cytokines such as IP-10, IL-12, IL-6 and TNF [32,33] suggesting that PuM did express IFN-γ receptors but that transmission of signaling was ineffective as far as we could see only for the induction of intracellular killing. This is certainly noteworthy since PuM have the appropriate "killing machinery" successfully induced upon treatment with AEC-derived factors. Even if we do not have a formal explanation for this selective un-responsiveness of PuM to IFN-γ, it may be speculated that these cells can be more restricted in the activation by inflammatory mediators to prevent tissue damage in the lungs. Moreover, unresponsiveness of interstitial macrophages (PuM) to IFN-γ is intriguing and may open a new way of thinking in the current evaluation of the importance of IFN-γ in the defense against tuberculosis and other lung infections. PuM may play a more regulatory role limiting local inflammatory responses [34-37] in contrast to AM, which are described to behave as a classical macrophage taking up most of the particulate material that is delivered to the alveolar space [35,38,39].

We have reported before [7] that AEC-secreted factors induce expression of MMR and promote an elongated shape on macrophages, both aspects compatible with the phenotype of alternatively activated M2 macrophages and different from the M1 phenotype induced by IFN-γ. M2 macrophages are associated

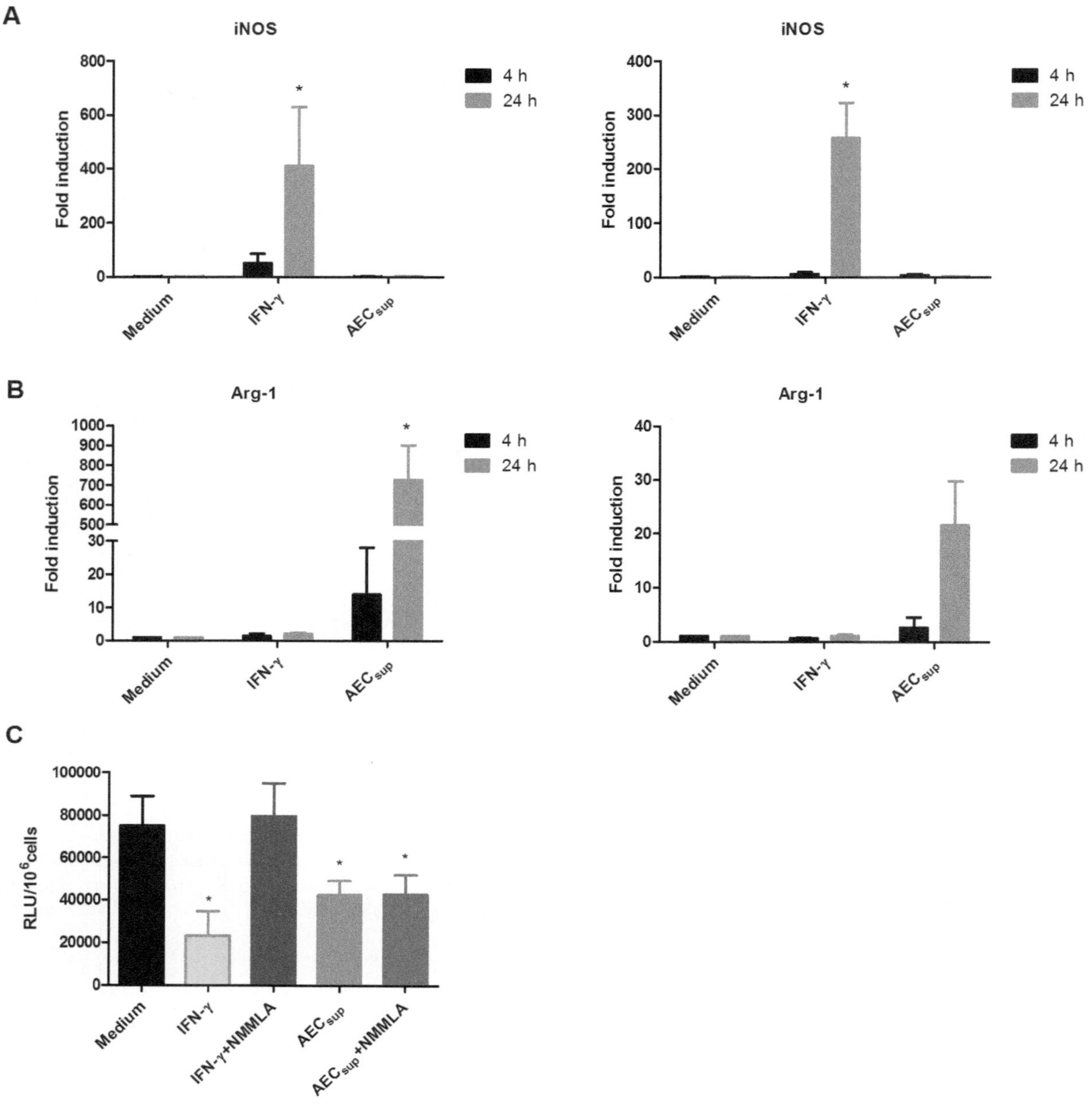

Figure 5. AEC_{sup} has opposite effects to IFN-γ on iNOS, and Arg-1 expression in BMM and PuM and mediates killing through an iNOS independent mechanism. BMM and PuM were infected with GFP-BCG. After infection, cells were treated with either IFN-γ or AEC_{sup} or 1 mM of NG-monomethyl-L-arginine (NMMLA) together with either IFN-γ or AEC_{sup} or left untreated. Total RNA was extracted from both cell-types after 4 and 24 h of treatment and the mean-fold accumulation of a) iNOS, b) Arg-1 ± SD from 3-4 experiments. In c) the effect of the iNOS inhibitor NMMLA on intracellular growth of BCG is shown. Bacterial growth was evaluated by determining RLU. Data are expressed as mean ± SD. The differences between groups of BMM and PuM were analyzed using non-parametric, one-way ANOVA (Kruskal-Wallis) with Dunn's post-test. * significantly different from medium control, $P<0.05$

with Th2 immune responses and are considered to play more immune modulatory than microbicidal roles [40-43]. However, in contrast to this classic M2 phenotype, the macrophage type induced by AEC factors has also potent microbicidal properties suggesting a type of intermediate M1/M2 macrophage possibly required in tissue responses as has been proposed by others [44,45]. Supporting this notion, an intermediate phenotype, M2b, was found to be important in the response and clearance of the intracellular pathogen *Leishmania infantum* [46]. In the same study, C-type lectins such as Dectin-1and MR were found to be crucial for the microbicidal response against this pathogen through ROS and caspase 1 induced IL-1β secretion pathways [46].

Although, IFN-γ is important in mediating macrophage activation, uncontrolled stimulation will not be beneficial to the host. Therefore, IFN-γ signaling is regulated by proteins such as SOCS1 identified as inducible feedback inhibitors of cytokines and consequently important for the regulation of inflammatory responses [25]. Thus, we evaluated whether SOCS1 proteins

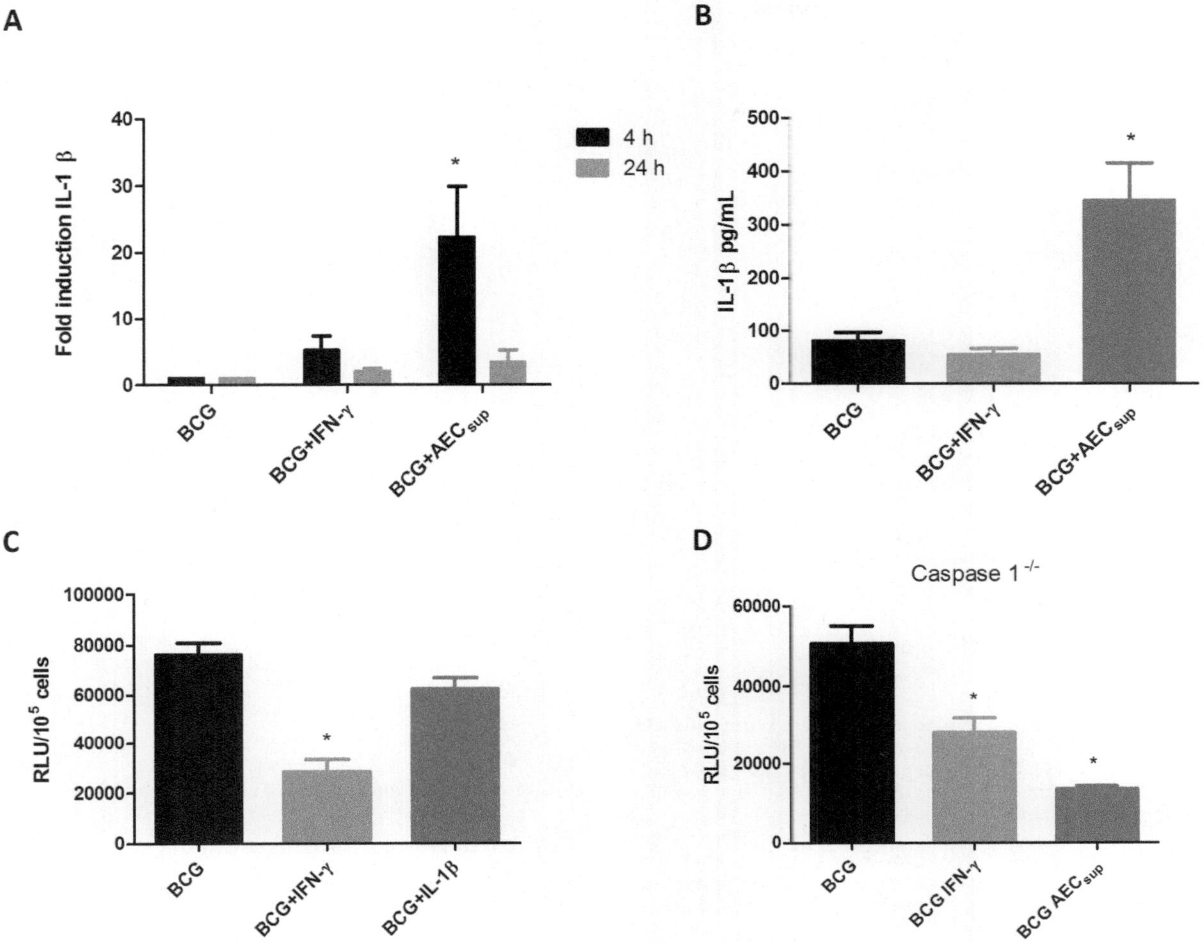

Figure 6. AEC_{sup} induces IL-1β expression and secretion without an effect of intracellular killing of BCG. BMM and PuM were infected with GFP-BCG. After infection, cells were treated with either IFN-γ or AEC_{sup} or left untreated. a) Total RNA was extracted from both cell-types after 4 and 24 h of treatment and the mean-fold accumulation of IL-1β transcripts shown. Data shows the average of 3 independent experiments. b) Cell culture supernatants were collected 24 h after infection and IL-1β measured with ELISA. Data shows the average of 6 independent experiments. c) The effect of incubating cells with IL-1β (1000 pg/ml) on intracellular growth of BCG. Data representative of 3 independent experiments d) The effect of $AECs_{up}$ on intracellular growth of BCG in BMM from caspase1$^{-/-}$ mice. Bacterial growth was evaluated by determining RLU. Values are expressed as means ± SD from 5 wells. The data is representative of 2 independent experiments. The differences between groups of BMM were analyzed using a one-way ANOVA followed by Bonferroni's Multiple Comparison Test. * significantly different from medium control, $P<0.05$.

were involved in the un-responsiveness of PuM to IFN-γ by testing the response in IFN-γ$^{-/-}$SOCS1$^{-/-}$ mice. Interestingly, PuM from these mice responded to IFN-γ and efficiently controlled the intracellular bacteria. This showed not only that the transmission of IFN-γ signals was regulated by SOCS1 but again that PuM were capable of responding to IFN-γ.

Since iNOS and Arg-1 are co-regulated, we followed their induction on PuM and BMM after treatment with IFN-γ or AEC-derived media. AEC_{sup} induced Arg-1 on both cell-types but iNOS was only detected after treatment with IFN-γ. This lack of upregulation of iNOS expression, together with the lack of effect of NMMLA, a NOS inhibitor [47], all strongly indicate that mechanisms independent of the iNOS pathway were involved in the macrophage microbicidal activation by AEC factors. The profiles were also different regarding expression of SOCS1, only induced upon IFN-γ activation in both cell types. SOCS1 have been shown to be important in determining macrophage phenotype and function and regulate the balance between Arg-1 and iNOS activity. SOCS1 are also associated with the induction of Arg-1 in IL-4 treated macrophages [48]. Contrary to the IL-4-induced M2 macrophages described by Whyte et al. we show that AEC-derived factors induced Arg-1 in the absence of SOCS1 expression. Most important, AEC-derived factors promoted control of intracellular bacterial growth in PuM in a SOCS1 independent manner demonstrating that none of the factors secreted by AEC were involved in the induction of SOCS1 despite the fact that AEC-derived factors induced high expression of MMR and Arg-1 similar to M2-like macrophages [7].

To explore other possible mechanisms for the control of intracellular bacterial growth promoted by AEC factors, we looked at the formation of reactive oxygen species ROS and to IL-1β involved in autophagy related mechanisms. AEC factors were efficient inducers of IL-1β but, in contrast to what was reported by other groups [27], IL-1β was not responsible for the induced intracellular killing in our experimental setup. In BMM, the intracellular BCG growth was not affected by addition of exogenous IL-1β to the cultures. Moreover, BMM from caspase

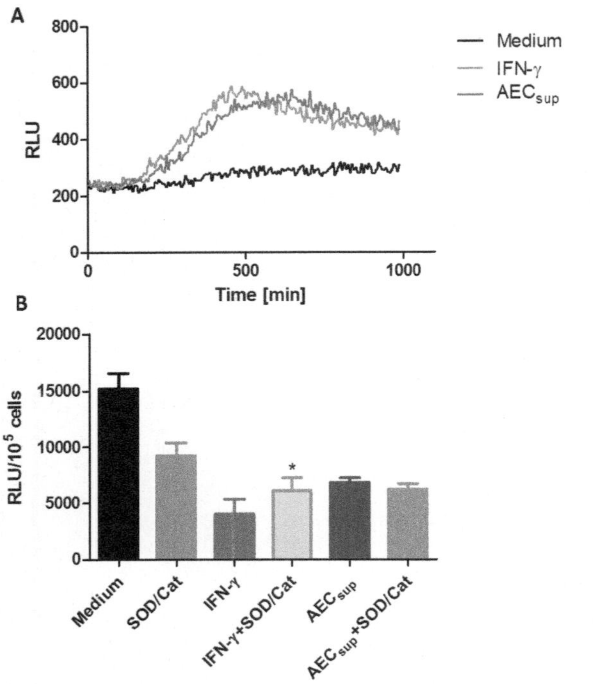

Figure 7. IFN-γ and AEC_{sup} induce an oxidative burst following infection with BCG. BMM were treated with IFN-γ (20 ng/ml) or AEC_{sup} 24 h prior to infection and then infected with GFP-BCG. a) ROS production was monitored in the cultures for 16 h by measuring luminescence with luminol as substrate. b) The effect of incubating cells with superoxide dismutase (SOD) and catalase on intracellular killing of BCG is shown. Values are expressed as means ± SD from 5 wells. Differences between treatments with or without SOD/catalase were analyzed using a paired t-test. * significant effect of SOD/catalase treatment, $P<0.05$.

$1^{-/-}$ mice, unable to produce mature IL-1β, efficiently responded to stimulation by AEC factors.

ROS formation may also be important in the control of BCG infection through the oxidative burst [28]. To a similar extent as IFN-γ, AEC factors were efficient inducers of ROS but by themselves did not contribute to the control of intracellular BCG growth. Finally, we investigated a possible effect on the phagosomal compartment by treating BMM with chloroquine, an inhibitor of phagosomal acidification but we could not observe a major difference between the treated and the untreated groups.

We have been unable to directly ascribe the effect of AEC factors to the most common mechanisms operating separately. Thus, it may be that other mechanisms, not considered herein are responsible for these effects. An alternative explanation is that the combination of different microbicidal events and pathways could be operating in mycobacterial killing as described by Lefevre et al. where ROS contribute partially to IL-1β secretion in the responses to *L. infantum* [46].

Collectively, our data show that PuM are restricted in inflammatory responses mediated by IFN-γ through SOCS1 and that AEC-secreted factors enhance the microbicidal capacities of macrophages by iNOS-independent mechanisms that may be linked with to an intermediate M1-M2-like macrophage.

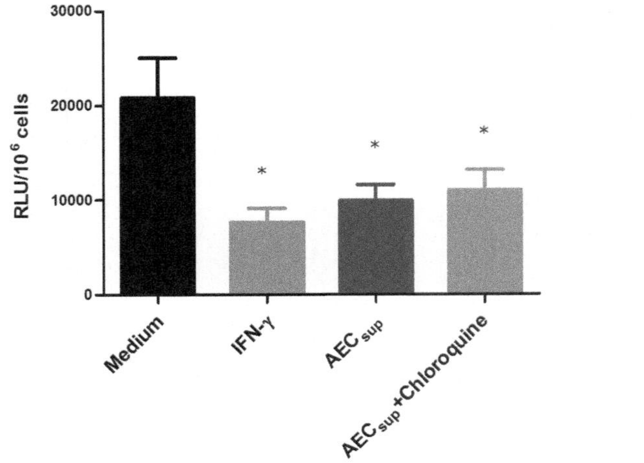

Figure 8. Inhibiting phagosomal acidification does not affect intracellular killing of BCG. BMM and PuM were infected with GFP-BCG. After infection, cells were treated with either IFN-γ or AEC_{sup} with or without chloroquine (10 μM). Bacterial growth was evaluated by determining RLU. Values are expressed as means ± SD from 5 wells. The data is representative of 2 independent experiments. Differences between treatments analyzed using a one-way ANOVA followed by Bonferroni's Multiple Comparison Test. * significantly different from medium control, $P<0.05$.

Supporting Information

Figure S1 Effect of either pretreating PuM with IFN-γ before (IFN-γ Pre Inf), or after infection (IFN-γ Post Inf). PuM were either pretreated with IFN-γ (20 ng/ml) for 24 h before infection or left untreated. Cells were then infected with GFP-BCG for 4 h. After infection, cells were thoroughly washed and treated with gentamicin for 30 min. After additional washing, cells were either cultured in complete medium (Medium) or in medium with IFN-γ (20 ng/ml) or in medium with AEC-derived supernatant (AEC_{sup} 1:2 diluted) for 48 h. Bacterial growth was evaluated by determining RLU in cell lysates. Data are shown as RLU/10^6. Values are means ± SD of the mean from 2 independent experiments. Differences were analyzed with a one-way ANOVA. * significantly different from Medium.

Acknowledgments

The authors wish to thank Martin Rottenberg (MTC, Karonlinska Institutet, Stockholm, Sweden) for supplying the IFN-$\gamma^{-/-}$SOCS$1^{-/-}$ mice and Sara Musciol and B. Heriques-Normark (MTC, Karolinska Institutet, Stockholm, Sweden) for supplying the Caspase-$1^{-/-}$ BMM.

Author Contributions

Conceived and designed the experiments: DHP ODC RF CF. Performed the experiments: DHP ODC RF. Analyzed the data: DHP ODC RF CF. Contributed reagents/materials/analysis tools: CF. Wrote the paper: DHP ODC RF CF.

References

1. Fehrenbach H (2012) Alveolar epithelial type II cells from embryonic stem cells: Knights in shining armour? Eur Respir J 39: 240–241.
2. Debbabi H, Ghosh S, Kamath AB, Alt J, Demello DE, et al. (2005) Primary type II alveolar epithelial cells present microbial antigens to antigen-specific CD4+ T cells. Am J Physiol Lung Cell Mol Physiol 289: L274–9.

3. Pechkovsky DV, Goldmann T, Ludwig C, Prasse A, Vollmer E, et al. (2005) CCR2 and CXCR3 agonistic chemokines are differently expressed and regulated in human alveolar epithelial cells type II. Respir Res 6: 75.
4. Lo B, Hansen S, Evans K, Heath JK, Wright JR (2008) Alveolar epithelial type II cells induce T cell tolerance to specific antigen. J Immunol 180: 881–888.
5. Witherden IR, Vanden Bon EJ, Goldstraw P, Ratcliffe C, Pastorino U, et al. (2004) Primary human alveolar type II epithelial cell chemokine release: Effects of cigarette smoke and neutrophil elastase. Am J Respir Cell Mol Biol 30: 500–509.
6. Thorley AJ, Ford PA, Giembycz MA, Goldstraw P, Young A, et al. (2007) Differential regulation of cytokine release and leukocyte migration by lipopolysaccharide-stimulated primary human lung alveolar type II epithelial cells and macrophages. J Immunol 178: 463–473.
7. Chuquimia OD, Petursdottir DH, Periolo N, Fernandez C (2013) Alveolar epithelial cells are critical in protection of the respiratory tract by secretion of factors able to modulate the activity of pulmonary macrophages and directly control bacterial growth. Infect Immun 81: 381–389.
8. Lawrence T, Natoli G (2011) Transcriptional regulation of macrophage polarization: Enabling diversity with identity. Nat Rev Immunol 11: 750–761.
9. Cassetta L, Cassol E, Poli G (2011) Macrophage polarization in health and disease. ScientificWorldJournal 11: 2391–2402.
10. Biswas SK, Chittezhath M, Shalova IN, Lim JY (2012) Macrophage polarization and plasticity in health and disease. Immunol Res 53: 11–24.
11. Martinez FO, Sica A, Mantovani A, Locati M (2008) Macrophage activation and polarization. Front Biosci 13: 453–461.
12. Vereyken EJ, Heijnen PD, Baron W, de Vries EH, Dijkstra CD, et al. (2011) Classically and alternatively activated bone marrow derived macrophages differ in cytoskeletal functions and migration towards specific CNS cell types. J Neuroinflammation 8: 58-2094-8-58.
13. Stein M, Keshav S, Harris N, Gordon S (1992) Interleukin 4 potently enhances murine macrophage mannose receptor activity: A marker of alternative immunologic macrophage activation. J Exp Med 176: 287–292.
14. Brysha M, Zhang JG, Bertolino P, Corbin JE, Alexander WS, et al. (2001) Suppressor of cytokine signaling-1 attenuates the duration of interferon gamma signal transduction in vitro and in vivo. J Biol Chem 276: 22086–22089.
15. Humphreys IR, Stewart GR, Turner DJ, Patel J, Karamanou D, et al. (2006) A role for dendritic cells in the dissemination of mycobacterial infection. Microbes Infect 8: 1339–1346.
16. Chuquimia OD, Petursdottir DH, Rahman MJ, Hartl K, Singh M, et al. (2012) The role of alveolar epithelial cells in initiating and shaping pulmonary immune responses: Communication between innate and adaptive immune systems. PLoS One 7: e32125.
17. Rahman MJ, Chuquimia OD, Petursdottir DH, Periolo N, Singh M, et al. (2011) Impact of toll-like receptor 2 deficiency on immune responses to mycobacterial antigens. Infect Immun 79: 4649–4656.
18. Corti M, Brody AR, Harrison JH (1996) Isolation and primary culture of murine alveolar type II cells. Am J Respir Cell Mol Biol 14: 309–315.
19. Carow B, Ye X, Gavier-Widen D, Bhuju S, Oehlmann W, et al. (2011) Silencing suppressor of cytokine signaling-1 (SOCS1) in macrophages improves mycobacterium tuberculosis control in an interferon-gamma (IFN-gamma)-dependent manner. J Biol Chem 286: 26873–26887.
20. Zhang X, Goncalves R, Mosser DM (2008) The isolation and characterization of murine macrophages. Curr Protoc Immunol Chapter 14: Unit 14.1.
21. Rothfuchs AG, Gigliotti D, Palmblad K, Andersson U, Wigzell H, et al. (2001) IFN-alpha beta-dependent, IFN-gamma secretion by bone marrow-derived macrophages controls an intracellular bacterial infection. J Immunol 167: 6453–6461.
22. Barrera LF, Kramnik I, Skamene E, Radzioch D (1994) Nitrite production by macrophages derived from BCG-resistant and -susceptible congenic mouse strains in response to IFN-gamma and infection with BCG. Immunology 82: 457–464.
23. Sato K, Akaki T, Tomioka H (1998) Differential potentiation of anti-mycobacterial activity and reactive nitrogen intermediate-producing ability of murine peritoneal macrophages activated by interferon-gamma (IFN-gamma) and tumour necrosis factor-alpha (TNF-alpha). Clin Exp Immunol 112: 63–68.
24. Flesch IE, Kaufmann SH (1991) Mechanisms involved in mycobacterial growth inhibition by gamma interferon-activated bone marrow macrophages: Role of reactive nitrogen intermediates. Infect Immun 59: 3213–3218.
25. Alexander WS, Starr R, Fenner JE, Scott CL, Handman E, et al. (1999) SOCS1 is a critical inhibitor of interferon gamma signaling and prevents the potentially fatal neonatal actions of this cytokine. Cell 98: 597–608.
26. Bronte V, Zanovello P (2005) Regulation of immune responses by L-arginine metabolism. Nat Rev Immunol 5: 641–654.
27. Pilli M, Arko-Mensah J, Ponpuak M, Roberts E, Master S, et al. (2012) TBK-1 promotes autophagy-mediated antimicrobial defense by controlling autophagosome maturation. Immunity 37: 223–234.
28. Ehrt S, Schnappinger D (2009) Mycobacterial survival strategies in the phagosome: Defence against host stresses. Cell Microbiol 11: 1170–1178.
29. Yu K, Mitchell C, Xing Y, Magliozzo RS, Bloom BR, et al. (1999) Toxicity of nitrogen oxides and related oxidants on mycobacteria: M. tuberculosis is resistant to peroxynitrite anion. Tuber Lung Dis 79: 191–198.
30. Gwinn MR, Vallyathan V (2006) Respiratory burst: Role in signal transduction in alveolar macrophages. J Toxicol Environ Health B Crit Rev 9: 27–39.
31. Alvarez E, Leiro J, Orallo F (2002) Effect of (-)-epigallocatechin-3-gallate on respiratory burst of rat macrophages. Int Immunopharmacol 2: 849–855.
32. Benoit M, Desnues B, Mege JL (2008) Macrophage polarization in bacterial infections. J Immunol 181: 3733–3739.
33. Taub DD, Key ML, Clark D, Turcovski-Corrales SM (1995) Chemotaxis of T lymphocytes on extracellular matrix proteins. analysis of the in vitro method to quantitate chemotaxis of human T cells. J Immunol Methods 184: 187–198.
34. Tschernig T, Pabst R (2009) What is the clinical relevance of different lung compartments? BMC Pulm Med 9: 39-2466-9-39.
35. Laskin DL, Weinberger B, Laskin JD (2001) Functional heterogeneity in liver and lung macrophages. J Leukoc Biol 70: 163–170.
36. Schneberger D, Aharonson-Raz K, Singh B (2011) Monocyte and macrophage heterogeneity and toll-like receptors in the lung. Cell Tissue Res 343: 97–106.
37. Steinmuller C, Franke-Ullmann G, Lohmann-Matthes ML, Emmendorffer A (2000) Local activation of nonspecific defense against a respiratory model infection by application of interferon-gamma: Comparison between rat alveolar and interstitial lung macrophages. Am J Respir Cell Mol Biol 22: 481–490.
38. Lohmann-Matthes ML, Steinmuller C, Franke-Ullmann G (1994) Pulmonary macrophages. Eur Respir J 7: 1678–1689.
39. Gordon SB, Read RC (2002) Macrophage defences against respiratory tract infections. Br Med Bull 61: 45–61.
40. Gordon S, Martinez FO (2010) Alternative activation of macrophages: Mechanism and functions. Immunity 32: 593–604.
41. Mantovani A, Locati M (2009) Orchestration of macrophage polarization. Blood 114: 3135–3136.
42. Anders HJ, Ryu M (2011) Renal microenvironments and macrophage phenotypes determine progression or resolution of renal inflammation and fibrosis. Kidney Int 80: 915–925.
43. Orme J, Mohan C (2012) Macrophage subpopulations in systemic lupus erythematosus. Discov Med 13: 151–158.
44. Edwards JP, Zhang X, Frauwirth KA, Mosser DM (2006) Biochemical and functional characterization of three activated macrophage populations. J Leukoc Biol 80: 1298–1307.
45. Sironi M, Martinez FO, D'Ambrosio D, Gattorno M, Polentarutti N, et al. (2006) Differential regulation of chemokine production by fcgamma receptor engagement in human monocytes: Association of CCL1 with a distinct form of M2 monocyte activation (M2b, type 2). J Leukoc Biol 80: 342–349.
46. Lefevre L, Lugo-Villarino G, Meunier E, Valentin A, Olagnier D, et al. (2013) The C-type lectin receptors dectin-1, MR, and SIGNR3 contribute both positively and negatively to the macrophage response to leishmania infantum. Immunity 38: 1038–1049.
47. Hibbs JB,Jr, Vavrin Z, Taintor RR (1987) L-arginine is required for expression of the activated macrophage effector mechanism causing selective metabolic inhibition in target cells. J Immunol 138: 550–565.
48. Whyte CS, Bishop ET, Ruckerl D, Gaspar-Pereira S, Barker RN, et al. (2011) Suppressor of cytokine signaling (SOCS)1 is a key determinant of differential macrophage activation and function. J Leukoc Biol 90: 845–854.

3

Overlapped Sequence Types (STs) and Serogroups of Avian Pathogenic (APEC) and Human Extra-Intestinal Pathogenic (ExPEC) *Escherichia coli* Isolated in Brazil

Renato Pariz Maluta[1], Catherine Mary Logue[2], Monique Ribeiro Tiba Casas[3], Ting Meng[2], Elisabete Aparecida Lopes Guastalli[4], Thaís Cabrera Galvão Rojas[1], Augusto Cezar Montelli[5], Teruê Sadatsune[6], Marcelo de Carvalho Ramos[7], Lisa Kay Nolan[2], Wanderley Dias da Silveira[1]*

1 Bacterial Molecular Biology Laboratory, Department of Genetics, Evolution and Bioagents, Institute of Biology, State University of Campinas (UNICAMP), Campinas, SP, Brazil, **2** Department of Veterinary Microbiology and Preventive Medicine, College of Veterinary Medicine, Iowa State University, Ames, Iowa, United States of America, **3** Instituto Adolfo Lutz, São Paulo, SP, Brazil, **4** Instituto Biológico, CAPTAA, Unidade de Pesquisa e Desenvolvimento de Bastos, Bastos, SP, Brazil, **5** Departamento de Clínica Médica, Faculdade de Medicina, Universidade Estadual Paulista, Botucatu, SP, Brazil, **6** Departamento de Microbiologia e Imunologia, Instituto de Biociências, Universidade Estadual Paulista, Botucatu, SP, Brazil, **7** Department of Internal Medicine, UNICAMP, Campinas, Brazil

Abstract

Avian pathogenic *Escherichia coli* (APEC) strains belong to a category that is associated with colibacillosis, a serious illness in the poultry industry worldwide. Additionally, some APEC groups have recently been described as potential zoonotic agents. In this work, we compared APEC strains with extraintestinal pathogenic *E. coli* (ExPEC) strains isolated from clinical cases of humans with extra-intestinal diseases such as urinary tract infections (UTI) and bacteremia. PCR results showed that genes usually found in the ColV plasmid (*tsh*, *iuc*A, *iss*, and *hly*F) were associated with APEC strains while *fyu*A, *irp*-2, *fep*C *sit*D$_{chrom}$, *fim*H, *crl*, *csg*A, *afa*, *iha*, *sat*, *hly*A, *hra*, *cnf*1, *kps*MTII, *clpV*$_{Sakai}$ and *mal*X were associated with human ExPEC. Both categories shared nine serogroups (O2, O6, O7, O8, O11, O19, O25, O73 and O153) and seven sequence types (ST10, ST88, ST93, ST117, ST131, ST155, ST359, ST648 and ST1011). Interestingly, ST95, which is associated with the zoonotic potential of APEC and is spread in avian *E. coli* of North America and Europe, was not detected among 76 APEC strains. When the strains were clustered based on the presence of virulence genes, most ExPEC strains (71.7%) were contained in one cluster while most APEC strains (63.2%) segregated to another. In general, the strains showed distinct genetic and fingerprint patterns, but avian and human strains of ST359, or ST23 clonal complex (CC), presented more than 70% of similarity by PFGE. The results demonstrate that some "zoonotic-related" STs (ST117, ST131, ST10CC, ST23CC) are present in Brazil. Also, the presence of moderate fingerprint similarities between ST359 *E. coli* of avian and human origin indicates that strains of this ST are candidates for having zoonotic potential.

Editor: Yung-Fu Chang, Cornell University, United States of America

Funding: RPM (2012/05073-3) and TCGR (2013/09167-5) are fellows from Fundação de Amparo à Pesquisa do Estado de São Paulo (FAPESP). This work was funded by Coordenação de Aperfeiçoamento de Pessoal de Nível Superior (PNPD/CAPES 23038.042588/2008-11) and FAPESP (2012/04931-6). WDS has a scientific fellowship from Conselho Nacional de Desenvolvimento Científico e Tecnológico (CNPq). CML was supported by the Dean's Office College of Veterinary Medicine, Iowa State University (ISU) and VPR Office, ISU. The funders had no role in study design, data collection and analysis, decision to publish, or preparation of the manuscript.

Competing Interests: The authors have declared that no competing interests.

* Email: wds@unicamp.br

Introduction

Strains of Avian Pathogenic *Escherichia coli* (APEC) are associated with colibacillosis, a disease of significant economic importance in poultry [1]. Besides the impact on animal health, it has been hypothesized that a subset of APEC may constitute a potential zoonotic risk [2–5]. While the chief source of *E. coli* causing extraintestinal disease in human beings is thought to be the human intestine [6,7], it is not clear how they come to be there [4]. In humans, *E. coli* strains associated with extra-intestinal diseases are termed ExPEC (Extra-intestinal pathogenic *E. coli*) [8]. ExPEC strains related to urinary tract infections (UTIs) are known as UPEC (Uropathogenic *E. coli*) and those causing neonatal meningitis, NMEC. Human ExPEC are also important causes of sepsis [9].

The hypotheses that poultry may be a vehicle of *E. coli* able to cause foodborne diseases in humans, or avian *E. coli* a reservoir of virulence and resistance genes, have been established previously. This was due to overlapping characteristics between APEC and UPEC, such as serogroups, virulence genotypes and assignments to ECOR phylogenetic groups [4]. Subsequent research comparing APEC and *E. coli* isolated from different clinical samples, including strains associated with other human disease syndromes, such as UPEC, NMEC [2] and *E. coli* causing septicemia [3], supported these hypotheses. Indeed, a number of APEC and ExPEC strains share the same phylogenetic background, as determined by multilocus sequence typing (MLST). The overlap is particularly striking for the ST95 and ST23 clonal complexes [10,11]. Also, the finding that *E. coli* from retail chicken and human UTI can share very close or even indistinguishable PFGE patterns [12] provides robust evidence supporting the hypothesis that APEC-contaminated poultry is a source of ExPEC in human disease. Recently, APEC strains have been shown to induce fluid accumulation in the mammalian gut [13].

Most published research demonstrating similarities between APEC and human EXPEC have been performed in North America and Europe, however due to the intense diversity found in extra-intestinal *E. coli*, either from poultry or humans, surveys from different regions of the world would be useful for further research studies, including vaccine production. For this reason, we compared a collection of APEC and ExPEC from Brazil, a large poultry producer, using a wide range of techniques in order to deduce if a subset of APEC strains shares characteristics with human ExPEC in Brazil.

Materials and Methods

E. coli strains

All *E. coli* strains analyzed in this work belongs to the collection of the Laboratory of Bacterial Molecular Biology (LBMB) from the Institute of Biology, State University of Campinas (UNICAMP), in Campinas, SP, Brazil. They were isolated and identified using standard methods [14]. A total of 81 APEC strains provided by three Brazilian reference laboratories (Microbiology Laboratory of UNICAMP, Campinas, SP, Brazil; Microbiology Laboratory of State University of Londrina, Londrina, PR, Brazil and Biologic Institute, Bastos, SP, Brazil) were used. The human ExPEC strains (n = 53) were isolated either in the UNICAMP Clinics Hospital in Campinas or the São Paulo State University (UNESP) Clinics Hospital in Botucatu, SP, Brazil. These were isolated from cases of bacteremia or UTIs. All *E. coli* strains (n = 134) were analyzed by serotyping and pulsed-field gel electrophoresis (PFGE). PCR, phylogenetic typing and multilocus sequence typing (MLST) were performed with 76 APEC and the 53 ExPEC strains.

Serotyping

All *E. coli* strains were serotyped at the *E. coli* Reference Center, Pennsylvania State University, USA. Somatic antigen (O) typing was performed as described previously [15]. Flagellar typing was performed via restriction fragment length polymorphism (RFLP) patterns of *fliC* [16].

Pulsed-field gel electrophoresis (PFGE)

Chromosomal DNA embedded in agarose gel, from all *E. coli* strains, was digested with *Xba*I. Electrophoresis conditions consisted of an initial time of 2.2 seconds, a final time of 54.2 seconds at a gradient of 6 V cm-1 and an included angle of 120°. The gels were electrophoresed for 18 h. The similarities of fragments were compared using a Dice coefficient at 1% tolerance and 0.5% optimization, and a dendrogram was constructed with the UPGMA clustering method using the BioNumerics software (version 6.6) (Applied Maths, Austin, TX). Clusters were established using the cutoff value of 70%.

Virotyping and Phylogenetic typing by PCR

APEC strains (n = 76) and ExPEC strains (n = 53) were assessed for their possession of 43 virulence genes using multiplex PCR. Isolates were cultured in LB at 37°C overnight. These cultures were used for the preparation of DNA templates using a thermal lysis procedure (http://www.apzec.ca/en/APZEC/Protocols/pdfs/ECL_PCR_Protocol.pdf). After preparation of the DNA templates, PCR was performed as described previously [17] using primers listed in Table S1. The PCR products were visualized by SYBR Safe DNA Gel Stain (Invitrogen) staining after agarose gel (1.5%) electrophoresis. Phylogenetic typing was established by the dichotomous tree of the PCR amplification of *chu*A, *yja*A and TSPE4.C2 [18]. The positive controls are shown in Table S1. All positive control strains were obtained from the Laboratory of Bacterial Molecular Biology except strains FVL16 and FVL35, which were kindly supplied by Dr. D. S. Leite, and strain PCAY, which was kindly provided by Dr. T. Yano, both from UNICAMP.

Gene clustering was based in the presence and absence of virulence genes. A binary matrix was used to determine the similarity using the Pearson correlation (centered) and the isolates were clustered by the complete linkage method using Gene Cluster 3.0 [19]. A distance tree was visualized using the software Java TreeView 1.1.6r4 [20].

Multilocus sequence typing (MLST)

MLST was performed with 76 APEC and 53 ExPEC strains using *E. coli* Achtman's scheme (http://mlst.ucc.ie/mlst/dbs/Ecoli) [21]. This scheme is based on the sequencing of the PCR amplification products of *adk*, *fum*C, *gyr*B, *icd*, *mdh*, *pur*A, and *rec*A (Table S1). DNA template preparation and PCR was done as described in the previous section. Sequencing was performed at "Centro de Biologia Molecular e Engenharia Genética (CBMEG)" or "Laboratório Central de Tecnologias de Alto Desempenho em Ciências da Vida (LaCTAD)", both from UNICAMP. For sequencing, amplicons were purified with a column based kit (Pure Link Quick PCR Purification Kit, Invitrogen, Germany). The purified product was sequenced using Sanger methodology using an ABI PRISM 3700 DNA Analyzer (Applied Biosystems with the program PCR-BD3700).

The sequences were concatenated in order to generate contiguous sequences that were aligned by the neighbor-joining clustering method using MUSCLE. The evolutionary history was inferred by using the Maximum Likelihood method, based on the Tamura-Nei model, with 1000 bootstrap replications [22]. All positions containing gaps and missing data were eliminated. A dendrogram of the clustering based on the alignments was created with MEGA5 [23]. The sequence type (ST) of each allele was attributed with the Achtman's scheme (http://mlst.ucc.ie/mlst/dbs/Ecoli). Novel alleles and STs were deposited in the MLST database (ST4131-37 and ST4139-41).

Statistics

The frequency of each *E. coli* virulence gene was compared between the groups by the Fisher's exact test using the Prism for Windows version 6.01 from GraphPad Software, Inc.

Results

Genes commonly found in the plasmid ColV were frequent and associated with APEC

The PCR results showed that the individual distribution of 20/43 genes examined in this work (46.5%) was significantly distinct between APEC (n = 76) and human ExPEC (n = 53) (Table 1). A positive correlation occurred more in ExPEC (n = 16 genes) than in APEC (n = 4 genes). Genes that were significantly associated with APEC included *tsh*, *iuc*A, *iss*, and *hly*F. Genes significantly associated with human ExPEC included those related to iron-acquistion (*fyu*A, *irp-2*, *fep*C *sit*D$_{chrom}$), adhesins (*fim*H, *crl*, *csg*A, *afa*, *iha*), autotransporter (*sat*), toxins (*hly*A, *hra*, *cnf*1), escape from

Table 1. Frequencies of 43 virulence genes tested in APEC (n = 76) and ExPEC (n = 53) strains from Brazil.

Gene	Category	APEC		Human ExPEC		p value	positive correlation to
		N	(%)	n	(%)		
tsh	SPATE[1]	28	(36.8)	4	(7.5)	p<0.01	APEC
hlyF	Toxin	51	(67.1)	17	(32.1)	p<0.01	APEC
iss	EHD[2]	39	(51.3)	16	(30.2)	p<0.05	APEC
iucA	Iron-acquisition	39	(51.3)	16	(30.2)	p<0.05	APEC
crl	Adhesin	51	(67.1)	49	(92.5)	p<0.01	ExPEC
sat	SPATE	-	(0)	8	(15.1)	p<0.01	ExPEC
irp-2	Iron-acquisition	18	(23.7)	28	(52.8)	p<0.01	ExPEC
fyuA	Iron-acquisition	19	(25.0)	28	(52.8)	p<0.01	ExPEC
fepC	Iron-acquisition	23	(30.3)	31	(58.5)	p<0.01	ExPEC
malX	UPEC PAI marker	8	(10.5)	16	(30.2)	p<0.01	ExPEC
iha	Adhesin	1	(1.3)	7	(13.2)	p<0.01	ExPEC
hlyA	Toxin	-	(0)	5	(9.4)	p<0.05	ExPEC
hra	Toxin	6	(7.9)	13	(24.5)	p<0.05	ExPEC
$clpV_{Sakai}$	T6SS[3]	10	(13.2)	17	(32.1)	p<0.05	ExPEC
$sitD_{chrom}$	Iron-acquisition	8	(10.5)	15	(28.3)	p<0.05	ExPEC
afa	Adhesin	-	(0)	4	(7.5)	p<0.05	ExPEC
cnf1	Toxin	-	(0)	4	(7.5)	p<0.05	ExPEC
kpsMTII	EHD	11	(14.5)	17	(32.1)	p<0.05	ExPEC
csgA	Adhesin	46	(60.5)	42	(79.2)	p<0.05	ExPEC
fimH	Adhesin	62	(81.6)	50	(94.3)	p<0.05	ExPEC
traT	EHD	41	(53.9)	36	(67.9)	NS	-
$sitD_{epi}$	Iron-acquisition	38	(50.0)	19	(35.8)	NS	-
ompT	EHD	35	(46.1)	16	(30.2)	NS	-
iucD	Iron-acquisition	35	(46.1)	19	(35.8)	NS	-
iroN	Iron-acquisition	33	(43.4)	20	(37.7)	NS	-
$vgrG_{Sakai}$	T6SS	32	(42.1)	26	(49.1)	NS	-
iutA	Iron-acquisition	31	(40.8)	24	(45.3)	NS	-
ireA	Iron-acquisition	24	(31.6)	10	(18.9)	NS	-
cvi/cva	EHD	22	(28.9)	11	(20.8)	NS	-
astA	Toxin	16	(21.1)	5	(9.4)	NS	-
tia	Invasin	16	(21.1)	7	(13.2)	NS	-
papC	Adhesin	15	(19.7)	13	(24.5)	NS	-
$icmF_{Sakai}$	T6SS	14	(18.4)	12	(22.6)	NS	-
$lpfA_{O157/OI-154}$	Adhesin	11	(14.5)	3	(5.7)	NS	-

Table 1. Cont.

Gene	Category	APEC		Human ExPEC		p value	positive correlation to
		N	(%)	n	(%)		
vat	Toxin	8	(10.5)	11	(20.8)	NS	-
pic	SPATE	5	(6.6)	5	(9.4)	NS	-
kpsMTIII	EHD	4	(5.3)	2	(3.8)	NS	-
lpfA$_{O157/OI-141}$	Adhesin	4	(5.3)	3	(5.7)	NS	-
neuC	EHD	4	(5.3)	4	(7.5)	NS	-
ibeA	Invasin	2	(2.6)	2	(3.8)	NS	-
hcp$_{Sakai}$	T6SS	2	(2.6)	3	(5.7)	NS	-
sfa	Adhesin	2	(2.6)	4	(7.5)	NS	-
gimB	Invasin	-	(0)	2	(3.8)	NS	-

[1]Serine protease autotransporter.
[2]Escape from host defenses.
[3]Type VI secretion system.

host defenses (*kps*MTII), type six secretion system (*clpV*$_{Sakai}$) and UPEC PAI marker (*mal*X).

The genes present in more than half of the APEC strains were *fim*H, *hly*F, *crl*, *csg*A, *iuc*A, *tra*T, *iss*, *iuc*A and *sit*D$_{epi}$ (Table 1). For human ExPEC, genes present in more than 50% of strains were *fim*H, *crl*, *csg*A, *tra*T, *fep*C *fyu*A and *irp*-2. No APEC strain was positive either for *sat*, *hly*A, *afa*, *cnf*1 or *gim*B while all of these surveyed genes were detected at least in two human ExPEC strains.

APEC and human ExPEC shared 13 serogroups

More than 30% of APEC and human ExPEC strains were non-typable (NT) for the somatic antigen (O) (Table 2). Among the O-typable strains, serogroup O8 was detected in more than 10% of APEC. No other serogroup was found in more than 10% of the human ExPEC strains. Among the 37 distinct serogroups detected in this work, nine (O2, O6, O7, O8, O11, O19, O25, O73, and O153) were shared between APEC and ExPEC, 13 serogroups (O5, O9, O36, O54, O68, O100, O101, O103. O106, O109, O150, O156 and OX9) were found only in APEC and 15 serogroups (O1, O15, O16, O20, O29, O30, O32, O69, O86, O102, O114, O126, O149, O167 and O176) were identified only in human ExPEC strains.

Two clusters were identified based on the analysis for the presence of virulence genes only

When APEC (n = 76) and human ExPEC (n = 53) strains were analyzed based on the presence of virulence genes, all strains were segregated between two major gene clusters (A and B) (Figure 1). Although both clusters contained strains from either APEC or ExPEC, most human ExPEC strains (71.7%) were segregated into cluster A while most APEC strains were placed in cluster B (63.2%). Interestingly, strains of cluster B had a high frequency (range around 60–95%) of pColV genes, including *sit*D$_{epi}$, *iss*, *omp*T, *hly*F, *tra*T, iroN, *iuc*A, *iuc*D and *iut*A and to a less extent *cvi/cva* and *tsh* (frequencies of 47.6% and 41.3%, respectively) (Table S2).

On the other hand, strains of cluster A presented frequencies of genes usually present in pColV plasmids at a much lower level (< 20%), except *iut*A (25.8%) and *tra*T (47.0%). Genes associated with cluster A were associated ($P<0.01$) for *kps*MTII, *clpV*$_{Sakai}$, *sit*D$_{chrom}$ and *sat* or $P<0.05$ for *crl*, *mal*X and *sfa* (Table S2). However, in contrast with pColV genes of strains in cluster B, genes associated with strains of cluster A, in general, presented a frequency lower than 35%. The only exception was *crl*, with a frequency of 86.4%, but this gene was detected in cluster B isolates at a prevalence of 68.3%.

Some APEC and ExPEC strains showed more than 70% of similarity by PFGE

Following *Xba*I restriction and PFGE, strains in both categories were, in general, very diverse. No indistinguishable profile was detected between APEC and ExPEC strains. However, some APEC and human ExPEC strains presented more than 70% of fingerprint similarity (Figure S1). Five APEC strains (UEL01, UEL19, UEL199, UEL89 and UEL126) and one APEC strain (H8327/11) were not restricted with *Xba*I.

Two pairs of strains presented additional similarities, besides the resembling fingerprints: (i) APEC EC191 and ExPEC U-8337/12 presented the same ST clonal complex (ST23 CC) and serotype (O8:H9). Both strains presented a genetic profile with few virulence genes and were identified as ECOR A. (ii) APEC EC126 and ExPEC M10062 belonged to the same ST (ST 359).

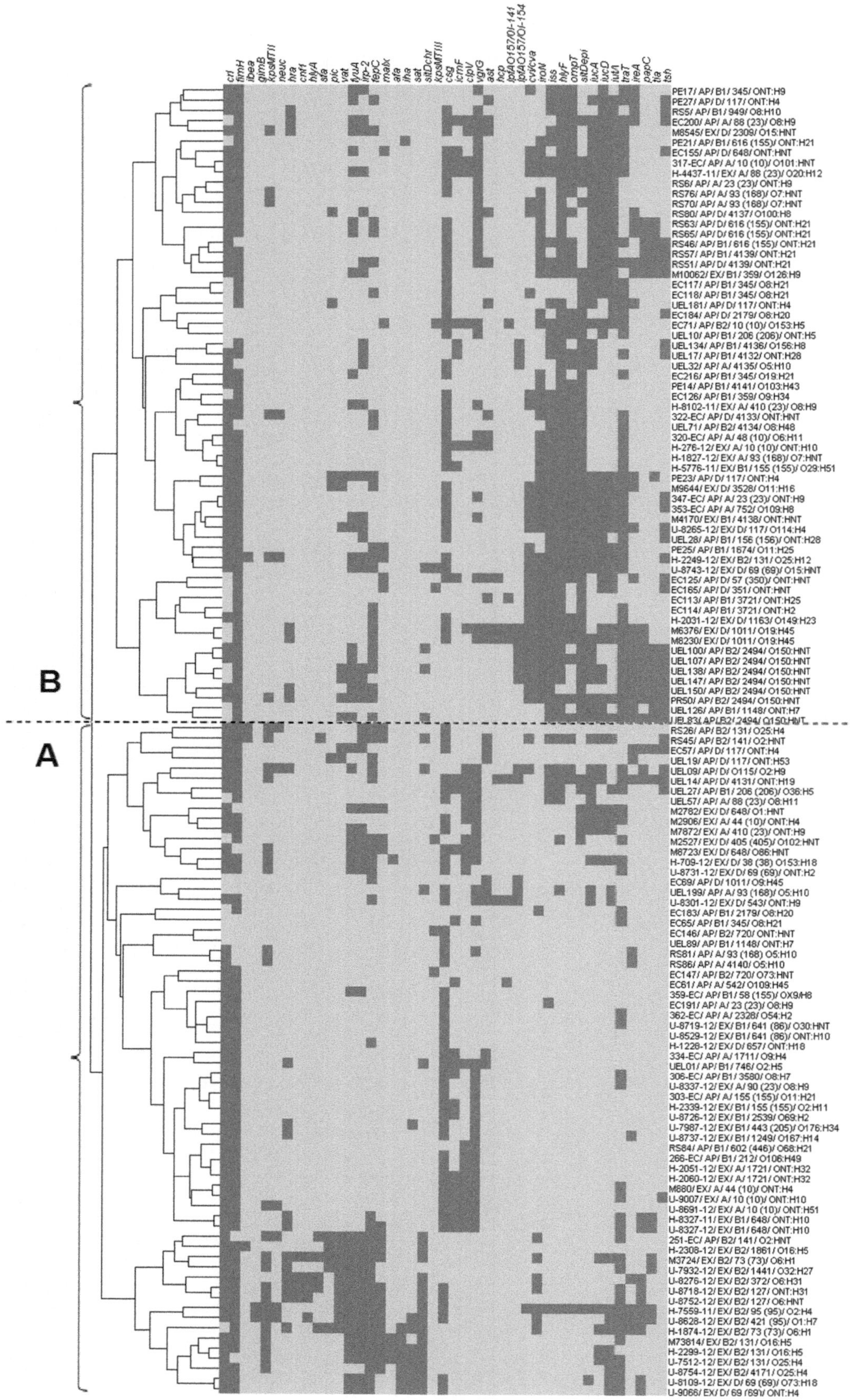
crl
fimH
ibea
gimB
kpsMTII
neuc
hra
cnf1
hlyA
sfa
pic
vat
fyuA
irp-2
fepC
malx
afa
iha
sat
sitDchr
kpsMTIII
csg
ccmF
clpV
vgrG
ast
hcp
lpfAO157/OI-141
lpfAO157/OI-154
cvi/cva
iroN
iss
hlyF
ompT
sitDepi
iucA
iucD
iutA
traT
ireA
papC
tia
tsh
B
A
PE17/ AP/ B1/ 345/ ONT:H9
PE27/ AP/ D/ 117/ ONT:H4
RS5/ AP/ B1/ 949/ O8:H10
EC200/ AP/ A/ 88 (23)/ O8:H9
M8545/ EX/ D/ 2309/ O15:HNT
PE21/ AP/ B1/ 616 (155)/ ONT:H21
EC155/ AP/ D/ 648/ ONT:HNT
317-EC/ AP/ A/ 10 (10)/ O101:HNT
H-4437-11/ EX/ A/ 88 (23)/ O20:H12
RS6/ AP/ A/ 23 (23)/ ONT:H9
RS76/ AP/ A/ 93 (168)/ O7:HNT
RS70/ AP/ A/ 93 (168)/ O7:HNT
RS80/ AP/ D/ 4137/ O100:H8
RS63/ AP/ D/ 616 (155)/ ONT:H21
RS65/ AP/ D/ 616 (155)/ ONT:H21
RS46/ AP/ B1/ 616 (155)/ ONT:H21
RS57/ AP/ B1/ 4139/ ONT:H21
RS51/ AP/ D/ 4139/ ONT:H21
M10062/ EX/ B1/ 359/ O126:H9
EC117/ AP/ B1/ 345/ O8:H21
EC118/ AP/ B1/ 345/ O8:H21
UEL181/ AP/ D/ 117/ ONT:H4
EC184/ AP/ D/ 2179/ O8:H20
EC71/ AP/ B2/ 10 (10)/ O153:H5
UEL10/ AP/ B1/ 206 (206)/ ONT:H5
UEL134/ AP/ B1/ 4136/ O156:H8
UEL17/ AP/ B1/ 4132/ ONT:H28
UEL32/ AP/ A/ 4135/ O5:H10
EC216/ AP/ B1/ 345/ O19:H21
PE14/ AP/ B1/ 4141/ O103:H43
EC126/ AP/ B1/ 359/ O9:H34
H-8102-11/ EX/ A/ 410 (23)/ O8:H9
322-EC/ AP/ D/ 4133/ ONT:HNT
UEL71/ AP/ B2/ 4134/ O8:H48
320-EC/ AP/ A/ 48 (10)/ O6:H11
H-276-12/ EX/ A/ 10 (10)/ ONT:H10
H-1827-12/ EX/ A/ 93 (168)/ O7:HNT
H-5776-11/ EX/ B1/ 155 (155)/ O29:H51
PE23/ AP/ D/ 117/ ONT:H4
M9644/ EX/ D/ 3528/ O11:H16
347-EC/ AP/ A/ 23 (23)/ ONT:H9
353-EC/ AP/ A/ 752/ O109:H8
M4170/ EX/ B1/ 4138/ ONT:HNT
U-8265-12/ EX/ D/ 117/ O114:H4
UEL28/ AP/ B1/ 156 (156)/ ONT:H28
PE25/ AP/ B1/ 1674/ O11:H25
H-2249-12/ EX/ B2/ 131/ O25:H12
U-8743-12/ EX/ D/ 69 (69)/ O15:HNT
EC125/ AP/ D/ 57 (350)/ ONT:HNT
EC165/ AP/ D/ 351/ ONT:HNT
EC113/ AP/ B1/ 3721/ ONT:H25
EC114/ AP/ B1/ 3721/ ONT:H2
H-2031-12/ EX/ D/ 1163/ O149:H23
M6376/ EX/ D/ 1011/ O19:H45
M8230/ EX/ D/ 1011/ O19:H45
UEL100/ AP/ B2/ 2494/ O150:HNT
UEL107/ AP/ B2/ 2494/ O150:HNT
UEL138/ AP/ B2/ 2494/ O150:HNT
UEL147/ AP/ B2/ 2494/ O150:HNT
UEL150/ AP/ B2/ 2494/ O150:HNT
PR50/ AP/ B2/ 2494/ O150:HNT
UEL126/ AP/ B1/ 1148/ ONT:H7
UEL83/ AP/ B2/ 2494/ O150:HNT
RS26/ AP/ B2/ 131/ O25:H4
RS45/ AP/ B2/ 141/ O2:HNT
EC57/ AP/ D/ 117/ ONT:H4
UEL19/ AP/ D/ 117/ ONT:H53
UEL09/ AP/ D/ O115/ O2:H9
UEL14/ AP/ D/ 4131/ ONT:H19
UEL27/ AP/ B1/ 206 (206)/ O36:H5
UEL57/ AP/ A/ 88 (23)/ O8:H11
M2782/ EX/ D/ 648/ O1:HNT
M2906/ EX/ A/ 44 (10)/ ONT:H4
M7872/ EX/ A/ 410 (23)/ ONT:H9
M2527/ EX/ D/ 405 (405)/ O102:HNT
M8723/ EX/ D/ 648/ O86:HNT
H-709-12/ EX/ D/ 38 (38) O153:H18
U-8731-12/ EX/ D/ 69 (69)/ ONT:H2
EC69/ AP/ D/ 1011/ O9:H45
UEL199/ AP/ A/ 93 (168)/ O5:H10
U-8301-12/ EX/ D/ 543/ ONT:H9
EC183/ AP/ B1/ 2179/ O8:H20
EC65/ AP/ B1/ 345/ O8:H21
EC146/ AP/ B2/ 720/ ONT:HNT
UEL89/ AP/ B1/ 1148/ ONT:H7
RS81/ AP/ A/ 93 (168) O5:H10
RS86/ AP/ A/ 4140/ O5:H10
EC147/ AP/ B2/ 720/ O73:HNT
EC61/ AP/ A/ 542/ O109:H45
359-EC/ AP/ B1/ 58 (155)/ OX9/H8
EC191/ AP/ A/ 23 (23)/ O8:H9
362-EC/ AP/ A/ 2328/ O54:H2
U-8719-12/ EX/ B1/ 641 (86)/ O30:HNT
U-8529-12/ EX/ B1/ 641 (86)/ ONT:H10
H-1228-12/ EX/ D/ 657/ ONT:H18
334-EC/ AP/ A/ 1711/ O9:H4
UEL01/ AP/ B1/ 746/ O2:H5
306-EC/ AP/ B1/ 3580/ O8:H7
U-8337-12/ EX/ A/ 90 (23)/ O8:H9
303-EC/ AP/ A/ 155 (155)/ O11:H21
H-2339-12/ EX/ B1/ 155 (155)/ O2:H11
U-8726-12/ EX/ B1/ 2539/ O69:H2
U-7987-12/ EX/ B1/ 443 (205)/ O176:H34
U-8737-12/ EX/ B1/ 1249/ O167:H14
RS84/ AP/ B1/ 602 (446)/ O68:H21
266-EC/ AP/ B1/ 212/ O106:H49
H-2051-12/ EX/ A/ 1721/ ONT:H32
H-2060-12/ EX/ A/ 1721/ ONT:H32
M880/ EX/ A/ 44 (10)/ ONT:H4
U-9007/ EX/ A/ 10 (10)/ ONT:H10
U-8691-12/ EX/ A/ 10 (10)/ ONT:H51
H-8327-11/ EX/ B1/ 648/ ONT:H10
U-8327-12/ EX/ B1/ 648/ ONT:H10
251-EC/ AP/ B2/ 141/ O2:HNT
H-2308-12/ EX/ B2/ 1861/ O16:H5
M3724/ EX/ B2/ 73 (73)/ O6:H1
U-7932-12/ EX/ B2/ 1441/ O32:H27
U-8276-12/ EX/ B2/ 372/ O6:H31
U-8718-12/ EX/ B2/ 127/ ONT:H31
U-8752-12/ EX/ B2/ 127/ O6:HNT
H-7559-11/ EX/ B2/ 95 (95)/ O2:H4
U-8628-12/ EX/ B2/ 421 (95)/ O1:H7
H-1874-12/ EX/ B2/ 73 (73)/ O6:H1
M73814/ EX/ B2/ 131/ O16:H5
H-2299-12/ EX/ B2/ 131/ O16:H5
U-7512-12/ EX/ B2/ 131/ O25:H4
U-8754-12/ EX/ B2/ 4171/ O25:H4
U-8109-12/ EX/ D/ 69 (69)/ O73:H18
U-9066/ EX/ D/ 69 (69)/ ONT:H4

Figure 1. Dendrogram showing similarity relationship among APEC (n = 76) and human ExPEC (n = 53). Similarity was established by the presence of virulence genes, using the Pearson correlation (centered). Isolates were clustered by the complete linkage method. Legends adopt the following pattern: STRAIN ID/CATEGORY (either APEC or human ExPEC)/ECOR/ST (ST COMPLEX – if applicable)/SEROTYPE. Darker spots indicate the presence of the referred genes.

This APEC strain was positive for eight virulence genes while the ExPEC was positive for 17, however, seven genes overlapped between these strains which belonged to ECOR B1.

APEC and ExPEC strains were uniformly distributed among the ECOR phylogenetic groups

When classified into phylogenetic groups (ECOR), the most frequent groups among 76 APEC strains were B1 (n = 27, 35.5%), followed by A (n = 18, 23.7%), D (n = 17, 22.4%) and B2 (n = 14, 18.4%). On the other hand, among the 53 human ExPEC strains, groups D (n = 16, 30.2%), B2 (n = 14, 26.4%), A (n = 12, 22.6%) and B1 (n = 11, 20.8%) were, the most prevalent. Overall, the phylogenetic groups most frequently detected among 129 *E. coli* strains analyzed were B1 (n = 38, 29.5%), D (n = 33, 25.6%), A (n = 30, 23.3%) and B2 (n = 28, 21.7%).

Nine STs overlapped among APEC and human ExPEC

Novel STs reported in this work include ST4131-37 and ST4139-41, found in APEC strains, and ST4139 and ST4171, detected in human ExPEC isolates. Seventy-six APEC strains were classified into 45 distinct sequence types (ST) while the 53 ExPEC strains were contained in 34 different STs (Table 3). Overall, 70 different STs were detected in 129 *E. coli* strains. No ST was detected in more than 10% of the strains of each category (APEC or human ExPEC). The STs more frequently found in APEC were ST2494, followed by ST117, ST345, ST93, and ST616. In human ExPEC, the most frequent STs were ST69, ST131, and ST648. A total of nine, among the 71 different STs, were shared between APEC and human ExPEC (ST10, ST88, ST93, ST117, ST131, ST155, ST359, ST648 and ST1011). APEC and human ExPEC strains included, respectively, 36 and 25 exclusive STs. In APEC and ExPEC strains that shared the same serotype, the same ST or ST complex was also detected (O7:HNT, ST93; O25:H4, ST131; and O8:H9, ST complex 23) (Figure 1).

The STs correlated well with the ECOR phylogenetic groups. Only seven different STs were represented by more than one phylogenetic group (ST10 - A and B2; ST155 - A and B1; ST616, ST648, ST2179, ST4139 - B1 and D) (Figure 1). The tree formed by the concatenated MLST alleles presented several clusters, most of them containing both APEC and human ExPEC strains. The number of clusters demonstrated that extra-intestinal *E. coli* isolated either from poultry or humans in Brazil, present a complex evolutionary history (Figure 2).

Discussion

Although previous work has suggested that a subset of avian and human ExPEC may harbor overlapping characteristics, the zoonotic potential of APEC is still questionable. Here, the comparison of several characteristics found in avian and human *E. coli* from Brazil revealed that, despite the diversity among extra-intestinal *E. coli*, many characteristics were shared.

Some overlapping serogroups detected in this work, included O2, O6, O7 and O25, which are commonly detected in human ExPEC [24]. Previous comparisons have demonstrated the overlapping of O2, O6 [2,4] and O25 [4] between APEC and human ExPEC. This work demonstrates that O7 may also be shared. Three O7:HNT avian or human *E. coli* strains (RS76, RS70 and H-1827/12) detected in this work also belonged to ST93 (ST complex 168) group A and were clustered together (cluster B). However, since the human ExPEC and the APEC strains did not present indistinguishable or closely related PFGE profiles, it seems reasonable to speculate that either no transmission has occurred or if it has occurred, there was a genotypic divergence (represented by the dissimilar PFGE patterns) after that event.

In general, genes usually found on plasmid pColV (*tsh*, *iss* and *hly*F) were associated with APEC. On the other hand, ExPEC strains showed lower frequencies or no statistical difference in the frequencies of genes usually found in plasmid pColV, except that that may also possess a chromosomal copy of a similar cluster of virulence genes (*sit*D). The same trend has been reported with APEC and ExPEC of North America [4,10] and Europe [2]. It seems that, despite the diversity found, *E. coli* strains that present some virulence determinants (i. e., pColV genes) are more adapted to produce disease in chickens worldwide, as suggested previously [1,25].

E. coli strains isolated in Europe and North America, from ST117 and phylogenetic group D, of avian and human disease origin, may be closely related [12,26,27]. This ST is spread in APEC populations in several countries, such as the USA [11], Sri Lanka [28], Egypt [29] and Denmark [30]. In the present work, the presence of ST117 and phylogroup D was found in more than 5% of the APEC and one human ExPEC. This human strain (U8265/12) also presented a genetic profile with many genes usually found in ColV plasmids and was grouped into the major APEC cluster. It would be very interesting to determine if ST117 strains isolated worldwide also have ColV related plasmids. These hypothetical hybrid strains could have the potential to infect humans and birds.

The sharing of ST10 clonal complex strains detected in this work are of interest since strains with this ST are emerging pathogens [26]. This ST has been detected in APEC [28,30] and has been shared among strains isolated from poultry and human cases of UTI and sepsis [31]. Thus, suggesting this complex seems to be an important zoonotic-associated ST with a worldwide range. Moreover, ST155 should also be examined carefully since it was previously detected in avian strains [28,29,31] and in this work it was shared between APEC and human ExPEC.

The detection of APEC and ExPEC isolates (RS26, U8752/12, H2299/12, M73814 and H2249/12) belonging to the serogroup O25 or O16, phylogroup B2 and ST131 warrants further attention because strains with these serogroups and STs are found in animals and food, and constitute a risk associated with extra-intestinal diseases in humans [32,33], including UTI and meningitis cases in Brazil [34,35]. *E. coli* ST131 O25:H4 has also been detected in poultry meat [12,31,36,37]. Interestingly, all the strains with these characteristics isolated in the present work clustered into phylogroup A and were not marked by the presence of pColV genes, contrasting with APEC strains isolated previously [32]. The only exception was a human ExPEC strain (H2249/12). Although the ST131 APEC strain did not present a pColV gene repertoire, it was positive for *ibe*A, similarly to the ST131 O25:H4 APEC strains isolated in Spain [37]. It was suggested that the

Table 2. Frequencies of O types among APEC (n = 81) and ExPEC (n = 53) strains isolated in Brazil.

Serogroup	APEC		Human ExPEC	
	n	(%)	n	(%)
NT	29	(35.8)	17	(32.1)
1	-	(0)	2	(3.8)
2	4	(4.9)	2	(3.8)
5	4	(4.9)	0	(0)
6	1	(1.2)	4	(7.5)
7	3	(3.7)	1	(1.9)
8	11	(13.6)	2	(3.8)
9	3	(3.7)	-	(0)
11	2	(2.5)	1	(1.9)
15	-	(0)	2	(3.8)
16	-	(0)	3	(5.7)
19	1	(1.2)	2	(3.8)
20	-	(0.0)	1	(1.9)
25	1	(1.2)	3	(5.7)
29	-	(0)	1	(1.9)
30	-	(0)	1	(1.9)
32	-	(0)	1	(1.9)
36	1	(1.2)	-	(0)
54	1	(1.2)	-	(0)
68	1	(1.2)	-	(0)
69	-	(0)	1	(1.9)
73	1	(1.2)	1	(1.9)
86	-	(0)	1	(1.9)
100	3	(3.7)	-	(0)
101	1	(1.2)	-	(0)
102	-	(0)	1	(1.9)
103	1	(1.2)	-	(0)
106	1	(1.2)	-	(0)
109	2	(2.5)	-	(0)
114	-	(0)	1	(1.9)
126	-	(0)	1	(1.9)
149	-	(0)	1	(1.9)
150	7	(8.6)	-	(0)
153	1	(1.2)	1	(1.9)
156	1	(1.2)	-	(0)
167	-	(0)	1	(1.9)
176	-	(0)	1	(1.9)
OX9	1	(1.2)	-	(0)
Total	81	(100)	53	(100)

*NT – Non-typable (including autoagglutination, multiple positives, rough and negative results).

APEC strains isolated in Spain display zoonotic potential due to the presence of very close PFGE patterns between them and human strains. In a similar fashion to this work, some APEC from ST23 clonal complex of a previous work also presented few virulence genes [11]. Also, ST23 clonal complex phylogroup A was previously detected both in APEC and human ExPEC strains [31]. This clonal complex has been detected in phylogroup A *E. coli* isolated from UTI cases in Brazil [34]. The present work shows that ST23 clonal complex, phylogroup A, *E. coli* of avian and human origin may share more than 70% of similarity by PFGE. These findings reinforce that avian strains of ST23 clonal complex are strong candidates demonstrating zoonotic potential.

Interestingly, ST359 was present in APEC EC126 and ExPEC M10062, strains that displayed more than 70% similarity by

Table 3. Frequencies of sequence types (STs) among APEC (n = 76) and ExPEC (n = 53) strains isolated in Brazil.

Sequence types (ST complex)	APEC		ExPEC	
	n	(%)	n	(%)
10 (10)	2	(2.5)	3	(5.7)
23 (23)	3	(3.7)	-	(0)
38 (38)	-	(0)	1	(1.9)
44 (10)	-	(0)	2	(3.8)
48 (10)	1	(1.2)	-	(0)
57 (350)	1	(1.2)	-	(0)
58 (155)	1	(1.2)	-	(0)
69 (69)	-	(0)	4	(7.5)
73 (73)	-	(0)	2	(3.8)
88 (23)	2	(2.5)	1	(1.9)
90 (23)	-	(0)	1	(1.9)
93 (168)	4	(4.9)	1	(1.9)
95 (95)	-	(0)	1	(1.9)
115	1	(1.2)	-	(0)
117	5	(6.2)	1	(1.9)
127	-	(0)	2	(3.8)
131	1	(1.2)	4	(7.5)
141	2	(2.5)	-	(0)
155 (155)	1	(1.2)	2	(3.8)
156	1	(1.2)	-	(0)
206 (206)	2	(2.5)	-	(0)
212	1	(1.2)	-	(0)
345	5	(6.2)	-	(0)
351	1	(1.2)	-	(0)
359	1	(1.2)	1	(1.9)
372	-	(0)	1	(1.9)
405 (405)	-	(0)	1	(1.9)
410 (23)	-	(0)	2	(3.8)
421 (95)	-	(0)	1	(1.9)
443 (205)	-	(0)	1	(1.9)
542	1	(1.2)	-	(0)
543	-	(0)	1	(1.9)
602 (446)	1	(1.2)	-	(0)
616 (155)	4	(4.9)	-	(0)
641 (86)	-	(0)	2	(3.8)
648	1	(1.2)	4	(7.5)
657	-	(0)	1	(1.9)
720	2	(2.5)	-	(0)
746	1	(1.2)	-	(0)
752	1	(1.2)	-	(0)
949	1	(1.2)	-	(0)
1011	1	(1.2)	2	(3.8)
1148	2	(2.5)	-	(0)
1163	-	(0)	1	(1.9)
1249	-	(0)	1	(1.9)
1441	-	(0)	1	(1.9)
1674	1	(1.2)	-	(0)
1711	1	(1.2)	-	(0)

Table 3. Cont.

Sequence types (ST complex)	APEC		ExPEC	
	n	(%)	n	(%)
1721	-	(0)	2	(3.8)
1861	-	(0)	1	(1.9)
2179	2	(2.5)	-	(0)
2309	-	(0)	1	(1.9)
2328	1	(1.2)	-	(0)
2494	7	(8.6)	-	(0)
2539	-	(0)	1	(1.9)
3528	-	(0)	1	(1.9)
3580	1	(1.2)	-	(0)
3721	2	(2.5)	-	(0)
4131	1	(1.2)	-	(0)
4132	1	(1.2)	-	(0)
4133	1	(1.2)	-	(0)
4134	1	(1.2)	-	(0)
4135	1	(1.2)	-	(0)
4136	1	(1.2)	-	(0)
4137	1	(1.2)	-	(0)
4138	-	(0)	1	(1.9)
4139	2	(2.5)	-	(0)
4140	1	(1.2)	-	(0)
4141	1	(1.2)	-	(0)
4171	-	(0)	1	(1.9)
Total	76	(100)	53	(100)

PFGE, in much the same way as strains of the ST23 clonal complex described above. Both strains shared some genes from ColV plasmid (e. g., *iss*, *hly*F, *iroN*) but the ExPEC strain also presented genes they may be found in other genetic elements, such as the operons *iuc/iut* and *pap*. To our knowledge, such similarities have not been described before in ST359 strains [11,31]. A previous work has detected this ST in strains associated with UTIs in humans in Brazil [38] and Spain [39]. In Spain, strains were of the phylogroup B1, similar to this work. Thus, avian ST359 strains deserve further attention in order to detect if they could represent a zoonotic concern.

Avian strains belonging to the ST95 group may present remarkable similarities with human strains and are present in APEC populations in North America [10,11] and Europe [30]. *E. coli* of this ST is suggested to present zoonotic potential [26]. Interestingly, this ST was found in only one human ExPEC and no APEC in this work. Although the lack of detection of this ST may be due to lack of a larger sampling, the absence of this ST in another work in Sri Lanka [28] suggests that this ST might be restricted to clones present in specific regions of the world.

Some important human ExPEC STs are not known to present food animal reservoirs to date (ST14, ST73, ST393 and ST405) [26]. In Brazil, at least one of these STs (ST404) has been found in *E. coli* isolated from human UTI and meningitis cases [35]. The lack of detection of these STs in the APEC studied here should be evaluated carefully and does not mean that these STs are not present in poultry in Brazil but may have been limited by the size of the study and number of samples tested.

Other STs shared among human and avian *E. coli* strains in this research included ST648 and ST1011 group D. These STs have been detected in group D human ExPEC strains also [31], but were not detected in a comparison between avian and human ExPEC strains in the USA [11]. ST648 group D APEC has recently been isolated from poultry in Denmark [30] indicating a hypothetical zoonotic potential. ST1011 avian strains should be evaluated further for the zoonotic potential.

The overlapping phylogenetic backgrounds (represented by similar STs) with virulence gene clusters detected in avian and human extra-intestinal *E. coli* in different regions of the world may reflect similar challenges that this pathogen has to overcome to colonize both hosts. The evolution of some extra-intestinal *E. coli* strains is suggested to be recent [26]. This possibility is favored by the fact that many novel STs detected in this work and previous work of others are derived from single mutations of previously known STs [11,28–30].

All in all, our results demonstrated that *E. coli* of avian and human origin may present overlapping characteristics in South America, although no indistinguishable fingerprint patterns were detected between APEC and human ExPEC. Some STs are commonly shared worldwide (ST131, ST117, ST 10 clonal complex, ST23 clonal complex). However, ST95, which is linked to APEC and ExPEC, and suggested to present zoonotic potential,

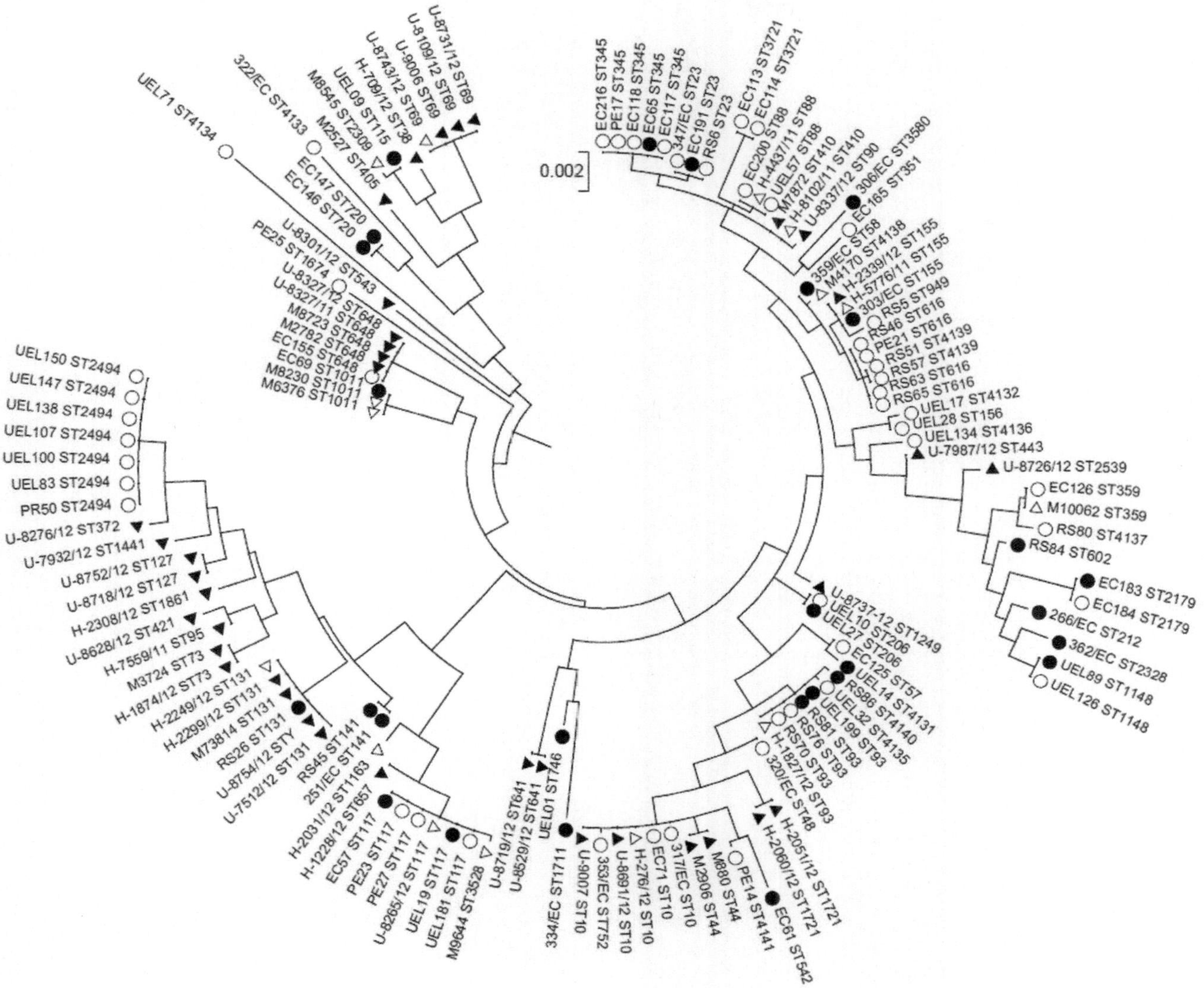

Figure 2. Molecular Phylogenetic analysis of APEC (n = 76) and human ExPEC (n = 53) based on concatenated MLST alleles. The evolutionary history was inferred by using the Maximum Likelihood method based on the Tamura-Nei model [22]. The tree with the highest log likelihood is shown. Circles represent APEC strains and triangles human ExPEC strains. Black geometrical shapes indicate that strains belonged to the major cluster A and white ones show that they are contained in major cluster B (see Figure 1 for more details).

was not detected in APEC in this study. The presence of shared ST359 strains with moderate fingerprint similarity indicates that strains of this ST should be scrutinized for zoonotic potential.

Supporting Information

Figure S1 Dendrogram showing similarity relationship established by PFGE based on the Dice coefficient and clustering by UPGMA. Legends adopt the following pattern: STRAIN ID/CATEGORY (either APEC or human ExPEC)/ ECOR/ST (ST COMPLEX – if applicable)/SEROTYPE. The vertical dotted line indicates the breakpoint for 70% of similarity. Dotted oval forms indicate APEC and ExPEC strains presenting more than 70% of similarity.

Table S1 Primers used to characterize the *E. coli* strains.

Table S2 Frequencies of 43 virulence genes tested in *E. coli* strains clustered in group A (n = 66) and B (n = 63). See Figure 1 for more details.

Author Contributions

Conceived and designed the experiments: WDS. Performed the experiments: RPM CML MRTC TM EALG ACM TS MCR WDS. Analyzed the data: RPM CML TCGR LKN WDS. Contributed reagents/ materials/analysis tools: CML EALG WDS. Contributed to the writing of the manuscript: RPM CML LKN WDS.

References

1. Dziva F, Stevens M (2008) Colibacillosis in poultry: unravelling the molecular basis of virulence of avian pathogenic *Escherichia coli* in their natural hosts. Avian Pathology 37: 355–366.
2. Ewers C, Li G, Wilking H, Kieling S, Alt K, et al. (2007) Avian pathogenic, uropathogenic, and newborn meningitis-causing *Escherichia coli*: How closely related are they? International Journal of Medical Microbiology 297: 163–176.
3. Moulin-Schouleur M, Répérant M, Laurent S, Brée A, Mignon-Grasteau S, et al. (2007) Extraintestinal Pathogenic *Escherichia coli* Strains of Avian and Human Origin: Link between Phylogenetic Relationships and Common Virulence Patterns. Journal of Clinical Microbiology 45: 3366–3376.
4. Rodriguez-Siek KE, Giddings CW, Doetkott C, Johnson TJ, Fakhr MK, et al. (2005) Comparison of *Escherichia coli* isolates implicated in human urinary tract infection and avian colibacillosis. Microbiology 151: 2097–2110.
5. Mellata M (2013) Human and avian extraintestinal pathogenic *Escherichia coli*: infections, zoonotic risks, and antibiotic resistance trends. Foodborne Pathog Dis 10: 916–932.
6. Moreno E, Andreu A, Pérez T, Sabaté M, Johnson JR, et al. (2006) Relationship between *Escherichia coli* strains causing urinary tract infection in women and the dominant faecal flora of the same hosts. Epidemiol Infect 134: 1015–1023.
7. Yamamoto S, Tsukamoto T, Terai A, Kurazono H, Takeda Y, et al. (1997) Genetic evidence supporting the fecal-perineal-urethral hypothesis in cystitis caused by *Escherichia coli*. J Urol 157: 1127–1129.
8. Russo TA, Johnson JR (2000) Proposal for a new inclusive designation for extraintestinal pathogenic isolates of *Escherichia coli*: ExPEC. Journal of Infectious Diseases 181: 1753–1754.
9. Kaper JB, Nataro JP, Mobley HLT (2004) Pathogenic *Escherichia coli*. Nature Reviews Microbiology 2: 123–140.
10. Johnson TJ, Wannemuehler Y, Johnson SJ, Stell AL, Doetkott C, et al. (2008) Comparison of extraintestinal pathogenic *Escherichia coli* strains from human and avian sources reveals a mixed subset representing potential zoonotic pathogens. Appl Environ Microbiol 74: 7043–7050.
11. Danzeisen JL, Wannemuehler Y, Nolan LK, Johnson TJ (2013) Comparison of multilocus sequence analysis and virulence genotyping of *Escherichia coli* from live birds, retail poultry meat, and human extraintestinal infection. Avian Dis 57: 104–108.
12. Vincent C, Boerlin P, Daignault D, Dozois CM, Dutil L, et al. (2010) Food reservoir for *Escherichia coli* causing urinary tract infections. Emerg Infect Dis 16: 88–95.
13. Maluta RP, Gatti MSV, Joazeiro PP, de Paiva JB, Rojas TCG, et al. (2014) Avian extra-intestinal *Escherichia coli*exhibits enterotoxigenic-like activity in the *in vivo* rabbit ligated ileal loop assay. Foodborne Pathogens and Disease 11: 484–489.
14. Arbeit DJ (1995) Laboratory procedures for the epidemiologic analysis of microorganisms. In: Murray PR, Baron EJ, Pfaller MA, Tenover FC, Yolken RH, editors. Manual of clinical microbiology. 6th ed. Washington, D.C.: American Society for Microbiology. 190–208.
15. Orskov F, Orskov I (1984) Serotyping of *Escherichia coli*. Methods Microbiol 14: 43–112.
16. Machado J, Grimont F, Grimont PA (2000) Identification of *Escherichia coli* flagellar types by restriction of the amplified fliC gene. Res Microbiol 151: 535–546.
17. Borges CA, Beraldo LG, Maluta RP, Cardozo MV, Guth BEC, et al. (2012) Shiga Toxigenic and Atypical Enteropathogenic *Escherichia coli* in the Feces and Carcasses of Slaughtered Pigs. Foodborne Pathogens and Disease 9: 1119–1125.
18. Clermont O, Bonacorsi S, Bingen E (2000) Rapid and simple determination of the *Escherichia coli* phylogenetic group. Appl Environ Microbiol 66: 4555–4558.
19. de Hoon MJL, Imoto S, Nolan J, Miyano S (2004) Open source clustering software. Bioinformatics 20: 1453–1454.
20. Saldanha AJ (2004) Java Treeview–extensible visualization of microarray data. Bioinformatics 20: 3246–3248.
21. Wirth T, Falush D, Lan R, Colles F, Mensa P, et al. (2006) Sex and virulence in *Escherichia coli*: an evolutionary perspective. Mol Microbiol 60: 1136–1151.
22. Tamura K, Nei M (1993) Estimation of the number of nucleotide substitutions in the control region of mitochondrial DNA in humans and chimpanzees. Mol Biol Evol 10: 512–526.
23. Tamura K, Peterson D, Peterson N, Stecher G, Nei M, et al. (2011) MEGA5: molecular evolutionary genetics analysis using maximum likelihood, evolutionary distance, and maximum parsimony methods. Mol Biol Evol 28: 2731–2739.
24. Smith JL, Fratamico PM, Gunther NW (2007) Extraintestinal Pathogenic *Escherichia coli*. Foodborne Pathogens and Disease 4: 134–163.
25. Ginns CA, Benham ML, Adams LM, Whithear KG, Bettelheim KA, et al. (2000) Colonization of the Respiratory Tract by a Virulent Strain of Avian *Escherichia coli* Requires Carriage of a Conjugative Plasmid. Infection and Immunity 68: 1535–1541.
26. Manges AR, Johnson JR (2012) Food-borne origins of *Escherichia coli* causing extraintestinal infections. Clin Infect Dis 55: 712–719.
27. Mora A, López C, Herrera A, Viso S, Mamani R, et al. (2012) Emerging avian pathogenic *Escherichia coli* strains belonging to clonal groups O111:H4-D-ST2085 and O111:H4-D-ST117 with high virulence-gene content and zoonotic potential. Vet Microbiol 156: 347–352.
28. Dissanayake DR, Octavia S, Lan R (2014) Population structure and virulence content of avian pathogenic *Escherichia coli* isolated from outbreaks in Sri Lanka. Vet Microbiol 168: 403–412.
29. Hussein AH, Ghanem IA, Eid AA, Ali MA, Sherwood JS, et al. (2013) Molecular and phenotypic characterization of *Escherichia coli* isolated from broiler chicken flocks in Egypt. Avian Dis 57: 602–611.
30. Pires-dos-Santos T, Bisgaard M, Christensen H (2013) Genetic diversity and virulence profiles of *Escherichia coli* causing salpingitis and peritonitis in broiler breeders. Vet Microbiol 162: 873–880.
31. Giufrè M, Graziani C, Accogli M, Luzzi I, Busani L, et al. (2012) *Escherichia coli* of human and avian origin: detection of clonal groups associated with fluoroquinolone and multidrug resistance in Italy. J Antimicrob Chemother 67: 860–867.
32. Platell JL, Johnson JR, Cobbold RN, Trott DJ (2011) Multidrug-resistant extraintestinal pathogenic *Escherichia coli* of sequence type ST131 in animals and foods. Vet Microbiol 153: 99–108.
33. Rogers BA, Sidjabat HE, Paterson DL (2011) *Escherichia coli* O25b-ST131: a pandemic, multiresistant, community-associated strain. J Antimicrob Chemother 66: 1–14.
34. Peirano G, Asensi MD, Pitondo-Silva A, Pitout JD (2011) Molecular characteristics of extended-spectrum β-lactamase-producing *Escherichia coli* from Rio de Janeiro, Brazil. Clin Microbiol Infect 17: 1039–1043.
35. Berman H, Barberino MG, Moreira ED, Riley L, Reis JN (2014) Distribution of strain type and antimicrobial susceptibility of *Escherichia coli* causing meningitis in a large urban setting in Brazil. Journal of Clinical Microbiology.
36. Kawamura K, Goto K, Nakane K, Arakawa Y (2014) Molecular Epidemiology of Extended-Spectrum β-Lactamases and *Escherichia coli* Isolated from Retail Foods Including Chicken Meat in Japan. Foodborne Pathog Dis 11: 104–110.
37. Mora A, Herrera A, Mamani R, López C, Alonso MP, et al. (2010) Recent Emergence of Clonal Group O25b:K1:H4-B2-ST131 ibeA Strains among *Escherichia coli* Poultry Isolates, Including CTX-M-9-Producing Strains, and Comparison with Clinical Human Isolates. Applied and Environmental Microbiology 76: 6991–6997.
38. Minarini LA, Camargo IL, Pitondo-Silva A, Darini AL (2007) Multilocus sequence typing of uropathogenic ESBL-producing *Escherichia coli* isolated in a Brazilian community. Curr Microbiol 55: 524–529.
39. Valverde A, Cantón R, Garcillán-Barcia MP, Novais Â, Galán JC, et al. (2009) Spread of blaCTX-M-14 Is Driven Mainly by IncK Plasmids Disseminated among *Escherichia coli* Phylogroups A, B1, and D in Spain. Antimicrobial Agents and Chemotherapy 53: 5204–5212.

4

Population Dynamics of a Salmonella Lytic Phage and Its Host: Implications of the Host Bacterial Growth Rate in Modelling

Sílvio B. Santos, Carla Carvalho, Joana Azeredo, Eugénio C. Ferreira*

Centre of Biological Engineering, Universidade do Minho, Braga, Portugal

Abstract

The prevalence and impact of bacteriophages in the ecology of bacterial communities coupled with their ability to control pathogens turn essential to understand and predict the dynamics between phage and bacteria populations. To achieve this knowledge it is essential to develop mathematical models able to explain and simulate the population dynamics of phage and bacteria. We have developed an unstructured mathematical model using delay-differential equations to predict the interactions between a broad-host-range *Salmonella* phage and its pathogenic host. The model takes into consideration the main biological parameters that rule phage-bacteria interactions likewise the adsorption rate, latent period, burst size, bacterial growth rate, and substrate uptake rate, among others. The experimental validation of the model was performed with data from phage-interaction studies in a 5 L bioreactor. The key and innovative aspect of the model was the introduction of variations in the latent period and adsorption rate values that are considered as constants in previous developed models. By modelling the latent period as a normal distribution of values and the adsorption rate as a function of the bacterial growth rate it was possible to accurately predict the behaviour of the phage-bacteria population. The model was shown to predict simulated data with a good agreement with the experimental observations and explains how a lytic phage and its host bacteria are able to coexist.

Editor: Julio Vera, University of Erlangen-Nuremberg, Germany

Funding: Financial support was received through the Strategic Project PEst-OE/EQB/LA0023/2013 from the FCT-Fundação para a Ciência e Tecnologia (http://www.fct.pt) and the projects "BioHealth - Biotechnology and Bioengineering approaches to improve health quality", Ref. NORTE-07-0124 FEDER-000027, co-funded by the Programa Operacional Regional do Norte (ON.2 – O Novo Norte), QREN, FEDER and "Consolidating Research Expertise and Resources on Cellular and Molecular Biotechnology at CEB/IBB", Ref. FCOMP-01-0124-FEDER-027462. Sílvio B. Santos was supported by the grant SFRH/BPD/75311/2010 and Carla Carvalho was supported by the grant SFRH/BPD/79365/2011 both from the FCT-Fundação para a Ciência e Tecnologia (http://www.fct.pt). The funders had no role in study design, data collection and analysis, decision to publish, or preparation of the manuscript.

Competing Interests: The authors have declared that no competing interests exist.

* Email: ecferreira@deb.uminho.pt

Introduction

(Bacterio)phages, or bacterial viruses, can be found wherever bacteria exist which includes every conceivable habitat as a result of their bacterial parasitism. Their presence in the biosphere is especially predominant in coastal water and in the oceans presenting an excess of 10^7–10^8 phage particles per millilitre and even higher concentrations in lakes. Phages are also abundant in pelagic marine environments and pointed as responsible for a significant loss of bacterioplankton. Roughly 10^9 phages per gram can be found in marine sediments and comparably high numbers in other sources like sewage and faeces, soil, sediments, deep thermal vents and in natural bodies of water. Phages are an extremely diversified group and it has been estimated that ten bacteriophage particles exist for each bacterial cell, accounting for an estimated size of the global phage population to be approximately 10^{31} particles, making phages the most abundant living entities on earth [1–12]. As a consequence of this high prevalence and ubiquity, even rare phage-induced events will be represented at a high frequency at the global level [7]. This fact, coupled with phage primitiveness and ability to infect bacteria, together with the accumulating data from both phage and bacterial genome sequencing projects, have highlighted their ecological impact acting as agents in the recycling of organic matter (including cells), their key role in the adaptive evolution of bacteria and also as tools in molecular biology and epidemiology [2,13–16]. The existing interactions between phages and bacteria, as well as all the predator–prey dynamics, have long been recognized to play a central role in the structure of ecological communities, or, in this particular case, in the relative proportions of different bacterial species or strains in a community [7,17].

The increasing problem of antibiotic-resistance bacteria together with the environmental costs caused by the use of chemotherapeutic agents, have motivated the renewed interest in the use of phages as alternative antimicrobial agents in pathogens control [18–21]. For an efficient therapeutic use of phages it is also vital to understand the phage-bacteria interactions in order to design efficient protocols, optimize phage production, determine optimal dose concentration and predict the outcome of phage therapy.

In order to understand phage ecology it is imperative to study the interaction between phages and their bacterial hosts making use of (and developing) mathematical models. However, the knowledge of the interactions between virus–host systems in their complex natural environments is still limited [20,22]. To improve this knowledge it is generally accepted that the development of mathematical and computer simulation models is essential. Such explicit models deal with phages density-dependent replication characteristics and have been used to study the population and evolutionary dynamics of phages [22–25]. It is thus critical to first obtain basic information on the behavior of specific virus–host systems during controlled conditions in the laboratory which may disclose some of the peculiar kinetics present even in complex environments, consequently predicting the outcome of an encounter between phage and bacteria [22,26]. Such studies will enable to fill the existing gap between mathematical models and natural communities by allowing the comparison between the outputs of mathematical models and the experimental data obtained from controlled, reproducible and easily manipulated biological systems, before conclusions can be extrapolated and applied to uncontrolled and complex natural systems [27]. If experimental results do not reflect the simulated ones, then the residues will be used to refine the model. These iterative experimental tests and refinement of the simulations allow for the understanding of relationships otherwise difficult to observe [25].

The first model which attempted to explain the kinetics of phage-bacteria interactions was developed by Levin (1977) from which subsequent models were adapted from [24]. According to the models developed so far, phage-bacteria population dynamics are typically modelled as three interacting populations: susceptible uninfected bacteria, phage-infected bacteria, and free-phage, which depend on phage growth kinetic parameters such as burst size, latent period, and adsorption rate [20,26].

The work presented herein aims at developing a mathematical model able to predict and explain the basic behaviour of the phage-bacteria population dynamics of lytic phages based on fundamental phage-bacteria biological parameters. Such a model was here developed and the output was found to mimic the experimental data. The key innovative aspects of the model are the incorporation of a normal distribution equation that rules the values of the latent period and the integration of a function that describes the variation of the adsorption rate according to the bacteria growth rate.

Materials and Methods

Media

LB broth, Miller (Sigma-Aldrich Inc., St. Louis MO, USA) was prepared according to the manufacturer's instructions. Agar plates were prepared by adding 1.2% of agar (Applichem, Darmstadt, Germany) to the LB broth and Molten Agar was prepared by adding 0.6% of agar to the LB broth. Minimal Medium was prepared based on M9 medium [28] using the following: Na_2HPO_4 6 g/L, KH_2PO_4 3 g/L, NH_4Cl 1 g/L, NaCl 0.5 g/L, $MgSO_4$ 0.12 g/L, and glucose 5 g/L.

Bacteriophage and Bacteria

The *Salmonella* phage PVP-SE1 isolated from a Regensburg (Germany) wastewater plant as part of the European Project "PhageVet-P"(FP6) was used along with its host, *Salmonella enterica* serovar Enteritidis strain S1400 which belongs to the University of Bristol private collection [29,30].

Determination of Phage Titre

The phage titre, expressed as plaque forming units (PFU), was determined using the Plaque Assay Modified with Antibiotic (PAMA) in LB medium as described by Santos *et al.* [31].

Single-Step Growth Curve Experiments

These experiments were carried out at 37°C, 120 rpm in minimal medium as described by Sambrook and Russell [28], using an overnight pre-inoculum of the bacteria in the same medium. The experiments were made in duplicate and repeated in two different occasions.

Adsorption Rate

Adsorption rate (δ) was determined based on Kropinski [32] using cells grown in minimal medium. Briefly, phage was added to an initial log phase of bacteria suspension at a multiplicity of infection (MOI) below 0.01 to assure that each phage has the opportunity to bind to one, and only one, bacterium. At fixed time intervals, 0.1 ml sample was taken and immediately diluted in 0.89 ml SM buffer (100 mM NaCl, 8 mM $MgSO_4$, 50 mM Tris-HCl at pH = 7.5) and 0.01 ml chloroform. Dilution was immediately vortexed and centrifuged at 9000 g for 5 min. Serial dilutions were carried in SM buffer and phage titre determined. The adsorption rate was calculated by the least squares fitting of the data to equation 1 [33].

$$\ln\left(\frac{P(t)}{P_0}\right) = -\delta\left(\frac{X_{S_0}}{\mu(S(t))}\right)\left(e^{\mu(S)t} - 1\right) \qquad (1)$$

where $\mu(S)$ is the bacteria multiplication rate, δ is the adsorption rate, X_S is the susceptible uninfected bacteria concentration, P is the phage concentration and t is the time. The subscript 0 refers to concentration at initial time.

This equation was obtained by solving the differential equations 2 (valid for $t <$ latent period) and 3, representing the phage and host dynamics respectively:

$$\frac{dP(t)}{dt} = -\delta X_S(t)P(t) \qquad (2)$$

$$\frac{dX_S(t)}{dt} = \mu(S)X_S(t) \qquad (3)$$

The bacteria multiplication rate depends on the substrate concentration and is modelled by the Monod kinetics equation 4:

$$\mu(S) = \frac{\mu_{max} \times S(t)}{S(t) + K_s} \qquad (4)$$

To determine the maximum rate of exponential growth (μ_{max}) and the half-saturation constant (K_s), an overnight pre-inoculum of the bacteria in minimal medium was transferred to fresh liquid minimal medium containing different concentrations of glucose (0.01–8 g/L of glucose) in order to obtain an Optical Density (OD) at 600 nm of 0.07 and incubated aerobically at 37°C with 120 rpm agitation. Successive determinations of bacterial cell concentration were accomplished every 20 min by reading the OD at 600 nm. The determination of the two parameters mentioned above was made by the least squares fitting of the data to the Monod's model (equation 4). Glucose (substrate) concentration was measured using the kit Glucose-TR (Spinreact, Spain) according to the manufacturer's instructions. The amount of substrate needed for a new bacterium (α) was determined using equation 5.

$$\alpha = \frac{S(t)}{X_s(t)} \tag{5}$$

Infection Assays

In these assays an overnight pre-inoculum of the bacteria in minimal medium was transferred to fresh liquid minimal medium in order to obtain an OD at 600 nm of 0.07 and incubated aerobically at 37°C with 120 rpm of agitation. Phages were added to the bacterial suspension during initial log phase in order to obtain the desired final concentration. The resultant suspension was incubated aerobically at 37°C with agitation (120 rpm) and samples were collected at fixed intervals to determine bacteria concentration through OD, phage concentration through PAMA and glucose (substrate) concentration using the kit Glucose-TR. These experiments were carried on in a 250 ml Erlenmeyer flask and in a 5 L bioreactor operating in batch mode and using an air inflow of 1 volume/volume of medium.

Model simulations

The DDEs were solved in Matlab (The Mathworks, Natick, MA, USA) using the *dde23* function/solver. The *fminsearch* function in Matlab was also used for least squares parameter fitting.

To assess the differences between the simulations obtained by the model and the experimental data, we have calculated the root-mean-square deviation (RMSD) normalized by the range of observed values of the simulated variable (NRMSD). Values are presented in percentage. Given the data range of each variable, NRMSD was calculated using the log10 values of the data from total bacteria and free phage to avoid favouring the simulated lower values. In the case of substrate the natural values were used instead of log10. Since the NRMSD is based on the residuals, the closer the value is to zero the best is the simulation fitting to the experimental data. It is important to mention that besides its good accuracy NRMSD cannot be used to compare between variables, but only for a particular variable since it is scale dependent.

A relative NRMSD (r.NRMSD) was calculated when assessing the deviation of the model output to a base-case, corresponding to the quotient of the obtained NRSMD due to the parameter variation and the NRMSD of the base-case. To easily interpret the values, when r.NRMSD is below the unity the value presented corresponds to $-1/r.NRMSD$. This way, a value of 2 means that the NRMSD is twice higher the NRMSD of the base-case, while a value of -2 means that the NRMSD is twice lower the NRMSD of the base-case and corresponds in this case to a better fitting of the simulation to the experimental data.

Results

Phage and bacteria interact with each other in a similar way to that of a predator and its prey. The simplest classical model explaining such behaviour is that of Lotka-Volterra. In this simple model the number of preys increase in the absence of predators and decreases as a function of their number. On the other hand, the number of predators decreases in the absence of their substrate (preys) and increases in its presence, proportionally [17,20,25].

We have constructed a mathematical model based in Levin *et al.* (1977) [24], a modification of the classical Lotka-Volterra model. The model developed includes the essential characteristics that rule the phage-bacteria behaviour. One of the most important features included is temporal heterogeneity, the existence of a time delay in the conversion of bacteria predation in phage progeny, an intrinsic characteristic of these systems. The proposed model is composed by the following five delay differential equations (DDEs):

$$\frac{dS(t)}{dt} = -\alpha\mu(S)X_t(t) \tag{6}$$

$$\frac{dX_s(t)}{dt} = \mu(S)X_s(t) - \delta X_s(t)P(t) \tag{7}$$

$$\frac{dX_i(t)}{dt} = \delta X_s(t)P(t) - \delta X_s(t-\tau)P(t-\tau) \tag{8}$$

$$\frac{dX_r(t)}{dt} = \mu(S)X_r(t) \tag{9}$$

$$\frac{dP(t)}{dt} = \beta\delta X_s(t-\tau)P(t-\tau) - \delta X_s(t)P(t) \tag{10}$$

and

$$\delta X_s(t-\tau)P(t-\tau) = 0, \text{ while } t < \tau \tag{11}$$

State variables: S is the substrate concentration, X_s is the susceptible uninfected bacteria concentration, X_i is the infected bacteria concentration, X_r is the resistant bacteria concentration, and P is the free phage concentration.

Parameters: $\mu(S)$ is the bacteria multiplication rate, α is the substrate needed for a new bacterium, δ is the adsorption rate, τ is the latent period, and β is the burst size.

Total bacteria (X_t) is calculated as:

$$X_t(t) = X_s(t) + X_i(t) + X_r(t) \tag{12}$$

The substrate S is uptake by the susceptible uninfected bacteria X_s, infected X_i and resistant bacteria X_r (i.e. the total bacteria X_t) by an amount of α for each newly formed bacterium. The grow rate $\mu(S)$ of resistant bacteria is assumed to be the same as the sensitive bacteria. The life cycle of the phage starts with its irreversible adsorption to the sensitive bacteria X_s at an infection rate δ, proportional to the product of their concentrations given by the law of mass action, and resistant bacteria are obviously not infected by the phage. The phage injects its DNA into the sensitive bacterium (X_s), which becomes infected (X_i), driving the cell to phage replication in order to produce their progeny, and thus it is assumed that the infected bacteria does not grow but still compete for substrate. The phage progeny is not released until the time needed for phage genome injection into the host progeny, phage proteins production and assembly have elapsed, the latent period τ, after which the cell lyses releasing an average number of β (burst size) new phages able to start a new cycle. This is ensured in the model by using delays in the variables of the differential equations. When $t<\tau$, the delayed variables $X_s(t-\tau)$ and $P(t-\tau)$ in equations 8 and 10 are set to zero and turning the term $\delta X_s(t-\tau)P(t-\tau)$ null (equation 11). Consequently, before the latent period, the infected bacteria (equation 8) will only vary as a

Table 1. Phage-bacteria interaction parameters determined experimentally.

Parameter	Symbol	Values	Units
Maximum bacterial growth rate	μ_{max}	0.356	h^{-1}
Half-saturation constant	K_s	390	mg L^{-1}
Substrate for a new bacterium	α	1.23×10^{-3}	mg L^{-1}
Burst size	β	58	PFU CFU^{-1}
Phage adsorption rate	δ_r	1.00×10^{-9}	ml CFU^{-1} PFU^{-1} h^{-1}
Latent period	τ	0.5	h
Rise period	ρ	0.55	h
lambda parameter (eq. 15)	λ	160	
bacterial growth at δ_r	mS_r	0.326	h^{-1}

consequence of phage adsorption to sensitive bacteria, increasing this way the concentration of infected bacteria. On the other hand, free phages will decrease at the same rate and equation 10 will assume the form of equation 2. This behaviour is in accordance with the biological data. As a consequence of the time needed for a bacterium to produce new phages, the population dynamics of a phage-bacteria system does not depend only on the phage and bacteria concentration at time t but also on their concentration τ hours earlier.

The parameters bacterial growth rate, phage adsorption rate, burst size, latent period, and substrate amount needed for a new bacterium, were determined through independent experiments as described in the section "Materials and Methods" and are listed in Table 1. The model was used to simulate the data obtained during the one-step growth experiments and assess the validity of the determined parameters (Figure 1).

It was observed that using the model as presented above, the data generated showed an overgrowth of the phage population in

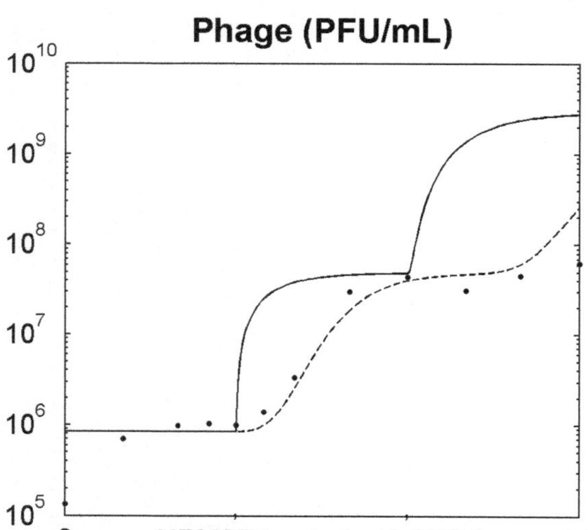

Figure 1. Simulating the one step growth curve using a normal distribution function (equations 6–14). Legend: • experimental data; ... data from the model using the average value of the latent period (equations 6–12); ___ data from the model using a distribution of values of the latent period (introduction of equations 13 and 14). The x axis represents time in hours.

comparison to the experimental data due to a rapid increase of phage titre right after the latent period. In order to overcome this limitation, a rise period (ρ) was introduced in the model to soften the predicted release of the phage by its host. This was accomplished by introducing a normal distribution function (equation 13) with an average $\bar{x}=\tau+\rho/2$ and standard deviation $\sigma=\rho/6$.

$$N_d(x)=\frac{1}{\sigma\sqrt{2\pi}}exp\left[-\frac{1}{2}\left(\frac{x-\bar{x}}{\sigma}\right)^2\right]\Delta x \qquad (13)$$

with $x\in[\tau,\tau+\rho]$ and $\Delta x=\rho/100$.

The normal distribution function when multiplied by Δx, the increment in x, will rule the distribution of the latent period determining the proportion of bacteria that will burst at each time t and which were infected at an elapsed time period between τ and $\tau+\rho$ (or x). The proportion of bacteria that will burst with a latent period x will then be multiplied by the number of bacteria infected x hours earlier giving the number of bacteria that will burst with a latent period of x in each iteration step (equation 14):

$$B_b(x)=\delta X_s(t-x)P(t-x)N_d(x) \qquad (14)$$

with $x\in[\tau,\tau+\rho]$.

The sum of all bacteria bursting with a latent period between τ and $\tau+\rho$, that is $\sum_{x=\tau}^{\tau+\rho} B_b(x)$, will give the number of bacteria that will burst at each time t, and thus will replace the term $\delta X_s(t-\tau)P(t-\tau)$ in equations 8 and 10. The introduction of such a function enabled a better agreement between the simulated and the experimental data (Figure 1).

After assessing the model in a single phage cycle, the model was tested to explain the phage-bacteria behaviour during a longer period. For that, an initial log phase bacterial culture (100 ml) was infected with phage at a MOI of 1.8×10^{-2} in a 250 ml Erlenmeyer flask and the phage, bacteria, and substrate concentrations were determined. This assay was carried out using minimal medium with no addition of substrate during the experiment. This means that the amount of substrate during the experiment will change over time due to bacterial consumption, influencing bacterial physiology. The infection assay was scaled up using a 5 L bioreactor operating in a batch mode with 2 L of medium, where phages were able to infect an initial log phase bacterial culture at a MOI of 5.1×10^{-4}. The data obtained was used to assess the suitability of the model presented and the behaviour of the phage-bacterium system at these two scale volumes (Figure 2 and Figure 3).

The data simulated by the model showed a good agreement with the experimental data but only during the first three hours of the experiment. This might suggest that at least one of the parameters in the model may not be constant but instead should vary in the course of the experiment. During the one-step growth experiments it was observed that the time at which the phage was added to the bacterial culture influenced the value of such parameters. In fact, phage infection of stationary grown cultures gives rise to lower burst sizes and higher latent periods than infection of cells growing exponentially (data not shown).

The variation of the burst size was included in the model but no significant improvement was achieved. Since the bacteria size and phage receptors in the cell wall vary with the different growth phases we have thus included a new equation (equation 15) to define the adsorption rate that allows it to vary along the experiment.

$$\delta(\mu(S)) = A * exp^{(\lambda * \mu(S))} \quad (15)$$

with $A = 2.22 * 10^{-32}, \lambda = 160$

This equation decreases the phage adsorption rate in function of the bacterial growth rate. The function passes through the experimental determined value of the adsorption rate of 1×10^{-9} for a bacterial growth rate of 0.326 (log phase) and decreases to a value close to zero for a very low bacterial growth rate (stationary cells). Taking into account the experimental coordinates, the parameter A was written in order of λ which was fitted using the Matlab *fminsearch* function with the experimental data from Figure 2. Since the adsorption is no longer constant, the term $\delta X_s(t-\tau)P(t-\tau)$ in equation 14 will be replaced by $\delta(t-\tau)X_s(t-\tau)P(t-\tau)$. The inclusion of this term in the model allowed a better agreement of the simulated data during all the experiment time periods (Figure 4 and Figure 5). The model was further tested with additional experimental data from an infection assay in the same conditions as the one from Figure 5 but at a

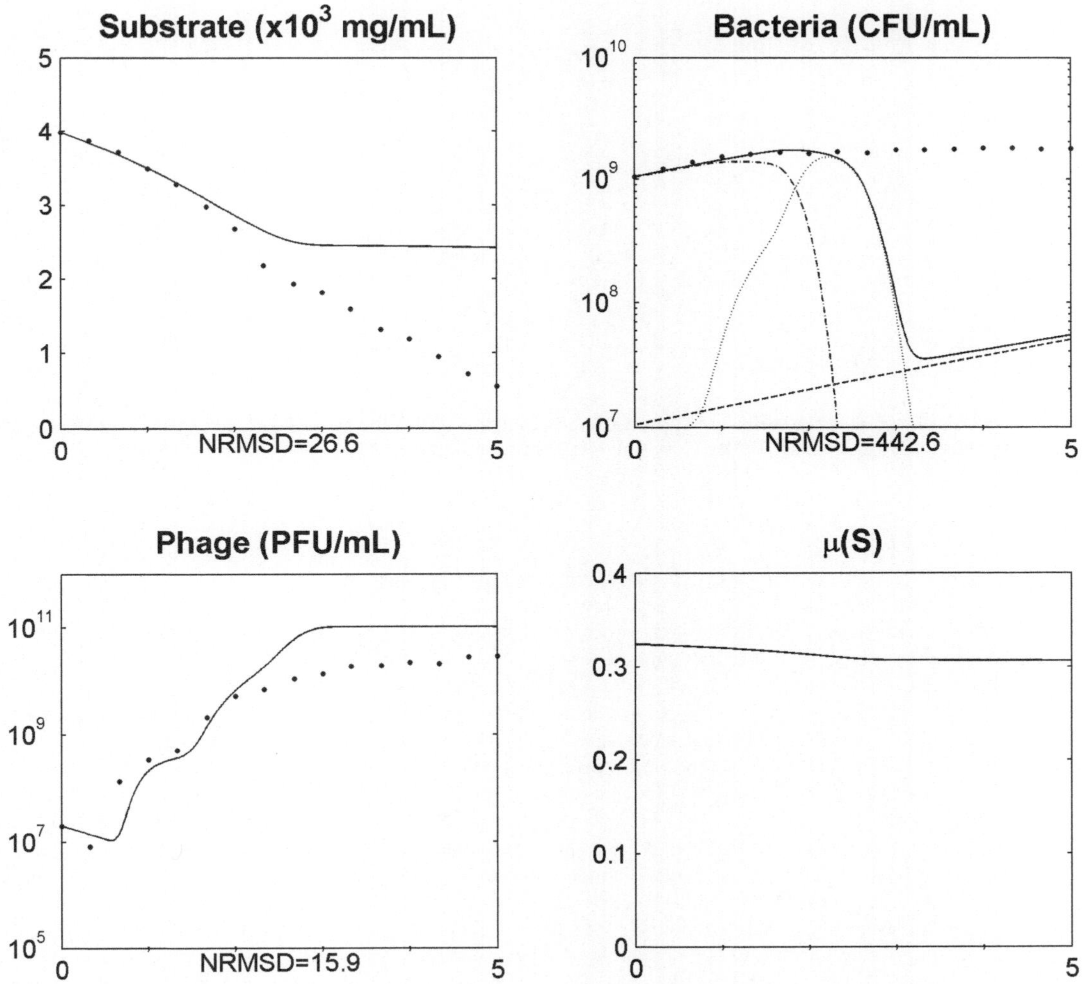

Figure 2. Simulating the phage and bacteria population dynamics using a distribution of values of the latent period (equations 6–14) in a 250 ml Erlenmeyer flask for a MOI = 1.8E10^{-2}. Legend: • experimental data; ___ model simulation; - . - model simulation of susceptible uninfected bacteria (X_s); ... model simulation of infected bacteria (X_i); _ _ model simulation of resistant bacteria (X_r). The x axis represents time in hours.

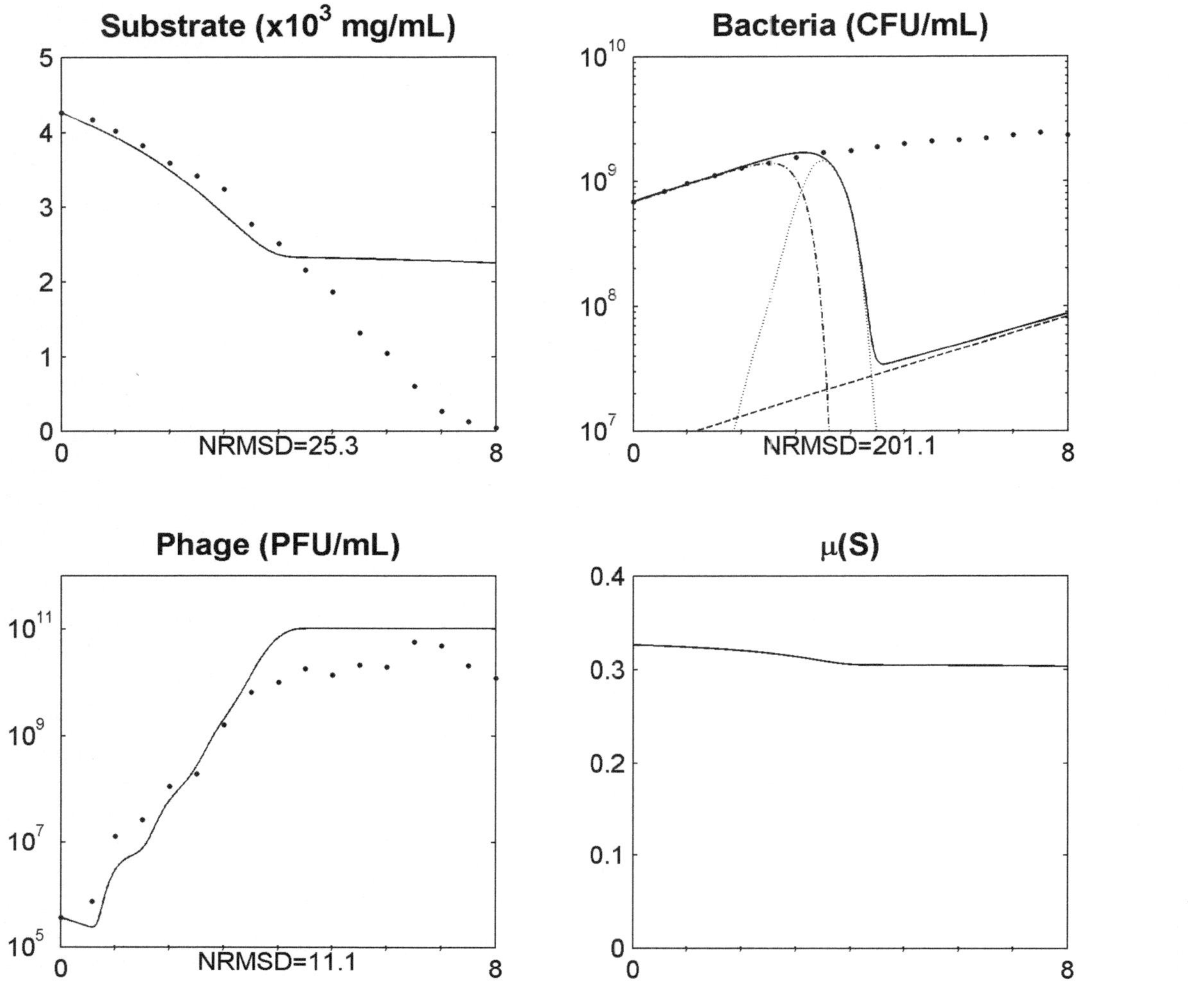

Figure 3. Simulating the phage and bacteria population dynamics using a distribution of values of the latent period (equations 5 and 13) in a 5 L bioreactor for a MOI = 5.1E10^{-4}. Legend: • experimental data; ___ model simulation; - . - model simulation of susceptible uninfected bacteria (X_s); ... model simulation of infected bacteria (X_i); _ _ model simulation of resistant bacteria (X_r). The x axis represents time in hours.

MOI of 2.8×10^{-3} (Figure 6). Also in this case, the simulation values presented a good agreement with the experimental data.

In order to evaluate the influence of the different parameters (inputs) on the model variables (outputs), a sensitivity analysis was conducted using the OAT (one-at-a-time) method, that is, changing the value of each parameter one at a time for a magnitude of 20% ($\pm$10%). This local sensitivity analysis was carried using as the base-case the simulation and data from Figure 5 and all the parameters included in the model were subjected to analysis (Figure 7).

For the sensitivity analysis we have determined a relative NRMSD (r.NRMSD) for each parameter variation which includes in its calculation the value of NRMSD of the base-case (as described in the section "Materials and Methods"). The r.NRMSD enables the determination of the parameter that causes the greatest deviation to the base-case and consequently the parameter for which the model is more sensitive.

To assess the outcome of a variation in the state variables we have modelled the outcome of the *Salmonella* phage and its host by changing each of the state variables alone in $\pm$10% (Figure 8), using as the base-case the same data from Figure 5.

Discussion

Modelling the phage–bacteria interactions

The unpredictability of the population dynamics of the phage–bacteria interactions is related to the intrinsic self-replicating nature of phages. Therefore, the awareness and understanding of kinetics and dynamics of phages and bacteria is important to predict the outcome of the interaction between these two populations in a given environment. To accomplish this, it is crucial to develop mathematical models able to explain and reveal the interaction between phages and bacteria [25]. When these models produce simulations with a good agreement with experimental data they can be powerful tools for the optimization of phage infection parameters towards the improvement of phage production or for the assessment of phage therapy outcome. Nevertheless, it is important to be aware that it is not possible to include all factors accounting for the population and evolutionary dynamics of the interactions between bacteria and phage in a mathematical model [20].

Structured models (models which consider the behaviour of intracellular metabolic pathways in response to environmental

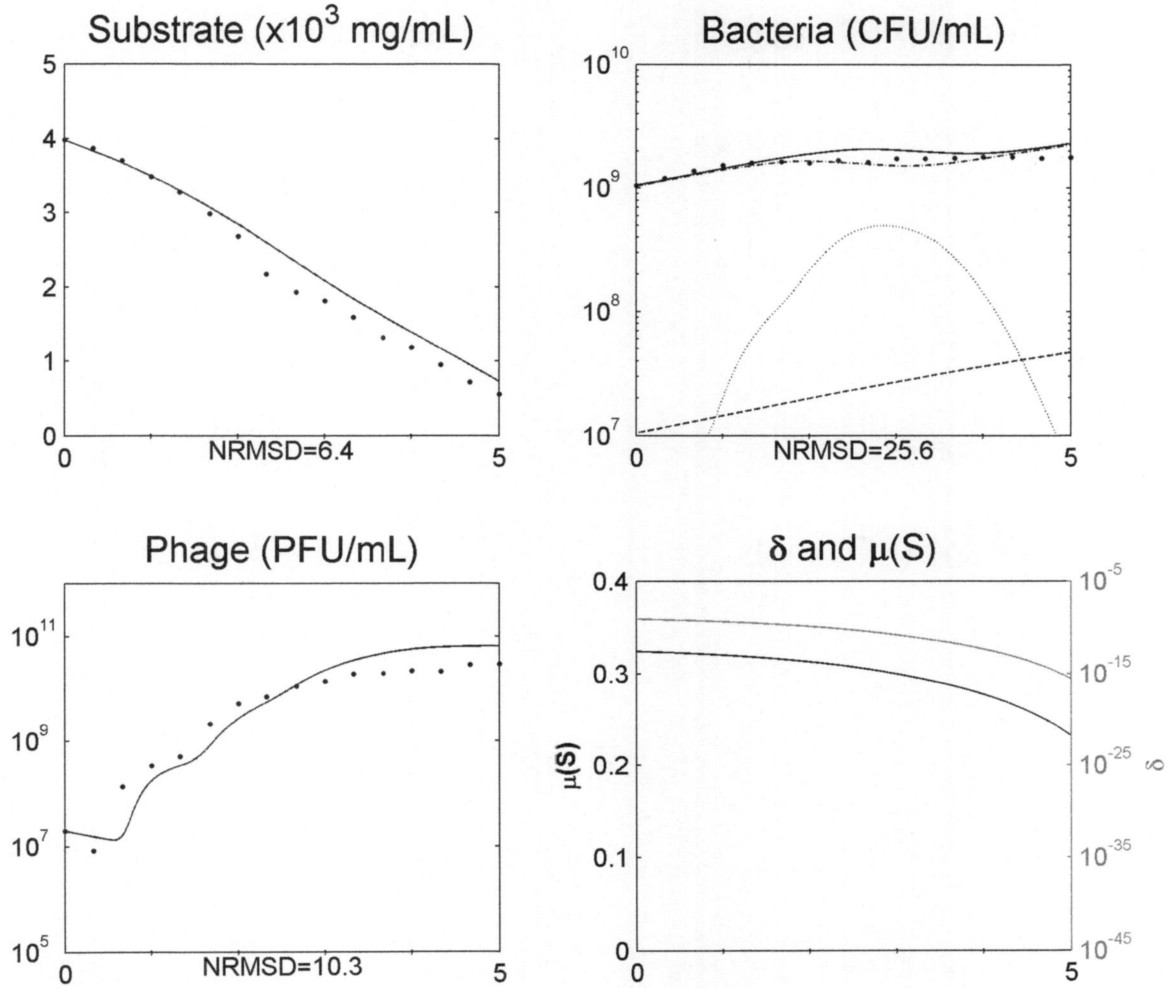

Figure 4. Simulating the phage and bacteria population dynamics using a distribution of values of the latent period and a variation of the adsorption constant (δ) as a function of the bacterial growth rate (μ) in a 250 ml Erlenmeyer flask for a MOI = 1.8E10^{-2}. Legend: • experimental data; ___ model simulation; - . - model simulation of susceptible uninfected bacteria (X_s); ... model simulation of infected bacteria (X_i); _ _ model simulation of resistant bacteria (X_r). The x axis represents time in hours.

changes) provide a more comprehensive description of microbial behaviour than unstructured models, however they are limited, complex and the majority of reported models are not validated with experimental data, in part due to difficulties for the knowledge *in vivo* of the reaction rates of the implied enzymes [34]. On the other hand, unstructured models describe the production of biomass being more simple but equally useful and descriptive of the experimental reality. Essentially, unstructured models of viruses keep track of virus as well of uninfected and infected host populations [35]. The results generated by such unstructured models can be useful in interpreting large data sets and identifying useful new experiments to perform [36].

Populations present in the model

We have here developed an unstructured model useful to understand the generic kinetic phenomena of the self-replicating nature of phages. Models describing phage-bacteria population dynamics are characterized by the interaction of three different populations: susceptible uninfected bacteria, phage-infected bacteria, and free phage [37]. Variations in these populations represent the core of a phage-bacteria unstructured model [35]. Another important population considered in the model developed herein consists of resistant bacteria. Even though this population do not interact with phages and generally do not contribute to the other two, its prevalence in a given environment will vary as a result of the interaction between phages and sensitive bacteria. Moreover, in a scenario where resistant bacteria are able to overcome the susceptible bacteria they will obviously compete for substrate with up to the same fitness. Nevertheless, due to the low proportion of resistant cells in the bacterial population and the time length of the experiments, resistant cells did not produce a significant impact in the results. This may constitute a particular case of the phage used in the experimental assays since it presents a broad lytic spectrum (a characteristic found more frequently in *Myoviridae* phages than in other *Caudovirales*) and thus bacteria may have more difficulty in developing resistance against this phage [7,29,38].

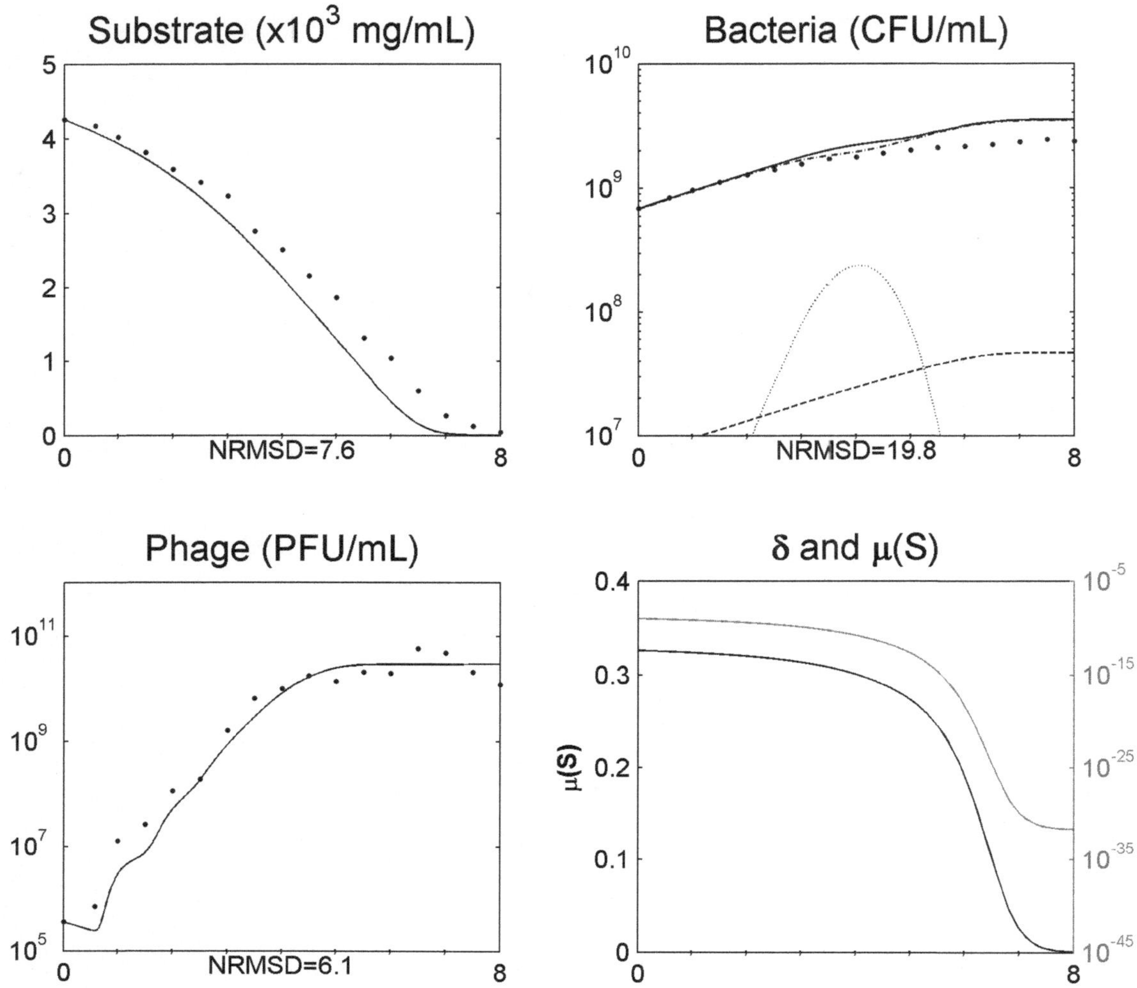

Figure 5. Simulating the phage and bacteria population dynamics using a distribution of values of the latent period and a variation of the adsorption constant (δ) as a function of the bacterial growth rate (μ) in a 5 l bioreactor for a MOI = 5.1E10^{-4}. Legend: • experimental data; ___ model simulation; - . - model simulation of susceptible uninfected bacteria (X_s); ... model simulation of infected bacteria (X_i); _ _ model simulation of resistant bacteria (X_r). The x axis represents time in hours.

Assumptions of the model

Some assumptions were made in the development of the model. The suitability of the model is only for lytic phages and not for temperate phages which can follow the lysogenic cycle not leading to the death of the bacterial cell and thus not acting always as a predator [39]. Phages encounter bacteria at random and will only adsorb to susceptible bacteria which can be infected by only one phage, since infection of the bacterium by a phage can inactivate the remaining receptors [23]. Consequently, the super-infection phenomenon that may exist when the multiplicity of infection (MOI) is high was not included. The infected bacteria do not grow by cell division, a commonly and more accurate assumption in phage-bacteria models, since after infection, phages take control of the bacterial machinery driving the cell only for phage production [23,37,40].

Resistant bacteria are assumed to not present a fitness disadvantage over sensitive bacteria and thus their grow rate $\mu(S)$ is assumed to be the same as the latter [18]. As infected bacterial cells are still metabolically active and requires a significant commitment of host resources in producing phage progeny, it was assumed that infected bacteria compete for substrate with the resistant and susceptible bacteria, an assumption found in other models like the one from Beretta and Kuang [41]. This assumption is corroborated by the work of Middelboe that showed that bacteria continue to uptake thymidine after cell infection until cell lysis, that is, during the lytic cycle of the phage [22]. Also, different studies found that viral production may even increase the rate of glycolysis of infected cells [42–45]. The work from Jain and Srivastava compared the major pathways in the infected cell metabolome with those in the uninfected cell. Even though that host metabolism is shifted for phage production diverting resources away from the TCA cycle and from the synthesis of cell wall components in order to upregulate the pentose phosphate pathway, the glucose uptake rates of uninfected and infected cells were found to be similar [36]. Conversely, the model developed by Jain and Srivastava to describe the interaction between the MS2 phage and *E. coli* predicted that the host cell do not grow since the value of the flux corresponding to the biomass production was zero [46]. Moreover, modelling infected bacteria as a growing population increases the phage growth rate, resulting

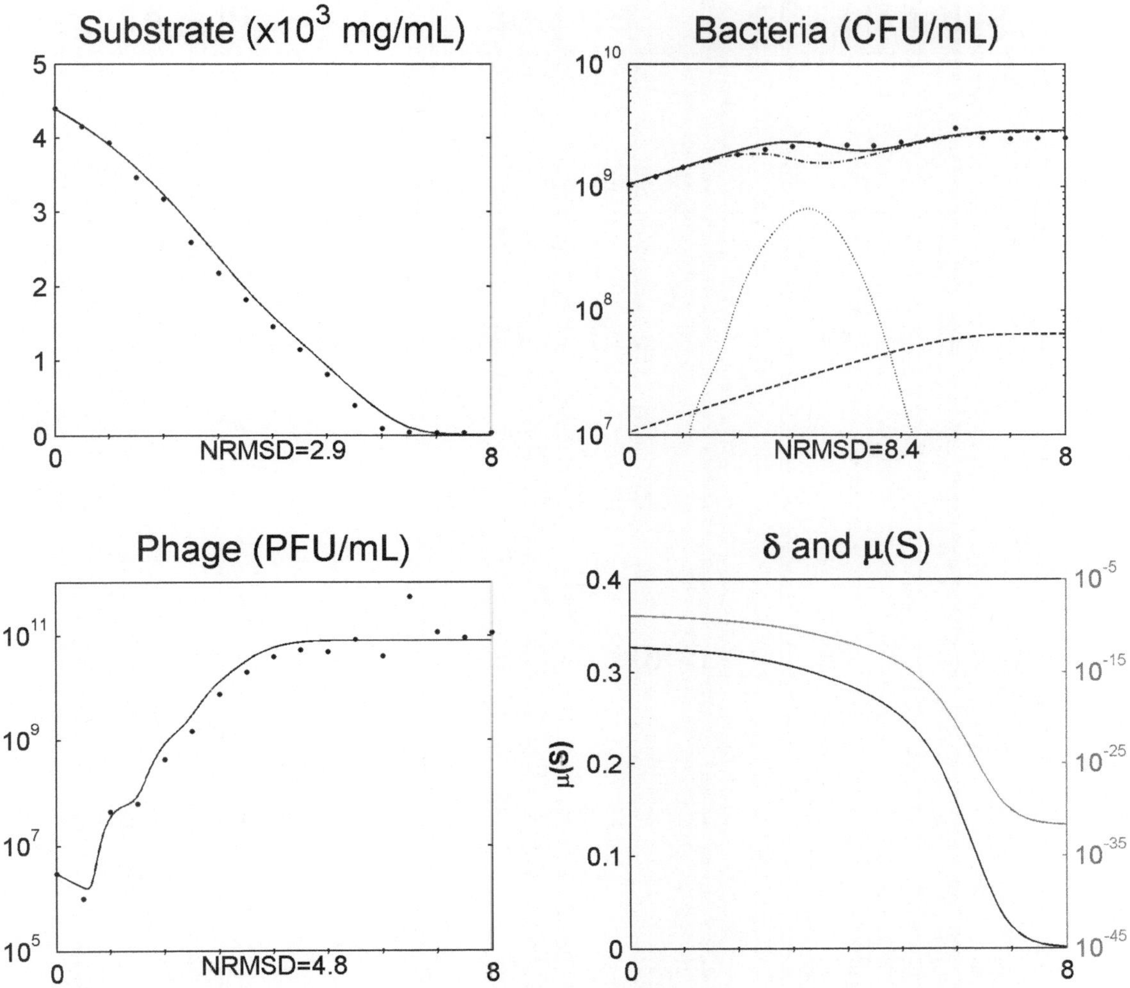

Figure 6. Simulating the phage and bacteria population dynamics using a distribution of values of the latent period and a variation of the adsorption constant (δ) as a function of the bacterial growth rate (μ) in a 5 l bioreactor for a MOI = 2.8E10^{-3}. Legend: • experimental data; ___ model simulation; - . - model simulation of susceptible uninfected bacteria (X_s); ... model simulation of infected bacteria (X_i); _ _ model simulation of resistant bacteria (X_r). The x axis represents time in hours.

in an increased distance to the observed biological data [37], fact that we also observed.

Acquired resistance of bacterial cells during the experiment time period and the possibility of bacteria to use cell contents of lysed cells as substrate were not included in this model since they are not relevant during the time scale used in this study [18,22].

Setting the latent period and inclusion of asynchronous bursts

The factors known to primarily rule phage-bacteria population dynamics are bacteria multiplication rate and substrate consumption, phage adsorption rate, burst size and latent period and thus these factors constitute the model parameters. The distribution of variation in these parameters also plays a critical role [20]. This last parameter, latent period, is an intrinsic characteristic of the phage-bacteria system (within a given environment) and has been neglected in the model presented by Levin and Bull (1996) and treated by Payne and Jansen (2001) as a lysis rate [47,48]. Without this parameter, simulated data will result in a higher phage growth due to an early release of phage progeny, leading to the appearance of new phages right after phage infection and before the latent period has passed. In the model presented herein, delay differential equations were used to include this temporal heterogeneity by treating the latent period as it is, a time period without phage release.

Experimental determination of the latent period is made through the single-step growth curve and corresponds to the time elapsed between infection and the time at which the phage concentration starts to rise. After this time point, the phage concentration continues to rise during a time period (the rise period) until a plateau is reached, showing that the infected bacteria do not all burst at the same time. These asynchronous bursts are due, in a lesser extent, to the different time at which phages adsorb (which depends on the adsorption rate that is usually very fast) but mainly to different latent periods within the infected-bacteria population, fact that will be reflected in the length of the rise period (ρ). Observing the single-step growth curve of a phage (as the one from Figure 1), the higher slope of the curve during the rise period, usually at its midpoint, reflects the majority of the infected cells bursting and releasing the progeny

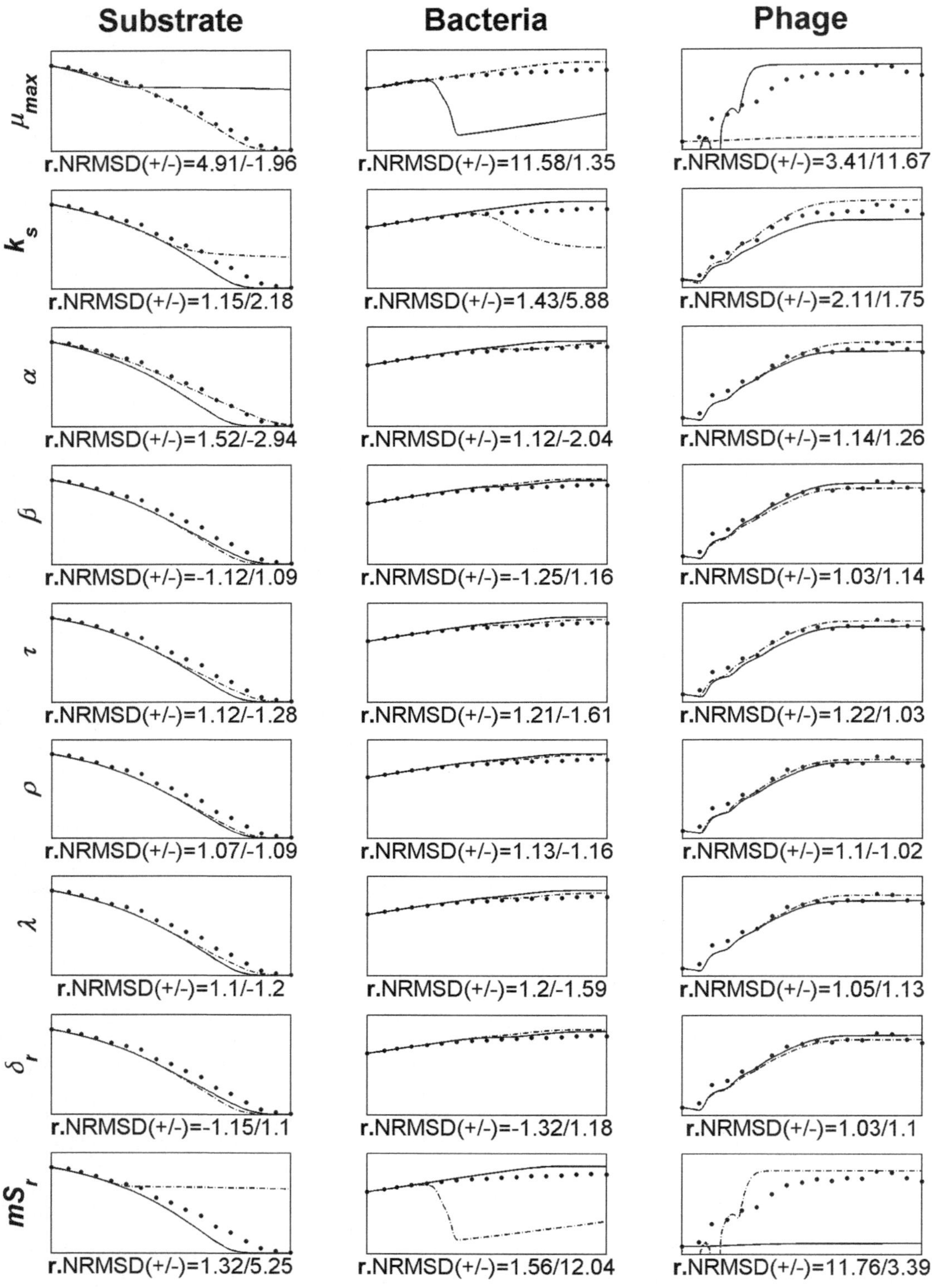

Figure 7. OAT sensitivity analysis of model parameters using as the base-case data from Figure 5. Legend: • experimental data; ___ model simulation when increasing the parameter in 10%; - . - model simulation when decreasing the parameter in 10%. The text on the left side of each line graph identifies the parameter being analysed. μ_{max}: maximum rate of exponential growth, K_s: half-saturation constant, α: substrate needed for a new bacterium, β: burst size, τ: the latent period, ρ: rise period, λ: lambda parameter used in equation 15, δ_r: experimental determined adsorption rate, mS_r: bacterial growth rate at which the adsorption rate was calculated, The x- and y-axis are the same as Figure 5 (x: time [1 8] hours, $y_{Substrate}$: [0 5000]; $y_{Bacteria}$: [10^7 10^{10}]; y_{Phage}: [10^5 10^{12}]).

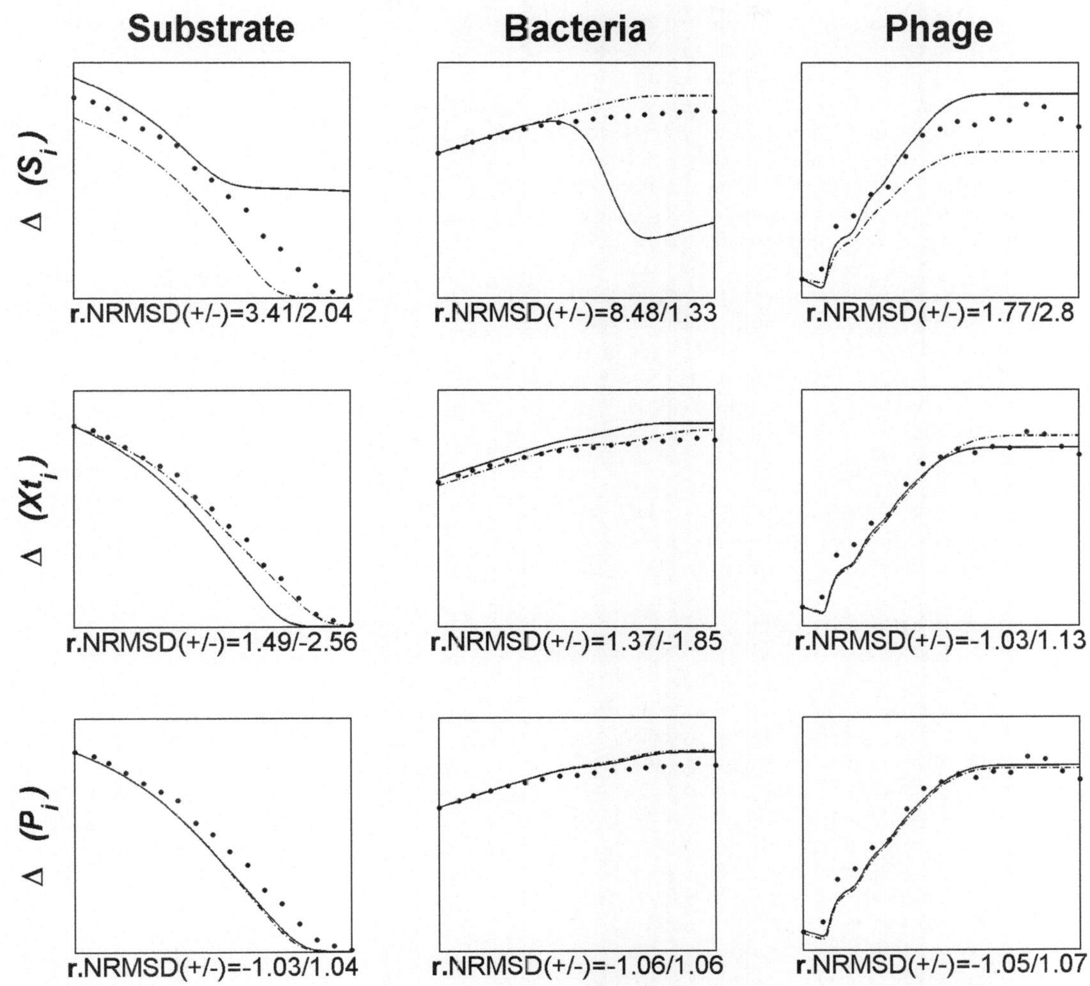

Figure 8. Variations in the model simulation when varying each of the state variables alone in 10% (using as the base case the data from Figure 5). Legend: • experimental data; ___ model simulation when increasing the variable in 10%; - . - model simulation when decreasing the variable in 10%. The text on the left side of each line graph identifies the variable being analysed. $\Delta(S_i)$: variation in initial substrate concentration, $\Delta(Xt_i)$: variation in initial total bacteria concentration, $\Delta(P_i)$: variation in initial free phage concentration. The x- and y-axis are the same as Figure 5 (x: time [1 8] hours, $y_{Substrate}$: [0 5000]; $y_{Bacteria}$: [10^7 10^{10}]; y_{Phage}: [10^5 10^{12}]).

phages. The non-inclusion of these asynchronous bursts in the model is common practice in modelling phage-bacteria populations but it will produce a faster increase in the phage release right after the latent period, enabling phages to start a new cycle much sooner, with the consequent overgrowth of simulated phage concentration when compared to the experimental data and leading to a larger decrease of the sensitive bacteria when compared with the real data [37]. Such behaviour was also observed in our model (Figure 1) as a result of the non-inclusion, in a first approach, of these asynchronous bursts. Figure 1 shows an overgrowth - more than two logs in only one phage cycle - of the simulated phage concentration in comparison to the experimental data. The model was readjusted to deal with the asynchronous bursts by including a normal distribution function (equation 13). The centre of the distribution curve is $\tau+\rho/2$, the midpoint of the rise period, which corresponds to the higher number of infected bacteria bursting. Using $\rho/6$ as the value of σ enables to include the limits of the distribution curve inside the rise period time length. The inclusion of this normal distribution function will rule the latent period distribution along the rise period by determining the proportion of infected bacteria that will burst at each possible time inside the limits τ and $\tau+\rho$ during the experiment. In contrast to what was observed by Weld and colleagues [37], this modification to the model enabled a much better agreement between the simulated and experimental data, as can be seen in Figure 1. Weld and colleagues distributed the burst over 10 min around the end of the latent period, apparently in the same way as Schrag and Mittler [49], as an even distribution over that period and not as a normal distribution, which seems to be the reason for this divergence.

Non-eradication of the bacterial population

Understanding the predator nature of phages and their ability to replicate each time a bacterium is infected lead us to speculate that the interaction between these two populations will result in a rapid growth of the phage population and the extinction of all sensitive bacterial population. Indeed, using the model with constant parameters and allowing for a normal distribution of the latent period to simulate the population and evolutionary dynamics of the phage–bacteria interactions, this behaviour is observed

(Figure 2 and Figure 3). This has also been reported in other models [17,20,24,47,49]. Nevertheless, experimental data reported herein and described by several other authors [22,49,50] shows that the phage concentration rapidly increases but the sensitive bacteria are not eradicated and are still found in the reactor even after the phage concentration have stopped increasing (Figure 2 and Figure 3). Two reasons have been pointed out to the non-eradication of the sensitive bacteria: the presence of surfaces, biofilms or others that may act as physical shelters to bacteria, consequently preventing phage adsorption [20,49]; and the physiological state of bacteria, since that stationary cells (due to its reduced metabolism) may preclude phage infection and replication [7,22]. This means that the bacterial cell needs to be accessible to the phage and in an appropriate physiological state to produce phage progeny [7]. As in the reactor experiment reported herein, the majority of bacteria are suspended and most likely not enclosed in physical shelters, the not eradication of sensitive bacteria is probably due to cells that are not in an appropriate physiological state to produce phage progeny. Indeed, we have observed that as the bacteria growth rate decreases, the rate of phage production also decreases (Figure 2 and Figure 3).

Physiological changes of stationary cells

The physiology of the bacterial cell will change through the different growth phases. When entering into the stationary phase, which is usually due to the depletion of available main nutritive components, the cell metabolic activity is greatly reduced, the bacterial growth declines due to reduced cell division, the outer membrane presents high degree of local charge heterogeneity, the receptors in the cell membrane may change in quantity and morphology, the bacterial motility is reduced and cells become smaller, in some cases changing their morphology from rod-shaped to cocci [51–53]. These changes on the physiology of the bacteria will induce variations in the parameters that rule the interaction between the phage and bacteria. Phage production and cell lysis were found to present a positive correlation with the bacterial growth rate of hosts in bioreactors suggesting a strong dependence of phage-bacteria population dynamics on the physiological state of bacteria [22].

Dependence of the adsorption rate on the cells physiology

The first step in the growth cycle of a phage is its adsorption to specific receptors on the susceptible bacterial surface representing the fitness of the phage in capturing its host. This characteristic has been found to vary to a great extent with the physiological state of the host cell [20,23,54]. An increase in the cell surface, a large number and density of receptors in the host cell surface and a higher motility of bacteria will be reflected in the adsorption rate constant increasing its value [17,54]. The collisions between the phage and bacteria happen at random Brownian motion and the adsorption rate is dependent on the concentration of both intervenients being typically modelled by the law of mass action [23]. The experimental determined adsorption rate constant will thus represent the density of phage receptors in the host cell surface, the diffusion rate in the medium, the size of both phage and bacteria and the efficiency of phage infection in relation to collisions [55]. Therefore, the adsorption rate constant will strongly depend on the physiological state of the bacterial cells which will in turn produce variations in that parameter during the time period of the experiment. This fact will also produce not only variations in the determined value of the constant between experiments carried at different physiological states of bacteria, but also difficulties in its determination since even slight physiological variations in the host cells may occur during an experiment [54].

Modelling phage-bacterial populations as a function of the growth rate

Although bacterial cells in stationary phase do not produce, or present minimal production of phage progeny, the resumption of cell growth by addition of nutrients results in an immediate lysis of host cells and phage production [7,22]. This fact suggests that phage production yield, and consequently the kinetic parameters, do not depend on the cell density or quorum sensing but rather on bacterial growth rate which can be determined by the amount of substrate available following a kinetic model such as that described in Monod's equation (equation 4). Consequently, the phage population (predators) starts to be primarily limited by the number of available preys but when the bacterial growth rate decreases it will be primarily limited by the host physiology (in this case as a consequence of resource availability). Hurley and colleagues [56] also found that the parameters that significantly influenced the simulation outcomes of their *in vivo* model for the SP6 phage-*Salmonella* system were phage adsorption rate and bacterial growth rate, however these authors did not related these two parameters and assumed a constant value in each simulation. Levin and Bull also concluded that a variation in phage adsorption and burst size values in time would eliminate the inconsistencies between the experimental results and their model, which predicted that the phage (combined with the immune defences) should eradicate the bacteria [47]. Even so, those variations were not implemented to avoid complexity of the model. Moreover, Hadas and colleagues observed that the rate of the irreversible step in T4 phage adsorption increased with the host growth rate [57].

Along these lines, we have modelled the adsorption rate as a function of the bacterial growth rate. Thus, phage adsorption will present its maximum when the growth rate equals the maximum bacterial growth rate and will decrease as the bacterial growth rate decreases until reaches a value of zero. This variation in the adsorption rate enabled a good fitting of the simulations with the experimental data for phage, bacteria and available substrate and was able to explain why (and how) sensitive uninfected bacteria are able to coexist with lytic phages (Figure 4 and Figure 5) as has been observed not only in therapy but also in environmental and ecological niches [17,24,49].

Like other authors we have also found dependence of the latent period and burst size on the host growth rate. The burst size seems to present a direct variation with that of the growth rate while the latent period presents an opposite variation, that is, the latent period is lengthened and the burst size reduced when the growth rate decreases, reducing the growth rate of phage progeny. Although the burst size presents a wide variation reaching the value of 1 when cells achieve stationary phase, the variation of the latent period is at a lesser extent. The variation of these two parameters as a response to growth rate was already observed [7,22,37,58]. The value of 1 for burst size of stationary cells reinforces the conclusion that infected stationary cells release progeny phage only after resumption of cell growth, which in this case, happens only after plating the cells, since no substrate was added during the experiments. These variations have been introduced in the model but no significant improvement in the simulations was observed. The wide variation of the burst size may be a result of a low adsorption of phage to cells that was already taken into account in the model, which may explain why these variations did not significantly improved the model simulations. Consequently, and due to the increasing complexity that it produces, we did not maintained these modifications for the sake of simplicity.

The dependence of phage fitness, represented by its biological parameters, on the bacterial growth rate will certainly have implications in environmental simulations and in phage therapy. *In vitro* (or in laboratory controlled conditions) studies of the population dynamics of phage and bacteria are usually carried at optimal conditions of bacterial growth which do not happen in the bacteria natural environments or organisms subjected to phage therapy. Therefore, kinetic parameters determined *in vitro* using exponential growth cells in optimal physiological state for phage infection cannot be used straightforward to model phage bacteria interactions in real environments since they will change while bacteria grow, due to variations in the cells physiological state.

Accordingly, the presence of a phage and a sensitive bacterium is not sufficient for infection to occur.

Sensitivity analysis

The complexity of the model and the number of parameters included turns difficult to understand the existing relationships between the inputs (parameters) and the outputs. Having this in mind we have carried a sensitivity analysis using the OAT method. This enables to vary a parameter and observe its effect on the model output, unambiguously showing the contribution of that particular parameter on the simulation. From this analysis (Figure 7), using the data from Figure 5 as the base-case, we have observed that the parameters that produce a wider deviation from the base case are the bacterial growth rate (mS_r) at which the adsorption rate was calculated (δ_r), the maximum rate of exponential growth (μ_{max}) and the half-saturation constant (Ks). The analysis showed that the Monod's parameters, and specially its output (bacterial growth rate), exert major influence in the model output. All these three parameters are directly related with the bacterial growth and confirm the influence that it has on the model. As a result, the identified parameters should be the focus of attention and need to be reliably determined. More important than determining the adsorption rate of reference (the one determined experimentally to define the adsorption equation, equation 15) is assessing the growth rate at which it was determined. Furthermore, the influence of the bacterial growth rate, and consequently the phage adsorption to the bacteria, is even higher than the burst size, a parameter commonly used to select for phage fitness. Using the values of the r.NRMSD (the higher absolute value from the ±10% variation) the parameters can be ranked in order of its higher impact on the model output: bacterial growth rate at which the adsorption rate was calculated (mS_r), the maximum rate of exponential growth (μ_{max}), the half-saturation constant (Ks) and at a lesser extent the substrate needed for a new bacterium (α), the latent period (τ), the lambda parameter used in equation 15 (λ), the substrate needed for a new bacterium (α), the experimental determined adsorption rate (δ_r), the burst size (β) and the rise period (ρ).

Variations in the state variables

Phage, bacterium and substrate concentrations are key variables when setting up conditions to produce phage, to use phages as antimicrobial agents or to understand changes in a phage-bacterium community. To assess the outcome of a variation in these state variables and consequently its influence in a phage-bacterium population we have modelled the outcome of the *Salmonella* phage and its host by changing each of the state variables alone in ±10% (Figure 8). The determination of the r.NRMSD was made using as the base-case the same data from Figure 5):

i) Substrate (Figure 8, line 1) – by increasing the initial substrate concentration in only 10% the sensitive bacteria are almost completely eradicated decreasing the total bacteria population (composed mostly by infected and resistant bacteria). The final phage concentration in the other hand is increased. At a first glance the increase in the substrate should favour bacterial growth but this will also increase the number of bacteria susceptible to the phage with a higher growth rate. As stated before, according to our model, the increased bacteria growth rate will increase phage infectiveness through a higher adsorption rate and therefore an equally increased phage progeny. The decrease in the substrate concentration does not have a great impact when comparing with the base-case. There is a slight increase in the bacteria population that will stop their growth sooner due to the depletion of substrate. Also, a slight increase in the phage concentration is observed.

ii) Bacteria (Figure 8, line 2) – increasing the total bacteria concentration will increase the number of sensitive bacterium able to be infected by phages and consequently it would be expectable an increase in the final phage concentration. Even though, the final concentration of phage decreased. Such behaviour is explained by the decrease in the substrate concentration in the medium due to the increase in bacterial population. This decrease in substrate concentration will lower the growth rate of the bacteria (calculated through the Monod equation, equation 4) that in turn will decline the adsorption rate leading to less phage infections and consequently less phage progeny. Moreover, the growth rate for which the phage adsorption is close to zero will be achieved sooner and thus the phage concentration will also stagnate sooner. When decreasing the total bacteria, beside the lower number of susceptible bacteria, the final phage concentration achieved will be higher than the positive variation of this state variable. This is a consequence of the higher availability of substrate per bacterium, since the substrate was maintained at its base line, which will lead to a higher growth rate of bacteria and in turn a higher phage infectiveness, as explained above. The phage concentration is primarily limited by the lower number of susceptible bacteria (until half the time) but then, the higher growth rate of cells leads to a higher phage population.

iii) Phage (Figure 8, line 3) - the variation of ±10% in the initial phage concentration produced a very small variation in the final phage and bacteria concentration when compared with the base case. These changes are almost imperceptible showing that small differences (at least up to 10%) in the initial phage concentration will not compromise/improve phage production and phage effectiveness in the control of the bacterial host;

This analysis showed that variations in the initial conditions of the state variables are not those that would be the most expectable. Increase in the bacterial substrate will lead to bacteria reduction and phage increase while decrease in the substrate may enable bacteria survival when confronted with a phage attack and an initial higher population of bacteria, if not supplemented with a higher amount of substrate, will lead to lower phage production. Only the increase of phage titre seems to have the expectable output, a higher final phage concentration and a lower bacterial population, but even so, for a population that reproduces exponentially, the variation is almost imperceptible.

From these variations it is possible to observe that the initial substrate concentration was the state variable that produced the most pronounced changes. From a clinical point of view or for phage production purposes it seems that, in contrary to what would be expected, to reduce the bacterial population or increase the final phage concentration it is advisable to increase the substrate available for the bacteria. Consequently, changes in the substrate may have dramatic impact in bacteria control with consequences for the success of therapy and phage production.

The proposed model clearly shows the major role of the physiological state of the bacteria in the population dynamics of a lytic phage and its respective host. The model contributed to give further insights about the role of the growth rate (as well the parameters that mostly can influence it, as is the case of the maximum rate of exponential growth - μ_{max} - and the half-saturation constant - K_s - and the state variable initial resource concentration - S_i) in the course and outcome of the interaction between a lytic phages and its respective host.

Conclusions

In conclusion, an unstructured mathematical model has been developed using delay differential equations to predict and explain the basic behaviour of the phage-bacteria population dynamics of lytic phages based on fundamental biological parameters which rule these interactions. The model is directed to lytic phages only, due to their significance in microbial ecology and in their therapeutic use as antimicrobial agents. The results have shown that the growth rate of the bacterial host plays a critical role in defining the value of the biological parameters of the phage-bacteria population dynamics, mainly the adsorption rate, and consequently in predicting the extent at which phages will grow and control the sensitive bacteria, i.e., the population dynamics of phages and bacteria. Although only the phage adsorption was modelled as a function of the growth rate, the variation of the phage adsorption may also reflect changes in the burst size and latency time that may be a consequence (or not) of that variation. This dependence is able to explain why phages and sensitive bacteria are able to coexist and may be the reason why phages are so successful in controlling bacteria in *in vitro* studies and are not able to achieve such a high performance *in vivo* (where substrate can be limiting). As a consequence, it is essential to understand the influence of environmental conditions (including, but not limited to, substrate availability) on bacterial growth to predict the phage-bacteria behaviour. This dependence will have implications on the use of phages therapeutically and also on their production in bioreactors.

Although the modification of simple models to allow accurate predictions of the *in vivo* interactions of phage and bacteria is complex, the construction of more accurate and precise models will for sure improve our understanding on the role of phages in natural systems and in therapy, diminishing the existing gap between *in vitro* and *in vivo* modelling. Therefore the presented model intends to be a step forward in this domain by introducing variations in the parameters as a function of the bacterial growth rate and also by treating the latent period as a mean value following a normal distribution and not as a single value valid to all the phage-bacteria present, an approximation never included in previous models.

Models explaining the phage-bacteria population dynamics are developed almost exclusively to understand the evolution of these populations in ecological studies and therapy. Nevertheless, in a time where the interest in phages as commercial products is growing exponentially, these models can be of extremely importance in monitoring, control and optimize phage production at large scales.

Acknowledgments

The authors acknowledge Prof. L. F. Shampine (Southern Methodist University, Texas) for its help on setting up dde23 in Matlab.

Author Contributions

Conceived and designed the experiments: SBS CMC JA ECF. Performed the experiments: SBS CMC. Analyzed the data: SBS CMC JA ECF. Contributed reagents/materials/analysis tools: JA ECF. Wrote the paper: SBS CMC JA ECF.

References

1. Ackermann HW (2006) Classification of Bacteriophages. In: Calendar R, editors. The Bacteriophages. New York: Oxford University Press. pp. 8–16.
2. Ackermann HW (2007) 5500 Phages examined in the electron microscope. Archives of Virology 152: 227–243.
3. Ashelford KE, Day MJ, Fry JC (2003) Elevated abundance of bacteriophage infecting bacteria in soil. Appl Environ Microbiol 69: 285–289.
4. Bergh O, Borsheim KY, Bratbak G, Heldal M (1989) High abundance of viruses found in aquatic environments. Nature 340: 467–468.
5. Breitbart M, Felts B, Kelley S, Mahaffy JM, Nulton J *et al.* (2004) Diversity and population structure of a near-shore marine-sediment viral community. Proc Biol Sci 271: 565–574.
6. Breitbart M, Wegley L, Leeds S, Schoenfeld T, Rohwer F (2004) Phage community dynamics in hot springs. Appl Environ Microbiol 70: 1633–1640.
7. Chibani-Chennoufi S, Bruttin A, Dillmann ML, Brussow H (2004) Phage-host interaction: an ecological perspective. J Bacteriol 186: 3677–3686.
8. Danovaro R, Manini E, Dell'anno A (2002) Higher abundance of bacteria than of viruses in deep Mediterranean sediments. Appl Environ Microbiol 68: 1468–1472.
9. Kepner RL Jr, Wharton RA Jr, Suttle CA (1998) Viruses in Antarctic lakes. Limnol Oceanogr 43: 1754–1761.
10. Rohwer F (2003) Global phage diversity. Cell 113: 141.
11. Suttle CA (1994) The significance of viruses to mortality in aquatic microbial communities. Microbial Ecology 28: 237–243.
12. Wommack KE, Colwell RR (2000) Virioplankton: viruses in aquatic ecosystems. Microbiol Mol Biol Rev 64: 69–114.
13. Boyd EF, Brussow H (2002) Common themes among bacteriophage-encoded virulence factors and diversity among the bacteriophages involved. Trends Microbiol 10: 521–529.
14. Fuhrman JA (1999) Marine viruses and their biogeochemical and ecological effects. Nature 399: 541–548.
15. Kropinski AM (2006) Phage therapy - everything old is new again. Can J Infect Dis Med Microbiol 17: 297–306.
16. Petty NK, Evans TJ, Fineran PC, Salmond GPC (2007) Biotechnological exploitation of bacteriophage research. Trends in Biotechnology 25: 7–15.
17. Bohannan BJM, Lenski RE (1997) Effect of Resource Enrichment on a Chemostat Community of Bacteria and Bacteriophage. Ecology 78: 2303–2315.
18. Cairns BJ, Timms AR, Jansen VA, Connerton IF, Payne RJ (2009) Quantitative models of in vitro bacteriophage-host dynamics and their application to phage therapy. PLoS Pathog 5: e1000253.
19. Clark JR, March JB (2006) Bacteriophages and biotechnology: vaccines, gene therapy and antibacterials. Trends Biotechnol 24: 212–218.
20. Levin BR, Bull JJ (2004) Population and evolutionary dynamics of phage therapy. Nat Rev Microbiol 2: 166–173.
21. Merril CR, Scholl D, Adhya SL (2003) The prospect for bacteriophage therapy in Western medicine. Nat Rev Drug Discov 2: 489–497.
22. Middelboe M (2000) Bacterial Growth Rate and Marine Virus-Host Dynamics. Microb Ecol 40: 114–124.
23. Lenski RE (1988) Dynamics of interactions between bacteria and virulent bacteriophage. In: Marshall KC, editors. Advances in microbial ecology. New York: Plenum Press. pp. 1–44.
24. Levin BR, Stewart FM, Chao L (1977) Resource-Limited Growth, Competition, and Predation: A Model and Experimental Studies with Bacteria and Bacteriophage. The American Naturalist 111: 3–24.
25. Payne RJ, Jansen VA (2003) Pharmacokinetic principles of bacteriophage therapy. Clin Pharmacokinet 42: 315–325.
26. Payne RJ, Jansen VA (2001) Understanding bacteriophage therapy as a density-dependent kinetic process. J Theor Biol 208: 37–48.
27. Lawton JH (1995) Ecological experiments with model systems. Science 269: 328–331.
28. Sambrook J, Russell DW (2001) Molecular Cloning: A Laboratory Manual. New York: Cold Spring Harbor Laboratory Press, Cold Spring Harbor.

29. Santos SB, Fernandes E, Carvalho CM, Sillankorva S, Krylov VN *et al.* (2010) Selection and Characterization of a Multivalent Salmonella Phage and Its Production in a Nonpathogenic Escherichia coli Strain. Appl Environ Microbiol .
30. Sillankorva S, Pleteneva E, Shaburova O, Santos S, Carvalho C *et al.* (2010) *Salmonella* Enteritidis bacteriophage candidates for phage therapy of poultry. J Appl Microbiol 108: 1175–1186.
31. Santos SB, Carvalho CM, Sillankorva S, Nicolau A, Ferreira EC *et al.* (2009) The use of antibiotics to improve phage detection and enumeration by the double-layer agar technique. BMC Microbiol 9: 148.
32. Kropinski AM (2009) Measurement of the Rate of Attachment of Bacteriophage to Cells. In: Clokie MRJ, Kropinski AM, editors. Bacteriophages. Humana Press. pp. 151–155.
33. Shao Y, Wang IN (2008) Bacteriophage adsorption rate and optimal lysis time. Genetics 180: 471–482.
34. Esener AA, Veerman T, Roels JA, Kossen NW (1982) Modeling of bacterial growth; formulation and evaluation of a structured model. Biotechnol Bioeng 24: 1749–1764.
35. Ellner SP, Guckenheimer J (2006) Dynamic models in biology. Princeton, N.J: Princeton University Press. 352 p.
36. Maynard ND, Gutschow MV, Birch EW, Covert MW (2010) The virus as metabolic engineer. Biotechnol J 5: 686–694.
37. Weld RJ, Butts C, Heinemann JA (2004) Models of phage growth and their applicability to phage therapy. J Theor Biol 227: 1–11.
38. Santos SB, Kropinski AM, Ceyssens PJ, Ackermann HW, Villegas A *et al.* (2011) Genomic and proteomic characterization of the broad-host-range Salmonella phage PVP-SE1: creation of a new phage genus. J Virol 85: 11265–11273.
39. Merabishvili M, Pirnay JP, Verbeken G, Chanishvili N, Tediashvili M *et al.* (2009) Quality-controlled small-scale production of a well-defined bacteriophage cocktail for use in human clinical trials. PLoS One 4: e4944.
40. Miller ES, Kutter E, Mosig G, Arisaka F, Kunisawa T *et al.* (2003) Bacteriophage T4 genome. Microbiol Mol Biol Rev 67: 86–156, table.
41. Beretta E, Kuang Y (1998) Modeling and analysis of a marine bacteriophage infection. Math Biosci 149: 57–76.
42. Singh VN, Singh M, August JT, Horecker BL (1974) Alterations in glucose metabolism in chick-embryo cells transformed by Rous sarcoma virus: intracellular levels of glycolytic intermediates. Proc Natl Acad Sci U S A 71: 4129–4132.
43. Bardell D, Essex M (1974) Glycolysis during early infection of feline and human cells with feline leukemia virus. Infect Immun 9: 824–827.
44. Green M, Henle G, Deinhardt F (1958) Respiration and glycolysis of human cells grown in tissue culture. Virology 5: 206–219.
45. Levy HB, Baron S (1957) The effect of animal viruses on host cell metabolism. II. Effect of poliomyelitis virus on glycolysis and uptake of glycine by monkey kidney tissue cultures. J Infect Dis 100: 109–118.
46. Jain R, Srivastava R (2009) Metabolic investigation of host/pathogen interaction using MS2-infected Escherichia coli. BMC Syst Biol 3: 121.
47. Levin BR, Bull JJ (1996) Phage Therapy Revisited: The Population Biology of a Bacterial Infection and Its Treatment with Bacteriophage and Antibiotics. The American Naturalist 147: 881–898.
48. Payne RJ, Phil D, Jansen VA (2000) Phage therapy: the peculiar kinetics of self-replicating pharmaceuticals. Clin Pharmacol Ther 68: 225–230.
49. Schrag SJ, Mittler JE (1996) Host-Parasite Coexistence: The Role of Spatial Refuges in Stabilizing Bacteria-Phage Interactions. The American Naturalist 148: 348–377.
50. Fischer CR, Yoichi M, Unno H, Tanji Y (2004) The coexistence of Escherichia coli serotype O157:H7 and its specific bacteriophage in continuous culture. FEMS Microbiol Lett 241: 171–177.
51. Akerlund T, Nordstrom K, Bernander R (1995) Analysis of cell size and DNA content in exponentially growing and stationary-phase batch cultures of Escherichia coli. J Bacteriol 177: 6791–6797.
52. Kolter R, Siegele DA, Tormo A (1993) The Stationary Phase of The Bacterial Life Cycle. Annual Review of Microbiology 47: 855–874.
53. Walker SL, Hill JE, Redman JA, Elimelech M (2005) Influence of growth phase on adhesion kinetics of Escherichia coli D21g. Appl Environ Microbiol 71: 3093–3099.
54. Delbruck M (1940) Adsorption of bacteriophage under various physiological conditions of the host. J Gen Physiol 23: 631–642.
55. Kasman LM, Kasman A, Westwater C, Dolan J, Schmidt MG *et al.* (2002) Overcoming the phage replication threshold: a mathematical model with implications for phage therapy. J Virol 76: 5557–5564.
56. Hurley A, Maurer JJ, Lee MD (2008) Using bacteriophages to modulate Salmonella colonization of the chicken's gastrointestinal tract: lessons learned from in silico and in vivo modeling. Avian Dis 52: 599–607.
57. Hadas H, Einav M, Fishov I, Zaritsky A (1997) Bacteriophage T4 development depends on the physiology of its host Escherichia coli. Microbiology 143 (Pt 1): 179–185.
58. Delbruck M (1945) The Burst Size Distribution in the Growth of Bacterial Viruses (Bacteriophages). J Bacteriol 50: 131–135.

5

Guava Leaf Extract Inhibits Quorum-Sensing and *Chromobacterium violaceum* Induced Lysis of Human Hepatoma Cells: Whole Transcriptome Analysis Reveals Differential Gene Expression

Runu Ghosh*⁹, Bipransh Kumar Tiwary, Anoop Kumar, Ranadhir Chakraborty⁹

OMICS Laboratory, Department of Biotechnology, University of North Bengal, Siliguri, West Bengal, India

Abstract

Quorum sensing (QS) is a process mediated via small molecules termed autoinducers (AI) that allow bacteria to respond and adjust according to the cell population density by altering the expression of multitudinous genes. Since QS governs numerous bioprocesses in bacteria, including virulence, its inhibition promises to be an ideal target for the development of novel therapeutics. We found that the aqueous leaf extract of *Psidium guajava* (GLE) exhibited anti-QS properties as evidenced by inhibition of violacein production in *Chromobacterium violaceum* and swarming motility of *Pseudomonas aeruginosa*. The gram-negative bacterium, *C. violaceum* is a rare pathogen with high mortality rate. In this study, perhaps for the first time, we identified the target genes of GLE in *C. violaceum* MTCC 2656 by whole transcriptome analysis on Ion Torrent. Our data revealed that GLE significantly down-regulated 816 genes at least three fold, with p value≤0.01, which comprises 19% of the *C. violaceum* MTCC 2656 genome. These genes were distributed throughout the genome and were associated with virulence, motility and other cellular processes, many of which have been described as quorum regulated in *C. violaceum* and other gram negative bacteria. Interestingly, GLE did not affect the growth of the bacteria. However, consistent with the gene expression pattern, GLE treated *C. violaceum* cells were restrained from causing lysis of human hepatoma cell line, HepG2, indicating a positive relationship between the QS-regulated genes and pathogenicity. Overall, our study proposes GLE as a QS inhibitor (QSI) with the ability to attenuate virulence without affecting growth. To the best of our knowledge, this is the first report which provides with a plausible set of candidate genes regulated by the QS system in the neglected pathogen *C. violaceum*.

Editor: George Dimopoulos, Johns Hopkins University, Bloomberg School of Public Health, United States of America

Funding: This study was funded by the Department of Biotechnology, Government of India (BT/Bio-CARe/05/141/2010-11) [Project Investigator- RG; Junior Research Fellow- BKT]. The funders had no role in study design, data collection and analysis, decision to publish, or preparation of the manuscript.

Competing Interests: The authors have declared that no competing interests exist.

* Email: disanru@yahoo.com

⁹ These authors contributed equally to this work.

Introduction

With the increase in the number of multi-drug-resistant pathogenic bacteria worldwide, there is a dire need for developing strategies to fight bacterial infections. The indiscriminate use of novel antibiotics that interfere with the metabolism of bacteria have only added to this number. Since Quorum Sensing (QS) regulates many virulence determinants of various pathogens, it has emerged as an attractive target to control their pathogenicity [1,2,3]. QS is a cell-to-cell communication mechanism, regulated by small diffusible signalling molecules termed autoinducers (AI), which allows bacteria to respond and adjust their needs in a population density-dependent manner by altering the expression of multitudinous genes [4,5,6]. The AIs used by Gram-negative bacteria are known as *N*-acyl homoserine lactones (AHLs), while Gram-positive bacteria utilize post-translationally modified oligopeptides as signaling molecules [7,8]. In most Gram-negative bacteria, QS systems are based on LuxI/LuxR homologues. The LuxI homologues encode an AHL synthetase involved in the synthesis of signal molecules, and the LuxR homologues encode the transcription regulatory protein which, upon binding of the cognate signal molecules, activates the transcription of the QS target genes [9].

Chromobacterium violaceum, an opportunistic pathogen, is a free-living, gram-negative, facultative anaerobic β-proteobacterium commonly found in water and soil in the tropical and subtropical regions [10]. In human, *C. violaceum* infection is rare, but this may be attributed to under-reporting of such cases in areas where the risks of exposure are high and diagnostic facilities are scarce [11]. In spite of this, more than 150 cases of infection were reported in tropical and subtropical regions, including India, where *C. violaceum* is normally found [12]. This rare infection is associated with a high mortality rate, between 60% and 80%, if not diagnosed at an early stage or treated correctly [11]. A recent

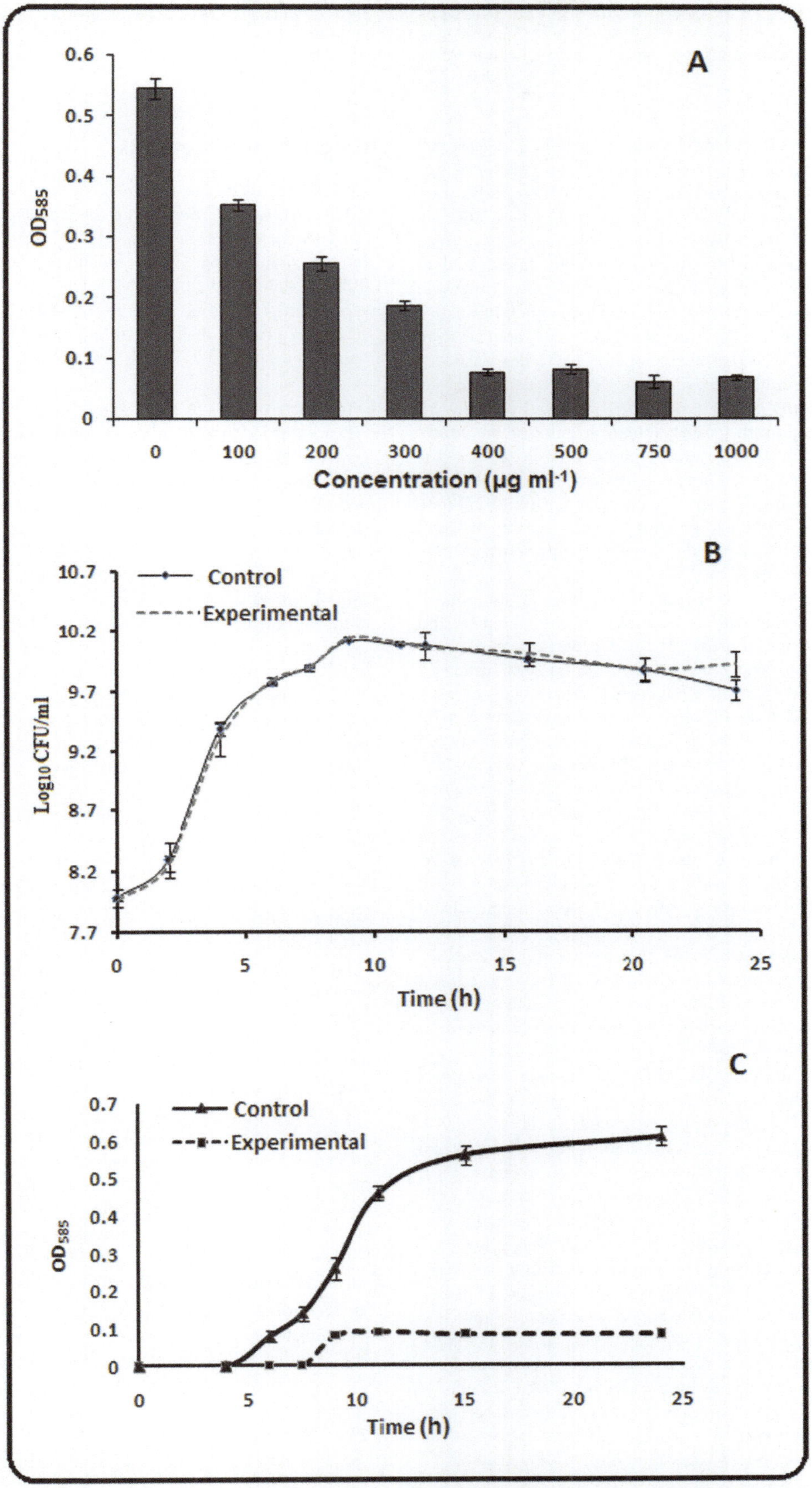
A
0.6
0.5
0.4
0.3
0.2
0.1
0
OD585
0
100
200
300
400
500
750
1000
Concentration (µg ml-1)
B
Control
Experimental
10.7
10.2
9.7
9.2
8.7
8.2
7.7
Log10 CFU/ml
0
5
10
15
20
25
Time (h)
C
Control
Experimental
0.7
0.6
0.5
0.4
0.3
0.2
0.1
0
OD585
0
5
10
15
20
25
Time (h)

Figure 1. Effect of GLE on violacein production and growth of wild-type *C. violaceum* MTCC 2656 cells. A. Violacein production in the presence of different concentrations of GLE. **B.** Viable cell number in batch culture grown without (Control) or with (Experimental) supplementation of 400 μg ml^{-1}GLE. **C.** Quantitation of violacein in batch culture grown without (Control) or with (Experimental) supplementation of 400 μg ml^{-1} GLE.

case of a man from South India with septicemic *C. violaceum* infection and septic arthritis, who had a fatal outcome, was reported [13]. A likely explanation for the high mortality rate could be the resistance of *C. violaceum* to a wide range of antibiotics and to other mech-anisms that pump out the cytotoxic drugs [14]. Thus, appropriate therapy is absolutely essential to control this neglected, potentially fatal infection. The strategy of controlling pathogens by interrupting its QS phenomenon was the prime focus of research in the recent years. The importance of quorum sensing in *C. violaceum* pathogenesis was demonstrated by the fact that QS-antagonist molecules protect the nematode *Caenorhabditis elegans* from *C. violaceum*-mediated killing [15]. The *C. violaceum* quorum-sensing system consists of the LuxI/LuxR homologues CviI/CviR, which controls virulence and the production of a variety of phenotypic characteristics that includes the production of the purple pigment, violacein, cyanide, chitinase and the antibiotic phenazine. The complete genomic sequence of *C. violaceum* ATCC 12472 has also revealed the presence of these QS-associated genes [16].

Inhibition of QS by some chemically synthesized compounds was identified but most of the QS inhibitors were isolated and characterized from plant sources [17,18]. Crude extracts of many plant parts were shown to possess anti-QS activity using *C. violaceum* as a model bacterium [19,20,21]. Crude plant extracts are often found to be more effective than isolated constituents at an equivalent dose perhaps owing to positive interactions between components of whole plant extracts. This synergy may involve prevention of the active component from degradation by enzymes or facilitate transport across cell barriers that result in higher efficacy of the crude drug when compared with purified components [22,23]. Hence, it is lately realized that crude extracts may possibly be the right strategy to treat multi-drug resistant pathogens as compared to the purified compound isolated from the same extract. In fact, the use of traditional herbal medicines is sometimes considered more effective than conventional drugs for the treatment of disease such as malaria [24]. It is proposed that the new generation of phytopharmaceuticals may enable successful use of herbal drug combinations to treat diseases in comparison to single active component [25].

Psidium guajava L. (Guava), widely distributed throughout India, belongs to the family Myrtaceae and is a well known traditional medicinal plant widely used in folk medicine [26,27]. The leaf extracts of this plant were shown to possess anti-microbial [28], anti-inflammatory, antidiarrhoea [29], anti-oxidant [30], antimutagenic [31], anti- cancer [32], anti-diabetic [33] and anti-plaque [34] activities. However, no molecular mechanism of antimicrobial property of guava-leaf extract was explored.

To the best of our knowledge, this is the first attempt to reveal gene expression profile of *C. violaceum* with the aid of whole transcriptome analyses on Ion-Torrent in presence of guava-leaf extract (GLE). GLE inhibited QS-controlled genes and QS-regulated phenotypes without affecting the bacterial growth up to 24 h suggesting these effects to be unrelated to bacteriostatic or bactericidal effects. Furthermore, the down-regulation of the wide array of genes, including those encoding virulence factors, affect pathogenicity as revealed by the ability of GLE to arrest *C. violaceum* induced cell lysis of human hepatoma cells.

Materials and Methods

Bacterial strains

The *Chromobacterium violaceum* wild type strain MTCC 2656 and *Pseudomonas aeruginosa* MTCC 2297 were obtained from the Microbial Type Culture Collection Center (MTCC), IMTECH, Chandigarh, India. MTCC 2656 and MTCC 2297 cells were routinely cultured on Nutrient broth (NB, Hi-Media- M002) agar and Luria-Broth (LB, Hi-Media-M575) agar respectively and maintained at 37°C.

Extraction of guava leaves

Leaves of *Psidium guajava* L. (Guava) were collected from the Centre of Floriculture and Agro Buisness Management (COFAM), University of North Bengal and extracted following standard method [35]. The leaves were washed thoroughly with sterile distilled water and rinsed with 70% (v/v) ethanol. Washed leaves were dried under sun initially and finally in the oven at 50°C for 1 hour. The dried leaves were crushed to fine powder, passed through an 80 mesh sieve and stored in a sealed plastic bag. To 50 gm of powder, 500 ml of sterile distilled water was added and the mixture was heated at 70°C for 1 hour and incubated at 30°C for 72 hours with shaking at regular intervals. The extract was filtered through a Whatman No. 1 filter paper (pore size 11 μm) and centrifuged at 10,000 rpm for 10 min. The resultant supernatant was freeze-dried using a lyophilizer (IIC Industrial Corporation, India). The dried sample was reconstituted with water, filter-sterilized (0.45 μm pore size, Sartorius Stedim, Germany) and tested for its ability to modulate quorum sensing in *Chromobacterium violaceum* strain MTCC 2656.

Demonstration of Inhibitory effect of Guava leaf extract (GLE) on Quorum Sensing (QS) activity of *C. violaceum* MTCC 2656

The inhibition of QS-mediated violacein production in *C. violaceum,* by GLE, was studied by the agar well assay [36]. Sterile molten Nutrient agar (Hi-Media, India) was pour-plated with the cells of *C. violaceum* MTCC 2656. After solidification of the plates, wells were made in which 100 μl of the aqueous extract was placed, and incubated at 37°C for 24 h. In this assay system, inhibition of bacterial growth (if any) results in producing a clear circular zone around the well, while a positive result of quorum sensing inhibition is demonstrated by a colourless circular translucent zone, signifying growth of cells with no production of pigment, around the well, circumscribed by purple pigmented bacterial growth in the remaining part of the plate.

Demonstration of anti-QS property of GLE using *Pseudomonas aeruginosa* MTCC 2297

To further confirm and or validate the anti-QS property of the GLE, motility assays of *Pseudomonas aeruginosa* MTCC 2297 were undertaken in petri dishes [37]. For this, LB media supplemented with 5% glucose and 0.5% agar containing either water (control) or 400 μg ml^{-1} of GLE was poured in petri dishes. An inoculum of 10 μl of overnight grown MTCC 2297 was inoculated at the center of the plates. The inoculated plates were

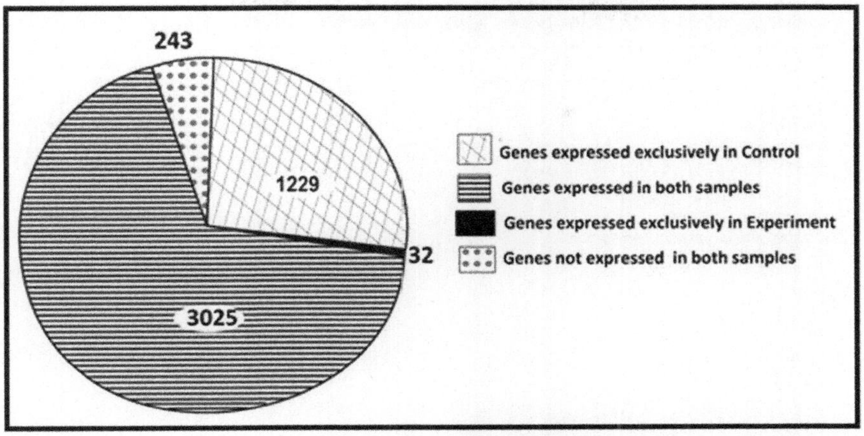

Figure 2. Profile of genome-wide gene expression in *C. violaceum* MTCC 2656 cells. Cells were grown in presence (experimental) and absence of 400 μg ml^{-1} GLE (control).

incubated at 37°C for 20 h and motility across the agar surface was visualized.

Determination of QS-inhibitory concentration of GLE with reference to violacein production

Log phase cells of MTCC 2656 (2.5×10^{6} CFU ml^{-1}) were inoculated (1.0% inocula) into 10 ml volumes of sterile nutrient broth in flasks containing different concentrations of the GLE (100 μg ml^{-1}to 1000 μg ml^{-1}) and incubated at 37°C for 24 h with shaking. The control set was devoid of GLE. The violacein pigment formation in the flask was quantified following the method of Choo *et al.* [38]. Briefly, 1 ml culture from each flask was centrifuged at 13,000 rpm for 10 min. The culture supernatant was discarded and 1 ml of DMSO was added to the pellet. The solution was vortexed vigorously to completely solubilize violacein and centrifuged at 10,000 rpm for 10 min to remove the cells. The absorbance of the supernatant was read at a wavelength of 585 nm in a digital spectrophotometer (ThermoSpectronic UV 1). The maximum O.D.$_{585}$ value observed in case of GLE-untreated cells was considered as 100% production of violacein.

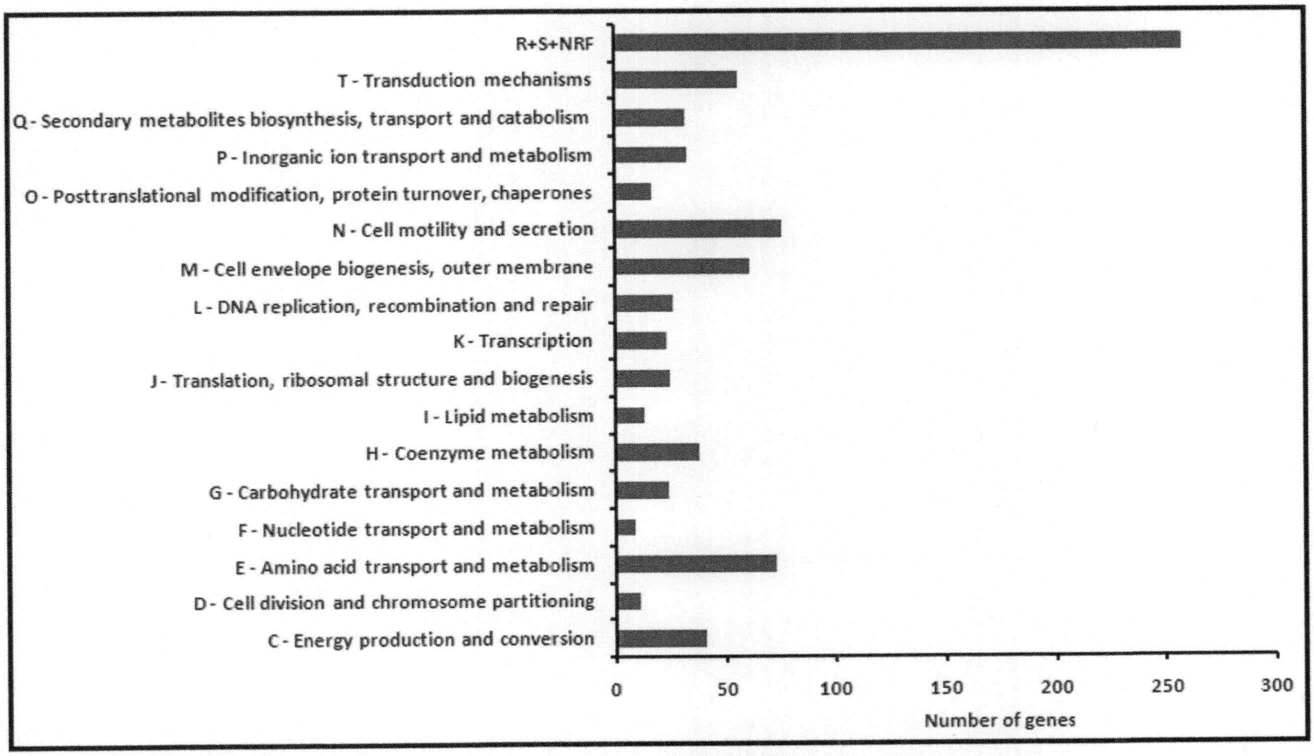

Figure 3. Differential expression of genes in *C. violaceum* in presence of GLE. The classification was based on Clusters of Orthologous Groups (COG) functional classification (R, General function prediction only; S, Function unknown; and NRF, No results found).

Table 1. Significantly downregulated *C. violaceum* genes associated with quorum-sensing and pathogenicity, in presence of GLE.

ORF no.	Gene name	Description	Control[1]	Expt[2]	$\log_2$(fold change)	p value
Constituent of Quorum sensing circuit						
CV_4091	*cviI*	N-acyl homoserine synthase; autoinducer synthase, quorum sensing controlled system	171.35	19.44	−3.14	5.39E-005
CV_4090	*cviR*	transcriptional activator, LuxR/UhpA family of regulators	194.51	14.75	−3.72	3.04E-006
Genes predicted to have cviR binding sites						
CV_1311		hypothetical	46.41	1.6	−4.86	0
CV_1329	*sbcB*	exonuclease I	15.82	0.66	−4.59	0
CV_1328		methyl-accepting chemotaxis protein	39.91	2.86	−3.8	3.86E-006
CV_1408	*sdaA2*	serine dehydratase	23.25	2.67	−3.12	0
CV_1444		hypothetical	14.81	0.94	−3.98	0.01
CV_2091		putative tetR-family transcriptional regulator	23.69	2.89	−3.03	0.01
CV_2321	*recN*	DNA repair protein	19.88	0.54	−5.19	0
CV_2434		hypothetical	56.1	1.93	−4.86	0
CV_2656		cytochrome P450 hydroxylase	96.94	8.63	−3.49	7.18E-006
CV_2716		hypothetical	52.45	5.47	−3.26	4.07E-005
CV_3062		enoyl-CoA hydratase	32.72	1.31	−4.64	0
CV_3300	*treB*	protein phosphohistidine-sugar phosphotransferase	8.82	0.55	−4	0.01
CV_4091	*cviI*	autoinducer synthase	171.85	19.44	−3.14	5.39E-005
CV_4092		aldehyde dehydrogenase	20.7	1.39	−3.9	0
CV_4142	*hoxX*	HoxX-like protein	7.99	0.53	−3.9	0.01
CV_4240		putative chitinase	42.51	2.82	−3.92	2.08E-006
CV_4379		hypothetical inside biotin synthesis operon	16.18	0.61	−4.74	1.65E-005
CV_4382	*comF*	competence protein F	26.97	1.71	−3.98	0.01
Quorum Sensing controlled genes						
hcn operon						
CV_1682	*hcnC*	hydrogen cyanide synthase HcnC	187.54	6.8	−4.78	7.43E-009
CV_1683	*hcnB*	hydrogen cyanide synthase HcnB	600.99	4.65	−7.01	0
CV_1684	*hcnA*	hydrogen cyanide synthase HcnA	600.99	4.65	−7.01	0
Protease						
CV_2571	*LasA*	LasA protease precursor	175.34	13.11	−3.74	1.13E-006
CV_0057	*LasB*	class 4 metalloprotease	211.37	20.93	−3.34	1.09E-006
Cellulose biosynthesis						
CV_2675	bscC	cellulose synthase, subunit C	9.3	0.22	−5.39	0
CV_2676	bscZ	endo-1,4-D-glucanase	9.3	0.22	−5.39	0
CV_2677	bcsB	cellulose synthase, subunit B	9.3	0.22	−5.39	0
Other genes related to virulence and pathogenicity						
CV_3824		conserved hypothetical protein	64.28	5.74	−3.49	6.81E-005
CV_3828	*pilB*	type 4 fimbrial biogenesis protein	63.95	5.32	−3.59	5.99E-006
CV_0829	*pilQ*	type 4 fimbrial biogenesis protein PilQ	96.39	2.11	−5.51	1.01E-013
CV_0830	*pilP*	type 4 fimbrial biogenesis protein PilP	96.39	2.11	−5.51	1.01E-013
CV_0832	*pilN*	type 4 fimbrial biogenesis protein PilN	85.03	6.41	−3.73	0
CV_0833	*pilM*	type 4 fimbrial biogenesis protein PilM	130.37	15.8	−3.04	5.79E-005
CV_0179	*pilT*	twitching motility protein PilT	53.97	6.7	−3.01	0
CV_0180	*pilU2*	twitching mobility protein transport	32.64	2.57	−3.67	0
CV_1458	*pilU1*	twitching motility protein	41.43	5.17	−3	0
CV_3112	*pilV*	type-4 fimbrial biogenesis PilV transmembrane protein	79.2	2.24	−5.14	0
CV_2618	*sipC*	cell invasion protein	17.62	0.92	−4.25	0.01
CV_2619	*sipB*	cell invasion protein	116.65	3.16	−5.21	6.23E-005

Table 1. Cont.

ORF no.	Gene name	Description	Control[1]	Expt[2]	log_2(fold change)	p value
CV_2620	*spaT*	surface presentation of antigens; secretory proteins	116.65	3.16	−5.21	6.23E-005
CV_2198	*pyrH*	uridylate kinase, Pyrimidine metabolism	40.96	4.63	−3.15	0
CV_2205	*ompH*	hypothetical protein, outer membrane protein	46.62	5.62	−3.05	0.01
CV_2206	*lpxD*	Lipopolysaccharide biosynthesis	41.56	4.12	−3.33	0.01
CV_2207	*fabZ*	3-hydroxyacyl-[acyl-carrier-protein] dehydratase, Fatty acid biosynthesis	41.56	4.12	−3.33	0.01
CV_2209	*lpxB*	lipid-A-disaccharide synthase, Lipopolysaccharide biosynthesis	29.44	3.69	−3	0
CV_2210	*rnhB*	ribonuclease HII	29.44	3.69	−3	0
CV_2212		hypothetical protein	561.06	41.05	−3.77	0
CV_4338	*ftsZ*	cell division protein	61.92	7.3	−3.08	9.58E-005
CV_4340	*ftsQ*	cell division transmembrane protein	41.07	1.13	−5.19	0
CV_4341	*ddlB*	D-alanine-D-alanine ligase, Peptidoglycan biosynthesis	41.07	1.13	−5.19	0
CV_4344	*ftsW*	cell division protein	25.93	3.05	−3.09	0
CV_4345	*murD*	Peptidoglycan biosynthesis	25.93	3.05	−3.09	0
CV_4346	*mraY*	Peptidoglycan biosynthesis	19.51	1.81	−3.43	0
CV_4349	*ftsI*	cell division protein FtsI	157.11	0.52	−8.25	3.40E-008
CV_4350		cell division protein	157.11	0.52	−8.25	3.40E-008
CV_4351	*mraW*	16 S rRNA (cytosine1402-N4)-methyltransferase	157.11	0.52	−8.25	3.40E-008
CV_4352		conserved hypothetical protein	157.11	0.52	−8.25	3.40E-008
CV_0516		Ca binding hemolysin	7.43	0.71	−3.39	3.36E-005
CV_0360		thermolabile hemolysin	25.86	2.28	−3.5	0
CV_1989		porin	257.04	24.41	−3.4	6.76E-006
CV_3104		porin	1207.93	95.97	−3.65	7.68E-006
CV_3829		porin	813.08	69.81	−3.54	6.04E-006
CV_3342		hemolysin III	274	26.88	−3.35	1.39E-005

[1]gene expression in cells grown for 24 h.
[2]gene expression in cells grown for 24 h in presence of 400 μg ml^{-1} of GLE.

The percentage (%) inhibition in violacein production was calculated as follows: % inhibition in violacein production = [(O.D.$_{585}$ value observed in the absence of GLE − O.D.$_{585}$ oberved in presence of a defined quantity of GLE)÷O.D.$_{585}$ value observed in absence of GLE] ×100.

Determination of growth and violacein production of *C. violaceum* MTCC 2656 in presence and absence of GLE

Growth of MTCC 2656 was quantified in terms of viable cell number present in the culture (dilution plating followed by counting CFUs) at different time interval [39]. Briefly, an overnight culture of MTCC 2656 (in NB medium), that had been inoculated with a freshly grown single colony, was diluted 100-fold into 10 ml NB medium and allowed to grow for 4 h to obtain log phase cells. The culture was then inoculated into 800 ml NB medium and divided into two portions, to one of which water (control) and to the other, GLE at a final concentration of 400 μg ml^{-1}, was added. These were finally distributed in 10 ml aliquots per pre-sterilized 100 ml Erlenmeyer flask, and incubated at 37°C with shaking. At different time intervals, cultures were withdrawn from time-defined flasks (both control and test) for dilution plating onto NB agar plates as well as for the estimation of violacein by the method described above.

Whole genome transcriptome analysis

The detail of the methodology is provided in File S1. In brief, RNA was isolated from control (cells grown without GLE) and experiment (cells grown with GLE) samples and cDNA Library was prepared using the Ion Total RNA-Seq Kit v2 (Catalog Number 4475936). Template preparation and enrichment was performed as per Ion OneTouch 200 Template Kit (Cat no. 4471263) and 200 base-read sequencing was performed using the Ion PGM 200 Sequencing Kit (Cat no 4474004) on ION TORRENT.

Infection and morphological assessment of HepG2 cells by phase-contrast inverted microscope

HepG2 cells (Human hepatocellular liver carcinoma cell line), obtained from Cell Culture Collection, NCCS, Pune, were grown in 100 mm polyvinyl coated plates, in DMEM (Dulbeco's Modified Eagle's Medium) media with 10% FCS at 37°C in a humidified atmosphere containing 5% CO_2. Cells (1×10^6) were seeded in 60 mm plates in DMEM medium supplemented with 10 U/ml penicillinG and 100 μg ml^{-1} streptomycin. After 24 h, the media was removed and cells were washed thrice with phosphate buffer (PBS, pH 7.2) and grown in DMEM media without antibiotics until *C. violaceum* infection. Bacteria (MTCC 2656) were grown overnight at 37°C in NB either without or with

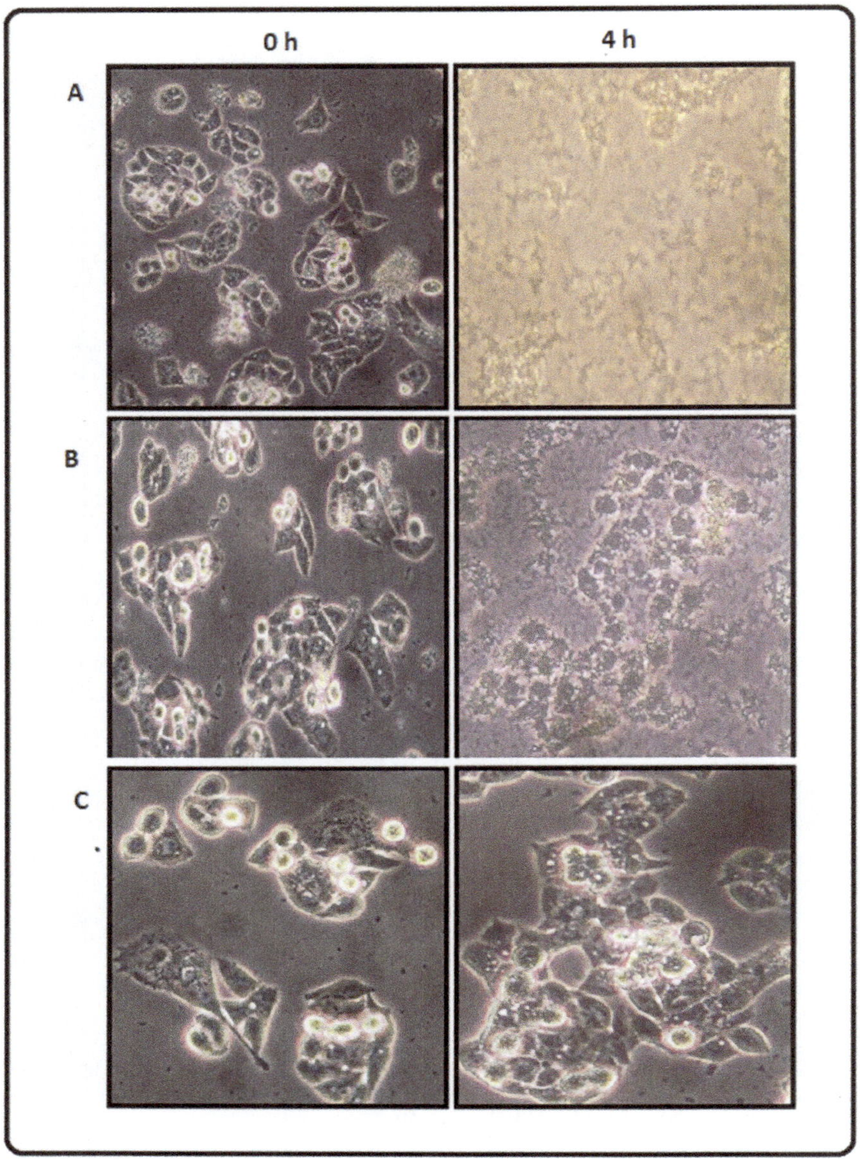

Figure 4. GLE inhibits *C. violaceum* induced cell-lysis of HepG2 cells. Phase contrast micrographs at 200X magnification showing (**A**) lysis of HepG2 cells after 4 h of infection with *C. violaceum* MTCC 2656 and (**B**) inhibition of lysis in presence of 400 μg ml^{-1} GLE. (**C**) growth of HepG2 cells in presence of GLE alone (to nullify any effect of GLE on HepG2 growth).

400 μg ml^{-1}GLE. The following day, the bacterial cells were grown further in fresh media under the same conditions for 3 h. Finally, the untreated or GLE treated *C. violaceum* cells were diluted in DMEM with 10% FCS supplemented either with water or GLE, and added to HepG2 cells at the infectivity ratio of 10:1. The plates were incubated at 37°C in an atmosphere of 5% CO_2 and observed under inverted microscope (Olympus, ck40-slp) at 0 and 4 h of infection at 200X magnification and photographed. HepG2 cells incubated in DMEM containing GLE served as control.

LDH cytotoxicity assay

Cytotoxicity induced in HepG2 cells was quantitated by measuring the release of the cytosolic enzyme lactate dehydrogenase (LDH) in the culture medium. For this, HepG2 cells were grown and exposed to *C. violaceum* as described above. After 4 h of incubation, supernatants were collected and evaluated for the presence of LDH using the LDH Cytotoxicity Assay Kit (Item No. 10008882, Cayman Chemical Company). The LDH activity (μU) was determined from the standard curve and the total LDH activity (μU ml^{-1}) was calculated as value from LDH activity assay (μU)/sample volume assayed (ml).

Statistical Analysis

All experiments were performed in triplicate. Results are expressed as mean value ± standard deviations (S.D) of three replicates and analyzed by using the software SPSS 15.0 for windows (SPSS Inc. Chicago, IL, USA). The statistical treatment of "Whole Transcriptome Analysis" data is elucidated in File S1.

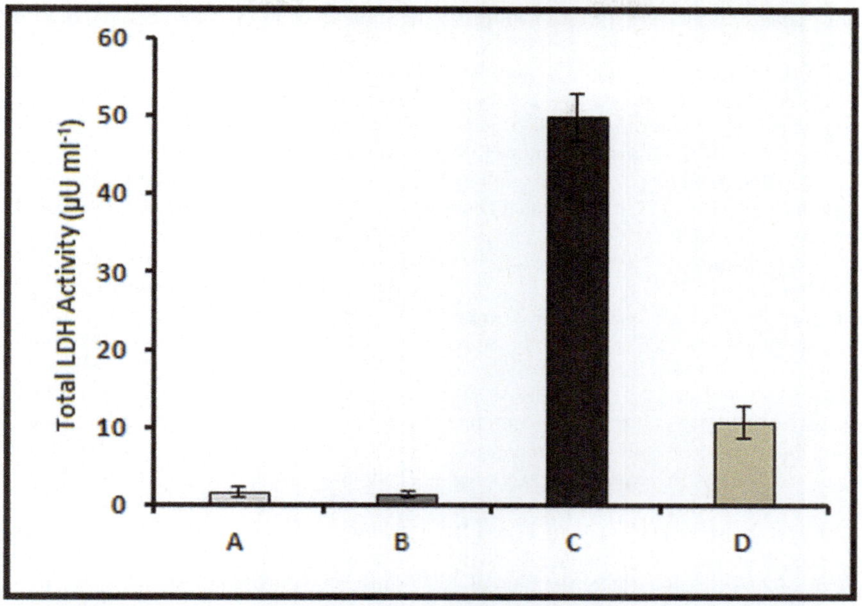

Figure 5. LDH activity in the culture medium of HepG2 cells. LDH assay was performed after 4 h treatment with: (A) only water; (B) only 400 μg ml^{-1} of GLE; (C) only *C. violaceum*; and (D) *C. violaceum* along with 400 μg ml^{-1} GLE.

Results and Discussion

Inhibition of violacein production in the presence of aqueous extracts of guava leaves

The best studied trait controlled by QS in *C. violaceum* is the production of the purple pigment, violacein [40]. To investigate QS inhibitory activity of aqueous extract of *Psidium guajava* L. leaves (GLE), its effect on violacein synthesis by *C. violaceum* MTCC 2656 was examined. A colourless transluscent zone around the zone of diffusion of GLE from the agar cup indicated growth of *C. violaceum* failing to produce violacein (purple pigment, Figure S1A). The production of violacein pigment at 24 h in the presence of varied concentration of GLE was quantified spectrophotometrically (Figure 1A). Violacein production was inhibited by 85% in presence of 400 μg ml^{-1} of GLE and the extent of inhibition remained similar at concentrations up to 1000 μg ml^{-1}. Thus, the concentration of GLE to be used for further studies was selected as 400 μg ml^{-1}. It has been documented in the Literature that nearly all parts of *Psidium guajava* L. tree, including fruits, leaves, bark, and roots, were used traditionally as a medicinal plant throughout the world for varied ailments [27]. The present study revealed that the guava-leaf extract was capable of inhibiting the quorum-dependent violacein production in *C. violaceum*.

The growth of *C. violaceum* is unaffected in presence of GLE

The number of viable cell enumerated from growth studies in presence and absence of the GLE (400 μg/ml) showed no significant difference (Figure 1B). On the contrary, the violacein production was significantly inhibited under similar growth conditions (Figure 1C). It has been noted by earlier authors that the toxicity of putative quorum sensing inhibitors towards bacterial cells may be assessed by addressing three principal issues; that the QS inhibitory effects occur below the minimum inhibitory concentration (if the inhibitor exhibits antimicrobial property), the quorum-sensing inhibitory concentration used in the study does not affect the final cell density after a certain period of incubation, and the QS inhibitor does not grossly affect the kinetics of growth [41]. Our results show that the growth of *C. violaceum* remained unaffected but the quorum-dependent production of violacein was significantly reduced in the presence of GLE. These observations led us to conclude that the decrease in violacein production was not due to any form of growth inhibition of the cells in the culture medium at least up to 24 h of incubation. Similar results with respect to concentration-dependent decrease in violacein production without affecting bacterial growth were reported with extracts of various medicinal plants [42,43]. Assuming that the quorum-inhibiting property of GLE should also be yielding in other bacteria that demonstrate quorum-dependent diverse phenotypes, we tested GLE for its activity using another test organism, *Pseudomonas aeruginosa*. The gram-negative bacterium *Pseudomonas aeruginosa* exhibits swarming motility which requires the bacteria to effectively work together via the process of quorum-sensing [37]. The cells of *P. aeruginosa,* MTCC 2297, in absence of GLE, formed tendrils migrating outwards from the point of bacterial inoculation, with continued branching as the bacteria moved farther from the center, while in presence of GLE, cells grew to form a localized colony in the center with no signs of swarming (Figure S1B). The ability of natural products to disrupt the quorum regulated swarming motility in *P. aeruginosa* was demonstrated earlier [44]. Thus, the manifestation of anti-QS property in the form of inhibition of swarming motility of *P. aeruginosa*, further provides evidence for the proposed quorum quenching property of GLE.

Impact of GLE on the genome-wide gene expression of *C. violaceum*

The quorum-regulated gene expression has been extensively studied in gram negative pathogenic bacteria such as *P. aeruginosa* [45] and *Escherichia coli* [46]. However, *C. violaceum* transcriptomic studies have not yet attracted the similar attention.

The increasing reports of *C. violaceum* cases [47] prompted us to unravel the status of gene expression in the presence or absence of GLE with a view to establish GLE as a QS inhibitory candidate for controlling the bacterial pathogenicity. Schuster *et al.* have observed that the timing affects the quorum-controlled gene expression in *P. aeruginosa* and most of the transcripts under QS-regulation were induced maximally at late log or stationary phase [48]. Thus, to get an insight into the effect of GLE on global gene expression of *C. violaceum*, we used 24 h grown cultures either in the absence (Control) or presence (Experimental) of 400 μg ml^{-1}GLE.

A total of 33,54,744 and 30,09,475 high quality Ion Torrent reads for Control and Experiment sample respectively were mapped on the reference genome of *Chromobacterium violaceum* ATCC 12472 with genome size of 4.75 Mb. There are 4,529 genes present in *C. violaceum* ATCC 12472 GFF file. The number of genes expressed in (i) both control and experiment; (ii) exclusively in control; (iii) exclusively in experiment; (iv) not expressed in both control and experiment were determined to be 3025, 1229, 32, and 243 respectively, as illustrated in Figure 2. Raw sequences were deposited at the NCBI Sequence Read Archive, under Bioproject accession number PRJNA243990 and SRA accession code SRP041018 (http://www.ncbi.nlm.nih.gov/bioproject/?term=PRJNA243990). The FPKM value for each gene was calculated for the combination of samples Control and Experiment. These FPKM values (FPKM_control and FPKM_experiment) were further used to calculate the log fold change as $\log_2$ (FPKM_experiment/FPKM_control). Moreover, uncorrected *p*-value of the test statistic for each gene was also computed. Using a stringent $\log_2$ threefold cut-off with p-value≤0.01, we observed that 816 genes, scattered throughout the *C. violaceum* MTCC 2656 genome, comprising about 19% of the total genes expressed, were significantly down regulated in the presence of GLE (Table S1). The 816 differentially expressed genes were categorised into 17 groups (Figure 3) based on the *C. violaceum* ATCC12472 database from the Brazilian Genome Virtual Institute of Genomic Research (BRGene) (http://www.brgene.lncc.br). The category representing the largest percentage, ~32% of the total 816 genes, encoded for the group of proteins assigned unclassified and unknown functions. The categories representing>6.5% belonged to the functional classes of proteins namely cell motility and secretion (9.31%), cell envelope biogenesis, outer membrane (7.47%), transduction mechanisms (6.86%) and amino acid transport and metabolism (8.94%). Among these, the first three groups of proteins are related to virulence in many pathogens since these processes mediate interaction of bacteria with their immediate environment and genes encoding virulence factors. GLE down-regulated genes like lyases (*argH, aspA*), dehydrogenase and hydratase (*usg, sdaA1*) and transporters (*sdaC, argT*), that encode products involved in metabolism and transport of amino acids. In addition, GLE down-regulated, about 4%, the genes involved in energy production and conversion, secondary metabolites biosynthesis, inorganic ion transport and metabolism, and coenzyme metabolism. The possibility of interference of GLE with one or several global regulators in *C. violaceum* resulting in differential expression of multiple genes cannot be ruled out. However, since GLE targeted virulent genes which are often QS controlled [49] and QS activated genes involved in uptake, synthesis and degradation of amino acids [50]; inhibited QS controlled phenotype, violacein production, without hampering growth of *C. violaceum* upto 24 h, pointed to the possible anti-QS property of GLE. Repression of genes related to cellular processes with no effect on growth was revealed by microarray analysis of *P. aeruginosa* gene expression in the presence of QS inhibitor furanone C-30 [51]. Earlier microarray studies with *E. coli* and *P. aeruginosa* have revealed that varied cellular functions are largely affected by the QS system [45,46]. Both the gram negative bacteria have more than one QS regulatory system. The cross-talk between different quorum signal pathways lead to either synergestic or antagonistic effects. It is thus tempting to assume that *C. violaceum* may also possess multiple QS systems and any interference would thus lead to a global differential expression of genes as observed with GLE. Previous transcriptomic studies have shown that as much as>10% of the total genes in the genome of *P. aeruginosa* are QS regulated [45], however, we observed a higher percentage (19%) of differential gene expression. This could be explained either due to the presence of multiple components of GLE or if we assume that these genes are QS-regulated, the higher number of genes revealed in this study may be due to the fact that earlier groups detected QS regulons using specific mutants but our studies involves revelation of QS regulons using a putative QS inhibitor on wild type cells. Thus, although the data obtained does not reflect a definitive analysis of gene regulation by quorum sensing, perhaps for the first time, our results provide with a set of probable candidate genes under QS regulation in *C. violaceum* for further investigation.

Differential expression of genes predicted to be under QS-regulation in *C. violaceum*

Cell to cell signaling in *C. violaceum* is based on the CviI/CviR circuit. At high cell density, the cytoplasmic quorum-sensing receptor, CviR, binds to the AHL autoinducer. The CviR-AHL complex then binds to DNA and activates transcription of a number of genes. The fact that GLE has the ability to induce break in the CviI/CviR circuitry is reflected phenotypically by the abolition of violacein production by *C. violaceum* in the growth medium. The gene responsible for violacein production, *vioA*, was $\log_2$>threefold repressed. Since the level of expression of *vioA* gene (FPKM_control value) in absence of GLE was low (perhaps owing to the experimental conditions), the comparison with FPKM_experiment value (in presence of GLE) was concluded as statistically non-significant. Our data revealed that GLE down-regulated the expression of major virulence factors (Table 1) such as, *lasA* (CV_2571), *lasB* (CV_0057), *hcnABC* operon (CV_1682 to CV_1684) and chitinase (CV_4240). The quorum regulated proteases LasA and LasB are capable of inactivating a wide range of biological tissues and immunological agents to allow the bacteria to invade and colonize host tissues [52,53]. The *hcnABC* operon responsible for biosynthesis of hydrogen cyanide (HCN), a secondary metabolite and a potent inhibitor of cytochrome oxidase and several other metalloenzymes, is known to be under the control of quorum-sensing mechanism in *C. violaceum* [54]. The enzyme, chitinase, is hypothesized to be involved in blocking the growth of fungi with chitinous exoskeleton, allowing a competitive advantage to *C. violaceum* during colonization [55]. In addition to these secreted factors, transcription of gene encoding acyl carrier protein (ACP) synthase, *fabF* (CV_3412) and other ACP-encoding genes like *fabZ* (CV_2207) and CV_3413, were significantly repressed by GLE. These gene products are proposed to be involved in synthesis of AHLs [56]. Similar down-regulation of these genes in *P. aeruginosa* was observed in the presence of the QS inhibitor, furanone C-30 [51]. Moreover, C-30 also targets *P. aeruginosa* virulence but it does not affect the expression of gene clusters encoding the components of QS system. On the contrary, GLE not only affected virulence but significantly repressed the expression of both *cviI* (CV_4091) and *cviR* (CV_4090) genes, the basic constituents of *C. violaceum* QS system, suggesting direct transcriptional interference of GLE.

A putative cviR binding site was detected in the *cviI* promoter region and so its expression was expected to be down-regulated. In addition, a number of other *cviR* -regulated genes based on the presence of an ideal CviR binding site coupled with genome scanning were predicted [57]. GLE down-regulated a subset of these genes, which included CV_1328 (methyl-accepting chemotaxis protein), CV_2321 (recN, DNA repair protein) and CV_3062 (enoyl-CoA hydratase). The downregulation of QS-controlled genes with no effect on bacterial cell growth led us to hypothesize the anti-QS property of GLE to be responsible for the observed differential gene expression.

Status of expression of genes plausibly networked with quorum sensing mechanism

The pathogenicity of *C. violaceum* is still poorly understood. Cause and effect relationship between QS system and pathogenicity of *C. violaceum* is yet an enigma. On the basis of whole genome sequence of *C. violaceum*, a catalogue of genes encoding probable virulence factors was prepared [58]. These included the genes encoding *pil* proteins and genes involved in lipopolysaccharide (LPS) and peptidoglycan biosynthesis. The type IV pilus gene cluster encodes Type IV pili which are appendages emanating from the surfaces of several gram-negative bacteria that are associated with pathogenicity [59,60]. The expression of a majority of the *pil* genes found in *C. violaceum*, namely *pil B, T, U2, U1, V* and *pilQPNM*, were inhibited by GLE. Transcription of *lpxB* (CV_2209) and *lpxD* (CV_ 2206) genes, responsible for LPS biosynthesis and *murA* (CV_0440) and *murD* (CV_4345), genes required for peptidoglycan biosynthesis, were significantly down-regulated in the presence of GLE. As found in other gram negative bacteria, the LPS and peptidoglycan of *C. violaceum* are associated with activation of immune response in the host, which induces secretion of inflammatory cytokines resulting in septic shock [58]. Very recently, genes encoding potential secreted virulence factors of *C. violaceum* were identified [61]. Among these, GLE targeted the hemolysins (CV_0516, CV_0360 and CV_3342) and porins (CV_1989, CV_3104, and CV_3829). Increased expression of hemolysin and porins in presence of quorum signal autoinducer 2 (AI-2), mediator of second signaling pathway, has previously been documented in *E.coli* [46]. While unravelling the virulence determinant for *C. violaceum* induced cell lysis, Miki *et al.* observed that the formation of pore structures on the host cell membrane results in cell death in hepatocytes [62]. The pore formation involved CipB (*sipB*, CV_2619), a translocator of the *Chromobacterium* pathogenicity islands 1 and 1a (Cpi-1/-1a). GLE significantly down-regulated the expression of the *sipB* gene and two other linked genes, *spaT* (CV_2620) and *sipC* (CV_2618), cell invasion protein. Transcription of a vast number of genes involved in flagellar biosynthesis (*flhB1, flhF, fliL, fliE, fliM, fliN, fliC2, fliC3, flgK1, flaG, fliI1, fleN* and *fliA1*, RNA polymerase sigma factor for flagellar operon) and chemotaxis (*cheV1, cheR3, cheR1, cheA, cheA2, cheZ, trg, nahY, motA1, motB1* and *motB2*) were repressed by GLE. Regulation of flagellar genes by QS has been previously reported in other bacteria such as *Yersinia pseudotuberculosis* and *Vibrio harvei* [63,64]. A decreased expression of *motA* was observed in QS-mutant strains of *E. coli* [49]. The results indicate that down-regulation of the expression of the varied genes, responsible for virulence and cell motility, by GLE, may alter *C. violaceum* pathogenicity.

C. violaceum genome consists of four genes involved in cellulose production: *bcsA, bcsB, bcsZ* and *bcsC* [65]. Interestingly, in the presence of GLE, expression of *bsc CZ* and *bcsB* (CV_2675-CV_2677) were down-regulated indicative of a low level of cellulose production. Bacterial biofilms consist of bacteria colonies embedded in their own extracellular matrix composed of exopolysaccharides (EPS). Cellulose has been identified as an exopolysaccharide (EPS) in the extracellular matrix produced by several bacterial species. Decreased cellulose production in the presence of GLE will thus lead to inhibition in the formation of biofilms. This emphasises the proposed QS inhibitory property of GLE in light of the finding that QS has been implicated as an important step in biofilm formation [5,66].

Earlier studies on anti-QS activities of crude plant extracts in bacteria have focussed mainly on demonstrating inhibition of expression of well-established specific QS-induced gene(s) [67]. In *C. violaceum*, quantitative real-time -PCR assay for expression of specific genes, namely, AI synthetase gene (*cviI*) and one of the violacein biosynthetic genes (*vioB*), was performed in order to establish that sub-lethal concentration of antibiotics can inhibit QS in *C. violaceum* [68]. Validation of the claims of remedies resulting from consumption of crude plant extracts (e.g crude extracts of *Ginkgo biloba*) by transcriptome profiling of the affected organs (e.g liver of rats on a high-fat diet) have provided with the molecular basis of herbal therapy [69]. The outcome of the present transcriptomic data improvises quorum inhibitory property of crude extract of guava leaves, GLE, and provides with a list of probable quorum regulated genes in *C. violaceum* giving insights into the features of this versatile bacterium. The attenuation of virulence by GLE, as revealed by the differential gene expression data, compelled us to test the phenotypic manifestation of *C. violaceum* induced cytolysis of human hepatoma cell line in presence of GLE.

GLE inhibits *C. violaecum*-induced cytotoxicity in HepG2 cells

To analyze the effect of GLE on cell death induced by *C. violaecum*, human hepatocellular liver carcinoma cell line HepG2 was infected with the bacteria either in the absence or presence of GLE. The course of infection was monitored by phase contrast microscopy at different time points. It was observed that *C. violaecum* was able to induce cytotoxicity in HepG2 cells, characterized by shrinkage of the cells and reduction in cell density, with time (Figure. 4A). Whereas, no morphological changes were observed in the cells when infection was carried out in presence of GLE indicating that the extract could protect the HepG2 cells from *C. violaecum* infection (Figure 4B). Notably, GLE alone had no inhibitory effect on the growth of HepG2 cells (Figure 4C). Furthermore, *C. violaceum*-induced cytotoxicity was quantified by assaying the release of the cytosolic enzyme lactate dehydrogenase (LDH). The cell death observed after *C. violaceum* infection in hepatocytes is characterized by membrane rupture with subsequent release of LDH [62]. In agreement, after 4 h following *C. violaceum* infection of HepG2 cells, a significant increase in extracellular LDH activity was observed as compared to the non-infected cells (Fig. 5). Interestingly, the amount of LDH release from *C. violaceum*-infected HepG2 cells decreased by ~80% in the presence of GLE indicating the ability of GLE to hinder cell death instigated by *C. violaceum*. A similar decrease in cytotoxicity in different cell lines was demonstrated using mutant strains of *C. violaceum* [62]. Moreover, GLE itself had no effect on human hepatoma HepG2 cells showing its non-toxicity towards the host cell (Figure 5). The guava leaf paste was highly likely to be non-toxic since it is used immensely in folklore practices, including as toothpaste for the cure of dental caries [70]. This result shows that GLE regulates the pathogenicity of *C. violaceum* by preventing the initiation of the cascade of gene expression required for successful infection and establishment in the host which, in turn, may provide time to the immune system of the host to eliminate the pathogen. This effect of GLE may be attributed to its ability to either directly or indirectly inhibit the QS system.

Conclusions

Identification of QS inhibitors (QSIs) from natural products as an alternative to antibiotics is currently an area of intense interest. However, establishment of a real QSI and the phenomenon of resistance to QSIs are debatable aspects in this field of research [71]. The present study proposes guava leaf aqueous extract (GLE) as QSI since it inhibits quorum regulated phenotypes such as violacein production and swarming in the pathogenic bacteria *C. violaceum* and *P. aeruginosa* respectively and induces a global differential gene expression in *C. violaceum*, a pathogen with high mortality rate, without affecting its growth. The complexity of GLE prevents us from directly linking it with the QS system. However, reduction in QS gene expression was correlated to the attenuation of bacterial virulence resulting in prevention of *C. violaceum* induced lysis of host (HepG2) cells *in vitro*. More studies are required to establish the exact nature of the metabolic cross-talks in presence of GLE but this study provides with the platform to think on using the crude extract of plants to combat pathogenic bacteria. Overall, our results provide insights into the candidature of GLE as QSI and the identification of a set of probable quorum regulated genes in *C. violaceum*.

Supporting Information

Figure S1 Inhibition of QS-regulated phenotypes by GLE. A. Formation of a colourless translucent zone around the well containing GLE indicating absence of violacein production by *C. violaceum* cells. **B.** Inhibition of swarming motility of *P. aeruginosa* cells grown in presence of GLE.

Author Contributions

Conceived and designed the experiments: RG RC. Performed the experiments: BKT AK. Analyzed the data: RG RC. Contributed reagents/materials/analysis tools: RG RC BKT AK. Contributed to the writing of the manuscript: RG RC BKT AK.

References

1. Rasmussen TB, Givskov M (2006) Quorum-sensing inhibitors as anti-pathogenic drugs. Intl J Med Microbiol 296: 149–161.
2. Clatworthy AE, Pierson E, Hung DT (2007) Targeting virulence: a new paradigm for antimicrobial therapy. Nat Chem Biol 3: 541–548.
3. Kalia VC, Purohit HJ (2011) Quenching the quorum sensing system: potential antibacterial drug targets. Crit Rev Microbiol 37: 121–140.
4. Williams P (2007) Quorum sensing, communication and cross-kingdom signalling in the bacterial world. Microbiology 153: 3923–3938.
5. Sarkar S, Chakraborty R (2008) Quorum sensing in metal tolerance of *Acinetobacter junii* BB1A is associated with biofilm production. FEMS Microbiol. Lett 282: 160–165.
6. Dandekar AA, Chugani S, Greenberg EP (2012) Bacterial quorum sensing and metabolic incentives to cooperate. Science 338: 264–266.
7. Waters CM, Bassler BL (2005) Quorum sensing: cell-to-cell communication in bacteria. Annu Rev Cell Dev Biol 21: 319–346.
8. Schuster M, Sexton DJ, Diggle SP, Greenberg EP (2013) Acyl-homoserine lactone quorum sensing: From evolution to application. Annu Rev Microbiol 67: 43–63.
9. Rutherford ST, Bassler BL (2012) Bacterial quorum sensing: Its role in virulence and possibilities for its control. Cold Spring Harb Perspect Med 2: a01242.
10. Koburger JA, May SO (1982) Isolation of *Chromobacterium* spp. from foods, soil, and water. Appl Environ Microbiol 44: 1463–1465.
11. Orsetti AG, Markiewicz P, Epstein MG, Conceiçao OJG, D'Ippolito G, et al. (2013) Liver abscesses by *Chromobacterium violaceum*: a case report of a rare disease. OA Case Reports 28: 19–23.
12. Teoh AYB, Hui M, Ngo KY, Wong J, Lee KF, et al. (2006) Fatal septicaemia from *Chromobacterium violaceum*: Case reports and review of the literature. Hong Kong Med J 12: 228–231.
13. Karthik R, Pancharatnam P, Balaji V (2012) Case Report Fatal *Chromobacterium violaceum* septicemia in a South Indian adult. J Infect Dev Ctries 6: 751–755.
14. Fantinatti-Garboggini F, Almeida R, Portillo VA, Barbosa TA, Trevilato PB, et al. (2004) Drug resistance in *Chromobacterium violaceum*. Genet Mol Res 3: 134–147.
15. Swem LR, Swem DL, O'Loughlin CT, Gatmaitan R, Zhao B, et al. (2009) A quorum-sensing antagonist targets both membrane-bound and cytoplasmic receptors and controls bacterial pathogenicity. Mol Cell 35: 143–153.
16. Vasconcelos ATR, Almeida de DF, Hungria M, Guimarães CT, Antonio RV, et al. (2003) The complete genome sequence of *Chromobacterium violaceum* reveals remarkable and exploitable bacterial adaptability. Proc Natl Acad Sci USA 100: 11660–11665.
17. Martinelli D, Grossmann G, Sequin U, Brandl H, Bachofen R (2004) Effects of natural and chemically synthesized furanones on quorum sensing in *Chromobacterium violaceum*. BMC Microbiol 4: 25.
18. Kalia VC (2013) Quorum sensing inhibitors: An overview. Biotechnol Adv 31: 224–245.
19. Koh KH, Tham FY (2011) Screening of traditional Chinese medicinal plants for quorum-sensing inhibitors activity. J Microbiol Immunol Infect 44: 144–148.
20. Vasavi HS, Arun AB, Rekha PD (2013) Inhibition of quorum sensing in *Chromobacterium violaceum* by Syzygium cumini L. and Pimenta dioica L. Asian Pac J Trop Biomed 3: 954–959.
21. Priya K, Yin W-F, Chan K-GF (2013) Anti-Quorum Sensing Activity of the Traditional Chinese Herb, *Phyllanthus amarus*. Sensors 13: 14558–14569.
22. Gilbert B, Alves LF (2003) Synergy in Plant Medicines. Curr Med Chem 10: 13–20.
23. Rasoanaivo P, Wright CW, Willcox ML, Gilbert B (2011) Whole plant extracts versus single compounds for the treatment of malaria: synergy and positive interactions. Malar J (Suppl 1): S4.
24. Ginsburg H, Deharo E (2011) A call for using natural compounds in the development of new antimalarial treatments– an introduction. Malar J (Suppl 1): S1.
25. Wagner H, Ulrich-Merzenich G (2009) Synergy research: approaching a new generation of phytopharmaceuticals. Phytomedicine 16: 97–110.
26. Gutierrez RM, Mitchell S, Solis RV (2008) *Psidium guajava*: a review of its traditional uses, phytochemistry and pharmacology. J Ethnopharmacol 117: 1–27.
27. Barbalho SM, Farinazzi-Machado FMV, de Alvares GR, Brun-nati ACS, Otoboni AM, et al. (2012) *Psidium Guajava* (Guava): A plant of mul-tipurpose medicinal applications. Med Aromat Plants 1: 104–109.
28. Metwally AM, Omar AA, Harraz FM, El Sohafy SM (2010) Phytochemical investigation and antimicrobial activity of *Psidium guajava* L leaves. Pharmacogn Mag 6: 212–218.
29. Birdi T, Daswani P, Brijesh S, Tetali P, Natu A, et al. (2010) Newer insights into the mechanism of action of *Psidium guajava* L. leaves in infectious diarrhoea. BMC Complement Altern Med 10: 33.
30. Chen HY, Yen GC (2007) Antioxidant activity and free radical-scavenging capacity of extracts from guava (*Psidium guajava* L) leaves. Food Chem 101: 686–694.
31. Matsuo T, Hanamure N, Shimoi K, Nakamura Y, Tomita I (1994) Identification of (+)-gallocatechin as a bio-antimutagenic compound in *Psidium guava* leaves. Phytochemistry 36: 1027–1029.
32. Ryu NH, Park KR, Kim SM, Yun HM, Nam D, et al. (2012) A Hexane Fraction of Guava Leaves (*Psidium guajava* L.) Induces Anticancer Activity by Suppressing AKT/Mammalian Target of Rapamycin/Ribosomal p70 S6 Kinase in Human Prostate Cancer Cells. J Med Food 15: 231–241.
33. Deguchi Y, Miyazaki K (2010) Anti-hyperglycemic and anti-hyperlipidemic effects of guava leaf extract. Nutr Metab (Lond) 7: 9.
34. Prabu GR, Gnanamani A, Sadulla S (2006) Guaijaverin - a plant flavonoid as potential antiplaque agent against *Streptococcus mutans*. J Appl Microbiol 101: 487–495.
35. Quave CL, Plano LRW, Pantuso T, Bennett BC (2008) Effects of extracts from Italian medicinal plants on planktonic growth, biofilm formation and adherence of methicillin-resistant *Staphylococcus aureus*. J Ethnopharmacol 118: 418–428.
36. Al-Hussaini R, Mahasneh AM (2009) Microbial growth and quorum sensing antagonist activities of herbal plants extracts. Molecules 14: 3425–3435.

37. Tremblay J, Richardson AP, Lepine F, Deziel E (2007) Self-produced extracellular stimuli modulate the *Pseudomonas aeruginosa* swarming motility behaviour. Environ Microbiol 9: 2622–2630.
38. Choo JH, Rukayadi Y, Hwang J-K (2006). Inhibition of bacterial quorum sensing by vanilla extract. Lett Appl Microbiol 42: 637–641.
39. Miller JH (1972) Determination of viable cell counts: bacterial growth curves. Experiments in Molecular Genetics. Cold Spring Harbor Press: New York, pp 31–36.
40. McLean RJ, Pierson LS III, Fuqua C (2004) A simple screening protocol for the identification of quorum signal antagonists. J Microbiol Methods 58: 351–360.
41. Defoirdt T, Brackman G, Coenye T (2013) Quorum sensing inhibitors: how strong is the evidence? Trends Microbiol 21: 619–624.
42. Adonizio AL, Downum K, Bennett BC, Mathee K (2006) Anti-quorum sensing activity of medicinal plants in southern Florida. J Ethnopharmacol 105: 427–435.
43. Koh KH, Tham FY (2011) Screening of traditional Chinese medicinal plants for quorum-sensing inhibitors activity. J Microbiol Immunol Infect 44: 144–148.
44. May CO, Tufenkji N (2011) The swarming motility of *P. aeruginosa* is blocked by cranberry proanthocyanidins and other tannin containing materials. Appl Environ Microbiol 77: 3061–3067.
45. Wagner VE, Bushnell D, Passador L, Brooks AI, Iglewski BH (2003) Microarray analysis of *Pseudomonas aeruginosa* quorum-sensing regulons: effects of growth phase and environment. J Bacteriol 185: 2080–2095.
46. DeLisa MP, Wu CF, Wang L, Valdes JJ, Bentley WE (2001) DNA microarray-based identification of genes controlled by autoinducer 2-stimulated quorum sensing in *Escherichia coli*. J Bacteriol 183: 5239–5247.
47. Yang CH, Li YH (2011) *Chromobacterium violaceum* infection: a clinical review of an important but neglected infection. J Chin Med Assoc 74: 435–441.
48. Schuster M, Lostroh CP, Ogi T, Greenberg EP (2003) Identification, timing, and signal specificity of *Pseudomonas aeruginosa* quorum-controlled genes: a transcriptome analysis. J Bacteriol 185: 2066–2079.
49. Passador LJ, Cook M, Gambello MJ, Rust L, Iglewski BH (1993) Expression of *Pseudomonas aeruginosa* virulence genes requires cell-to-cell communication. Science 260: 1127–1130.
50. Baca-DeLancey RR, South MM, Ding X, Rather PN (1999) *Escherichia coli* genes regulated by cell-to-cell signaling. Proc Natl Acad Sci USA 96: 4610–4614.
51. Hentzer M, Wu H, Andersen JB, Riedel K, Rasmussen TB, et al. (2003) Attenuation of *Pseudomonas aeruginosa* virulence by quorum sensing inhibitors. EMBO J 22: 3803–3815.
52. Kessler E (1995) Beta-lytic endopeptidases. Methods Enzymol 248: 740–756.
53. Preston M, Seed P, Toder D, Iglewski B, Ohman D, et al. (1997) Contribution of proteases and LasR to the virulence of *Pseudomonas aeruginosa* during corneal infections. Infect Immun 65: 3086–3090.
54. Duran N, Menck CF (2001) *Chromobacterium violaceum*: a review of pharmacological and industiral perspectives. Crit Rev Microbiol 27: 201–222.
55. Chernin LS, Winson MK, Thompson JM, Haran S, Bycroft BW, et al. (1998) Chitinolytic activity in *Chromobacterium violaceum*: substrate analysis and regulation by quorum sensing. J Bacteriol 180: 4435–4441.
56. Schaefer AL, Val DL, Hanzelka BL, Cronan JE, Greenberg EP (1996) Generation of cell-to-cell signals in quorum sensing: acyl homoserine lactone synthase activity of a purified *Vibrio fischeri* LuxI protein. Proc Natl Acad Sci USA 93: 9505–9509.
57. Stauff DL, Bassler BL (2011) Quorum Sensing in *Chromobacterium violaceum*: DNA Recognition and Gene Regulation by the CviR Receptor. J Bact 193: 3871–3878.
58. Brito CFAD, Carvalho CMB, Santos FR, Gazzinelli RT, Oliveira SC, et al. (2004) *Chromobacterium violaceum* genome: molecular mechanisms associated with pathogenicity. Genet Mol Res 3: 148–161.
59. Bieber D, Ramer SW, Wu CY, Murray WJ, Tobe T, et al. (1998) Type IV pili, transient bacterial aggregates, and virulence of enteropathogenic *Escherichia coli*. Science 280: 2114–2118.
60. Kennan RM, Dhungyel OP, Whittington RJ, Egerton JR, Rood JI (2001) The type IV fimbrial subunit gene (fimA) of *Dichelobacter nodosus* is essential for virulence, protease secretion, and natural competence. J Bacteriol 183: 4451–4458.
61. Castro-Gomes T, Cardoso MS, DaRocha WD, Laibida LA, Nascimento AM, et al. (2014) Identification of secreted virulence factors of *Chromobacterium violaceum*. J Microbiol 52: 350–353.
62. Miki T, Iguchi M, Akiba K, Hosono M, Sobue T, et al. (2010) Chromobacterium pathogenicity island 1 type III secretion system is a major virulence determinant for *Chromobacterium violaceum*-induced cell death in hepatocytes. Mol Microbiol 77: 855–872.
63. Atkinson S, Throup JP, Stewart GSAB, Williams P (1999) A hierarchical quorum-sensing system in *Yersinia pseudotuberculosis* is involved in the regulation of motility and clumping. Mol Microbiol 33: 1267–1277.
64. Yang Q, Defoirdt T (2014) Quorum sensing positively regulates flagellar motility in pathogenic *Vibrio harveyi*. Env Microbiol doi:10.1111/1462-2920.12420 (in Press).
65. Recouvreux DO, Carminatti CA, Pitlovanciv AK, Rambo CR, Porto LM, et al. (2008) Cellulose biosynthesis by the beta-proteobacterium, *Chromobacterium violaceum*. Curr Microbiol 57: 469–476.
66. Wood TK (2009) Insights on *Escherichia coli* biofilm formation and inhibition from whole-transcriptome profiling. Environ Microbiol 11: 1–15.
67. Lee J-HJ, Kim Y-GY, Cho MHM, Wood TKT, Lee JJ (2011) Transcriptomic analysis for genetic mechanisms of the factors related to biofilm formation in *Escherichia coli* O157: H7. Curr Microbiol 62: 1321–1330.
68. Liu Z, Wang W, Zhu Y, Gong Q, Yu W, et al. (2013) Antibiotics at subinhibitory concentrations improve the quorum sensing behavior of *Chromobacterium violaceum*. FEMS Microbiol Lett 341: 37–44.
69. Gu X, Xie Z, Wang Q, Liiu G, Qu Y, et al. (2009) Transcriptome profiling analysis reveals multiple modulatory effects of Ginkgo biloba extract in the liver of rats on a high-fat diet. Febs J 276: 1450–1458.
70. Fathilah AR, Rahim ZHA (2003) The anti-adherence effect of *Piper betle* and *Psidium guajava* extracts on the adhesion of early settlers in dental plaque to saliva-coated glass surfaces. J Oral Sci 45: 201–206.
71. Defoirdt T, Boon N, Bossier P (2010) Can Bacteria Evolve Resistance to Quorum Sensing Disruption, PLoS Pathog 6: e1000989.

6

Francisella novicida Pathogenicity Island Encoded Proteins Were Secreted during Infection of Macrophage-Like Cells

Rebekah F. Hare[1], Karsten Hueffer[2*]

1 Department of Biology and Wildlife, Institute of Arctic Biology, University of Alaska Fairbanks, Fairbanks, Alaska, United States of America, **2** Department of Veterinary Medicine, University of Alaska Fairbanks, Fairbanks, Alaska, United States of America

Abstract

Intracellular pathogens and other organisms have evolved mechanisms to exploit host cells for their life cycles. Virulence genes of some intracellular bacteria responsible for these mechanisms are located in pathogenicity islands, such as secretion systems that secrete effector proteins. The *Francisella* pathogenicity island is required for phagosomal escape, intracellular replication, evasion of host immune responses, virulence, and encodes a type 6 secretion system. We hypothesize that some *Francisella novicida* pathogenicity island proteins are secreted during infection of host cells. To test this hypothesis, expression plasmids for all *Francisella novicida* FPI-encoded proteins with C-terminal and N-terminal epitope FLAG tags were developed. These plasmids expressed their respective epitope FLAG-tagged proteins at their predicted molecular weights. J774 murine macrophage-like cells were infected with *Francisella novicida* containing these plasmids. The FPI proteins expressed from these plasmids successfully restored the intramacrophage growth phenotype in mutants of the respective genes that were deficient for intramacrophage growth. Using these expression plasmids, the localization of the *Francisella* pathogenicity island proteins were examined via immuno-fluorescence microscopy within infected macrophage-like cells. Several *Francisella* pathogenicity island encoded proteins (IglABCDEFGHIJ, PdpACE, DotU and VgrG) were detected extracellularly and they were co-localized with the bacteria, while PdpBD and Anmk were not detected and thus remained inside bacteria. Proteins that were co-localized with bacteria had different patterns of localization. The localization of IglC was dependent on the type 6 secretion system. This suggests that some *Francisella* pathogenicity island proteins were secreted while others remain within the bacterium during infection of host cells as structural components of the secretion system and were necessary for secretion.

Editor: Yousef Abu Kwaik, University of Louisville, United States of America

Funding: Research reported in this publication was supported by the National Institute of General Medical Sciences of the National Institutes of Health under Award Number P20GM103395. The content is solely the responsibility of the authors and does not necessarily represent the official views of the National Institutes of Health. The funders had no role in study design, data collection and analysis, decision to publish, or preparation of the manuscript.

Competing Interests: The authors have declared that no competing interests exist.

* Email: khueffer@alaska.edu

Introduction

Pathogenicity islands exist in many pathogenic bacteria, are acquired via horizontal gene transfer, and encode genes that facilitate interactions with host cells [1]. Secretion systems in bacteria involve the transport or translocation of effector molecules from the interior of a bacterial cell through its membranes to the exterior. Protein secretion is an important mechanism for bacteria to adapt and survive in their environment, including within an infected host [2]. Effector proteins are enzymes or toxins that facilitate infection and are secreted by these secretion systems [3].

Francisella tularensis is an intracellular pathogen that possesses the *Francisella* pathogenicity island (FPI) [4]. The FPI is found in all *Francisella* species and strains, and is duplicated in all human-virulent biovars of *F. tularensis*. *F. novicida* and *F. philomiragia* harbor only one copy of the FPI, which makes these species attractive for creating isogenic FPI gene deletion mutants [4,5]. The molecular mechanisms contributing to the intracellular survival of *Francisella* are poorly understood, and FPI mutagenesis approaches are useful in identifying genes required for intracellular replication and virulence [4,6,7,8,9,10,11].

The FPI contains genes with homology to genes encoding type 6 secretion systems (T6SS) in other bacteria [12,13,14,15]. Bioinformatics, genetics, biochemical, and cell biology approaches provide evidence the FPI encodes a functional secretion system [12,13]. Homologues of *iglAB*, *pdpB*, *dotU*, and *vgrG* are found in most T6SS identified to date; therefore, some suspect the secretion system of the FPI is a T6SS, although this is debatable [15,16]. DotU and PdpB are inner membrane components that are homologous with the T6SS proteins DotU and IcmF, respectively [15]. IglA and IglB are IcmF-associated homologous proteins seen in *Rhizobium leguminosarum*, *Salmonella enterica*, and *Vibrio cholerae* [4,12,13,15,16,17,18]. The solubility properties of IglABC suggest these proteins could be part of the needle spanning through the bacterial membranes, and the protein-protein interactions of IglAB also suggests the auxiliary roles within *F. novicida* as described in other species [13]. Mutations in IglA and IglB result in bacteria that are unable to escape the phagosome and unable to replicate intracellularly [4,6,12,19,20]. In some species, these homologues are responsible for secretion of proteins, including Hcp and VgrG [16,18,21,22,23]. Recent studies suggest the T6SSs constitute important *Francisella* virulence, intracellular growth, or survival factors; however, only basic aspects of this system have been characterized [13,24,25].

Although the ability of *F. tularensis* to replicate within macrophages is multifactorial, our working hypothesis is that *F. tularensis* secretes FPI-encoded proteins that facilitate the organism's ability to escape the vacuole, enter the cytoplasm to replicate intracellularly, and down regulate the host immune cytokine response. If this is correct, then FPI-encoded proteins should be secreted during infection within host macrophages. Currently available genetic tools for studying the FPI-encoded proteins consist of green fluorescent protein (GFP) tags [8] and more recently reporter fusion tag systems [11]. Secretion of FPI-encoded proteins have previously been examined in the *Francisella* live vaccine strain (LVS) with a fusion β-lactamase, however, this system is not applicable to wild type *F. novicida* and was assessed in a β-lactamase gene mutant because *F. novicida* possesses native β-lactamase genes that exhibit the same activity toward the TEM substrate and interfere with the assay [11]. In the current study, FPI-encoded proteins were expressed as fusion proteins with the small triple FLAG tag and tracked within infected macrophage-like cells. The localization of IglC in a T6SS mutant was also assessed.

Materials and Methods

Bacterial and Cell Cultures

Bacterial strains and cell lines used in this study are listed in Table 1. *Escherichia coli* D10 (Invitrogen) was grown aerobically at 37°C on Luria-Bertani (LB) media, containing 50 μg/ml of ampicillin (LBA) when appropriate for selection and maintenance. *F. novicida* U112 (ATTC 15482) was cultured aerobically at 37°C on tryptic soy agar (TSA) or in tryptic soy broth (TSB) supplemented with 0.1% cysteine. When selecting for or maintaining transformants, *Francisella* was cultured on TSA containing 15 μg/ml of kanamycin. J774-1A murine macrophage-like cells were obtained from the American Type Culture Collection (ATCC, TIB 67, BALB/C macrophage). J774 cells were grown in flasks in Dulbecco's Modified Eagle Medium (DMEM) (GIBCO Invitrogen Grand Island, NY USA) supplemented with 10% newborn calf serum (NCS) and maintained at 37°C in a humidified 6.5% CO_2 incubator. The mosquito hemocyte like cells Sua-1B were grown in Schneider's Insect Medium (Sigma Aldrich St. Louis, MO USA) supplemented with 20% fetal bovine serum at 28°C and flasks were capped tightly [26].

DNA Manipulations

Restriction enzyme digests, sub-cloning, cloning, and DNA electrophoresis for *E. coli* was performed using standard cloning techniques [27] and the (Invitrogen Carlsbad, CA. USA) E-Gel Clonewell 0.8% SYBR Safe gel system. By cutting the *Francisella* expression plasmid groE-GFP- pFNLTP6 [8] with BamHI, the GFP insert was removed, leaving the groE promoter and the multiple cloning site (MCS) in place. Within the MCS a fragment containing *sopB* and a triple FLAG tag from the plasmid pSB2598 was inserted [28]. Next through quick-change mutagenesis, the second NcoI site in the plasmid's kanamycin resistance gene was removed without changing the coding sequence. By modifying this second NcoI site, most of the FPI genes have been inserted into plasmids using the restriction enzymes EcoRI and NcoI within the MCS; this allows for easy primer design and cloning of C-terminus triple FLAG plasmids. Primers used to construct the *Francisella* expression plasmids with the epitope tag on the C-terminus of FPI are listed in Table 2. After cloning all C-terminal tagged FPI genes, pKH8, IglA-FLAG, has been modified becoming the backbone for N-terminus tagged plasmids. These modifications involved removing *iglA*, leaving the triple FLAG tag down stream of the groE promoter and the Shine-Delgarno sequences of *iglA* yet upstream of the MCS. Primers were designed for N-terminus triple FLAG-tagged FPI genes to be inserted with the restriction enzymes XmaI and XhoI (Fig. S1). Primers used to construct the *Francisella* expression plasmids with the epitope tag on the N-terminus of FPI are listed in Table 3. PCR for cloning was done using Phusion High-Fidelity PCR (Finnzymes). Restriction enzyme digest was performed as described in New England Biolabs Catalog and Technical Reference (Ipswich, MA. USA). PCR products and restriction enzyme digest products were purified via Wizard SV Gel and PCR clean up System (Promega Madison, WI, USA). Ligations using T4 DNA ligase (Fisher Scientific) were done at 16°C for 14–16 h. Plasmids were recovered from *E. coli* through PureYield Plasmid Miniprep System (Promega Madison, WI, USA) for screening and PureYield Plasmid Midiprep System (Promega Madison, WI, USA) to collect a stock for transformations. Plasmids generated in this study are listed in Table 4. Plasmids were initially screened using restriction enzymes that were used for cloning and when applicable another restriction enzyme that would cut within the specific FPI gene. Plasmids were also screened for correct gene and triple FLAG sequence using standard Taq PCR. After plasmids and gene inserts were ligated, they were transformed into *E. coli* D10 using electroporation. Plasmids collected from *E. coli* were sequenced to confirm that the FPI gene sequence was not altered before being transformed into *F. novicida*.

Transforming *Francisella*

These newly constructed *Francisella* expression plasmids containing the FPI encoded ORFs were chemically transformed into *F. novicida* strain U112. A sub-culture of bacteria was grown aerobically at 37°C, shaking at 200 rpm until mid log phase or an $OD_{600\ nm}$ of 0.3–0.5. Cells were pelleted at 5,000×*G* for 5 min at room temperature. Cells were suspended in *Francisella* transformation buffer or transformation medium [29] and then 400 μl of cell suspension was mixed with DNA and incubated aerobically at 37°C with shaking at 90 rpm for 1 h. 56 μl of 10% glucose and one ml of TSB was then added per transformation and incubated overnight aerobically at 37°C with shaking at 150 rpm. Cells were plated in 100 μl aliquots on freshly prepared TSA containing 15 μg/ml of kanamycin. Colonies were picked, isolated, and then screened by PCR, restriction enzyme digest, and Western blotting for confirmation of successful transformation.

SDS-PAGE and Western Blotting

SDS-PAGE was performed using standard techniques [27]. Proteins were transferred to Immobilon-P membrane (Millipore Billerica, MA USA), and then blocked in 5% non-fat dry milk (NFDM) in Tris-Buffered Saline and Tween 20 (Fisher BioReagents Fair Lawn, NJ US) solution (TBST) containing 1 mM Tris, 15 mM NaCl, 2 mM KCl, and 0.1% Tween 20 for 1 h. To detect FLAG-tagged proteins, the blots were incubated with (1/5000) monoclonal M2 anti-FLAG antibodies (Sigma Aldrich St. Louis, MO USA) in 5% NFDM in TBST. For the detection of IglA, IglC, PdpA, and PdpC, polyclonal rabbit anti IglA, IglC,

Table 1. List of Strains and Plasmids.

Strain	Description	Reference
J774-1A	Murine Macrophages cell lines	ATTC
Sua-1B	Mosquito hemocytes cell lines	[26]
E. coli DH5α	Sub cloning competent cells	Invitrogen
F. novicida U112	*Francisella novicida* prototype strain	ATTC
F. novicida U112R 2008	U112 Δ restriction genes	[40]
Jlo	U112 with deletion of gene FTN1758	[36]
Δ*pdpA*	U112 Δ*pdpA*	[32]
Δ*pdpB*	U112 Δ*pdpB*	[13]
Δ*iglE*	U112 Δ*iglE*	[13]
Δ*vgrG*	U112 Δ*vgrG*	[13]
Δ*iglF*	U112 Δ*iglF*	[13]
Δ*iglG*	U112 Δ*iglF*	[13]
Δ*iglH*	U112 Δ*iglH*	[13]
Δ*dotU*	U112 Δ*dotU*	[13]
Δ*iglI*	U112 Δ*iglI*	[13]
Δ*iglJ*	U112 Δ*iglJ*	[13]
Δ*pdpC*	U112 Δ*pdpC*	[13]
Δ*pdpE*	U112 Δ*pdpE*	[13]
Δ*iglD*	U112 Δ*iglD*	[13]
Δ*iglC*	U112 Δ*iglC*	[12]
Δ*iglB*	U112 Δ*iglB*	[13]
Δ*iglA*	U112 Δ*iglA*	[13]
Δ*pdpD*	U112 Δ*pdpD*	[36]

PdpA, and PdpC antibodies were used 1/2000. To detect bound antibodies, blots were incubated with Peroxidase-Goat Anti-Mouse or Peroxidase-Goat Anti-Rabbit secondary antibodies (1/5000) (Zymed Laboratories Invitrogen Immundetection San Francisco, CA USA) in 5% NFDM in TBST. To visualize protein bands, blots were incubated with SuperSignal West Pico Chemiluminescent Substrate (Thermo Scientific Rockford, IL USA) prior to exposing and developing film.

Macrophage Growth Assay and Analysis

J774A.1 mouse macrophage like cells (ATCC TIB-67) were seeded in 24-well cell culture plates at 1.4×10^5 cells/well for 24 h in complete Dulbecco's Modified Eagle Medium (cDMEM) containing 10% newborn calf serum (NCS). In 4 independent experiments, cells were infected in triplicate with *F. novicida* strains at a multiplicity of infection (MOI) of 1:50 (bacterium-to-macrophage). To help promote bacterial uptake after bacteria have been added, the 24-well dishes containing infected macrophages were centrifuged at $600 \times G$ for 10 min. Infected monolayers were incubated for 2 h in DMEM to allow for phagocytosis to occur, washed five times in Hank's Phosphate Buffered Saline (HPBS) (GIBCO Invitrogen Grand Island, NY USA). At this time, the infection is at 0 h, and infected macrophages were then either lysed at this time or incubated at 37°C in 5% CO_2 for 24 and 48 h. To determine bacterial replication, infected macrophages were lysed in 0.1% dexoxycholate in HPBS at 0, 24, and 48 h post infection. The lysates were serially diluted in HPBS and plated on TSA and incubated at 37°C for 24 or 48 h. The colony forming units (cfu) were enumerated, and used to plot growth curves. To perform statistical analysis, the fold replication at 48 h was first determined (cfu 48 h/cfu 0 h), and then the log of the 48 h fold replication was used in a two-way ANOVA with XLSTAT to compare the means of each group in Tukey multiple comparisons ($\alpha = 0.05$).

Immuno-fluorescence Microscopy and Analysis

J774 murine macrophage-like cells or Sua-1B mosquito hemocyte-like cells were grown on coverslips and infected with *F. novicida* strains as indicated in each figure (Table 1) [13]. Cells were infected for 30 min at an MOI of 50:1 (bacteria per macrophage), washed with phosphate buffered saline (PBS), and incubated until the desired time point in DMEM containing 10% NCS. Cells were then fixed in 4% paraformaldehyde for 15 min at room temperature and rinsed three times with PBS. FLAG-tagged proteins and *Francisella* were detected with anti-FLAG M2 monoclonal antibodies and rabbit anti-*Francisella novicida* serum, respectively. The antibodies were diluted (1/500) in PBS containing 0.5% BSA and 0.1% saponin to permeate host cell membranes, while leaving the bacterial cell membranes intact [30]. Primary antibodies were detected with goat anti-mouse and goat anti-rabbit serum conjugated to Alexa Fluor 488 and 594, respectively (Invitrogen MOLECULAR PROBES Eugene, OR US). DNA was detected with DAPI (Invitrogen MOLECULAR PROBES Eugene, OR US). Coverslips were mounted using Prolong Gold Antifade reagent (Invitrogen MOLECULAR PROBES Eugene, OR US) and examined using an Olympus TE81 inverted fluorescent microscope with spinning disc confocal capabilities.

Table 2. C-term Primers Used in this Study.

C-terminal FLAG *Francisella* expression plasmid primers
pdpA_C_terminalFLAG_F_Nde1: ggcagCATATGctaattaagtagacaatgatagc
pdpA_C_terminalFLAG_B_Nco1: ggcagCCATGGGatttcctttgatttatat
pdpB_C_terminalFLAG_F_Kpnl1: agGGTACCcaaaaggaaattaaaagtatg
pdpB_C_terminalFLAG_B_Nco1: agCCATGGGttgtacattaacttctccttg
iglE_C_terminal_F_EcoR1: aggaGAATCCggcaaaacaaggagaagttaatg
iglE_C_terminal_B_Nco1: gacgCCATGGCatctttttctatgctgctatc
vgrG_C_terminal_F_EcoR1: gagaGAATTCgattaaggggatattcttatg
vgrG_C_terminal_B_Nco1: agagCCATGGCtccaaccattgttgctgcggaacc
iglF_C_terminal_F_Nde1: gcagCATATGcaatggttggataataatatg
iglF_C_terminal_B_Nco1: agcaCCATGGCattttccaataagcttcttgcttgc
iglG_C_terminal_F_EcoR1: agagGAATTCgaagcttattggaaaatttaaatg
iglG_C_terminal_B_Nco1: agagCCATGGCagatgtttttacatttatttg
iglH_C_terminal_F_EcoR1: agagGAATTCcttagaaggtcattatcatg
iglH_C_terminal_B_Nco1: agagCCATGGCtatagagttatttaaaacaatc
dotU_C_terminal_F_EcoR1: aggaGAATTCctatataaaggatattagaaatg
dotU_C_terminal_B_Nco1: aggaCCATGGCccagcttaataaaattag
iglI_C_terminal_F_EcoR1: cgagGAATTCgggtaagaggagatttatatg
iglI_C_terminal_B_Nco1: gaagCCATGGCtatgtcaaaaagatcttc
iglJ_C_terminal_F_EcoR1: caagGAATTCcaaatgagatagatg
iglJ_C_terminal_B_Nco1: agcgCCATGGCtaaattaaaataacc
pdpC_C_terminal_F_EcoR1: ggcgaGAATTCgataaattaaggaagtacatatg
pdpC_C_terminal_B_Nco1: ggcagCCATGGGtgacgatattttttaaaaaagtc
pdpE_C_terminal_F_EcoR1: gaagGAATTCcttaaggatgcaaaaatatg
pdpE_C_terminal_B_Nco1: aggcCCATGGCtattatagtaattttcttttc
iglD_C_terminal_F_EcoR1: ggcagGAATTCaagatcggagttgattctaatg
iglD_C_terminal_B_Nco1: ggcgaCCATGGGagaaaaggctataaagaaatc
iglC_C_terminal_F_EcoR1: gaGAATTCaaaggagaatgattatgagtgag
iglC_C_terminal_B_Nco1: gaCCATGGGtgcagctgcaatatatcc
iglB_C_terminal_F_Nde1: gagaCATATGgtagagaggattttgttatg
iglB_C_terminal_B_Nco1: agagCCATGGCgttattatttgtacc
iglA_C_terminal_F_EcoR1: agagGAATTCgtaaaaaaaggacaataagatg
iglA_C_terminal_B_Nco1: ggaaCCATGGCcttatcatctacttg
pdpD_C_terminal_F_EcoR1: ggcgaGAATTCgtaagagtagtaagtatggatcaag
pdpD_C_terminal_B_Nco1: ggcgaCCATGGGaacccagatcattggtctatac
anmK _C_terminal_F: agagGAATTCgaatataaatattgtgtaggaatcatg
anmK _C_terminal_B_Nco1: aggaCCATGGCaaagaaatttatttgacc

The description contains the FPI ORF nomenclature, which terminal the tag was fused to, direction of the primer, restriction enzyme used in cloning the respective ORF, the sequence of the primer in 5′-3′ direction, underlining represents in frame codon.

Table 3. N-term Primers Used in this Study.

N-terminal FLAG *Francisella* expression plasmid primers
pdpA_N_terminal_F_Xma1: acggCCCGGGgaatagcagtaaaagatataac
pdpA_N_terminal_B_Xho1: acggCTCGAGttaatttcctttgatttatatc
pdpB_N_terminal_F_Xma1: acggCCCGGGgaaattttattaaaaatcatc
pdpB_N_terminal_B_Xho1: acggCTCGAGttattgtacattaacttctccttg
iglE_N_terminal_F_Xma1: acggCCCGGGgatacaataaattattgaaaaatc
iglE_N_terminal_B_Xho1: acggCTCGAGttaatctttttctatgctgc
vgrG_N_terminal_F_Xma1: acggCCCGGGgatcaaaagcagaccatattttc
vgrG_N_terminal_B_Not1: caggGCGGCCGCttatccaaccattgttgctgcgg
iglF_N_terminal_F_Xma1: acggCCCGGGgaaataatgatattgataaatgg
iglF_N_terminal_B_Xho1: acggCTCGAGttaaattttccaataagcttcttgc
iglG_N_terminal_F_Xma1: acggCCCGGGgattaaatattataaatgactcc
iglG_N_terminal_B_Xho1: acggCTCGAGctaagatgtttttacatttatttgtcc
iglH_N_terminal_F_Xma1: acggCCCGGGatgaaaaaagaaaagatttaag
iglH_N_terminal_B_Xho1: acggCTCGAGttatatagagttatttaaaacaatc
dotU_N_terminal_F_Xma1: acggCCCGGGgaaaagactttaaagagatag
dotU_N_terminal_B_Xho1: acggCTCGAGttaccagcttaataaaattagtaagc
iglI_N_terminal_F_Xma1: acggCCCGGGgaagtcagataatatctacac
iglI_N_terminal_B_Xho1: acggCTCGAGttatatgtcaaaaagatcttc
iglJ_N_terminal_F_Xma1: acggCCCGGGgaaagactattttgaagatcttttg
iglJ_N_terminal_B_Xho1: acggCTCGAGtcataaattaaaataacctagatatatc
pdpC_N_terminal_F_Xma1: acggCCCGGGgaaacgacaaatatgaactaaatatc
pdpC_N_terminal_B_Xho1: acggCTCGAGctatgacgatattttttaaaaaag
pdpE_N_terminal_F_Xma1: acggCCCGGGgaagtaaaaaagtatttcaattattattaatatttg
pdpE_N_terminal_B_Xho1: acggCTCGAGttatattatagtaattttcttttc
iglD_N_terminal_F_Xma1: acggCCCGGGgatttctagaaaggatttattg
iglD_N_terminal_B_Xho1: acggCTCGAGttaagaaaaggctataaagaaatc
iglC_N_terminal_F_Xma1: acggCCCGGGgaattatgagtgagatgataacaag
iglC_N_terminal_B_Xho1: acggCTCGAGctatgcagctgcaatatatcc
iglB_N_terminal_F_Xma1: acggCCCGGGgaacaataaataaattaag
iglB_N_terminal_B_Xho1: acggCTCGAGttagttattatttgtaccg
iglA_N_terminal_F_Xma1: acggCCCGGGcaaaaaataaaatcccaaattc
iglA_N_terminal_B_Not1: caggGCGGCCGCctacttatcatctacttgttgattac
pdpD_N_terminal_F_Xma1: acggCCCGGGatcaagatatcaacgatttattatatg
pdpD_N_terminal_B_Xho1: acggCTCGAGttaaacccagatcattggtctatac
anmk_N_terminal_F_Xma1: acggCCCGGGgatctggaacatcactagatgg
anmk_N_terminal_B_Xho1: acggCTCGAGttaaaagaaatttatttgacc

The description contains the FPI ORF nomenclature, which terminal the tag was fused to, direction of the primer, restriction enzyme used in cloning the respective ORF, the sequence of the primer in 5′-3′ direction, underlining represents in frame codon.

Images containing a total of 10–60 infected cells and 100–600 bacteria for each of the 3 independent experiments were collected as Z-stacks and a projection image was generated using the Intelligent Imaging SlideBook software package. Exposure time and settings were constant for all slides in each experiment. Using SlideBook software, masks were generated for infected macrophage-like cells, bacteria, and FLAG-tagged protein signals. The percentage of bacterial masks that overlapped with FLAG masks was used to determine the percentage of bacteria associated or co-localized with FLAG-tagged protein. However, this did not account for FLAG-tagged protein that dispersed away from bacteria, therefore the percentage of infected macrophage masks containing FLAG-tagged masks were also determined for every FPI protein with each the C-terminal and N-terminal FLAG-tag. Three independent experiments were performed. The data were analyzed in XLSTAT with an ANOVA paired with a left sided Dennett's test comparing each FLAG-tagged proteins' mean to the mean of bacteria not containing a FLAG expressing plasmid for either the percentage of bacteria co-localized with FLAG signal or the percentages of infected macrophage-like cells containing FLAG signal. Significant differences were determined with an $\alpha = 0.05$. We also tested the data for correlations between the same proteins with different tags

Table 4. Plasmids used in this study.

Plasmid	Description	Reference
pFNLTP6-gro-gfp	groE-gfp; Km^r Ap^r	[8]
pKH1	Km^r Ap^r	This study
pSB2598	sopB-FLAG	[28]
pKH2	Km^r Ap^r	This study
pKH3	gro-sopB-FLAG; Km^r Ap^r	This study
pKH22	gro-pdpA-FLAG; Km^r Ap^r	This study
pKH24	gro-pdpB-FLAG; Km^r Ap^r	[13]
pKH9	gro-iglE-FLAG; Km^r Ap^r	[13]
pKH10	gro-vgrG-FLAG; Km^r Ap^r	[13]
pKH26	gro-iglF-FLAG; Km^r Ap^r	[13]
pKH11	gro-iglG-FLAG; Km^r Ap^r	[13]
pKH12	gro-iglH-FLAG; Km^r Ap^r	[13]
pKH13	gro-dotU-FLAG; Km^r Ap^r	[13]
pKH14	gro-iglI-FLAG; Km^r Ap^r	[13]
pKH15	gro-iglJ-FLAG; Km^r Ap^r	[13]
pKH5	gro-pdpC-FLAG; Km^r Ap^r	This study
pKH16	gro-pdpE-FLAG; Km^r Ap^r	This study
pKH6	gro-iglD-FLAG; Km^r Ap^r	This study
pKH4	gro-iglC-FLAG; Km^r Ap^r	[13]
pKH18	gro-iglB-FLAG; Km^r Ap^r	This study
pKH8	gro-iglA-FLAG; Km^r Ap^r	This study
pKH7	gro-pdpD-FLAG; Km^r Ap^r	This study
pKH25	gro-anmK-FLAG; Km^r Ap^r	This study
pKH40	gro-FLAG-pdpA; Km^r Ap^r	This study
pKH50	gro-FLAG-pdpB; Km^r Ap^r	This study
pKH34	gro-FLAG-iglE; Km^r Ap^r	This study
pKH35	gro-FLAG-vgrG; Km^r Ap^r	This study
pKH44	gro-FLAG-iglF; Km^r Ap^r	This study
pKH36	gro-FLAG-iglG; Km^r Ap^r	This study
pKH45	gro-FLAG-iglH; Km^r Ap^r	This study
pKH37	gro-FLAG-dotU; Km^r Ap^r	This study
pKH39	gro-FLAG-iglI; Km^r Ap^r	This study
pKH47	gro-FLAG-iglJ; Km^r Ap^r	This study
pKH41	gro-FLAG-pdpC; Km^r Ap^r	This study
pKH38	gro-LAG-pdpE; Km^r Ap^r	This study
pKH48	gro-FLAG-iglD; Km^r Ap^r	This study
pKH46	gro-FLAG-iglC; Km^r Ap^r	This study
pKH43	gro-FLAG-iglB; Km^r Ap^r	This study
pKH27	gro-FLAG-iglA; Km^r Ap^r	This study
pKH42	gro-FLAG-pdpD; Km^r Ap^r	This study
pKH49	gro-FLAG-anmK; Km^r Ap^r	This study

This table lists all the plasmids used in designing the *Francisella* expression plasmids and all of the *Francisella* exoression plasmids that were generated in this study.

using the Spearman correlation test in XLSTAT. Significant differences were determined with an $\alpha = 0.05$.

Results

FLAG-tagged FPI Protein Expression in *F. novicida*

Expression of C-terminal and N-terminal epitope tagged FPI proteins from *F. novicida* U112 was confirmed by Western blotting (Fig. 1). Western blotting showed the C-terminal and N-terminal tags do not interrupt FPI protein expression at their expected sizes (Fig. 1 and Table S1), with the exception of FLAG-PdpE. In addition to the expected sizes, lower intensity bands of different sizes were detected for some proteins (Fig. 1). IglG-FLAG expression was lower than the other FPI proteins and was not visible here (Fig. 1C); expression of IglG-FLAG was confirmed with longer exposure times causing over exposure with the other proteins (data not shown). Antibodies towards IglA, IglC, PdpA, and PdpC detected proteins of the same size as western blots detecting the FLAG-tag (Fig. S2).

Intramacrophage Growth Complementation

Since several FPI genes are needed for intracellular growth, the C-FLAG and N-FLAG-tagged proteins ability to complement respective knock out mutant strains were assessed (Table 1). As previously described, the FPI deletion mutants of *iglABCDEFHIJ, pdpAB, dotU*, and *vgrG* were defective for intramacrophage growth [8,12,30,32] (Fig. 2). Expression of C-terminal and N-terminal tagged FPI proteins, IglABCDEFHIJ, PdpAB, DotU, and VgrG, in FPI mutants increased growth rates, indicating that *Francisella* expression plasmids complemented their mutants. Genetic complementation of each deletion mutant with the C-FLAG and N-FLAG-tagged *Francisella* expression plasmids restored intramacrophage growth, and the growth of complemented mutants were significantly higher compared to their parental mutant (Fig. 2) ($P<0.05$). Expression of the tagged proteins did not always completely restore growth to that of the wild type; growth of 22 of the 26 complements were equivalent to that of the wild type ($P>0.05$). However, growth of the complements, FLAG-IglB, IglC-FLAG, IglJ-FLAG, and FLAG-IglJ, were significantly different compared to wild type and their respective mutant ($P<0.05$). Together, these data indicated that most of the plasmids expressed biologically functional proteins.

Unique Patterns of FPI-Encoded Proteins Co-localized with *F. novicida* within Infected Cells

FPI proteins with consistent FLAG detection had varying patterns of distribution of FLAG signal when compared to each other within infected murine macrophage-like cells (Fig. 3) and mosquito hemocyte-like cells (Fig. 4). IglA was localized with bacteria (Fig. 3 and Fig. 4). IglCE and VgrG were co-localized with bacteria and also extending beyond, completely surrounding bacteria (Fig. 3 and Fig. 4). IglD and PdpA were also localized with bacteria, and on occasion surrounding the bacteria (Fig. 3 and Fig. 4). PdpC was also detected both co-localized with bacteria and dispersing away from the bacteria (not shown). IglI was distinctly localized to the bacterium, it surrounded the bacterium uniformly (Fig. 3 and Fig. 4). The C-FLAG PdpE was studded around the bacterium, while the N-terminal tagged protein was not detected by immuno-fluorescent microscopy (Fig. S3).

C-terminal FLAG-tagged FPI Proteins Co-Localization with Bacteria During Cell Infection

The localization of FPI-encoded proteins was examined via immuno-fluorescent microscopy of infected murine macrophage-like J774 cells with bacteria expressing C-terminal fusion proteins. The percent of bacteria co-localized with FLAG signal within infected macrophages was determined for all 18 FPI-encoded proteins. At 30 min post-infection, bacteria expressing FLAG-tagged IglABDEH, PdpE, VgrG, and DotU were significantly more often co-localized with fluorescent signal compared to control bacteria not expressing epitope-tagged protein ($p \leq 0.038$) (Fig. 5A and Table S2). Bacteria expressing the remaining FLAG-

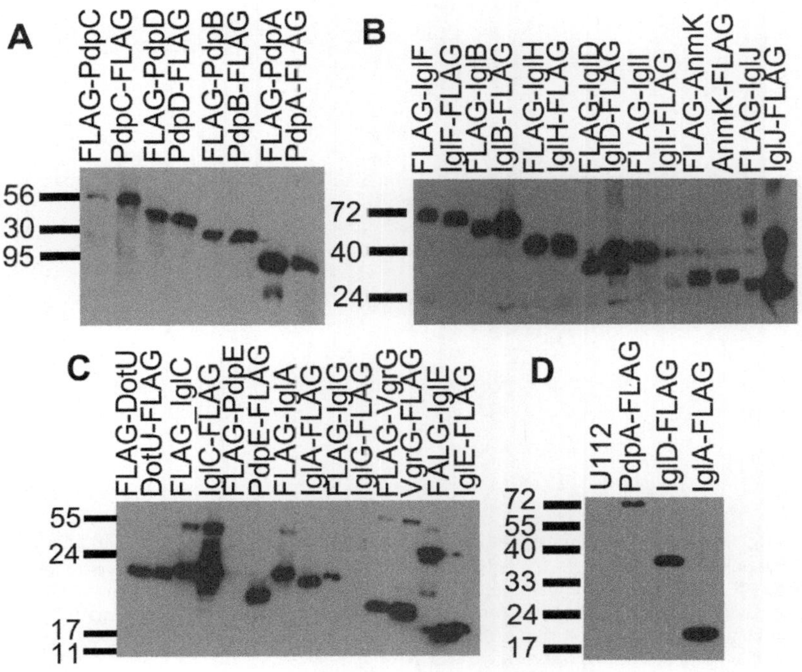

Figure 1. FPI FLAG-tagged Protein Expression in *F. novicida*. Whole cell lysates have been analyzed for production of N- terminal and C-terminal FLAG-tagged proteins by Western blotting. *F. novicida* U112 wild type and U112 expressing the respective FPI protein from the *Francisella* expression plasmid are labeled above each lane. C-terminally tagged proteins are referred to as protein-FLAG, and N-terminally tagged proteins are referred to as FLAG-protein. FPI encoded proteins have been grouped into similar predicted sizes (A) 156–95 kDa, (B) 67.6–30.9, (C) 24.6–14.5, and (D) a 10% gel for comparisons. FLAG-tagged proteins are detected with monoclonal mouse anti FLAG, goat anti mouse conjugated HRP, and chemiluminescent substrate.

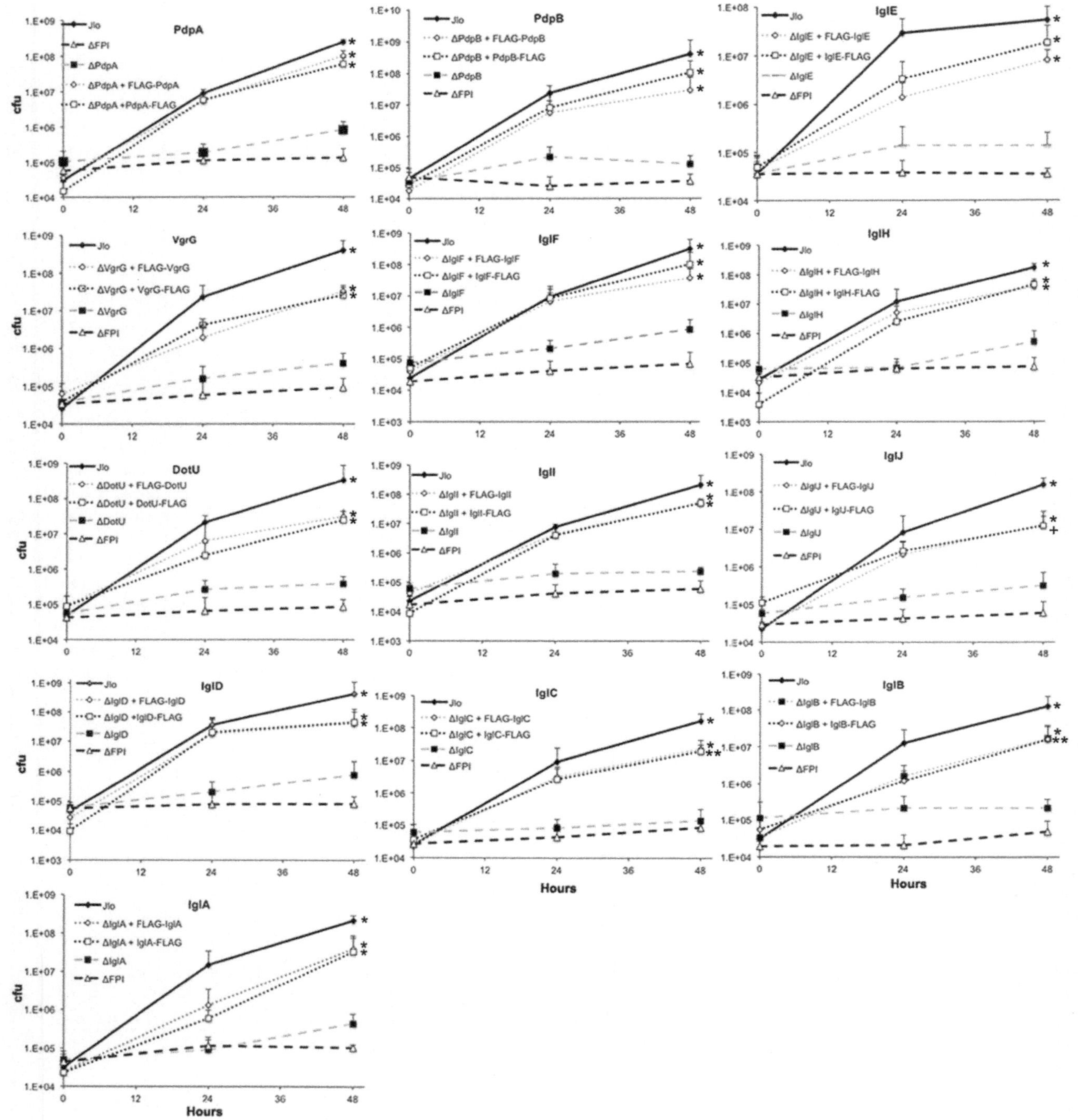

Figure 2. Intracellular Growth. J774 cells were infected with indicated strains and harvested at 0, 24 and 48 h post infection and the amount of intracellular bacteria at each time point was determined. The FPI mutant lacks all FPI genes and was defective for intracellular growth, as were each of the individual FPI gene mutants. All of the genetic complements express the cognate FPI gene as a C-terminal or an N-terminal triple FLAG-tagged version from a plasmid vector. Graphs are the average of four independent experiments and error bars show standard deviations, * indicate significance from mutants, ** partial complementation that was different form mutants and wild type, and + approaching significance at 0.054. P values generated by Tukey multiple comparisons via a two-way ANOVA.

tagged FPI-encoded proteins were not statistically different from the control bacteria at 30 min ($p \geq 0.368$) (Fig. 5A and Table S2). Bacteria expressing FLAG-tagged IglACEI, PdpA, VgrG, and DotU all had significantly more bacteria co-localized with fluorescent signal compared to control bacteria not expressing epitope-tagged protein ($p \leq 0.008$) at 4 h post-infection (Fig. 5B and Table S2). Also at 4 h, bacteria expressing tagged IglH were approaching statistical significance when compared to the control

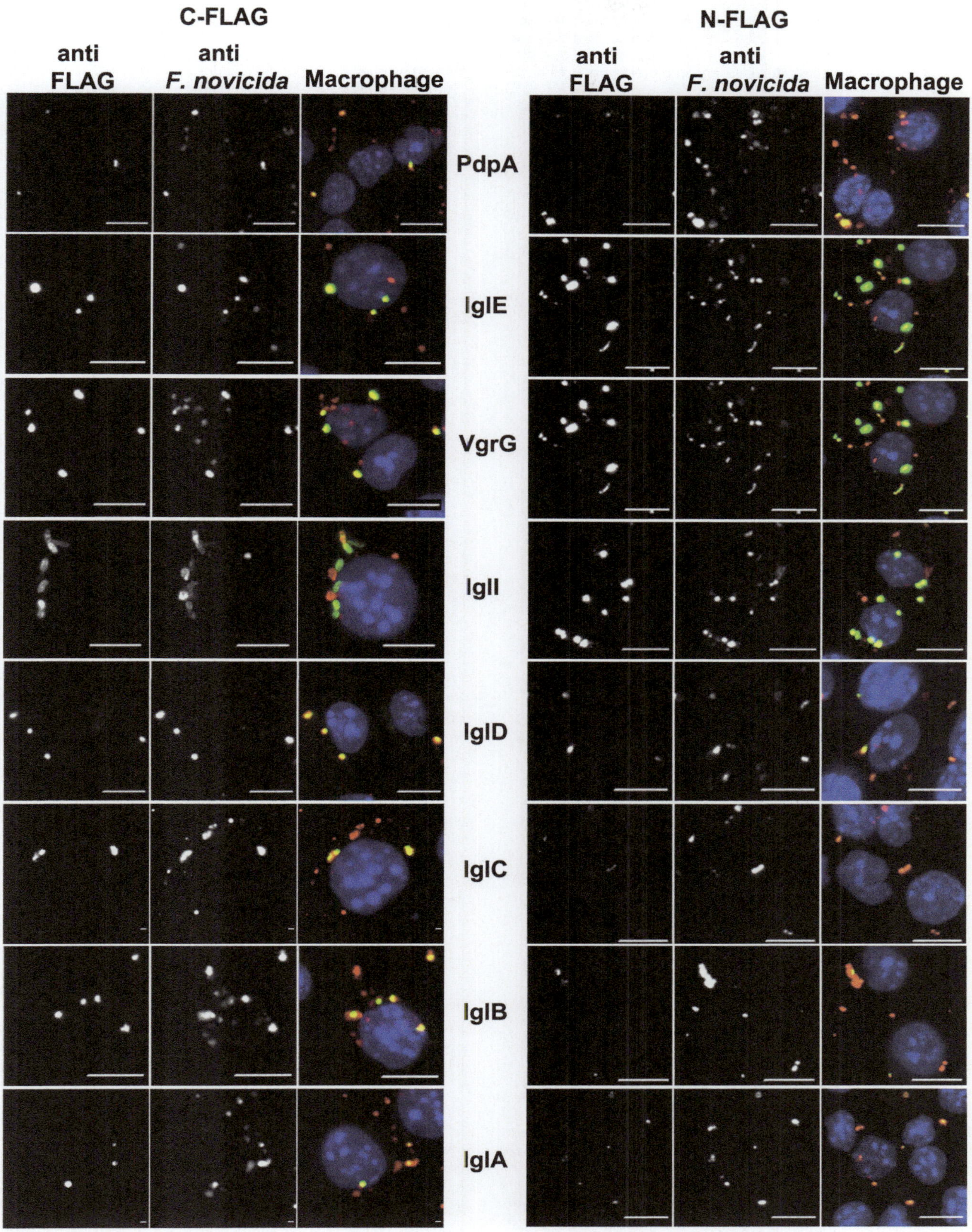
C-FLAG
anti FLAG
anti F. novicida
Macrophage
N-FLAG
anti FLAG
anti F. novicida
Macrophage
PdpA
IglE
VgrG
IglI
IglD
IglC
IglB
IglA

Figure 3. Localization of FPI Proteins within Infected Macrophages. J774 cells were infected with wild type bacteria without a plasmid and wild type containing the *Francisella* expression plasmids that express PdpA, IglABCDEDI and VgrG with either C-term FLAG tag (left) or N-term FLAG tag (right) at 4 h post-infection. Middle columns and the red in the merged right columns indicate bacteria. Left columns and the green in the merged right columns indicate FLAG-tagged proteins. Host cell nuclei are indicated by blue in merged right columns.

(p = 0.054) (Fig. 5B and Table S2). The bacteria expressing the rest of the FPI-tagged proteins were not different from control bacteria, when examining the bacteria co-localized with FLAG signal at 4 h (p≥0.814) (Fig. 5B and Table S2). 8 h into infection, bacteria expressing IglEHI and DotU had significantly more bacteria co-localized with fluorescent signal than the control bacteria (p≤0.029) (Fig. 5C and Table S2). Also at 8 h, bacteria expressing IglA had 55% of bacteria co-localized with FLAG signal, approaching statistical significance when compared to the control (p = 0.054) (Fig. 5C and Table S2). The other FLAG-tagged FPI-encoded proteins were not different from the controls (p≥0.274) (Fig. 5C and Table S2). More bacteria were co-localized with fluorescent signal from tagged IglE, DotU, PdpE, and IglA (p≤0.012) with bacteria expressing those tagged proteins compared to control bacteria at 12 h, while bacteria expressing

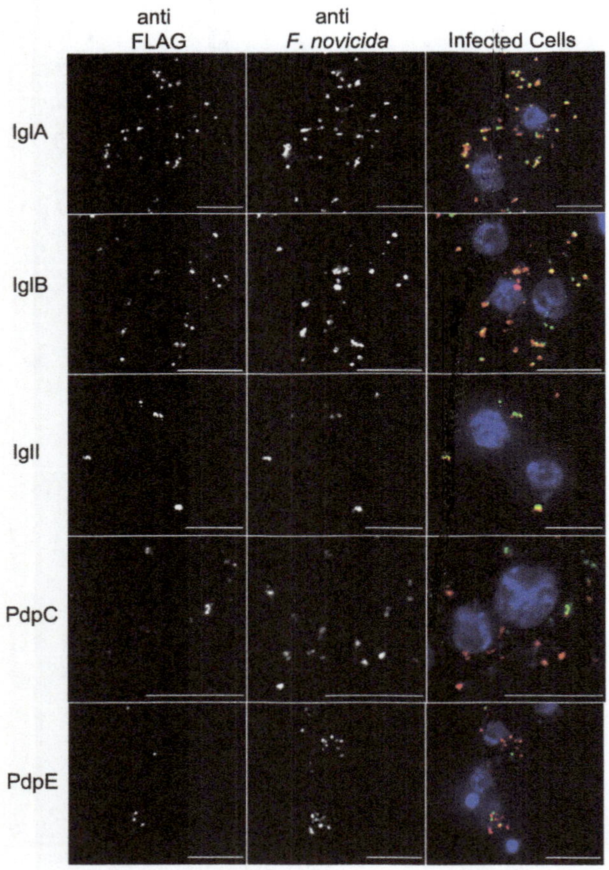

Figure 4. Localization of FPI Proteins During Infection of Sua-1B Cells. J774 cells infected with wild type containing the *Francisella* expression plasmids that express IglA, IglC, IglI, and PdpE for 4 h (A). Sua-1B cells infected with wild type containing the *Francisella* expression plasmids that express IglA, IglC, IglI, and PdpE for 4 h (B). Middle columns and red in the right column indicate bacteria. C-terminally FLAG-tagged proteins are shown in the left columns and are also green in the right column. Host cell nuclei are displayed as blue in the merged, right columns.

the other FPI-tagged proteins were not different than the control bacteria (p≥0.129) (Fig. 5D and Table S2).

C-term FLAG-tagged FPI Protein Localization within Infected Cells

The localization of C-term FLAG-tagged FPI proteins were alternatively assessed by calculating the percent of infected cells containing fluorescent signal to account for proteins that were secreted into the infected cells but did not co-localize with the bacteria expressing the epitope tagged proteins. Within 30 min of infection, the cells infected with bacteria expressing tagged IglABCDEHI, PdpACDE, VgrG, DotU, and Amnk all had significantly more infected cells containing fluorescent signal when compared to the control cells infected with bacteria not expressing FLAG-tagged proteins (p≤0.036) (Fig. 6A and Table S2). Also at 30 min into infection, the cells infected with bacteria expressing tagged PdpB and IglF were not different from cells infected with control bacteria (p≥0.232) (Fig. 6A and Table S2). At 4 h post infection, cells infected with bacteria expressing tagged IglEHI, VgrG, and DotU, all had significantly more infected cells containing fluorescent signal (p≤0.029) than the cells infected with bacteria not expressing FLAG-tagged proteins. However, the cells infected with tagged IglBCDFGJ, PdpABCE, and Anmk were not significantly different from cells infected with bacteria not expressing FLAG-tagged proteins (Fig. 6B and Table S2). When comparing the amount of infected cells containing FLAG signal at 8 h after infection, the infected cells that expressed IglABC-DEFGI, PdpACE VgrG, and DotU had significantly more infected cells containing FLAG signal (p≤0.009). While the cells infected with bacteria expressing tagged IglG, PdpBD and Anmk were not different from the control cells containing bacteria not expressing FLAG-tagged proteins (p≥0.142) (Fig. 6C and Table S2). More infected cells contained FLAG signal at 12 h within cells infected with bacteria expressing tagged PdpACE, IglABCDEJ, VgrG, DotU, and Anmk (p≤0.042). Additionally at 12 h, cells infected with bacteria expressing FLAG-tagged PdpBD and IglFGH were not statistically different (p≥0.103) when compared to cells infected with bacteria not expressing FLAG-tagged proteins (Fig. 6D and Table S2).

Bacteria Co-Localization with N-terminal FLAG-tagged FPI Proteins During Cell Infection

The localization of FPI-encoded proteins was also examined via immuno-fluorescent microscopy of infected murine macrophage-like cells with bacteria expressing N-terminal fusion proteins in order to assess effects of tags on the termini of FPI encoded proteins. The percent of bacteria co-localized with FLAG signal within infected macrophages was determined for all 18 FPI proteins with a N-terminus FLAG tag. At 30 min post-infection, bacteria expressing FLAG-tagged IglEI, VgrG, DotU, and PdpC had significantly more bacteria co-localized with FLAG signal compared to bacteria not expressing epitope-tagged protein (p≤ 0.047 (Fig. 7A and Table S2). Also at 30 min, bacteria expressing tagged IglD were approaching statistical significance with the amount of bacteria co-localized with FLAG-tagged protein when compared to the control bacteria (p = 0.072) (Fig. 7A and Table S2). The bacteria expressing the other tagged proteins at 30 min were not statistically different from the control bacteria not

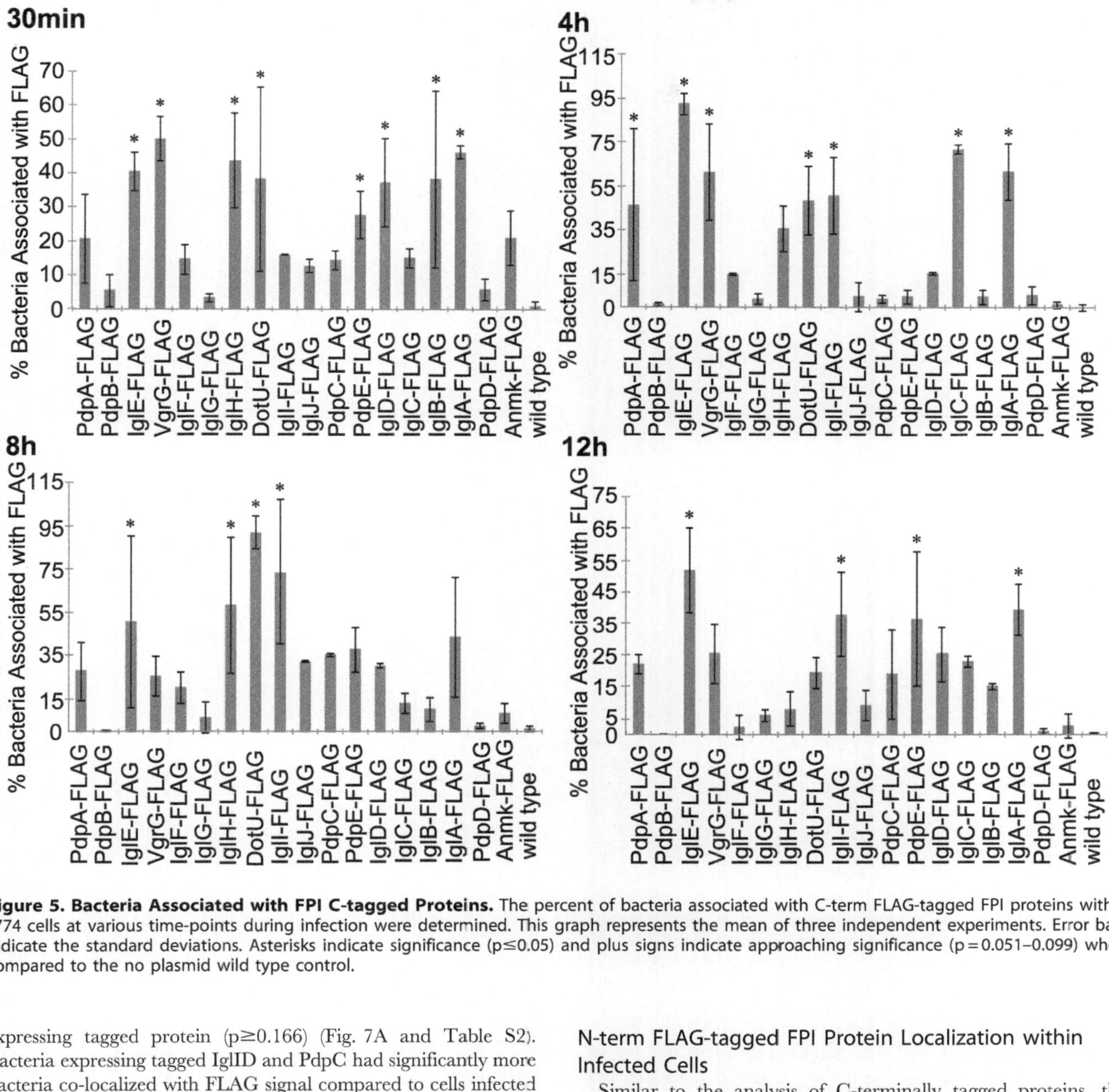

Figure 5. Bacteria Associated with FPI C-tagged Proteins. The percent of bacteria associated with C-term FLAG-tagged FPI proteins within J774 cells at various time-points during infection were determined. This graph represents the mean of three independent experiments. Error bars indicate the standard deviations. Asterisks indicate significance ($p \leq 0.05$) and plus signs indicate approaching significance ($p = 0.051$–0.099) when compared to the no plasmid wild type control.

expressing tagged protein ($p \geq 0.166$) (Fig. 7A and Table S2). Bacteria expressing tagged IglID and PdpC had significantly more bacteria co-localized with FLAG signal compared to cells infected with control bacteria not expressing epitope-tagged protein ($p \leq 0.021$) at 4 h post-infection (Fig. 7B and Table S2). Additionally at 4 h into infection, bacteria expressing the rest of the FPI-tagged proteins were not different from control bacteria when examining the bacteria co-localized with FLAG signal ($p \geq 0.347$) (Fig. 7B and Table S2). Bacteria expressing tagged IglEI and DotU had significantly more bacteria co-localized with FLAG signal than cells infected with the control bacteria at 8 h post infection ($p \leq 0.037$) (Fig. 7C and Table S2). Bacteria expressing the other FPI proteins were not different from the control ($p \geq 0.161$) (Fig. 7C). More bacteria were co-localized with FLAG signal from tagged VgrG, DotU, IglACEI, and PdpC ($p \leq 0.032$) when compared to control bacteria at 12 h, while bacteria expressing the other FPI-tagged proteins were not different than the control bacteria ($p \geq 0.115$) (Fig. 7D and Table S2).

N-term FLAG-tagged FPI Protein Localization within Infected Cells

Similar to the analysis of C-terminally tagged proteins, the localization of FPI proteins with N-term FLAG tags were also assessed by calculating the percent of infected cells containing FLAG signal. At 30 min into the infection, cells infected with bacteria expressing tagged PdpAC, IglAEI, DotU, and PdpC all had significantly more infected cells containing FLAG signal when compared to control cells infected with bacteria not expressing FLAG-tagged proteins ($p \leq 0.008$) (Fig. 8A and Table S2). Also at 30 min, cells infected with bacteria expressing tagged IglBCDFGHJ PdpB, VgrG, and PdpDE, and Amnk were not significant ($p \geq 0.135$) (Fig. 8A and Table S2). Cells infected with bacteria that expressed tagged IglCDE and PdpC all had significantly more infected cells containing FLAG signal ($p \leq 0.045$) than cells infected with bacteria not expressing FLAG-tagged proteins at 4 h post-infection (Fig. 8B and Table S2). Additionally at 4 h post infection, the cells infected with bacteria expressing tagged PdpABD, IglABIFGHJ, VgrG, DotU, and Anmk were not significantly different from the cells infected with

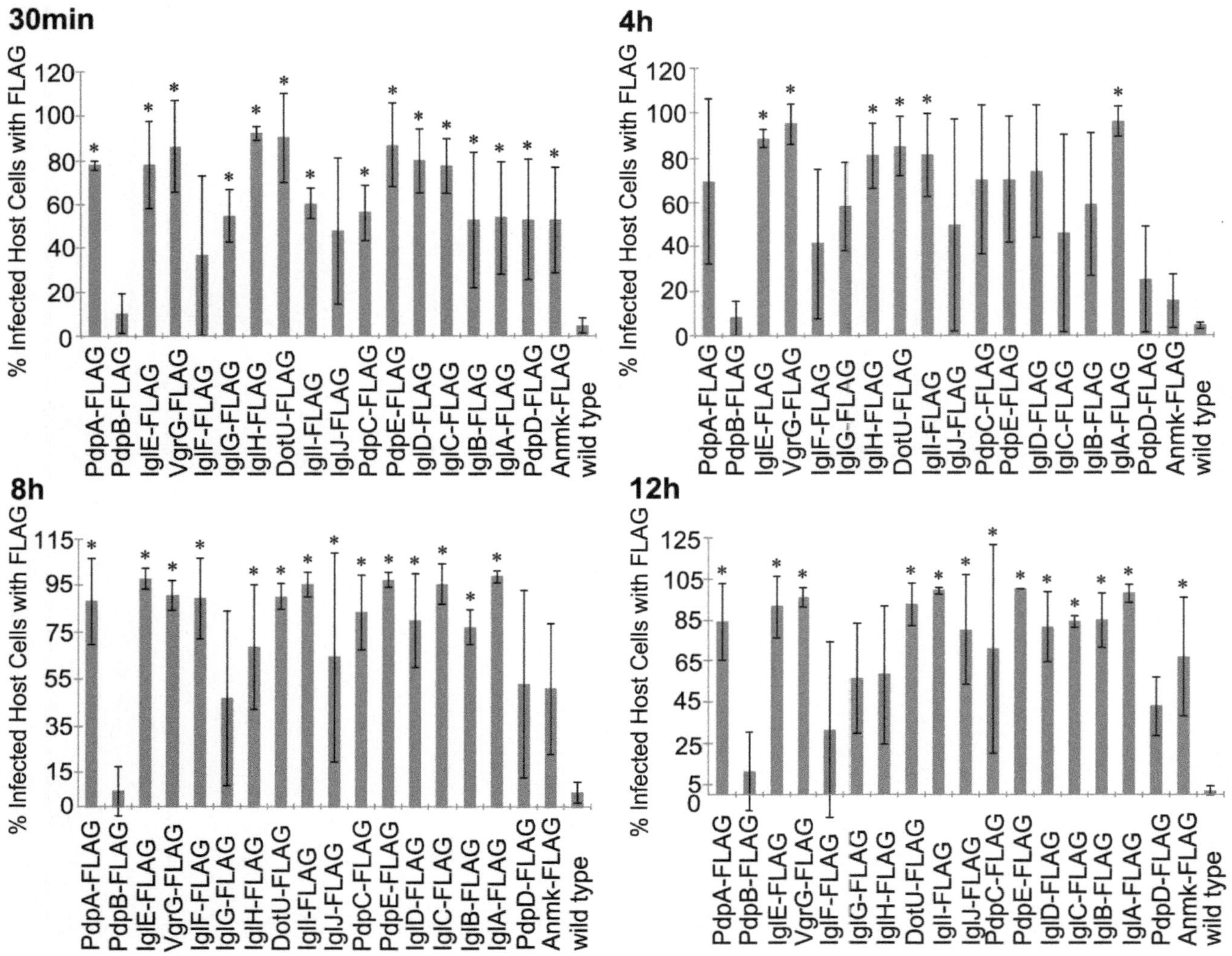

Figure 6. Infected Macrophages with FPI C-tagged Proteins. The percent of infected macrophages containing C-term FLAG-tagged FPI proteins within J774 cells at various time-points during infection were determined. This graph represents the mean of three independent experiments. Error bars indicate the standard deviations. Asterisks indicate significance ($p \leq 0.05$) and plus signs indicate approaching significance ($p = 0.051–0.099$) when compared to the no plasmid wild type control.

bacteria not expressing FLAG-tagged proteins ($p \geq 0.173$) (Fig. 8B and Table S2). When comparing the amount of infected cells containing FLAG signal at 8 h after infection, the cells infected with bacteria that expressed tagged FPI proteins were not different ($p \geq 0.132$) than infected cells containing bacteria not expressing FLAG-tagged proteins (Fig. 8C and Table S2). None of the infected cells were significant for containing FLAG signal in cells infected ($p \geq 0.105$) when compared to cells infected with bacteria not expressing FLAG-tagged proteins at 12 h into the infection (Fig. 8D and Table S2).

IglC Localization was Dependent on PdpB

Less FLAG was detected when IglC was expressed by the Δ*pdpB* mutant when compared to expression of tagged IglC by wild type bacteria (Fig. 9). Over 20% of wild type bacteria were associated with IglC, while less than 7% of Δ*pdpB* bacteria were associated with IglC (Fig. 9B). The differences between IglC association with wild type or Δ*pdpB* mutant bacteria were statistically significant. IglC was examined by Western blot to determine if the mutant expressed the tagged proteins. IglC-FLAG was expressed at similar levels in the wild type and Δ*pdpB* mutant backgrounds (Fig. 9C). Therefore, extracellular co-localization of IglC with bacteria was dependent on the T6SS.

Discussion

Genes within the FPI are required for a T6SS, intracellular growth, and virulence [13,24,25,32,33]. Many of the FPI-encoded proteins are part of a T6SS, therefore we hypothesized that some of the FPI encoded proteins would be directed for secretion by that secretion system, such as effector proteins or chaperones [11,33]. A recent study that utilized a fusion Temoniera (TEM) β-lactamase reporter in LVS identified IglCEFIJ, PdpAE, and VgrG as secreted proteins, and determined that secretion is dependent on the core components of the T6SS: IglCG, DotU, and VgrG [11]. Also within that same study, only IglCE and PdpAE are secreted from *F. novicida*, suggesting differences in the secretion of FPI proteins among subspecies of *Francisella* [11]. In another

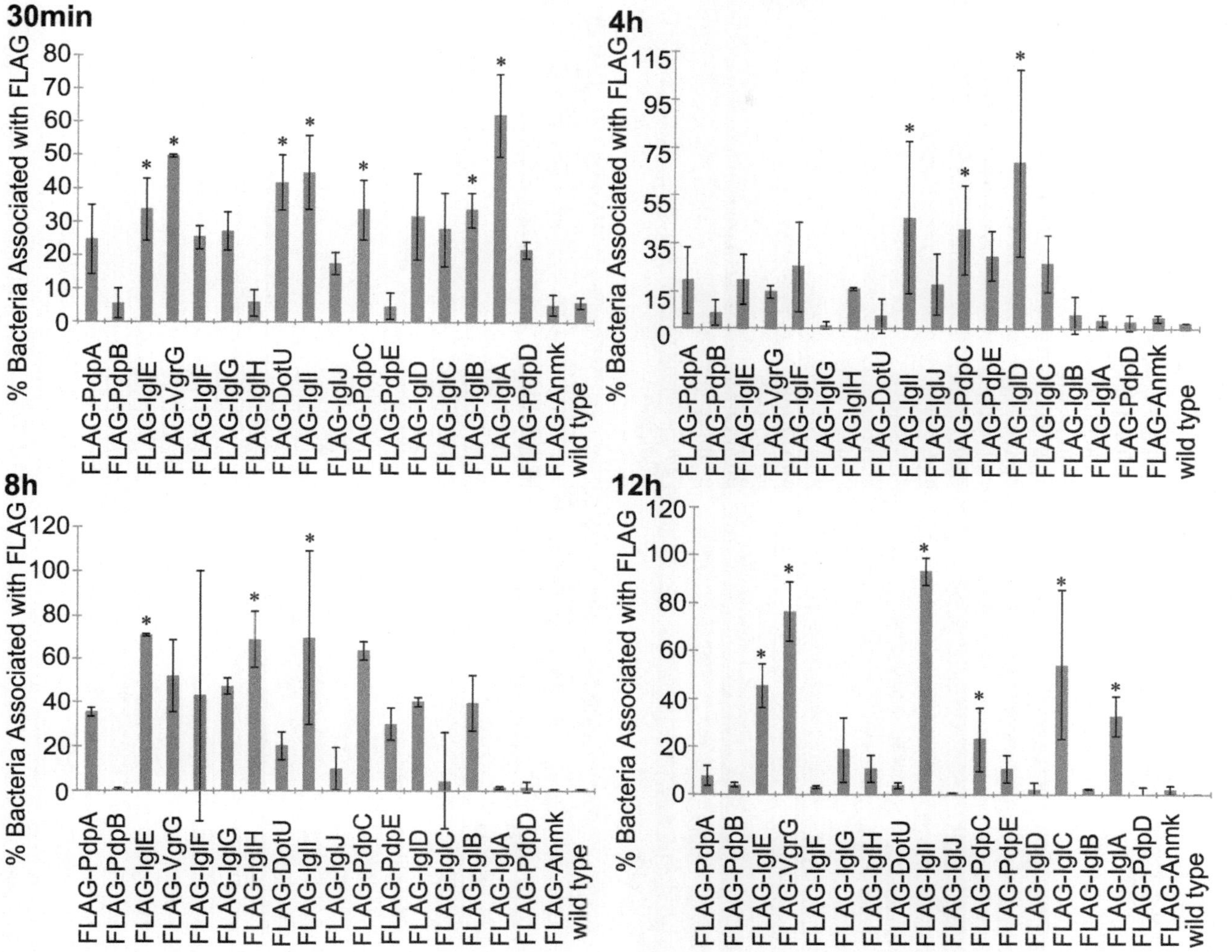

Figure 7. Bacteria Associated with FPI N-tagged Proteins. The percent of bacteria associated with N-term FLAG-tagged FPI proteins within J774 cells at various time-points during infection were determined. This graph represents the mean of three independent experiments. Error bars indicate the standard deviations. Asterisks indicate significance ($p \leq 0.05$) and plus signs indicate approaching significance ($p = 0.051–0.099$) when compared to the no plasmid wild type control.

study a CyaA reporter was used to show IglI and VgrG are secreted in both LVS and *F. novicida* [33].

Although these studies were the first to identify secreted FPI-encoded proteins, there were some limitations to the tools used in those experiments. The TEM β-lactamase reporters have adverse effects including low levels of protein expression; therefore results were only based on one time point at 18 h post-infection [11]. In addition to low expression, the TEM β-lactamase expresses functionless proteins that are unable to complement an intracellular growth phenotype [11]. The TEM β-lactamase assay was used in a *F. novicida* β-lactamase mutant because *F. novicida* possesses native β-lactamases that interfered with the fusion tag [11].

In the present study, we examined the localization of FPI proteins in *F. novicida* during infection of macrophage-like cells. We used *Francisella* expression plasmids that express all 18 of the FPI-encoded proteins from *F. novicida* with a C-terminal epitope FLAG tag as well as a N-terminal epitope FLAG tag. The *Francisella* expression plasmids were the first tools described for *Francisella novicida* that have epitope tags for both termini for the entire set of FPI proteins. Some secreted proteins possess a secretion signal on either the N- or C-terminus [16]. Consequently, adding amino acids at either end may block the secretion signal, disrupting the protein's localization and function. It was unlikely that both the N- and C- termini of the proteins were required for localization. Therefore, the *Francisella* expression plasmids were developed to contain epitope FLAG tags at both the N- and C-termini. Moreover, the triple FLAG tag is short; it was less likely to alter protein folding and function. Triple repeating sequences increase affinity of FLAG monoclonal antibodies and reduce background in Western blotting, immuno-fluorescence microscopy, and many other commercially available biochemical products specific for localizing the FLAG tag sequence. However, there were also some disadvantages associated with the *Francisella* expression plasmids. Adding the triple FLAG tag does increase the molecular weight and could alter protein processing as evident by altered migration pattern in SDS gels (Fig. 1 and Fig. S2). These proteins encoded by the plasmids were constitutively expressed, which may lead to over expression and could also explain the apparent altered processing for some proteins. Additionally the

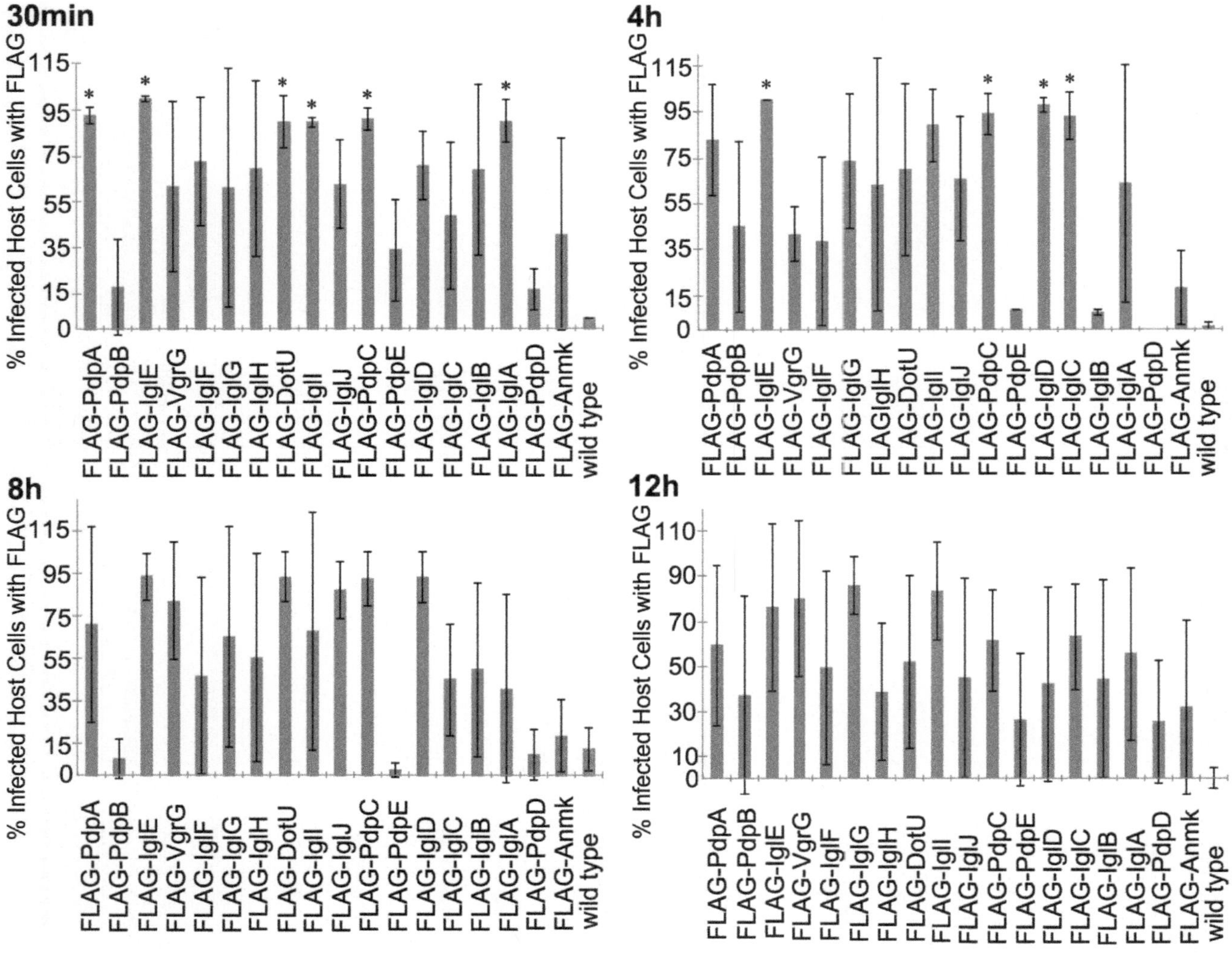

Figure 8. Infected Macrophages with FPI N-tagged Proteins. The percent of infected macrophages containing N-term FLAG-tagged FPI proteins within J774 cells at various time-points during infection were determined. This graph represents the mean of three independent experiments. Error bars indicate the standard deviations. Asterisks indicate significance ($p \leq 0.05$) and plus signs indicate approaching significance ($p = 0.051$–0.099) when compared to the no plasmid wild type control.

FPI blots were performed with similar amounts of bacteria and not adjusted for optimal visualization of individual proteins, which lead to over exposure of some proteins (Fig. 1). The lack of a wild type PdpC band visible with the PdpC antibody could be explained by low expression levels of native PdpC compared to the overexpression of PdpC-FLAG. Also it should be noted that different antibodies were used for detection of PdpC; the FLAG-tagged proteins contain three repeats of the FLAG sequences, which increases the avidity of the FLAG antibodies compared to the antibody raised against native PdpC (Fig. S2B).

To help discern the functions of FPI-encoded proteins, their localization within host cells were examined via microscopy. There are limitations associated with microscopy. Microscopy permeabilization techniques can result in false positives due to cell death or leakiness of membranes. In this study we used saponin to permeate host cell membranes, while leaving the bacterial cell membranes intact [30].

If the localization of a FPI protein was dependent on the expression of other proteins at a specific time, then we would have detected it because we examined four different time points during infection. These time points were chosen according to *Francisella's* intracellular life cycle [34,35]. By 30 min post-infection, internalized bacteria escape the phagosome and enter the host cell's cytoplasm [20,31]. Bacteria replicate intracellularly by four and 8 h bacteria post-infection [35]. By 12h post-infection bacteria manipulate host cells by avoiding immune responses and initiating autophagy [37]. However, since the plasmids constitutively express proteins inferences cannot be made on the timeline of secretion under natural transcriptional control.

In this study the secretion and localization of FPI tagged proteins during an infection of a macrophage-like cell line was determined by assessing the amount of fluorescence per cell. First, we generated a complete set of plasmids that contain either N- or C-terminal tags. These plasmids were used to examine the entire FPI for potentially secreted proteins, and internal controls within the FPI were used for this study. IglE is described as an outer-membrane protein, PdpB is an inner-membrane protein, and several FPI proteins have previously been identified as secreted in *F. tularensis* [11,13,33]. As an initial screen of the entire FPI, we hope this study inspires future investigations that further charac-

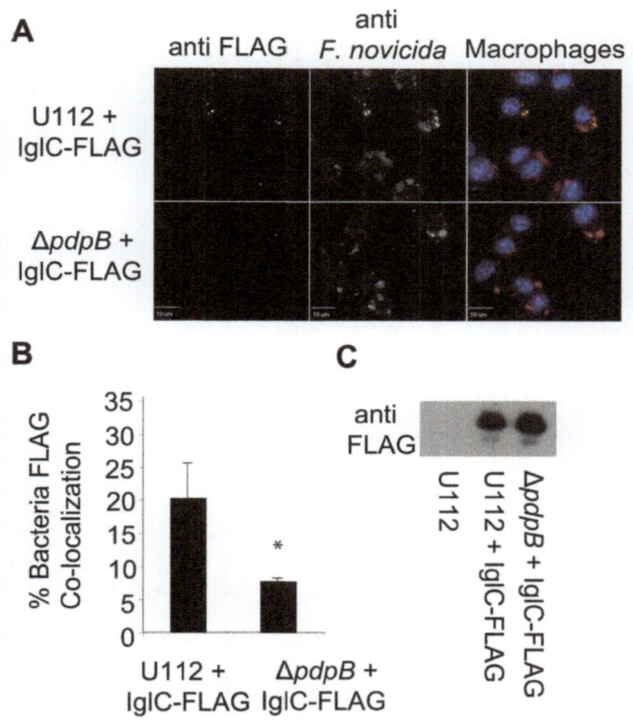

Figure 9. IglC Secretion was Dependent on T6SS. Microscopy of IglC-FLAG expressed in both wild type and Δ*pdpB* strains of U112 within J774 cells at 4 h post infection (A). Automated image analysis of bacterial cells associated with FLAG signal in wild type and Δ*pdpB* backgrounds shown in (B). Each bar represents the mean from three independent experiments of the percent of bacteria associated with FLAG signal, and error bars represent the standard deviation. An asterisk indicates the value was significantly different from that of the wild type in a Two-tailed t test ($p \leq 0.05$). Western blot shows FLAG expression of IglC-FLAG in wild type and Δ*pdpB* strains of U112 (C).

terize individual FPI proteins and their secretion pattern and localization within infected cells.

Using the *Francisella* expression plasmids in *F. novicida* IglABCDEFGHIJ, PdpACE, DotU, and VgrG were localized within macrophage-like cells on the outside of bacterial cell membranes. Proteins that were consistently detected in the cytoplasm of host cells were IglABCDEI and VgrG. There were high degrees of variation observed among time points, in the localization of proteins within infected host cells, among the N-terminal tagged proteins at, and between different tags of the same proteins. The detection of IglFGHJ, PdpACE, and DotU varied depending on time point. Since these proteins were constitutively expressed inferences based on time would not be valid. Although if some of the FPI proteins require the expression of other proteins for their delivery, secretion, or localization this may have been missed with only examining one time point. The variance in the cellular analyses can be explained by the fact that more area was assessed in the cell level analysis increasing total non-specific background fluorescence compared to bacterial association. Additionally, it was not possible to address how bacteria that were co-localized with FLAG were dispersed among infected cells. In addition to the increase in background signal in the analysis of host cells infected with FLAG, the variation among the N-tagged proteins can be explained by the fact that proteins are translated from N-terminus to C-terminus; therefore, the additional amino acid sequences from the FLAG tag may have an affect on folding and protein stability, however these proteins were able to restore the growth phenotype (Fig. 2). It was also not surprising that N-terminally tagged PdpE was not expressed, as PdpE is predicted to have an export signal sequence on the N-terminus [11]. In addition IglE also may possess an N-terminal signal and is an outer-membrane protein [11].

Since the C-terminal and the N-terminal-flagged proteins were individually analyzed for each protein, the expression levels of the two variants were statistically compared for correlation (Fig. S4). The two tags were significantly correlated at half of the time points; overall there was always a positive correlation trend. Reasons that detection of N- and C-terminally tagged proteins were not always correlated could include effects from the tag such as alterations in expression and processing, post translation modifications, masking transportation of proteins, or stability. To avoid artifacts from tagging one terminus of the individual FPI proteins, this study used two tags and examined several time points. It should also be noted that for a given protein there was also a wide variation from one time point to another (Fig. 5–8). This is not surprising since the stabilities of these proteins are not known and previous studies of stability of pathogen proteins in host cells can be altered by tagging effects or degraded by host cell [37].

Some proteins, while expressed, were not detected in the cytoplasm of host cells. The lack of detection of fluorescence signal for PdpBD and Anmk possessing FLAG tags were similar to that of wild type cells in bacteria, and within infected host cells. This suggests PdpBD and Anmk were not secreted from bacteria, which is consistent with other studies [13,38]. These non-secreted proteins were expressed at similar levels compared to most of the secreted proteins (Fig. 1). Our inability to detect PdpB confirms the appropriateness of using saponin-permeabilized cells to detect FLAG epitopes outside of bacteria while leaving the bacterial cell wall intact [30]. Previous fractionation of *F. novicida* expressing FLAG-tagged PdpB show this PdpB localized to the inner-membrane [13]. The current model for the T6SS in *Francisella* suggests PdpB is a transmembrane anchor protein, which spans the inner-membrane with parts extending into the periplasmic space [13]. DotU is an inner-membrane component of the secretion system of *Francisella* and all T6SS's where it interacts with PdpB [13,38]. Solubility properties have identified DotU as predominantly membrane-associated, partially soluble, and localized to the inner-membrane and periplasmic space where it stabilizes the secretion system [13,39]. The localization of DotU has not been visualized before; it is interesting that in this study, microscopy detected DotU as extracellular. DotU could be temporarily exposed to the extracellular space of bacterial cells during the contraction of the tube of the secretion system as proteins were secreted. Also the extracellular localization of DotU could be from effects of the FLAG tag.

In the current model of the T6SS in *Francisella novicida* as described by de Bruin, the inner-tube of the T6SS is speculated to be a polymer of IglC, which lies within the IglA and IglB polymer that contracts and drives IglC through the host cell membrane [13]. This contraction of IglAB could temporarily expose components of the secretion system (IglABC, DotU, and potentially other proteins) to extracellular staining. Also, IglA-IglB polymers span both the inner- and outer-membrane of *Francisella*, and thus were exposed extracellularly but not necessarily secreted. VgrG and PdpE were located on the point of the secretion channel-forming tube and would therefore appear outside of bacteria, as shown in this study [13].

Several secreted proteins were identified in this study that have previously been identified as secreted from LVS, including PdpAE, IglCEFIJ, and VgrG [11,33,39]. This study also identified DotU, IglABDGH, and PdpC as being localized to the outside of bacteria within infected host cells. Fractionation of *F. novicida* show IglABCD localized in all fractions of the bacterial cell [13], which supports their detection, outside of bacteria within macrophages. IglH, and PdpC might be secreted proteins, as it is not clear whether these proteins were secreted, localized to the outer-membrane of *Francisella*, or temporarily localized to the outer-membrane as components of the secretion system during transport of other secreted proteins. An alternative explanation is their detection in this study is leakage due to over expression from the *Francisella* expression plasmids. Detection observed in microscopy is not likely from dying cells that leak tagged proteins; since we have shown these tagged, plasmid-expressed proteins restored the intracellular growth phenotype (Fig. 2). In addition, PdpB, a protein localized to the inner bacterial membrane was not detected in microscopy while it was expressed (Fig. 1) and restored growth (Fig. 2). If leakage were a systematic problem with this study, we would expect PdpB to be detected despite its localization to the inner-membrane. However, we cannot exclude this possibility for other proteins. In any type of secretion assay, leakage from dead or dying bacteria is always a possibility that has to be considered in data interpretation.

To further examine localization of IglC, the IglC-FLAG plasmid was transformed into a Δ*pdpB* strain to test if localization was dependent on the T6SS. PdpB is homologous to IcmF, which is an inner-membrane component of the T6SS in *V. cholera* and is required for the secretion of Hcp [14]. PdpB was not detected through microscopy because it is an inner-membrane protein of *Francisella* [13]. The co-localization of IglC-FLAG with bacterial cells was significantly lower in Δ*pdpB* bacteria compared to wild type cells (Fig. 9B). The expression of IglC was examined to determine if the mutant was expressing IglC-FLAG, and both wild type and Δ*pdpB* expressed IglC at similar levels (Fig. 9C).

The relevance of the findings may not be generally applicable to other, virulent subspecies of *Francisella* since there are previous data showing that the secretion patterns differ between *F. novicida* and LVS [11,33]. Although, due to the similarities of the secreted proteins between LVS and *F. novicida*, this study confirms *F. novicida* as a valuable model to study the molecular mechanism employed by *F. tularensis* during infection of host cells.

This study describes the development of genetic tools to assess and elucidate the function of the complete set of FPI encoded proteins. These genetic tools include plasmids that contain an entire set of both N- and C-terminus epitope triple FLAG-tagged FPI genes that express FPI proteins. Western blotting of bacterial lysates reveals expression of 35 full-length epitope tagged FPI proteins. The *Francisella* expression plasmid expresses full-length functional proteins that restore the intramacrophage growth phenotype in respective mutants. Therefore the *Francisella* expression plasmids were genetically viable tools that can be used to further understand the intracellular life cycle of *F. tularensis* and elucidate potential intervention strategies. Overall these plasmids will contribute to a better understanding of the molecular mechanisms involved in the intracellular life cycle of *F. tularensis*.

Supporting Information

Figure S1 *Francisella* Expression Plasmids. Representative diagram of the *Francisella* expression plasmids, pKH4 containing *iglC*. with a C-terminal FLAG tag and pKH46 containing *iglC* with a N-terminal FLAG tag, are shown as examples of all 36 plasmids. All of the *Francisella* expression plasmids contain a groE promoter, a multiple cloning site (MCS), triple FLAG epitope tag, antibiotic cassettes, and an origin of replication. The MCS shows restriction enzyme sites used for insertion of FPI genes. Each of the FPI genes was individually inserted where *iglC* is depicted in the diagram. Arrows represent the direction of transcription and size of gene products.

Figure S2 Native and FLAG-tagged FPI Proteins. Western blot of *F. novicida* U112 wild type and U112 expressing the respective C-terminally tagged FPI proteins from the *Francisella* expression plasmid are labeled above each lane. (A) 15% gel with IglAC, (B) 8% gel with PdpAC. IglAC and PdpAC proteins were detected with polyclonal rabbit anti IglAC and PdpAC. FLAG-tagged proteins are detected with monoclonal mouse anti FLAG, goat anti mouse conjugated HRP, and chemiluminescent substrate.

Figure S3 Three-dimensional Reconstruction of PdpE. Three-dimensional reconstructions were comprised from a series of images that were taken through the macrophage cell infected with wild type containing *Francisella* expression plasmids. Bacteria in red, FLAG-tagged protein in green, and host cell nuclei in blue.

Figure S4 N- and C-tag Correlation. The data for N- and C-tagged proteins in both bacterial and cellular analyses were plotted against each other for each protein, at each time point. The data were subjected to a Spearman correlation tests. A best-fit trend line was inserted along with the slope, R^2 values, and the P value. Asterisks indicate significance ($p \leq 0.05$) of the Spearman's test for correlation. Specific analysis and times points are indicated on graphs.

Table S1 Molecular weights of FPI proteins. The molecular weights of *F. novicida* FPI encoded proteins.

Table S2 Means of FLAG with bacteria and FLAG within cells and significance. Each of the FPI proteins were examined for their localization with bacteria or within infected host cells, which is indicated as Bacterial or Cellular in the analysis column. Within each analysis proteins were examined via the FLAG tag on the N-terminus and the C-terminus. Values indicate the mean percentage of bacterial-FLAG co-localization or the mean percentage of infected cells containing FLAG from 3 independent experiments. Significance was determined with a left sided Dunett's test, * $p<0.05$, ** $p<0.001$, and *** $p<0.0001$.

Acknowledgments

Thank you Lisa McGilvary for providing excellent technical support. Thank you Thomas Zhart for pFNLTP-GFP-6. Thank you Fran Nano for supplying mutant strains and anti FPI protein antibodies.

Author Contributions

Conceived and designed the experiments: KH RFH. Performed the experiments: RFH. Analyzed the data: KH RFH. Contributed reagents/materials/analysis tools: KH RFH. Contributed to the writing of the manuscript: RFH KH.

References

1. Gal-Mor O, Finlay BB (2006) Pathogenicity islands: a molecular toolbox for bacterial virulence. Cell Microbiol 8(11): 1707–1719.
2. Galan JE, Wolf-Watz H (2006) Protein delivery into eukaryotic cells by type III secretion machines. Nature 444(7119): 567–573.
3. Mattoo S, Lee YM, Dixon JE (2007) Interactions of bacterial effector proteins with host proteins. Curr Opin Immunol 19(4): 392–401.
4. Nano FE, Zhang N, Cowley SC, Klose KE, Cheung KK, et al (2004) A *Francisella tularensis* pathogenicity island required for intramacrophage growth. J Bacteriol 186(19): 6430–6436.
5. Owen CR, Burker EO, Jellison WL, Lackman DB, Bell JF (1964) Comparative studies of *Francisella tularensis* and *Francisella novicida*. J Bacteriol 87: 676–83.
6. Gray CG, Cowley SC, Cheung KK, Nano FE (2002) The identification of five genetic loci of *Francisella novicida* associated with intracellular growth. FEMS Microbiol Lett 215(1): 53–56.
7. Tempel R, Lai XH, Crosa L, Kozlowicz B, Heffron F (2006) Attenuated *Francisella novicida* transposon mutants protect mice against wild type challenge. Infect Immun 74(9): 5095–5105.
8. Maier TM, Havig A, Casey M, Nano FE, Frank DW, et al (2004) Construction and characterization of a highly efficient *Francisella* shuttle plasmid. Appl Environ Microbiol 70(12): 7511–7519.
9. Nix EB, Cheung KK, Wang D, Zhang N, Burke RD, et al (2006) Virulence of *Francisella* spp. in chicken embryos. Infect Immun 74(8): 4809–4816.
10. Santic M, Molmeret M, Barker JR, Klose KE, Dekanic A, et al (2007) A *Francisella tularensis* pathogenicity island protein essential for bacterial proliferation within the host cell cytosol. Cell Microbiol 9(10): 2391–2403.
11. Bröms JE, Meyer L, Sun K, Lavander M, Sjöstedt A (2012) Unique substrates secreted by the type VI secretion system of *Francisella tularensis* during intramacrophage infection. PLOS ONE 7(11).
12. de Bruin OM, Ludu JS, Nano FE (2008) The *Francisella* pathogenicity island protein IglA localizes to the bacterial cytoplasm and is needed for intracellular growth. BMC Microbiol 7: 1.
13. de Bruin OM, Duplantis BN, Ludu JS, Hare RF, Nix EB, et al (2011) The biochemical properties of the *Francisella* pathogenicity island (FPI)-encoded proteins IglA, IglB, IglC, PdpB, and DotU suggest roles in type VI secretion. Microbiol 157: 3483–3491.
14. Barker JR, Chong A, Wehrly TD, Yu JJ, Rodriguez SA, et al (2009) The *Francisella tularensis* pathogenicity island encodes a secretion system that is required for phagosome escape and virulence. Mol Microbiol 74(6): 1459–1470.
15. Bingle LE, Bailey CM, Pallen MJ (2008) Type VI secretion: a beginner's guide. Curr Opin Microbiol 11(1): 3–8.
16. Filloux A, Hachani A, Bleves S (2008) The bacterial type VI secretion machine: yet another player for protein transport across membranes. Microbiol 154(2008): 1570–1583.
17. Bladergroen MR, Badelt K, Spaink HP (2003) Infection-blocking genes of a symbiotic *Rhizobium leguminosarum* strain that are involved in temperature-dependent protein secretion. Mol Plant Microbe Interact 16(1): 53–64.
18. Pukatzki S, Ma AT, Sturtevant D, Krastins B, Sarracinot D, et al (2006) Identification of a conserved bacterial protein secretion system in *Vibrio cholerae* using the *Dictyostelium* host model system. Proc Natl Acad Sci U S A 103(5): 1528–1533.
19. Lindgren H, Golovliov I, Baranov V, Ernst RK, Telepnev M, et al (2004) Factors affecting the escape of *Francisella tularensis* from the phagolysosome. J Med Microbiol 53(10): 953–958.
20. Santic M, Molmeret M, Klose KE, Jones S, Kwaik YA (2005) The *Francisella tularensis* pathogenicity island protein IglC and its regulator MglA are essential for modulating phagosome biogenesis and subsequent bacterial escape into the cytoplasm. Cell Microbiol 7(7): 969–979.
21. Mougous JD, Cuff ME, Raunser S, Shen A, Zhou M, et al. (2006) A virulence locus of *Pseudomonas aeruginosa* encodes a protein secretion apparatus. Science 312(5779): 1526–30.
22. Schell MA, Ulrich RL, Ribot WJ, Brueggemann EE, Hines HB, et al. (2007) Type VI secretion is a major virulence determinant in *Burkholderia mallei*. Mol Microbiol 64(6): 1466–1485.
23. Zheng J, Leung KY (2007) Dissection of a type VI secretion system in *Edwardsiella tarda*. Mol Microbiol 66(5): 1192–1206.
24. Lindgren M, Bröms J, Meyer L, Golovilov I, Sjöstedt A (2013) The *Francisella tularensis LVS Delta-pdpC* mutant exhibits a unique phenotype during intracellular infection. BMC Microbiol 13(20).
25. Lindgren M, Eneslatt K, Bröms J, Sjöstedt A (2013) Importance of PdpC, IglC, IglI, and IglG for modulation of a host cell death pathway induced by *Francisella tularensis*. Infect Immun 81(6): 2076–2084.
26. Read A, Sigrid J, Hueffer K, Gallagher L, Happ G (2008) *Francisella* genes required for replication in mosquito cells. J Med Entomol 45(6): 1108–1116.
27. Sambrook J, Fritsch EF, Maniatis T (1989) Molecular cloning: a laboratory manual, 2nd ed. Cold Spring Harbor Laboratory, Cold Spring Harbor, N.Y.
28. Patel JC, Hueffer K, Lam TT, Galan JE (2009) Diversification of a *Salmonella* virulence protein function by ubiquitin-dependent differential localization. Cell 137: 283–294.
29. Tyeryar FJ, Lawton WD (1970) Factors affecting transformation of *Pasteurella novicida*. J Bacteriol 104(3): 1312–1317.
30. Johnson MB, Criss AK (2013) Fluorescence microscopy methods for determining the viability of bacteria in association with mammalian cells. J Vis Exp 5(79).
31. Chong A, Wehrly TD, Nair V, Fischer ER, Barker JR, et al. (2008) The early phagosomal stage of *Francisella tularensis* determines optimal phagosomal escape and *Francisella* pathogenicity island protein expression. Infect Immun 76(12): 5488–5499.
32. Schmerk CL, Duplantis BN, Howard PL, Nano FE (2009) A *Francisella novicida pdpA* mutant exhibits limited intracellular replication and remains associated with the lysosomal marker LAMP-1. Microbiol 155(5): 1498–1504.
33. Bröms JE, Sjöstedt A, Lavander M (2010) The role of the *Francisella tularensis* pathogenicity island in type VI secretion, intracellular survival, and modulation of host cell signaling. Front Microbiol 1(136).
34. Clemens DL, Lee BY, Horwitz MA (2005) *Francisella tularensis* enters macrophages via a novel process involving pseudopod loops. Infect Immun 73(9): 5892–5902.
35. Santic M, Molmeret M, Klose KE, Kwaik YA (2006) *Francisella tularensis* travels a novel, twisted road within macrophages. Trends Microbiol 14(1): 37–44.
36. Checroun C, Wehrly TD, Fischer ER, Hayes SF, Celli J (2006) Autophagy-mediated reentry of *Francisella tularensis* into the endocytic compartment after cytoplasmic replication. Proc Natl Acad Sci U S A 103(39): 4578–4583.
37. Kubori T, Galan J (2003) Temporal regulation of *Salmonella* virulence effector function by proteasome-dependent protein degradation. Cell 115: 333–342.
38. Ludu JS, de Bruin OM, Duplantis BN, Schmerk CL, Chou AY, et al (2008) The *Francisella* pathogenicity island protein is PdpD is required for full virulence and associates with homologues of the type VI secretion system. J Bacteriol 190(13): 4584–4595.
39. Leiman PG, Basler M, Ramagopal UA, Bonanno JB, Sauder, et al (2009) Type VI secretion apparatus and phage tail-associated protein complexes share a common evolutionary origin. Proc Natl Acad Sci U S A 106(11): 4154–4159.
40. Robertson G, Child R, Ingle C, Celli J, Norgard M (2013) IglE is an outer membrane-associate lipoprotein essential for intracellular survival and murine virulence of Type A *Francisella tularensis*. Infect Immun 81(11): 4026–4040.

7

RecA Protein Plays a Role in the Chemotactic Response and Chemoreceptor Clustering of *Salmonella enterica*

Albert Mayola[1], Oihane Irazoki[1], Ignacio A. Martínez[2], Dmitri Petrov[2], Filippo Menolascina[3], Roman Stocker[3], José A. Reyes-Darias[4], Tino Krell[4], Jordi Barbé[1], Susana Campoy[1]*

1 Departament de Genètica i de Microbiologia, Universitat Autònoma de Barcelona, Bellaterra (Cerdanyola del Vallès), Spain, **2** ICFO-Institut de Ciències Fotòniques, Castelldefels, Spain, **3** Department of Civil and Environmental Engineering, Massachusetts Institute of Technology, Cambridge, Massachusetts, United States of America, **4** Department of Environmental Protection, Estación Experimental del Zaidín-Consejo Superior de Investigaciones Científicas, Granada, Spain

Abstract

The RecA protein is the main bacterial recombinase and the activator of the SOS system. In *Escherichia coli* and *Salmonella enterica* sv. Typhimurium, RecA is also essential for swarming, a flagellar-driven surface translocation mechanism widespread among bacteria. In this work, the direct interaction between RecA and the CheW coupling protein was confirmed, and the motility and chemotactic phenotype of a *S.* Typhimurium *ΔrecA* mutant was characterized through microfluidics, optical trapping, and quantitative capillary assays. The results demonstrate the tight association of RecA with the chemotaxis pathway and also its involvement in polar chemoreceptor cluster formation. RecA is therefore necessary for standard flagellar rotation switching, implying its essential role not only in swarming motility but also in the normal chemotactic response of *S.* Typhimurium.

Editor: Axel Cloeckaert, Institut National de la Recherche Agronomique, France

Funding: The authors acknowledge financial support from the Spanish Ministry of Science and Innovation (grant BFU2011-23478 to JB and FIS200800114 and FIS2011-24409 to DP), Generalitat de Catalunya (grant 2009SGR159 to DP and 2014SGR572 to JB), Fundació Privada Cellex Barcelona and the National Institute of Health (grant 1R01GM100473 to RS). The funders had no role in the design of the study, in data collection and analysis, in the decision to publish, or in the preparation of the manuscript.

Competing Interests: The authors have declared that no competing interests exist.

* Email: Susana.Campoy@uab.cat

Introduction

RecA is a DNA-dependent ATPase [1,2] present in almost all members of the Domain Bacteria [3,4]. As a protein that is highly conserved among bacterial species, it is commonly used in phylogenetic studies [5]. In addition, RecA is the main bacterial recombinase involved in the central steps of homologous recombination and recombinational DNA repair [6–8]. It is also the activator of the DNA damage response known as the SOS system [9]. In this system, RecA acts as a DNA damage sensor by binding to single-stranded DNA, which activates the protein and thereby its co-protease activity. Activated RecA (RecA*) prompts autocleavage of the LexA repressor, thus inducing its expression and that of other SOS genes, mostly those involved in DNA recombination and repair [9]. RecA* is also able to induce the autocleavage of other serine proteases such as UmuD [10] and several repressors of the bacteriophage lytic cycle [11–13]. Furthermore, RecA is directly associated with other repair pathways such as the activated error-prone DNA polymerase V [10,14], excision base repair [15], and RecN, involved in the repair of double-stranded DNA breaks [16].

Increasing its activity profile, a new role for RecA in swarming motility, has been added to its functional catalog [17,18]. Swarming is a specialized and highly coordinated form of flagellar-driven multicellular surface translocation [19,20] and the fastest mode of bacterial surface navigation [21]. In the absence of RecA, the swarming ability of both *Escherichia coli* and *Salmonella enterica* sv. Typhimurium is impaired [17,18]; interestingly, the same phenotype is observed in *S.* Typhimurium strains overexpressing RecA protein [18]. Thus, none of the as yet known RecA activities, i.e., SOS response activation, recombinational DNA repair, or genetic recombination, seem to be necessary for the control of swarming motility [17]. Nevertheless, a possible link between RecA and the chemotaxis pathway through the CheW protein has been suggested. The CheW coupling protein is essential for the formation of the ternary signaling complex that also contains the CheA autokinase and MCPs (methyl-accepting chemotaxis proteins) [22]. The *in vitro* interaction of RecA and CheW was shown in a large-scale genome-wide screen assay [23] and a balance between the intracellular concentrations of these two proteins was shown to be essential for swarming [18].

Swarming is not governed by chemotaxis, nevertheless it has been described that the chemosensory pathway is essential for the motility on solid surfaces [24]. Mutants in the chemotaxis (*che*) pathway fail to swarm because of defective colony hydration [25], which in turn is associated with the flagellar rotation bias present in *che* mutants. Thus, in the wild-type strain flagellar rotation switches from clockwise (CW) to counterclockwise (CCW), whereas *che* mutants have CW or CCW biases depending on the specific mutation [26]. During CCW rotation, all flagella of the bacterium form a bundle that falls apart when one or more flagellar motors turn in the CW direction. This switch promotes lubrication of the cell–surface interface, which is needed to overcome surface friction, both of which are critical requirements for swarming motility in temperate swarmers such as *S.* Typhimurium and *E. coli* [27]. The impaired swarming phenotype of *che* mutants can be restored by adding water or osmolytes to the semi-solid surface [25,27] or by restoring the normal flagellar rotation bias [28].

To further investigate the role of RecA in the control of swarming motility, in this work we examined the relationship between RecA and CheW in two-hybrid experiments and on co-immunoprecipitation assays, which confirmed the interaction of these two proteins. In addition, we explored the possibility that RecA affects the bacterial flagellar rotation pattern, by studying the swimming profile, flagellar motor rotation, and the chemotactic response of a *recA* knockout *S.* Typhimurium mutant (*ΔrecA*). The results confirmed a role for RecA in polar chemoreceptor cluster formation and therefore in flagellar rotation switching.

Materials and Methods

Bacterial strains, plasmids, and growth conditions

The bacterial strains and plasmids used in this study are listed in Table 1. Except when indicated, all strains of bacteria were grown at 37°C in Luria–Bertani (LB) broth or on LB plates. When necessary, ampicillin (100 μg/ml) or chloramphenicol (34 μg/ml) was added to the culture. The same cell growth conditions were used for the microfluidics and optical trap assays (described below). The bacteria were grown overnight in 2 ml of LB broth supplemented, when needed, with the appropriate antibiotic. Each culture was diluted 1:10 into LB broth without antibiotics but containing an oxygen scavenging system consisting of 100 μg glucose oxidase/ml and 20 μg catalase/ml (final concentrations) [29]. The added glucose is a substrate for the oxygen scavenging system and provides the energy needed for swimming under anaerobic conditions [29]. The cells were incubated at 37°C for 1 h and then diluted 100-fold in measuring medium (1% Bacto Tryptone, 0.8% NaCl, 2% glucose, 100 mM Tris-Cl, pH 7.5) containing the oxygen scavenging system. In all cases the scavenging system was added at least 2 h before the medium was used, to ensure a stable low level of oxygen.

Dead cells used for optical trap assays were prepared by the addition of 2% formaldehyde to the culture, with subsequent dilution steps carried out following the same protocol used for live bacterial cultures.

For chemotactic assays, overnight cultures of *S.* Typhimurium were grown in tryptone broth (1% Bacto Tryptone, 0.5% NaCl) supplemented, when required, with the appropriate antibiotics [30]. All strains used in the chemotactic assays (LT2, UA1928 (*ΔcheB*), UA1931 (*ΔrecA*) and UA1931/pUA1130) had similar growth kinetics in tryptone broth medium (data not shown). The cultures were then diluted 1:100 in the same medium but without antibiotics and incubated at 30°C with constant shaking until an OD_{600} of approximately 0.5 was reached. The culture was then harvested by centrifugation at 4500 g for 10 min at room temperature. The obtained cell pellet was washed twice in 1 ml of tempered tethering buffer (10 mM potassium-phosphate pH 7, 67 mM NaCl, 10 mM Na-lactate, 0.1 mM EDTA, and 0.001 mM L-methionine) and the resuspended cells were diluted to approximately 6×10^7 colony-forming units (cfu)/ml.

Construction of the *S.* Typhimurium mutant strains

The *S.* Typhimurium mutants were constructed using the one-step PCR-based gene replacement method [31], except when indicated. All DNA techniques were performed as described elsewhere [32]. The chloramphenicol resistance cassette from the plasmid pKD3 was amplified using suitable oligonucleotides containing 80-nucleotide stretches homologous to each of the insertion sites and the corresponding P1 and P2 sites of pKD3 (Table S1). The PCR products were digested with *Dpn*I and used to transform *S.* Typhimurium electrocompetent cells containing the pKOBEGA plasmid [33]. Following selection of the transformant clones, the latter plasmid was eliminated by taking advantage of its temperature sensitivity, incubating the clones at 42°C. Gene substitution was confirmed by PCR and sequencing using the appropriate primers.

To construct the *ΔcheRΔcheW* and *ΔcheRΔrecA* mutant strains, the chloramphenicol resistance cassette present in the *ΔcheR* strain was eliminated as previously described using the pCP20 plasmid [31]. Afterwards, either *ΔrecA* or *ΔcheW* mutations from UA1927 and UA1907 were transferred to the *ΔcheR* chloramphenicol-sensitive strain (UA1910) by transduction, using the P22 HT bacteriophage [34]. The same procedure was used to construct the *ΔrecA ΔcheW* strain. In this case, the chloramphenicol resistance cassette present in *ΔcheW* strain was removed and the *ΔrecA* mutation was transducted from the UA1927 strain. In all cases, the absence of the prophage in the selected clones was determined by streaking them onto green plates as previously described [35]. All of the resulting strains were verified by PCR and sequencing.

β-Galactosidase-based two-hybrid system

The two-hybrid assay was performed as described [36]. The *recA* and *cheW* genes were PCR-amplified using suitable oligonucleotides (Table S1) that included *Sph*I and *Bam*HI restriction sites in the amplicon. After their release from the plasmids by endonuclease digestion, the amplified genes were cloned in both pB2HΔα and pB2HΔω vectors. The same procedure was used to clone the *amyA* and *dnaE* genes into pB2HΔα and pB2HΔω, respectively. These constructs served as the non-interaction assay controls, as previously described [36]. All of the constructs were confirmed by sequencing.

To simultaneously express the two fusion proteins within a cell, electrocompetent *E. coli* MC1061 cells were co-transformed with the two plasmids of interest and the transformants were selected by adding chloramphenicol and ampicillin to the solid medium. The presence of both fusions was confirmed by PCR and sequencing.

For β-galactosidase assays, the selected clones were grown in LB supplemented with ampicillin and chloramphenicol at 37°C until an OD_{550} of 0.2 was reached. IPTG was then added to the culture to a final concentration of 20 nM and the cultures were incubated at 37°C. Samples were taken 5 h after IPTG addition, and β-galactosidase activity was assayed as described by Miller (1991) [37]. The relative expression of β-galactosidase in each strain was calculated as the enzyme's activity with respect to that of the non-interaction control strain, which expressed the ΔαAmyA and ΔωDnaE proteins [36]. The reported results are the means of at least three independent assays, each performed in triplicate.

Construction of RecA and CheW tagged proteins

Co-immunoprecipitation assays were carried out using RecA-6×His and CheW-FLAG tagged proteins. The *recA* and *cheW* genes were PCR-amplified using the appropriate oligonucleotide pair (Table S1). In both cases, the corresponding tag sequence was included at the 5′ end of the suitable oligonucleotides that also contained a 3× Gly linker between the tag and the corresponding

Table 1. Bacterial strains and plasmids used in this work.

Strain or plasmid	Relevant characteristic(s)	Source or reference
Strains		
DH5α	*E. coli supE4 ΔlacU169 (φ80 ΔlacZ ΔM15) hsdR17, recA1, endA1, gyrA96, thi-1, relA1*	Clontech
MC1061	*E. coli* F^- *Δ(ara-leu)7697 [araD139] B/r Δ(codB-lacI)3 galK16 galE15* λ^- $e14^-$ *mcrA0 relA1 rpsL150* (Str^R) *spoT1 mcrB1 hsdR2*(r^-m^+)	CGSC
BL21 (DE3) pLysS	*E. coli* F^- *dcm ompT lon hsdS*($r_B^-m_B^-$) *galλ(DE3)* carring pLysS plasmid, Cm^R	Stratagene
LT2	*S.* Typhimurium wild type strain	ATCC
UA1907	*S.* Typhimurium *ΔcheWΩcat*. Cm^R	This work
UA1908	*S.* Typhimurium *ΔcheW*	This work
UA1910	*S.* Typhimurium *ΔcheR*	This work
UA1915	*S.* Typhimurium *ΔcheR ΔcheW*	This work
UA1913	*S.* Typhimurium *ΔcheR ΔrecA*	This work
UA1927	*S.* Typhimurium *ΔrecAΩ,cat*. Cm^R	This study
UA1928	*S.* Typhimurium *ΔcheB*	[29]
UA1929	*S.* Typhimurium *ΔcheY*	[29]
UA1930	*S.* Typhimurium *ΔcheW ΔrecA*	This work
UA1931	*S.* Typhimurium *ΔrecA*	This work
Plasmids		
pKOBEGA	Vector containing the λ Red recombinase system, Amp^r, temperature sensitive	Generous gift of Prof. G. M. Ghigo, [33]
pKD3	Vector carrying FRT-Cm construction, Amp^R, Cm^R	[31]
pCP20	Vector carrying FLP system, $OriV_{ts}$, Amp^R	[31]
pGEM-T	Cloning Vector; Amp^R	Promega
pGEX 4T-1	Expression vector carrying the Ptac IPTG - inducible promoter and the *lacI*^{q} gene; GST fusion tag, Amp^R	Amersham Biosciences
pUA1108	pGEX 4T-1 derivative plasmid carrying without the GST fusion tag, carrying only the Ptac promoter and the lacIq gene; used as overexpression vector, Amp^R	This work
pUA1109	pUA1108 derivative containing the native *S.* Typhimurium *recA* gene under the control of the Ptac promoter, Amp^R.	This work
pUA1127	pUA1108 derivative vector carrying the eYFP::*cheR* fusion, Amp^R	This work
pUA1130	pUA1108 derivative containing the native *S.* Typhimurium *recA* gene under the control of the Ptac promoter, Amp^R.	This work
pUA1131	pUA1108 derivative overexpression vector carrying the *cheW*-FLAG gene	This work
pB2H*Δα*	pACYCDuet-1 derivative vector with the *E. coli Δα* β-galactosidase fragment under the control of the Ptac promoter	BCCM/LMBP, [36]
pB2H*Δω*	pACYCDuet-1 derivative vector with the *E. coli Δω* β-galactosidase fragment under the control of the Ptac promoter	BCCM/LMBP, [36]
pUA1114	pB2H*Δα* derivative vector contaning the *Δα-recA* fusion	This work
pUA1115	pB2H*Δω* derivative plasmid contaning the *Δω-recA* fusion	This work
pUA1116	pB2H*Δα* derivative vector contaning the *Δα-cheW* fusion	This work
pUA1117	pB2H*Δω* derivative plasmid contaning the *Δω-cheW* fusion	This work
pUA1118	pB2H*Δα* derivative plasmid contaning the *Δα-amyA* fusion, used as a control in two hybrid assays.	This work
pUA1119	pB2H*Δω* derivative plasmid contaning the *Δω-dnaE* fusion, used as a control in two hybrid assays.	This work

gene sequence (Table S1). The PCR products were digested with *Nde*I and *Bam*HI and cloned into pUA1108, with each tagged protein under the control of the Ptac promoter. The plasmids were transformed into *E. coli* DH5α and confirmed by sequencing. The confirmed plasmids were used in the electrotransformation of the *S.* Typhimurium *ΔrecA ΔcheW* strain, thereby ensuring that every RecA and CheW protein produced by that strain carried the specific tag. The selected transformants were confirmed again by PCR and sequencing.

Co-immunoprecipitation assays

The assays were performed as described by D'Ulisse and others [38] with modifications. Briefly, cultures of *S.* Typhimurium *ΔrecA ΔcheW* harboring the plasmids with constructs encoding the tagged proteins were grown in LB broth supplemented with ampicillin to an OD_{550} of 0.2. Expression of the tagged genes was induced by the addition of 1 mM IPTG. After 3 h of growth, the cultures were centrifuged and the resulting pellet was washed in TBS 1× buffer (1.5 M NaCl, 250 mM Tris, pH 7.3) and then resuspended in cold IP lysis buffer (1 × TBS, 15% glycerol, and 1% Triton X-100). The samples were incubated at 4°C for 40 min, with vortexing every 5 min. Finally, the samples were centrifuged at 4°C and the supernatant was collected. The protein concentration was determined by the Bradford method. As a control, cell lysates of *S.* Typhimurium *ΔrecA ΔcheW* containing the empty pUA1108 were also obtained following the same procedure.

Pure Proteome Protein A magnetic beads (Millipore) were used for immunoseparation. Either anti-6×His mouse IgG (Roche) or anti-FLAG mouse IgG (Acris) antibody was attached to the beads according to the manufacturer's instructions.

For co-immunoprecipitation the corresponding cell lysates were mixed at a ratio of 1:1 and incubated at 30°C for 1 h without shaking, to allow interaction of the proteins. The appropriate antibody-coated magnetic beads (either anti-6×His or anti-FLAG IgGs) were added to the mixture and the samples were incubated overnight at 4°C with gentle mixing on a shaker. The beads were recovered and washed three times with wash buffer (1×TBS, 15% glycerol, and 1% Triton X-100) and finally resuspended in 45 µl of Laemmli sample buffer and heated for 10 min at 90°C. The samples were separated by SDS-PAGE on a 15% polyacrylamide gel and analyzed by western blotting using as primary antibody either anti-6×His or anti-FLAG mouse IgG and as secondary antibody horseradish peroxidase (HRP)-coupled anti-mouse IgG goat IgG antibody (Acris) together with Luminata western HRP chemiluminescence substrates (Millipore). HRP-coated Precision Plus protein western C standard (BioRad) was used as the molecular mass marker.

Microfluidics assays

Microfluidics experiments were performed for each bacterial strain as previously described by Ahmed *et al.* [39]. After overnight incubation, cultures in log phase were diluted 1:50, to a final volume of 2 mL, in fresh medium and grown to an optical density $OD_{600} = 0.45 \pm 0.02$. In order to optimize trajectory identification, cells were then diluted 1:4 before injecting them in a 60 µm thick microfluidic channel hosting a microwell (∅ 2 mm) at its center. The focal plane was set to 30 µm above the glass bottom, at channel mid-depth, to minimize the interaction of bacteria with surfaces. Cells were imaged with a 20× objective (Nikon Pan Fluor ELWD, NA 0.45, WD 7.4 mm) in phase contrast at 25 frames per second, for a total of 20 s per experiment. Three biological replicates, each with 5 technical replicates, were carried out for each strain. All frames were segmented to obtain the cells' coordinates and cells were then tracked using in-house developed tracking routines based on the nearest neighbor method, implemented in MATLAB (The Mathworks, Natick, MA). Manual selection of high-quality tracks followed trajectory identification and allowed 300–500 trajectories per strain to be used for the analysis of motility.

Optical trapping assays

Optical trapping was carried out as previously described using a 1064-nm optical beam from a laser coupled to a single-mode fiber (Avanex) expanded up to 10 mm and then highly focused by an immersion oil objective (Nikon, CFI PL FL 100× NA 1.30 WD 0.16 mm) [29]. The oxygen scavenging system guaranteed a constant low level of oxygen and hence cell survival during the measurements [40].

As previously described [29], data for each bacterial strain were obtained from ten different randomly chosen cells of four distinct biological replicates; thus 40 cells per strain were analyzed.

The forward scattered light is collected by a 40X objective and projected into a quadrant photo diode (New Focus 2911). By this technique, the position of each trapped cell was acquired for 1000 s at 2 kHz of acquisition rate. The entire set of acquired data (1000 s) was then divided into 1-s-blocks. For each data block the angular velocity of the cell around the optical axis (Θ value) was calculated as described [29]. About 80% of the histograms showed very similar patterns. All plots shown below for the wild-type, mutant strains, and dead bacteria present the Θ histogram of one trapped cell either from the corresponding bacterial strain or from a dead cell control. In all cases, the selected histograms were within the above-mentioned 80%. During the experiments, videos were recorded using a CCD camera at the beginning (capture) and end (liberation) of the measurements.

Chemotaxis capillary assays

Chemotaxis assays were conducted as described by Adler [30], with some modifications. The chemotaxis chamber set was formed by placing three V-shaped bent needles (40 mm 18 G needle, Nipro) on the surface of an aseptic 140-mm Petri dish (Deltalab) and then covering them with a 24×65 mm microscope cover slip (Menzel-Glässer).

One-ml capillary tubes, 3 cm long (Microcaps, Drummond Scientific Co.), were used. One end of each tube was heat-sealed in a flame. After autoclaving, the sealed capillaries were filled with either tethering buffer or 10 mM L-aspartate dissolved in tethering buffer [41].

Approximately 2 ml of each cell suspension was placed in the chemotaxis chambers, which were then incubated for 1 h at 30°C. After the incubation, the exterior of the capillaries was rinsed under a stream of sterile double-distilled water. The sealed end of the capillaries was then broken off and the contents of the tube were emptied into a 1.5-ml microcentrifuge tube containing 0.9% NaCl. Suitable dilutions were plated on LB plates; after an overnight incubation, the cfu/ml were calculated.

Construction of the eYFP::*cheR* fusion

The enhanced yellow fluorescent protein (eYFP)::*cheR* fusion was constructed by the overlap extension procedure as follows. The *cheW* and *eYFP* genes (Clontech) were amplified using CheRstmF/CheRstmBamHI and eYFPNdeI/eYFPR oligonucleotide pairs, respectively, with eYFPR and cheRstmF containing complementary overhangs and a 3× Gly linker (Table S1). The resulting DNA fragments were annealed and amplified in a second round of PCR using eYFPNdeI and CheRstmBamHI to form the corresponding eYFP::*cheR* fusion. These outer primers contained *Nde*I and *Bam*HI restriction sites that were used to clone the fragments into the IPTG-inducible pUA1108 expression vector, giving rise to plasmid pUA1127, in which the eYFP::*cheR* fusion is under Plac promoter control. The fusion was confirmed by sequencing and the pUA1127 plasmid was transformed into several genetic backgrounds (*ΔcheR*, *ΔcheRΔrecA*, and *ΔcheRΔcheW*) to obtain the bacterial strains used in the chemoreceptor clustering assays.

Chemoreceptor clustering assay

Receptor clustering experiments were performed as described [22,42,43] with modifications. Briefly, overnight cultures of *S.* Typhimurium strains carrying the pUA1127 (eYFP::*cheR*) plasmid were grown at 30°C in tryptone broth supplemented with ampicillin under constant agitation. After 24 h of incubation, the cultures were diluted 1:100 in tryptone broth supplemented with ampicillin and 25 mM IPTG to induce eYFP::*cheR* fusion expression. The cultures were then incubated at 30°C until an OD_{600} of 0.5 was reached. The cells were harvested by low speed centrifugation (5300 g) for 15 min, washed once in cold tethering buffer, and finally resuspended in 100 ml of ice-cold tethering buffer. The cells were maintained on ice throughout the assay.

For fluorescence microscopy assays, the cells were immobilized and fixed at the same focal plane using thin 1% agarose pads in tethering buffer. Three µl of cells were applied on the pad, which was then covered with a clean cover slip. Fluorescence microscopy was performed using a Zeiss AxioImager M2 microscope (Carl Zeiss Microscopy) equipped with a Zeiss AxioCam MRm monochrome camera (Carl Zeiss Microscopy) and a filter set for eYFP (excitation BP 500/25; beam splitter FT 515; emission BP535/30). Cell fields were photographed and at least 250 cells were inspected by eye to determine the presence and type of clusters. All fluorescence images were obtained at 1000× magnification and were acquired under identical conditions. Each experiment was performed in triplicate using independent cultures. The images presented in the corresponding figure are representative of the entire images that are included as (Files S1, S2 and S3). ImageJ software (National Institutes of Health) was used to either quantify the number of clusters or to prepare images for publication.

Statistical analysis

The chemotaxis capillary and chemotaxis clustering assays were statistically evaluated using, respectively, a two-way or one-way analysis of variance (ANOVA) with Prism (GraphPad), as previously described [44,45]. In all cases, the analyses were followed by the Bonferroni multiple comparison post-hoc test, with $p<0.05$ defined as statistically significant. In all of the figures, the error bars indicate the standard deviation.

Results

The interaction of RecA and CheW proteins

Large-scale protein-protein *in vitro* interaction studies had previously identified RecA as a prey protein when CheW was used as bait, but not vice versa [23]. Thus, to ascertain the RecA and CheW interaction two-hybrid assays and co-immunoprecipitation experiments were carried out.

In the two-hybrid assay, the previously described pB2H$\Delta\alpha$/pB2H$\Delta\omega$ system [36] was used. RecA and CheW proteins were fused to the two non-functional but complementary β-galactosidase truncations ($\Delta\alpha$ and $\Delta\omega$) in the system. In the reporter strain, β-galactosidase activity is driven by protein-protein recognition between both non-β-galactosidase parts of the chimeras. As shown in Fig. 1A, relative β-galactosidase expression by strains co-expressing either the $\Delta\alpha$RecA/$\Delta\omega$CheW or the $\Delta\alpha$CheW/$\Delta\omega$RecA chimera pair was >7, indicating significantly higher β-galactosidase activity in these strains than in the non-interaction assay control strain [36].

To further confirm the two-hybrid assay results and thus obtain additional evidence for the interaction between RecA and CheW, co-immunoprecipitation assays were carried out using *S.* Typhimurium *ΔrecA ΔcheW* strains carrying the corresponding plasmids that overexpress either the RecA-6×His or the CheW-FLAG tagged proteins (Table 1). The immunoprecipitation was performed by using magnetic beads coated with either anti-6×His or anti-FLAG mouse IgG antibodies that specifically interact with the corresponding tagged protein and the recovered proteins were detected by Western blot. As seen in Fig. 1B, when both recombinant proteins were present in the protein mixture, and anti-6×His antibody coated beads were used, CheW-FLAG proteins were observed in the recovered supernatants. The same results were observed when anti-FLAG antibody coated beads were added to the mixture, then RecA-6×His proteins were recovered. All together, these data indicate that RecA-6×His was able to pull down CheW-FLAG and vice versa. Thus, the results of the two assays together confirm RecA–CheW pair formation and suggest the association of RecA with the chemotaxis pathway through its interaction with CheW.

The absence of RecA causes a decrease in swimming speed

To further understand the role of RecA in motility, the swimming speed of a *S.* Typhimurium *ΔrecA* mutant was evaluated through microfluidics assays and compared to that of the wild-type strain. In this assay, swimming speed computed over the whole set of identified trajectories was measured. As shown in Fig. 2, which depicts the relative frequency of swimming speeds for each strain, the absence of RecA prompted a change in the swimming profile. Specifically, at the highest relative frequency, the velocity of the mutant was lower than that of the wild-type strain under the same experimental conditions.

The *ΔrecA* mutant present a CW-bias of flagella rotation

To determine whether the slower swimming speed prompted by the absence of RecA was due to a bias in flagellar rotation, the flagellar rotation patterns of the *ΔrecA* mutant was studied using a single optical trap. Optical trapping is an excellent tool for analyzing the dynamic properties of bacteria [46–48]. It is based on the ability of an optical beam to trap a single cell because the refractive index of cells (and their constituents) is higher than that of the surrounding medium. Once trapped, the movement of that cell is measured, yielding information on its momentary position. In the case of rod shaped *S.* Typhimurium, a cell trapped in the single optical beam aligns itself along the optical axis. Thus, besides Brownian motion, the torque produced by flagellar rotation alters the dynamics of the cell [29]. The measurement of the rotation profile for each strain is expressed as the distribution of the change in the mean value of Θ, which is the angular velocity of the cell around the optical axis [29]. Using this technique we were able to distinguish CW to CCW switching of the flagella.

Figure 3 shows the flagellar rotation profile of both the *S.* Typhimurium *ΔrecA* mutant and the wild-type strain. Dead wild-type cells and *ΔcheB* and *ΔcheY* mutants were used as controls. There is no change in the angular velocity of dead cells, which exhibit only Brownian motion; thus, their Θ distribution pattern is centered at zero. Among the mutants, *ΔcheB* cells, described by their tumbling motility because of their CW flagellar rotation bias [49], displayed a Θ distribution pattern centered near zero. Thus, there was no change in the average angular velocity of the mutant cells but, as would be expected, the histogram was broader than that of the dead cells. Conversely, the flagella rotation profile of *ΔcheY* cells had a mean Θ that was not zero and was highly positive, consistent with the smooth swimming ability characteristic of this strain because of its CCW-biased flagellar rotation pattern [50,51]. In the living wild-type cells, the Θ histogram showed the anticipated two peaks, reflecting normal switching between CCW and CW flagellar rotations. The peak centered at zero along the x axis corresponded to the tumbling state (CW rotation) whereas the peak located around 7.5 was due to the running state (CCW rotation) (Fig. 3). By contrast, but similar to the *ΔcheB* mutant, *ΔrecA* cells showed only one peak, centered

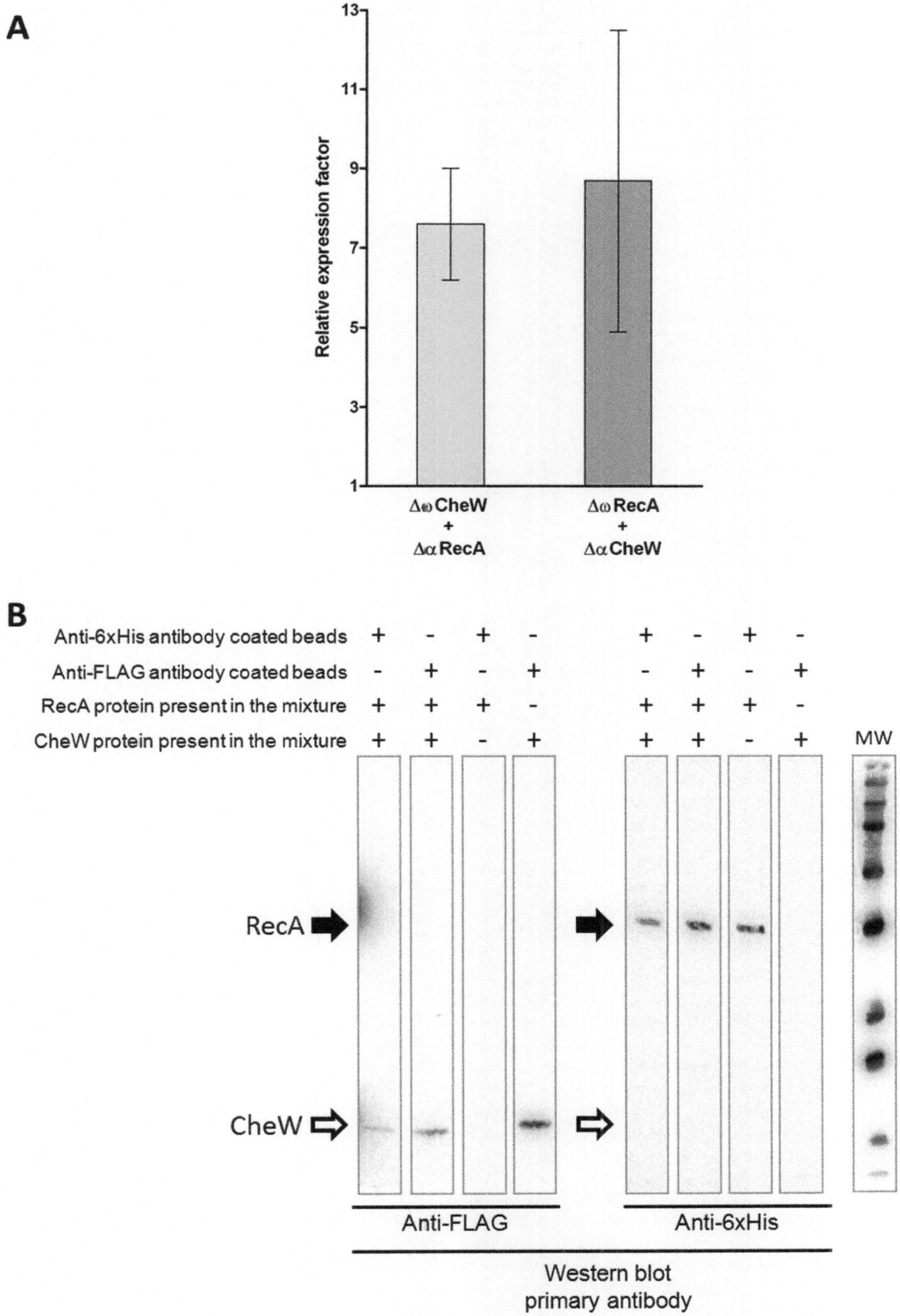

Figure 1. RecA protein directly interacts with CheW. A) Two-hybrid assay. Measurement of the β-galactosidase activity of strains co-expressing the chimera protein pairs ΔαRecA/ΔωCheW or ΔαCheW/ΔωRecA. The results are expressed relative to those obtained with the non-interacting control strain expressing ΔαAmyA and ΔωDnaE [36]. Measurements were made 5 h after the addition of 20 nM IPTG to the culture. In

each case the mean value from three independent experiments (performed in triplicate) is shown. Error bars indicate the standard deviation. **B) Co-immunoprecipitation assay.** Lysates prepared from cells overexpressing RecA-6×His and CheW-FLAG tagged proteins were mixed to allow the proteins to interact. Immunoprecipitation (IP) was performed by adding magnetic beads coated with either anti-6×His or anti-FLAG antibodies to the mixture and the attached proteins were recovered and separated by SDS-PAGE electrophoresis. The presence of the recombinant protein in the supernatants was assessed by western blotting (WB). As a control, co-IP assays were conducted using lysates from a *ΔrecAΔcheW S.* Typhimurium strain carrying an empty overexpression plasmid, thus expressing neither RecA-6×His nor CheW-FLAG proteins. The presence (+) or absence (−) of RecA, CheW, or both tagged proteins in the corresponding lysate mixture is indicated. Black and white arrows show the position of RecA-6×His and CheW-FLAG, respectively. IP indicates the antibody attached to the beads and WB the primary antibody used in western blotting. MW indicate the molecular mass marker.

near zero, in their Θ distribution pattern (Fig. 3). The absence of the second peak was indicative of the CW-biased rotation of Δ*recA* cells.

Chemotaxis response of *recA* mutants

To further confirm the switching defect of the *ΔrecA* mutant, its ability to move towards an L-aspartate source was evaluated using a classical capillary assay. The results are shown in Fig. 4. As expected, and in concordance with observations in other tumbling strains such as the *ΔcheB* mutant [49], the capillary assays clearly demonstrated that cells lacking RecA are unable to respond to the presence of L-aspartate. Furthermore, chemotaxis by the *ΔrecA* mutant was restored when the RecA deficiency was complemented by the presence of a plasmid containing the *recA* gene under the control of an IPTG- inducible promoter (Fig. 4) but not by the presence of the empty plasmid (data not shown). These results unequivocally showed that the presence of RecA is essential for a normal chemotactic response.

Chemotaxis receptor clustering

Based on our results, the phenotype of the *ΔrecA* mutant is similar to that of other *che* mutants, since in all cases the absence of RecA impairs not only swarming [17,18] but also the switching of flagellar rotation (Fig. 3) and the chemotactic response (Fig. 4). To elucidate the role of RecA in chemotaxis, and taking into consideration the direct interaction of RecA with the CheW coupling protein, which bridges the MCPs to histidine kinase CheA [43,52], we asked whether, like CheW, RecA was involved in chemoreceptor clustering. To investigate this possibility fluorescently tagged CheR (eYFP-CheR) was used as a specific reporter for chemoreceptor localization [43].

The eYFP-CheR fusion was constructed and cloned into pUA1108 vector under the control of an IPTG-inducible promoter (Table 1). For correct chemoreceptor localization, native CheR had to be removed; accordingly, the plasmid was included in the *S.* Typhimurium *ΔcheR ΔrecA* transformant. Additionally, it was also used to obtain the *ΔcheR* and *ΔcheR ΔcheW* mutants. The *ΔcheR* mutant served as the positive control strain since it exhibited normal polar clusters. The *ΔcheR ΔcheW* strain was used as the negative control strain since the absence of CheW inhibits polar cluster formation [22]. As expected, and in agreement with previous reports [43], in the positive control (the *ΔcheR* strain) single tight polar spots were seen in ~70% of the observed cells. These spots corresponded to the clustering of thousands of chemoreceptors at the cell pole (Fig. 5). However, in agreement with previous data [43], in the absence of CheW compact polar clusters were formed in only ~10% of the cells; instead, the presence of diffuse clusters (known as caps) was observed (Fig. 5A). Thus, according to our findings, the absence of RecA significantly impairs normal polar cluster formation, which occurred in only ~50% of the cells, and increases the presence of caps. Nonetheless, neither the reduction in polar spot formation nor the increase in caps was as high in the absence of RecA as in the absence of CheW.

Discussion

A role for RecA in controlling the swarming motility of both *E. coli* and *S.* Typhimurium was clearly shown in previous studies [17,18]. However, besides its possible connection to CheW [18,23], nothing was known about the mechanisms that link RecA to motility. To determine whether RecA is associated with the chemotaxis pathway and, specifically, with flagellar function, we examined its putative direct interaction with CheW, its role in swimming motility and in chemotaxis, as well as the flagellar switching pattern of cells lacking RecA.

Our results support a tight relationship between RecA and the chemotactic response. First, our results unequivocally confirmed the interaction of RecA with CheW through two widely used techniques. The results of the two-hybrid assays were in concordance with those previously obtained in a large-scale genome-wide screening assay [23] and suggested an interaction between RecA and CheW (Fig. 1A), which was definitively shown in the co-immunoprecipitation assays (Fig. 1B). Second, the motility phenotype of cells lacking RecA, as determined herein, was similar to that of some *che* mutants. The latter finding was supported by microfluidics assays, in which the average swimming speed of *S.* Typhimurium *ΔrecA* was lower than that of the wild-type strain (Fig. 2), and by the observed differences in the flagellar rotation patterns of these two strains (Fig. 3). Thus, the absence of

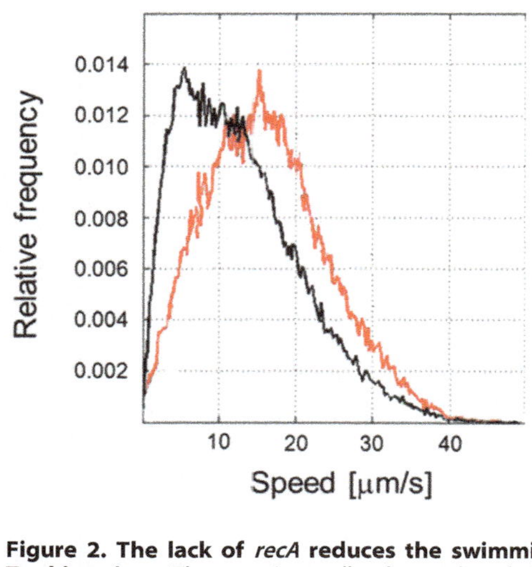

Figure 2. The lack of *recA* reduces the swimming speed of *S.* Typhimurium. The experimentally observed probability distribution of the swimming speeds within a population of wild-type *S.* Typhimurium (red line) and of the *ΔrecA* mutant (black line), assessed using a microfluidics assay [39]. Values are expressed as the relative frequency of a given speed within a cell population. For each strain, the results were obtained from three independent experiments supported by five technical replicates each, for a total of 300–500 cells tracked per strain.

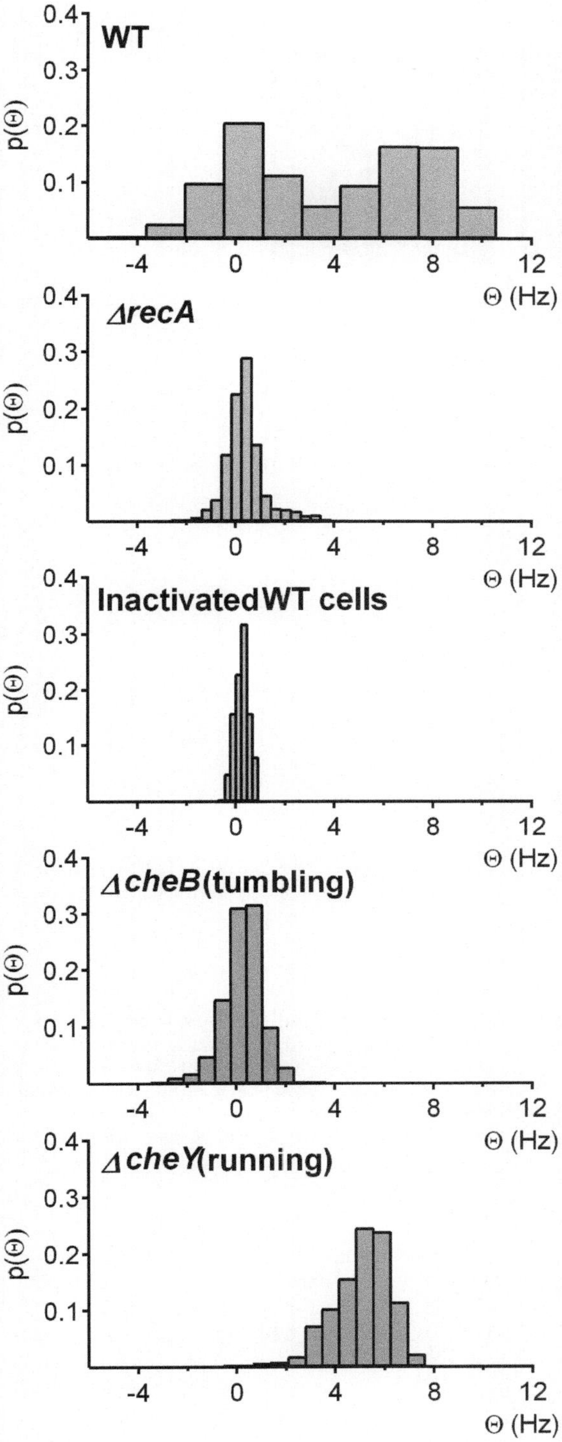

Figure 3. Flagellar rotation is CW-biased in the *S.* Typhimurium Δ*recA* mutant. The flagellar switching profiles of *S.* Typhimurium LT2 wild-type (WT), dead wild-type (dead), *ΔrecA*, *ΔcheB* (tumbling), and *ΔcheY* (running) cells were evaluated. The resulting histograms show the distribution of the change in the mean cellular angular velocity around the optical axis (Θ). A zero-centered peak, as displayed by dead cells and the *ΔcheB* tumbling mutant, is indicative of CW-biased flagellar rotation, and a peak with positive values, as displayed by the *ΔcheY* running mutant, CCW-biased rotation. The presence of two peaks, one zero-centered and the other centered at positive values, indicates a mixed population displaying both CW and CCW rotational patterns and thus a non-biased flagellar rotational pattern. For each strain, the results were obtained from four independent experiments of ten cells each.

RecA impaired flagellar switching, leading to a CW bias similar to that of other tumbling strains, like the *ΔcheB* mutants [49]. Furthermore, consistent with the tumbling phenotype of the *ΔrecA* mutant, our results demonstrate that RecA is essential for a normal chemotactic response. Specifically, in quantitative chemotaxis assays the *ΔrecA* mutant was unable to detect the presence of L-aspartate, a well-known chemoattractant (Fig. 4); instead, chemotaxis was restored only when the *recA* deficiency was complemented by a plasmid carrying a copy of the *recA* gene (Fig. 4). The slower-moving phenotype of the RecA-deficient mutants can be explained by the CW bias displayed by these cells. By being anchored in a tumbling state, without normal running, the *ΔrecA* mutant was slower than the wild-type strain. In a previous study, the inability to switch the direction of flagellar rotation was linked to defects in chemotaxis and to improper colony hydration, leading to an inability to swarm [25,27,28]. It was previously reported that an *E. coli recA1* mutant did not exhibit any apparent alterations in chemotaxis [53]. Nevertheless, this *recA1* strain was not a knockout mutant, as was the *ΔrecA* mutant used in this work, in which the *recA* gene was completely removed. Furthermore, the *recA1* allele is a single amino acid missense mutation that prevents RecA recombinatorial activity [54] but still allows normal binding to ssDNA as well as ATP-independent renaturation of comple-

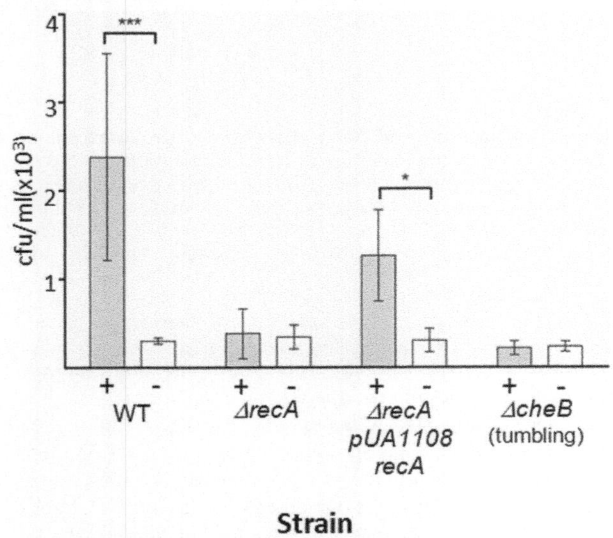

Figure 4. The chemotactic response of the *S.* Typhimurium Δ*recA* mutant is impaired. The chemotactic responses of *S.* Typhimurium wild-type (WT), Δ*recA*, Δ*recA* complemented (Δ*recA* pUA1109), and Δ*cheB* (tumbling) cells were assessed using Adler's capillary assay [30] with the modifications described in Materials and Methods. Values are expressed as the number of viable cells (in cfu/ml) in a capillary tube containing either 10 mM aspartate (+) or tethering buffer alone (−). The results are the mean of five independent experiments of three capillaries each. Error bars indicate the standard deviation. ***$p<0.001$ and *$p<0.05$ as determined by two-way ANOVA with Bonferroni correction.

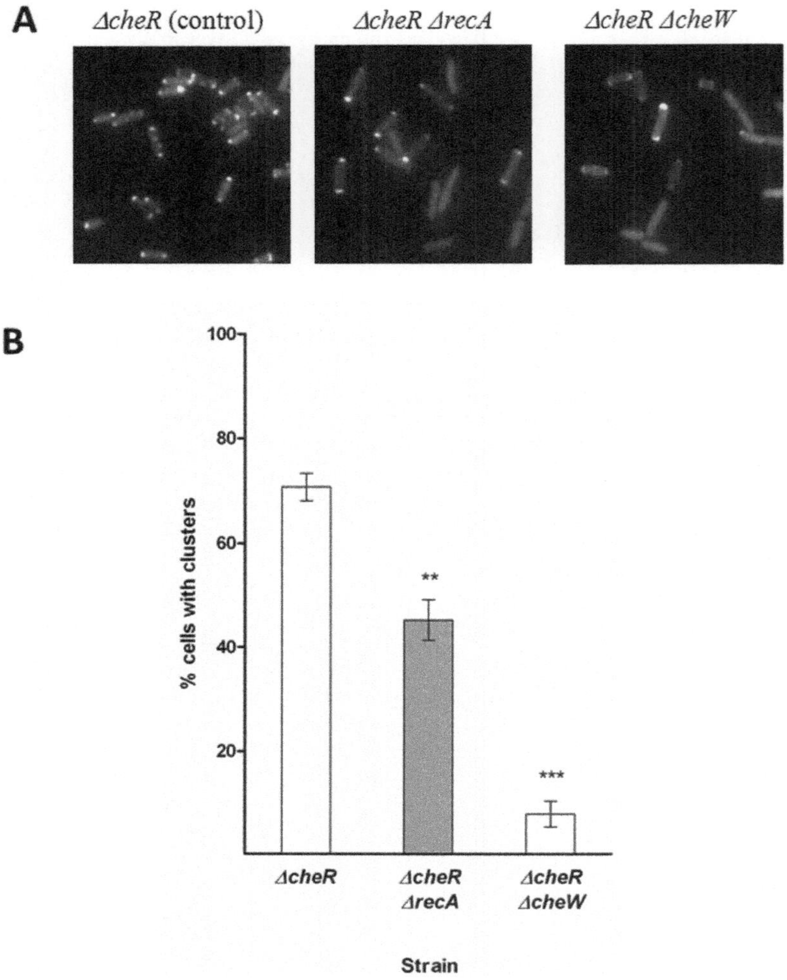

Figure 5. The formation of polar chemoreceptor clusters is altered in the absence of RecA protein. A) A representative fluorescence microscopy image of the *ΔcheR ΔrecA* strain harboring plasmid pUA1127, containing the inducible eYFP::*cheR* fusion. Images of the *ΔcheR* and *ΔcheRΔcheW* strains containing the gene fusion were also included as positive and negative controls of polar chemoreceptor cluster structuring, respectively. **B**) The fraction of cells with well-structured polar chemoreceptor clusters. The percentage of cells showing polar, round, and diffraction-limited spots (previously referred to as clusters; [43]) was quantified in each strain. The results are the mean of three independent imaging experiments. Error bars represent standard deviation. ***$p<0.001$ and *$p<0.01$ as determined by one-way ANOVA with Bonferroni correction.

mentary ssDNA molecules [55]. Thus, the results obtained with the two mutants cannot be compared.

Nevertheless, how RecA modulates flagellar rotation was unclear. In an earlier study, the absence of RecA had no effect on the expression of genes involved in either flagellar biosynthesis or chemotaxis [17], as shown for other proteins such as H-NS, which is required not only for flagellar motor function but also for flagellar biogenesis [56]. The direct association of RecA with the CheW coupling protein led us to ask whether RecA, like CheW, plays a role in the architecture of chemoreceptor arrays. CheW tethers CheA kinase to the MCPs forming the MCP-CheW-CheA ternary complexes and chemoreceptor arrays, enabling MCPs to modulate CheA autokinase activity [22,57] which, in turn, controls the level of phosphorylated CheY (CheY-P). Once activated by the MCPs, phosphorylated CheA transfers its phosphoryl group to CheY. CheY-P then promotes a switch in the direction of flagellar rotation, from CCW to CW. According to our observations, the RecA protein is necessary for the formation of normal polar chemoreceptors arrays. Although its role may not be the same as that of CheW, the absence of RecA significantly reduces the polar clustering of chemoreceptors (Fig. 5). The absence of MCPs, CheW, or CheA is known to impair chemoreceptor array formation, leading cells to run constantly because of the CCW bias of their flagellar rotation [51]. Conversely, and in addition to the demonstrated effect of RecA on chemoreceptor array formation, our results show that the absence of this protein results in a CW bias, similar to that observed in *cheZ, cheR*, or *cheB* null mutants. All of these Che proteins are associated with chemotactic response adaptation: CheZ phosphorylase returns phosphorylated Che (CheY-P) to its non-phosphorylated state (CheY), and CheB methylesterase and CheR methyltransferase control the MCP methylation state, adjusting it to the presence and concentration of external stimuli. Therefore, a similar function can be hypothesized for RecA in the chemotactic response adaptation. It is worth noting that although CheZ, CheR, and CheB co-localize with MCPs-CheW-CheA complexes at the cell poles [42], none of these proteins have an effect on polar cluster formation [43], unlike RecA. While further work is needed to determine the exact role of RecA in the chemotaxis pathway, our results clearly reveal previously unknown functions of RecA: its involvement in the control or modulation of flagellar rotation, and thus not only with swarming but also with swimming and chemotaxis, and its role in the architecture of polar chemoreceptor arrays.

Supporting Information

Table S1 Oligonucleotides used in this work.

File S1 ZIP File containing the entire set of images used to determine the presence and type of clusters found in the *S.* Typhimurium UA1910 (*ΔcheR*) strain carrying the pUA1127 (eYFP::*cheR*) plasmid.

File S2 ZIP File containing the entire set of images used to determine the presence and type of clusters found in the *S.* Typhimurium UA1915 (*ΔcheRΔcheW*) strain carrying the pUA1127 (eYFP::*cheR*) plasmid.

File S3 ZIP File containing the entire set of images used to determine the presence and type of clusters found in the *S.* Typhimurium UA1913 (*ΔcheRΔrecA*) strain carrying the pUA1127 (eYFP::*cheR*) plasmid.

Acknowledgments

IN MEMORIAM

We mourn the loss of Dmitri Petrov (1947–2014) a great colleague and friend.

Author Contributions

Conceived and designed the experiments: SC AM JB DP RS. Performed the experiments: AM OI IAM FM SC. Analyzed the data: SC AM OI JB DP IAM FM RS. Contributed reagents/materials/analysis tools: JB DP RS TK JAR. Contributed to the writing of the manuscript: SC JB.

References

1. Barbe J, Villaverde A, Cairo J, Guerrero R (1986) ATP hydrolysis during SOS induction in Escherichia coli. J Bacteriol 167: 1055–1057.
2. Bedale WA, Cox M (1996) Evidence for the coupling of ATP hydrolysis to the final (extension) phase of RecA protein-mediated DNA strand exchange. J Biol Chem 271: 5725–5732.
3. Moran NA, Wernegreen JJ (2000) Lifestyle evolution in symbiotic bacteria: insights from genomics. Trends Ecol Evol 15: 321–326.
4. Tamas I, Klasson L, Canback B, Naslund AK, Eriksson AS, et al. (2002) 50 million years of genomic stasis in endosymbiotic bacteria. Science 296: 2376–2379.
5. Eisen JA (1995) The RecA protein as a model molecule for molecular systematic studies of bacteria: comparison of trees of RecAs and 16S rRNAs from the same species. J Mol Evol 41: 1105–1123.
6. Roca AI, Cox MM (1997) RecA protein: structure, function, and role in recombinational DNA repair. Prog Nucleic Acid Res Mol Biol 56: 129–223.
7. Cox MM (1999) Recombinational DNA repair in bacteria and the RecA protein. Prog Nucleic Acid Res Mol Biol 63: 311–366.
8. Lusetti SL, Cox MM (2002) The bacterial RecA protein and the recombinational DNA repair of stalled replication forks. Annu Rev Biochem 71: 71–100.
9. Sassanfar M, Roberts JW (1990) Nature of the SOS-inducing signal in Escherichia coli. The involvement of DNA replication. J Mol Biol 212: 79–96.
10. Jiang Q, Karata K, Woodgate R, Cox MM, Goodman MF (2009) The active form of DNA polymerase V is UmuD'(2)C-RecA-ATP. Nature 460: 359–363
11. Roberts JW, Roberts CW (1975) Proteolytic cleavage of bacteriophage lambda repressor in induction. Proc Natl Acad Sci U S A 72: 147–151.
12. Sauer RT, Ross MJ, Ptashne M (1982) Cleavage of the lambda and P22 repressors by recA protein. J Biol Chem 257: 4458–4462.
13. Eguchi Y, Ogawa T, Ogawa H (1988) Cleavage of bacteriophage phi 80 CI repressor by RecA protein. J Mol Biol 202: 565–573.
14. Patel M, Jiang Q, Woodgate R, Cox MM, Goodman MF (2010) A new model for SOS-induced mutagenesis: how RecA protein activates DNA polymerase V. Crit Rev Biochem Mol Biol 45: 171–184.
15. Reuven NB, Arad G, Maor-Shoshani A, Livneh Z (1999) The mutagenesis protein UmuC is a DNA polymerase activated by UmuD', RecA, and SSB and is specialized for translesion replication. J Biol Chem 274: 31763–31766.
16. Keyamura K, Sakaguchi C, Kubota Y, Niki H, Hishida T (2013) RecA protein recruits structural maintenance of chromosomes (SMC)-like RecN protein to DNA double-strand breaks. J Biol Chem 288: 29229–29237.
17. Gomez-Gomez JM, Manfredi C, Alonso JC, Blazquez J (2007) A novel role for RecA under non-stress: promotion of swarming motility in Escherichia coli K-12. BMC Biol 5: 14.
18. Medina-Ruiz L, Campoy S, Latasa C, Cardenas P, Alonso JC, et al. (2010) Overexpression of the recA gene decreases oral but not intraperitoneal fitness of Salmonella enterica. Infect Immun 78: 3217–3225.
19. Harshey RM (2003) Bacterial motility on a surface: many ways to a common goal. Annu Rev Microbiol 57: 249–273.
20. Kearns DB (2010) A field guide to bacterial swarming motility. Nat Rev Microbiol 8: 634–644.
21. Henrichsen J (1972) Bacterial surface translocation: a survey and a classification. Bacteriol Rev 36: 478–503.
22. Cardozo MJ, Massazza DA, Parkinson JS, Studdert CA (2010) Disruption of chemoreceptor signalling arrays by high levels of CheW, the receptor-kinase coupling protein. Mol Microbiol 75: 1171–1181.
23. Arifuzzaman M, Maeda M, Itoh A, Nishikata K, Takita C, et al. (2006) Large-scale identification of protein-protein interaction of Escherichia coli K-12. Genome Res 16: 686–691.
24. Burkart M, Toguchi A, Harshey RM (1998) The chemotaxis system, but not chemotaxis, is essential for swarming motility in Escherichia coli. Proc Natl Acad Sci U S A 95: 2568–2573.
25. Wang Q, Suzuki A, Mariconda S, Porwollik S, Harshey RM (2005) Sensing wetness: a new role for the bacterial flagellum. EMBO J 24: 2034–2042.
26. Manson MD (1992) Bacterial motility and chemotaxis. Adv Microb Physiol 33: 277–346.
27. Partridge JD, Harshey RM (2013) More than motility: Salmonella flagella contribute to overriding friction and facilitating colony hydration during swarming. J Bacteriol 195: 919–929.
28. Mariconda S, Wang Q, Harshey RM (2006) A mechanical role for the chemotaxis system in swarming motility. Mol Microbiol 60: 1590–1602.
29. Martinez IA, Campoy S, Tort M, Llagostera M, Petrov D (2013) A simple technique based on a single optical trap for the determination of bacterial swimming pattern. PLoS One 8: e61630.
30. Adler J (1973) A method for measuring chemotaxis and use of the method to determine optimum conditions for chemotaxis by Escherichia coli. J Gen Microbiol 74: 77–91.
31. Datsenko KA, Wanner BL (2000) One-step inactivation of chromosomal genes in Escherichia coli K-12 using PCR products. Proc Natl Acad Sci U S A 97: 6640–6645.
32. Green MR, Sambrook J, Sambrook J (2012) Molecular cloning: a laboratory manual. Cold Spring Harbor, N.Y.: Cold Spring Harbor Laboratory Press.
33. Chaveroche MK, Ghigo JM, d'Enfert C (2000) A rapid method for efficient gene replacement in the filamentous fungus Aspergillus nidulans. Nucleic Acids Res 28: E97.
34. Campoy S, Jara M, Busquets N, Perez De Rozas AM, Badiola I, et al. (2002) Role of the high-affinity zinc uptake znuABC system in Salmonella enterica serovar typhimurium virulence. Infect Immun 70: 4721–4725.
35. Davis RW, Bostein D, Roth JR (1980) Advanced bacterial genetics. A manual for genetic engineering. In: Cold Spring Harbor Laboratory Press CSH, NY, editor.
36. Borloo J, De Smet L, Vergauwen B, Van Beeumen JJ, Devreese B (2007) A beta-galactosidase-based bacterial two-hybrid system to assess protein-protein interactions in the correct cellular environment. J Proteome Res 6: 2587–2595.
37. Miller JH (1972) Experiments in Molecular Genetics: Cold Spring Harbor Laboratory Press, Cold Spring Harbor, NY.
38. D'Ulisse V, Fagioli M, Ghelardini P, Paolozzi L (2007) Three functional subdomains of the Escherichia coli FtsQ protein are involved in its interaction with the other division proteins. Microbiology 153: 124–138.
39. Ahmed T, Shimizu TS, Stocker R (2010) Microfluidics for bacterial chemotaxis. Integr Biol (Camb) 2: 604–629.
40. Landry MP, McCall PM, Qi Z, Chemla YR (2009) Characterization of photoactivated singlet oxygen damage in single-molecule optical trap experiments. Biophys J 97: 2128–2136.
41. Block SM, Segall JE, Berg HC (1983) Adaptation kinetics in bacterial chemotaxis. J Bacteriol 154: 312–323.
42. Sourjik V, Berg HC (2000) Localization of components of the chemotaxis machinery of Escherichia coli using fluorescent protein fusions. Mol Microbiol 37: 740–751.

43. Kentner D, Thiem S, Hildenbeutel M, Sourjik V (2006) Determinants of chemoreceptor cluster formation in Escherichia coli. Mol Microbiol 61: 407–417.
44. Brennan CA, DeLoney-Marino CR, Mandel MJ (2013) Chemoreceptor VfcA mediates amino acid chemotaxis in Vibrio fischeri. Appl Environ Microbiol 79: 1889–1896.
45. Raterman EL, Welch RA (2013) Chemoreceptors of Escherichia coli CFT073 play redundant roles in chemotaxis toward urine. PLoS One 8: e54133.
46. Altindal T, Chattopadhyay S, Wu XL (2011) Bacterial chemotaxis in an optical trap. PLoS One 6: e18231.
47. Min TL, Mears PJ, Chubiz LM, Rao CV, Golding I, et al. (2009) High-resolution, long-term characterization of bacterial motility using optical tweezers. Nat Methods 6: 831–835.
48. Koch M, Rohrbach A (2012) Object-adapted optical trapping and shape-tracking of energy-switching helical bacteria. 6: 680–686.
49. DeFranco AL, Parkinson JS, Koshland DE Jr (1979) Functional homology of chemotaxis genes in Escherichia coli and Salmonella typhimurium. J Bacteriol 139: 107–114.
50. Warrick HM, Taylor BL, Koshland DE Jr (1977) Chemotactic mechanism of Salmonella typhimurium: preliminary mapping and characterization of mutants. J Bacteriol 130: 223–231.
51. Parkinson JS (1978) Complementation analysis and deletion mapping of Escherichia coli mutants defective in chemotaxis. J Bacteriol 135: 45–53.
52. Sourjik V (2004) Receptor clustering and signal processing in E. coli chemotaxis. Trends Microbiol 12: 569–576.
53. Smith RA, Parkinson JS (1980) Overlapping genes at the cheA locus of Escherichia coli. Proc Natl Acad Sci U S A 77: 5370–5374.
54. Kawashima H, Horii T, Ogawa T, Ogawa H (1984) Functional domains of Escherichia coli recA protein deduced from the mutational sites in the gene. Mol Gen Genet 193: 288–292.
55. Bryant FR, Lehman IR (1986) ATP-independent renaturation of complementary DNA strands by the mutant recA1 protein from Escherichia coli. J Biol Chem 261: 12988–12993.
56. Ko M, Park C (2000) Two novel flagellar components and H-NS are involved in the motor function of Escherichia coli. J Mol Biol 303: 371–382.
57. Liu JD, Parkinson JS (1989) Role of CheW protein in coupling membrane receptors to the intracellular signaling system of bacterial chemotaxis. Proc Natl Acad Sci U S A 86: 8703–8707.

8

Evaluation of Up-Converting Phosphor Technology-Based Lateral Flow Strips for Rapid Detection of *Bacillus anthracis* Spore, *Brucella* spp., and *Yersinia pestis*

Pingping Zhang[1]⁹, Xiao Liu[1,2]⁹, Chengbin Wang[3]⁹, Yong Zhao[1], Fei Hua[4], Chunfeng Li[3], Ruifu Yang[1]*, Lei Zhou[1]*

1 Laboratory of Analytical Microbiology, State Key Laboratory of Pathogen and Biosecurity, Beijing Institute of Microbiology and Epidemiology, Beijing, People's Republic of China, **2** School of Public Health and Management, Chongqing Medical University, Chongqing, People's Republic of China, **3** Department of Clinical Laboratory, Chinese People's Liberation Army General Hospital, Beijing, People's Republic of China, **4** Department of Etiology, Taishan Medical University, Tai'an, People's Republic of China

Abstract

Bacillus anthracis, *Brucella* spp., and *Yersinia pestis* are zoonotic pathogens and biowarfare- or bioterrorism-associated agents that must be detected rapidly on-site from various samples (e.g., viscera and powders). An up-converting phosphor technology-based lateral flow (UPT–LF) strip was developed as a point-of-care testing (POCT) to satisfy the requirements of first-level emergency response. We developed UPT–LF POCT to quantitatively detect the three pathogens within 15 min. Sample and operation-error tolerances of the assay were comprehensively evaluated. The sensitivity of UPT–LF assay to bacterial detection reached 10^4 cfu·mL^{-1} (100 cfu/test), with a linear quantitative range of 4 to 6 orders of magnitude. Results revealed that the UPT–LF assay exhibited a high specificity with the absence of false-positive results even at 10^9 cfu·mL^{-1} of non-specific bacterial contamination. The assay could tolerate samples with a wide pH range (2 to 12), high ion strengths (≥4 mol·L^{-1} of NaCl), high viscosities (≤25 mg·mL^{-1} of PEG20000 or ≥20% of glycerol), and high concentrations of bio-macromolecule (≤200 mg·mL^{-1} of bovine serum albumin or ≥80 mg·mL^{-1} of casein). The influence of various types of powders and viscera (fresh and decomposed) on the performance of UPT–LF assay was determined. The operational error of liquid measurement exhibited few effects on sensitivity and specificity. The developed UPT–LF POCT assay is applicable under field conditions with excellent tolerance to sample complexity and operational error.

Editor: Mónica V. Cunha, INIAV, I.P.- National Institute of Agriculture and Veterinary Research, Portugal

Funding: This study was supported by the National Natural Science Foundation of China (Grant No. 81000774), the National High Technology Research and Development Program of China (Grant No. 2013AA032205) and the Major National Science and Technology Programs of China (Grant Nos. 2011ZX10004 and 2012ZX1004801). The funders had no role in study design, data collection and analysis, decision to publish, or preparation of the manuscript.

Competing Interests: The authors have declared that no competing interests exist.

* Email: ammszhoulei@aliyun.com (LZ); ruifuyang@gmail.com (RY)

⁹ These authors contributed equally to this work.

Introduction

Anthrax, brucellosis, and plague, which are caused by *Bacillus anthracis*, *Brucella* spp., and *Yersinia pestis*, respectively, are infectious zoonotic diseases with high mortality rates. These diseases have persisted in several countries with a wide range of animal hosts, susceptible animals, vectors, or soils as reservoirs in nature foci [1–3]. Moreover, anthrax, brucellosis, and plague have become threats to public health because their corresponding pathogens are easily transmitted from contaminated animal products or environment materials to humans through gastrointestinal, respiratory, and integumentary systems [1–4]. The pathogens have been used in biological warfare (biowarfare) and biological terrorism (bioterrorism) because of their wide distribution, simple culture needs, and aerosol dissemination capability [2,5–7]. Pathogen spread by human-to-human respiratory transmission has worsened the problem [2,8].

Point-of-care testing (POCT) assays are necessary for surveillance in nature foci and first-level emergency responses in biowarfare and bioterrorism through on-site detection of the pathogens (e.g., viscera and decomposed viscera, blood, excretion, soil, water, and food). The simplicity and error tolerance of detection methods should also be considered for nonprofessional operations. Polymerase chain reaction (PCR) [4] and other techniques [9–13] have been developed, but these methods cannot be used as POCT assays because of expensive equipment, complex sample pretreatment, and complicated and time-consuming operations. Lateral flow (LF) immunoassay is a widely used and comprehensively applied POCT technology [14,15] to rapidly detect various pathogens, such as *Y. pestis* [16], *B. anthracis* [17], and *Brucella* spp. [18]. However, traditional LF assays that use colloidal gold as label have the following limitations under potential bioterrorism scenarios with mixed bacteria and complex samples: (i) absence of reactions with different strains of the same pathogen species, (ii) low sensitivities caused by macroscopic observations, (iii) results cannot be analyzed because of serious interferences by colored samples such as whole blood [16], and (iv) inapplicability in extreme conditions (e.g., acid, basic, saline, viscose, and protein-rich solutions) because of the interference in conjugation between antibodies and gold particles that depend on physical adsorption. To simultaneously detect different strains of the same pathogen species, specific antigens are used to prepare antibodies, and two or more antibodies are mixed in LF assays to efficiently recognize species of *B. anthracis* [19], *Brucella* [20,21], and *Y. pestis* [16,22] under field conditions. However, only a few works have focused on the rapid detection of pathogens from complex samples using LF assay to meet the requirements of POCT diagnostic and system assessments using extreme reagents and on-site samples. The rarity of studies on LF assay-based pathogen detection is attributed to the limitation of colloidal gold as a biolabel.

An up-converting phosphor technology-based lateral flow (UPT–LF) assay was developed with at least 10-fold improvement using a nanometer-sized luminous up-converting phosphor (UCP) particle as the biolabel [23]. Given the unique optical characteristics of UCP and the covalent bond between UCP and biomolecules, UPT–LF assay exhibits robust performance to detect blood [24–26] and saliva [27] samples. UPT–LF assays have also been successfully used to quantitatively detect bacteria [28–30], viruses [31], biomolecules [25,32], and drugs of abuse [28]. The assays can potentially be applied to POCT diagnostics upon improvement because of their less background noise, high sensitivity, low-cost operation, and stable photochemical properties [26,33].

This study aims to improve the UPT–LF assay for point-of-care applications. The assay performance was comprehensively evaluated to rapidly detect *B. anthracis*, *Brucella* spp., and *Y. pestis* under the interference of various factors. With this method, surveillance in nature foci and first-level emergency response in bioterrorism can be performed.

Materials and Methods

Ethic statement

Eight-week-old female Balb/c mice were obtained from the Laboratory Animal Research Center, Academy of Military Medical Sciences (China). Mice acquisition was given license by the Ministry of Health in the General Logistics Department of Chinese People's Liberation Army (Permit No. SCXK-2007-004). All experiments were approved by the Committee of the Welfare and Ethics of Laboratory Animals, Beijing Institute of Microbiology and Epidemiology (Beijing, China). All mice were housed in an accredited research animal facility fully staffed with trained personnel. Humane endpoints were strictly observed. Balb/c mice were humanely culled by cervical dislocation.

Reagents and materials

UCP ($NaYF_4$:Yb^{3+},Er^{3+}) with a diameter of approximately 50 nm was prepared and provided by Dr. Yan Zheng from Shanghai Kerune Phosphor Technology Co., Ltd. (Shanghai, China). The excitation and emission spectrum peaks of UCP were 980 nm and 541.5 nm, respectively.

Nitrocellulose membrane (SHF 1350225) and glass fiber (GFCP20300) were obtained from Millipore Corp. (Bedford, MA, USA). Papers (Nos. 470 and 903) were purchased from Schleicher & Schuell, Inc. (Keene, NH, USA). Plastic cartridges and laminating cards were manufactured by Shenzhen Jincanhua Industry Co. (Shenzhen, China) and Shanghai Liangxin Biotechnology Co. (Shanghai, China), respectively; both materials were designed by our group.

Agar powder, bovine serum albumin V (BSA), casein, tryptone, yeast extract powder, $Ca(NO_3)_2$, HCl, $FeSO_4$, glycerin, $MgSO_4$, $MnCl_2$, PEG20000, KCl, NaN_3, NaCl, and NaOH were all purchased from Sigma–Aldrich (St. Louis, MO, USA). All other reagents were of analytical grade, used without further purification, and supplied by Sinopharm Chemical Reagent Co., Ltd. (Shanghai, China) unless otherwise specified. Acid instant drink, alkaline putty powder, flour, gourmet powder, milk powder, and sugar were purchased from a local supermarket. Viscera of Balb/c mouse, including heart, liver, lung and spleen, were homogenized and divided into two parts. One part was stored at $-20°C$ as fresh specimens, and the other was incubated at 37°C for two weeks as decomposed specimens.

Bacterial cultures and monoclonal antibody preparation

B. anthracis, *B. melitensis* M55009, and *Y. pestis* were used for specific detection. The following bacteria were used to evaluate the specificity: *Bacillus atrophaeus* (two isolates), *B. cereus* (three isolates), *B. subtilis* (three isolates), *B. thuringiensis* (two isolates), *Escherichia coli* O157:H7, *Salmonella choleraesuis*, *S. enteritidis*, *S. paratyphi* A, *S. paratyphi* B, *S. paratyphi* C, *S. typhi*, *S. typhimurium*, *Vibrio cholerae* O1, *V. cholerae* O139, *Y. aldovae*, *Y. enterocolitica*, *Y. intermedia*, *Y. kristensenii*, *Y. mollaretii*, *Y. pseudotuberculosis*, *Y. rohdei*, and *Y. ruckeri*.

To prepare the spores, *B. anthracis*, *B. atrophaeus* (two isolates), *B. cereus* (three isolates), *B. subtilis* (three isolates), and *B. thuringiensis* (two isolates) were grown on nutrient agar plates at 37°C for 7 d. The plates contained 6 $g \cdot L^{-1}$ tryptone, 3 $g \cdot L^{-1}$ yeast extract powder, 10 $g \cdot L^{-1}$ NaCl, 1 $g \cdot L^{-1}$ KCl, 0.122 $g \cdot L^{-1}$ $MgSO_4$, 0.23 $g \cdot L^{-1}$ $Ca(NO_3)_2$, 0.197 $g \cdot L^{-1}$ $MnCl_2$, 0.0002 $g \cdot L^{-1}$ $FeSO_4$, and 15 $g \cdot L^{-1}$ agar powder. Vegetative cells were placed in sterilized distilled water and lysed by osmotic pressure. The spores were collected by scraping the cells into sterilized saline and pelleted at 7,000 rpm for 15 min. The pellets were washed thrice through resuspension and centrifugation in sterilized saline. The resultant spores exhibited 99% purity through microscopic observation; spore population was determined by Luria–Bertani (LB) agar plate count. *V. cholerae* and other bacteria were cultured in alkaline peptone water and LB broth at 37°C, respectively. Pure cultures were collected at the logarithmic phase through centrifugation and resuspension in sterilized saline. The population was determined by LB plate count.

To prepare monoclonal antibodies (mAb), eight-week-old female Balb/c mice were immunized every two weeks for two months by subcutaneous injection. To prepare mAb against *Y. pestis*, each mouse was injected with 25 μg F1 antigens [30]. Subsequently, 2.5×10^8 cfu of Sterne strain, a vaccine strain of *B. anthracis*, was inactivated by 4% formaldehyde solution for immunization. Several mAbs were obtained following preliminary preparation. The antibodies were then screened by ELISA coated by Sterne and 170044 stains. The mAbs positive for the two strains in the ELISA assay were chosen [34]. To prepare mAbs against *Brucella*, 2.5×10^8 cfu of three strains, namely, *B. abortus* S19, *B. suis* S1, and *B. melitensis* M55009 strains, was mixed and inactivated for immunization. Then, preliminary mAbs were screened by ELISA assay coated by the three strains [29].

Establishment of UPT–LF assay

UPT–LF strips were prepared as previously described [29–31] to detect *B. anthracis*, *Brucella spp.*, and *Y. pestis*, and the strips were named Ban-UPT–LF, Bru-UPT–LF, and Ype-UPT–LF, respectively. The mAb coating corresponding to each bacterium (1.0 mg/mL) and goat anti-mouse IgG (2 mg/mL) was dispensed on a nitrocellulose membrane at a rate of 1 μL/cm as the test (T) and control (C) lines, respectively. Subsequently, UCP-conjugating mAb conjugate (1 mg/mL, 30 μL/cm) was fixed in the glass fiber as the conjugate pad. The conjugate pad and membrane prepared by various candidate mAbs were paired; the optimized pair was selected for strip fabrication. The assembled strip was placed in the cartridge, with the sample-adding and result-scanning windows located above the sample pad and nitrocellulose membrane, respectively (Figure 1). During testing, the sample was initially diluted 10 times with the sample-treating buffer [0.03 M phosphate buffer (PB) that contains 1% BSA, 0.5% SDS, 0.1 M NaCl, and 0.01% NaN_3, pH 7.2]. Exactly 100 µL of the diluted sample was applied to each strip. After 15 min, the strip was scanned with a UPT biosensor (UPT-3A) to obtain the results.

UPT-3A is similar to a previously described single-channel biosensor [30], which is equipped with a laser diode to emit 980 nm excited light and a photomultiplier tube to transform a 541.5 nm UCP-sourced emitted light to an electric signal (Figure 1). UPT-3A with an upgraded design was used to improve operational convenience. For the hardware, integrated keys were fabricated on the front panel to replace the mouse and keyboard. A platform was installed in the biosensor to settle the strips. When the "measure" key (all keys were in Chinese) was first pressed, the platform was stretched out, and the strips were settled on the platform. The button was pressed again, and the platform was drawn back and began to scan automatically. A radiofrequency identification device (RFID) was installed to automatically receive all information on a specific test from the built-in RFID chip of the UPT–LF strip (Figure 1). The information included the test item, cut-off threshold, quantitation parameters, and expiry date of the strip. For the software, a mathematical algorithm was designed to automatically fix the highest point and two inflection points of a signal peak. The integral value of that area was determined, and the value was used as the specific signal of the immunological reaction. Peak areas corresponding to the test and control lines were referred to as T and C values, respectively; the ratio of T/C

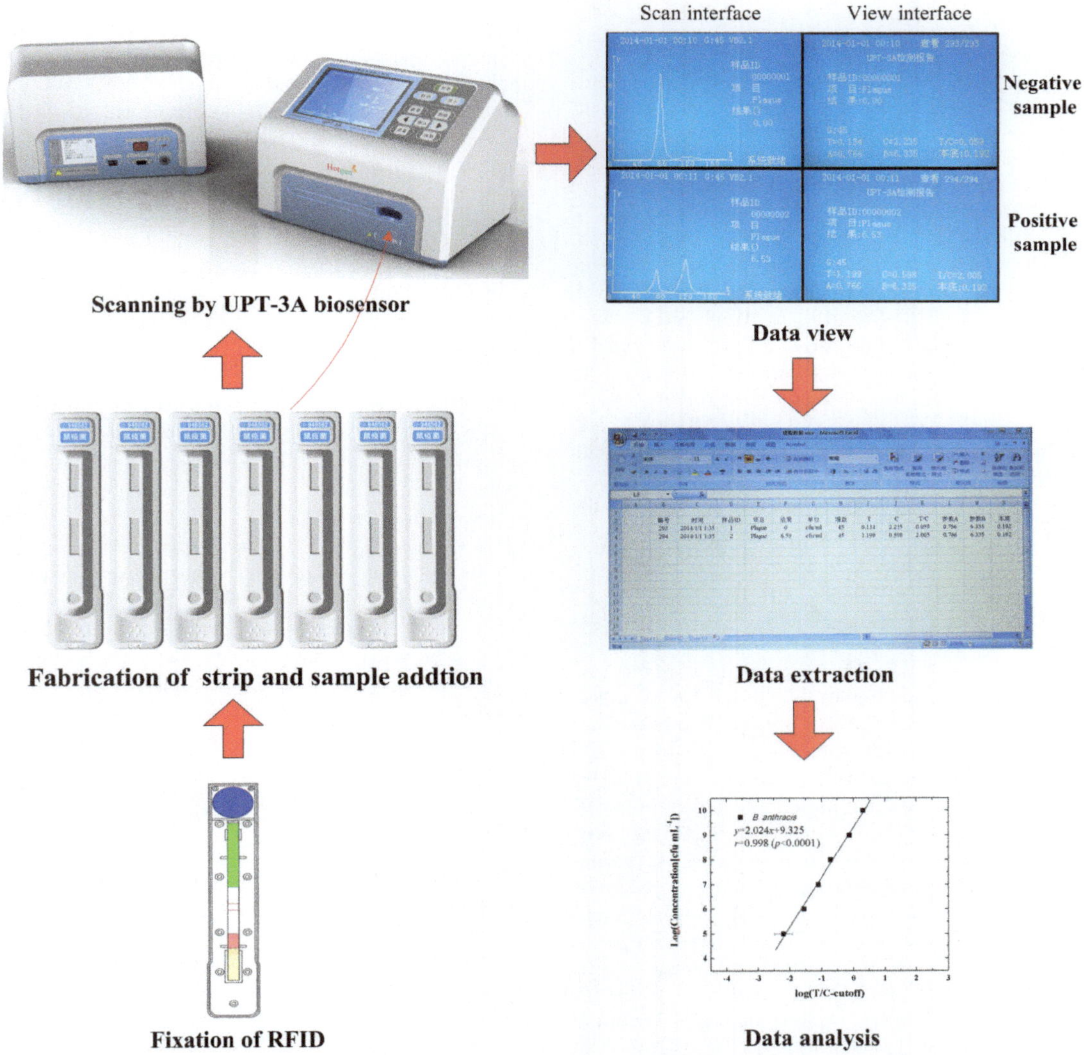

Figure 1. Workflow for pathogen detection using the UPT–LF biosensor. UPT–LF strips fixed with RFID are fabricated first. The samples are added to each strip every 30 s. Afterwards, the strips are scanned by an upgraded UPT-3A biosensor. The dimension of the upgraded UPT biosensor is 32 cm×21 cm×20 cm. The power sockets, powder switch, and USB interface are installed on the rear panel (left), and the integrated keys are installed on the front panel (right). When the "measure" key (marked in Chinese) at the top of panel is pressed, the scanning image is shown in the window. After the "view" key is pressed, the results are shown in detail. The data are exported for analysis using Excel through the USB interface to perform data analysis.

was the result of the measurement [23,35]. Samples with T/C ratios higher than the cutoff threshold (mean +3SD of T/C ratios corresponding to the blank) were considered positive and vice versa. Scanning and analysis were accomplished within 30 s by pressing a single key. To detect a series of samples, multiple strips were arranged in parallel, and the samples were added to the corresponding strip every 30 s. Each strip could be scanned following the addition of the sample for 15 min.

Evaluation of sensitivity, linearity, and precision

PB was used as the blank sample and tested 10 times by each type of UPT–LF strip. Subsequently, the mean +3SD of T/C values were set as the cut-off thresholds that corresponded to the detected bacteria. Standard samples of *B. anthracis* spore, *B. melitensis* M55009, and *Y. pestis*, with concentrations ranging from 10^3 cfu·mL^{-1} to 10^{10} cfu·mL^{-1}, were prepared by serial dilution with PB. Each sample was detected in triplicate by the corresponding strip.

Specificity evaluation

The specificity of the Ban-UPT–LF strip was determined using 12 bacterial strains [*B. atrophaeus* (two isolates) spore, *B. cereus* (three isolates) spore, *B. subtilis* (three isolates) spore, *B. thuringiensis* (two isolates) spore, *B. melitensis* M55009, and *Y. pestis*] with concentrations ranging from 10^5 cfu·mL^{-1} to 10^9 cfu·mL^{-1}. Thirteen bacterial strains (*E. coli* O157:H7, *S. choleraesuis*, *S. enteritidis*, *S. paratyphi* A, *S. paratyphi* B, *S. paratyphi* C, *S. typhi*, *S. typhimurium*, *V. cholerae* O1, *V. cholerae* O139, *Y. enterocolitica*, *B. anthracis* spore, and *Y. pestis*), with concentrations ranging from10^6 cfu·mL^{-1} to 10^9 cfu·mL^{-1}, were used to evaluate the specificity of the Bru-UPT–LF strip. Finally, the specificity of Ype-UPT–LF was determined by using 10 bacterial strains (*Y. aldovae*, *Y. enterocolitica*, *Y. intermedia*, *Y. kristensenii*, *Y. mollaretii*, *Y. pseudotuberculosis*, *Y. rohdei*, *Y. ruckeri*, *B. anthracis* spore, and *B. melitensis* M55009), with concentrations ranging from 10^4 cfu·mL^{-1} to 10^9 cfu·mL^{-1}. Specific targets with the same concentrations were set as the positive controls.

Evaluation of sample tolerance

Evaluation of single-factor theoretical tolerance. Immunological assays based on the capillarities of microporous materials cause the pH, ion strength, viscosity, and biological matrices (biomacromolecules) of the sample to noticeably interfere with the accuracy of the test. pH tolerance was determined by HCl and NaOH solutions, and ion strength tolerance was obtained from KCl–NaCl saline solutions. Glycerol and PEG20000 were selected as the representative micromolecules and macromolecules, respectively, to change the liquid viscosity. BSA and casein were selected as the representatives of serum-derived albumin and milk-derived phosphoproteins, respectively, to block immune reactions as biological matrices. All solutions were prepared with distilled water. Specific concentration points for interference reagents were determined according to the performance of each UPT–LF strip. *B. anthracis* spore, *B. melitensis* M55009, and *Y. pestis* were spiked into each solution of a specific reagent with a final concentration of 10^5 cfu·mL^{-1} to 10^9 cfu·mL^{-1}, 10^5 cfu·mL^{-1} to 10^9 cfu·mL^{-1}, and 10^4 cfu·mL^{-1} to 10^8 cfu·mL^{-1}, respectively. Each simulated sample was tested by the corresponding UPT–LF strip in triplicate.

Evaluation of real sample tolerance. Various kinds of powder and viscera were selected as potential samples for anti-bioterrorism and zoonotic investigation. Seven kinds of powder, namely, flour, fruit juice, gourmet powder, milk powder, putty powder, soil, and sucrose, were combined with 0.03 mol·L^{-1} of PB to obtain solutions. Decomposed and fresh viscera of Balb/c mice were ground and homogenized in PB. The concentration of each solution was high and nearly close to the saturation point. *B. anthracis* spore, *B. melitensis* M55009, and *Y. pestis* were spiked into the aforementioned specimens at 10^5 cfu·mL^{-1}, 10^5 cfu·mL^{-1} to 10^6 cfu·mL^{-1}, and 10^4 cfu·mL^{-1}, respectively. Each simulated sample was tested thrice using the corresponding UPT–LF strip.

Evaluation of operation error tolerance

The standard operation was set as the control. Exactly 10 μL of the sample was mixed with 90 μL of the sample-treating buffer to obtain 100 μL of loading mixture, which was applied per strip. Each step of the liquid measurement procedure was analyzed to evaluate the operation error tolerance. To evaluate the tolerance of sample measure error, sample and sample-treating buffer were mixed at ratios of 5:90, 20:90, and 30:90, and then 100 μL of the loading mixture was applied per strip. To evaluate the tolerance of the sample-treating buffer measure error, sample and sample-treating buffer were mixed at ratios of 10:70, 10:110, and 10:130, and then 100 μL of the loading mixture was applied per strip. To evaluate the tolerance of the loading mixture measure error, sample and sample-treating buffer were mixed at standard ratios of 10:90, and 70, 110, and 130 μL of the loading mixtures were applied per strip, respectively. *B. anthracis* spore (10^5 cfu·mL^{-1}), *B. melitensis* M55009 (10^6 cfu·mL^{-1}), and *Y. pestis* (10^4 cfu·mL^{-1}) were used as positive samples. Each sample was tested thrice using the corresponding UPT–LF strip.

Results and Discussion

Sensitivity, linearity, and precision of the UPT–LF assay

Performance of the UPT–LF assay was initially evaluated using bacteria diluted in PB. The results gathered from the performance evaluation were used as standard in the evaluation of different interference factors. Sensitivity was set to the lowest concentration of dilution, with the T/C ratio higher than the cut-off threshold. For the detection of *B. anthracis* spore, *Brucella* spp., and *Y. pestis*, the sensitivity of the UPT–LF assay was 10^5, 10^6, and 10^4 cfu·mL^{-1}, respectively (Figure 2). The correlation coefficient (r) of the linear correlation analysis and the formula of the linear regression analysis, with the logarithm of T/C-cutoff as x and the logarithm of concentration as y, were used to evaluate linearity and achieve bacteria quantitation, respectively. The r of the linear correlation analysis was larger than 0.99, with linear quantitative values ranging from 10^5 cfu·mL^{-1} to 10^{10} cfu·mL^{-1}, 10^6 cfu·mL^{-1} to 10^9 cfu·mL^{-1}, and 10^4 cfu·mL^{-1} to 10^8 cfu·mL^{-1} for *B. anthracis* spores, *Brucella* spp. and *Y. pestis*, respectively. The coefficients of the variation in the three tests corresponding to each sample were all less than 15%, guaranteeing the quantitation precision.

In this study, SDS was used as the surface-active agent in the sample-treating buffer to reduce the non-specific reaction by improving the hydrophilicity of biological macromolecules and also to facilitate the release of UCP particles from the conjugate pad. The covalent bonds between UCP particles and biological macromolecules are stable and free from the negative influence of SDS. The SDS concentration must be carefully optimized to reduce its potential negative influence. The bacteriolytic effect of SDS may also slightly improve the sensitivity of the test by increasing the amount of dissociative antigenic determinants.

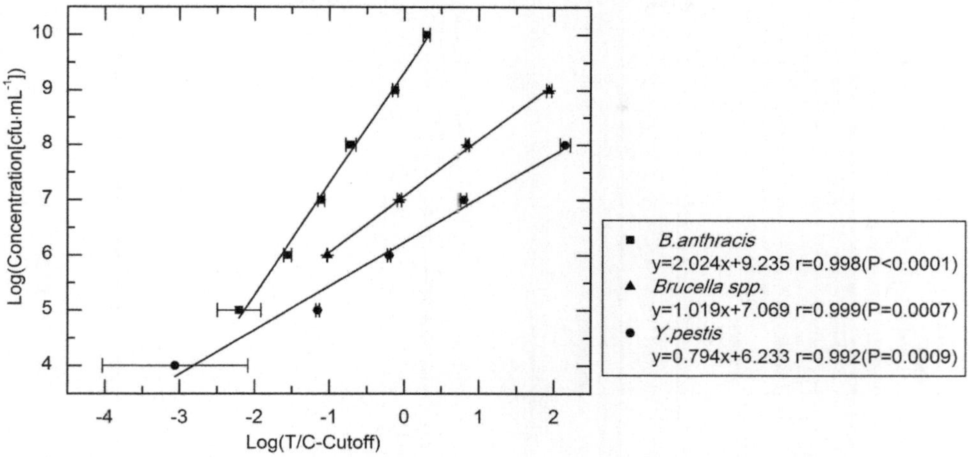

Figure 2. Sensitivity, linearity, and precision of the UPT–LF assay. For the detection of *B. anthracis* spore, *Brucella* spp., and *Y. pestis*, the sensitivity of the UPT–LF assay was 10^5, 10^6, and 10^4 cfu·mL^{-1}, respectively. The coefficients of variation (CV) of the three tests corresponding to each sample were all less than 15%.

Specificity of the UPT–LF assay

Bacteria that shared close genetic relatedness and similar transmission routes with the target were selected to evaluate the specificity of each kind of strip. Figures 3B and C show that the Bru-UPT-LF strip and the Ype-UPT-LF strip had high specificity and did not produce false-positive results even when the concentration of nonspecific bacteria was very high (10^9 cfu·mL^{-1}). However, the specificity of the Ban-UPT-LF strip was slightly disappointing (Figure 3A). Although significant differences were found between specific and nonspecific signals, *B. cereus* (isolate No. 1) and *B. subtilis* (isolate No. 2) spores began to cause a false-positive result at 10^7 cfu·mL^{-1}. No cross reaction occurred when the contaminating non-specific bacteria were at a lower concentration. The cross-reaction was caused by the monomorphic nature of *Bacillus* spp. that creates spore cell surface antigens [36] highly similar to those of some strains of *B. cereus*. Thus, the strains were even indistinguishable by PCR based on plasmids *pXO1* and *pXO2* encoding the toxin and capsule antigen, respectively [37].

Sample tolerance of the UPT–LF assay

B. anthracis spore, *Brucella* spp., and *Y. pestis* are pathogens with dual identities, namely, zoonotic pathogens and biowarfare/bioterrorism agents. Given their dual identities, the targets of these species may exist in different samples (typically viscera and "white powder") with various properties, which can be roughly classified into four: pH, ion strength, viscosity, and biological matrix (bio-macromolecule). UPT–LF assay is a microporous material-based immunological reaction. pH, ion strength, and biological matrix (blocking effect of bio-macromolecules) have important functions in the specificity and sensitivity of immunological reactions. Viscosity matters remarkably change the capillarity of microporous material. In this study, two experiments were carried out. The first experiment investigated the theoretical tolerance of the UPT–LF assay to single-factor interference. The second experiment determined the actual ability of the UPT–LF assay to detect real samples, including various viscera and powders. The results corresponding to the three kinds of strips were gathered to demonstrate the performance of the UPT–LF assay.

Theoretical tolerance to single-factor interference in the model samples. The serially diluted standard samples with PB were used as the controls for various interference factors. The T/C ratios of each strip under the interference of various factors with different concentrations (corresponding to the quantitative range) did not show significant change, except 0.1 mol·L^{-1} NaOH (pH 13), which altered the immunological reaction and led to a noticeable signal decrease (data not shown). This result indicates that the holistic signal response of the UPT–LF assay is relatively stable and promises accurate quantitation.

The T/C ratios of negative and weak positive samples (with concentration reaching the sensitivity) demonstrate the specificity and sensitivity of the test under the influence of interference factors (Figure 4). The highest concentration of an interference factor that does not decrease the specificity (i.e., no false-positive result for the negative sample test) and the sensitivity (i.e., no false-negative result for the weak positive sample test) of the test is defined as the tolerance limit. The tolerance limits of the three kinds of UPT–LF strips under the influence of various interference factors were determined (Table 1).

The tolerance of the UPT–LF assay to pH ranged from 2 to 12, with extreme acid and basic backgrounds leading to false-positive and false-negative results, respectively. Ion strength showed a similar effect with NaOH solution. Saline solution with a high concentration resulted in the simultaneous decrease in the T/C ratios of negative and positive samples. This simultaneous decrease is a phenomenon that agrees well with the notion that ion strength can eliminate the nonspecific binding of immunological reaction at a proper concentration but damage the specific binding at an unsuitably high concentration. The maximum tolerated concentration (tolerance limit) of Ban-UPT-LF strip was 0.25 mol·L^{-1}, whereas that of Bru-UPT-LF strip was more than 4 mol·L^{-1}. In viscosity adjustment, the maximum tolerance limits of the UPT–LF assay to PEG20000 and glycerol were 25 mg·mL^{-1} and more than 20% (V/V), respectively. The macromolecule and the micromolecule exerted different influences on the assay. PEG20000 noticeably increased the signal, resulting in a false-positive result (i.e., destroyed the specificity of the test), whereas glycerol mainly decreased the signal. The tolerance limits of the three kinds of strip to PEG20000 were similar, but the tolerance of the three kinds of strip to glycerol was significantly different from that of Bru-UPT-LF, which was below 5%, and from that of Ban-UPT-LF, which was above 20%. As a rough representative of the

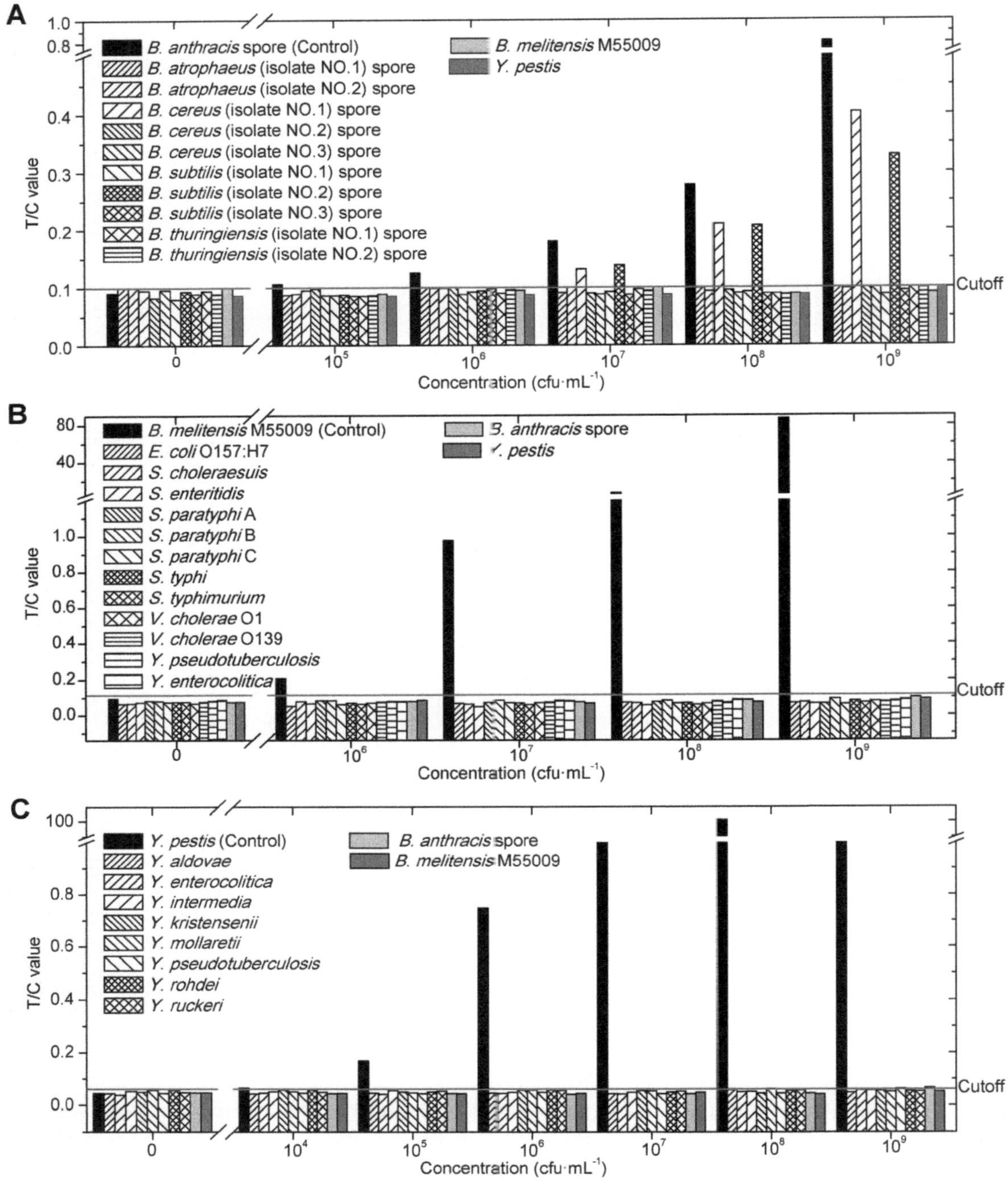

Figure 3. Specificity of the UPT–LF assay.

bio-macromolecules in the biological matrix, the influence of BSA and casein was similar to that of PEG20000 and glycerol. The former simultaneously increased nonspecific and specific binding, whereas the latter decreased them. Therefore, specificity and sensitivity were damaged. The maximum tolerance limits to BSA and casein were 200 mg·mL^{-1} and more than 80 mg·mL^{-1}, respectively.

Under the interference of various factors, the stability of the UCP optical characteristic and the UCP–biomolecule conjugate was critical. The properties of the antibodies in the UPT–LF strips were also important and especially noticeable in Bru-UPT-LF. The sensitivity and/or specificity of Ban-UPT-LF and Ype-UPT-LF were threatened by interference factors. The tolerance limits were determined as the highest concentration at which two kinds of strip could maintain the same specificity and sensitivity as the PB control. However, for Bru-UPT-LF, most interference factors (except glycerol) significantly improved the sensitivity. The sensitivity of Bru-UPT-LF changed from 10^6 cfu·mL^{-1} to 10^4 cfu·mL^{-1}, but specificity was not damaged. Many reasons can account for this phenomenon. For the Bru-UPT-LF assay

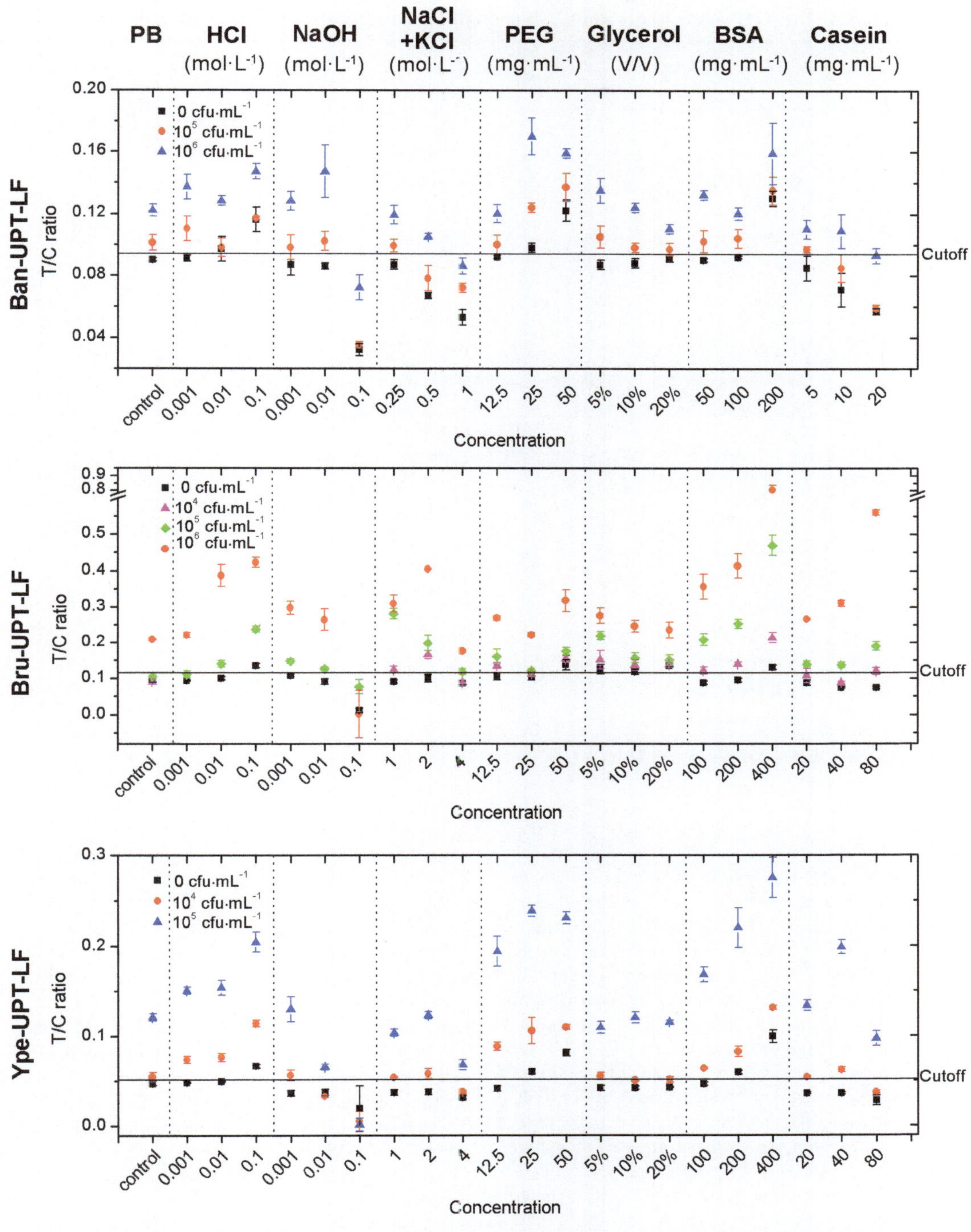

Figure 4. Tolerance of the UPT–LF strips to single-factor interference. The UPT–LF strip maintained sensitivity and specificity [T/C ratio of the positive sample (red circles) is higher than the cut-off threshold and T/C ratio of the negative sample (black squares) is lower than the cut-off threshold] and even improved the sensitivity at certain concentrations of the interference factors [T/C ratio of the positive sample (purple triangles, green diamonds) is higher than the cut-off threshold].

under the interference of NaOH, ion strength and casein, T/C ratios of the negative samples decreased, whereas those of the positive samples increased. However, the T/C ratios of both samples simultaneously decreased for Ban-UPT-LF and Ype-UPT-LF, thus generating false-negative results. For the Bru-UPT-LF assay under the interference of HCl, PEG20000, and BSA, T/C ratios of the positive samples increased, whereas those of the negative samples remained unchanged. However, the T/C ratios of both samples increased simultaneously for Ban-UPT-LF and Ype-UPT-LF, and false-positive results were generated.

Tolerance to real sample interference. The T/C ratios of the negative and weak positive samples under the influence of real

Table 1. Tolerance limits of UPT–LF assay to pH value, ion strength, viscosity, bio-macromolecule, and real samples.

Interference Factor		Unit	Ban-UPT–LF	Bru-UPT–LF	Ype-UPT–LF
pH value	HCl	mol·L^{-1}	≤0.001 (pH 3)	≤0.01 (pH 2)**	≤0.01 (pH 2)
	NaOH	mol·L^{-1}	≤0.01 (pH 12)	≤0.01 (pH 12)**	≤0.001 (pH 11)
Ion strength	NaCl+KCl	mol·L^{-1}	≤0.25	≥4**	≤2
Viscosity	PEG20000	mg·mL^{-1}	≤12.5	≤25**	≤12.5
	Glycerol	%(V/V)	≥20%	<5%	≤5%
Bio-macromolecule	BSA	mg·mL^{-1}	≤100	≤200**	≤100
	Casein	mg·mL^{-1}	≤5	≥80**	≤40
Power	Flour	mg·mL^{-1}	≤100	≥200	≤50
	Fruit juice	mg·mL^{-1}	≤100	≤50**	≤50
	Gourmet powder	mg·mL^{-1}	≥400	≥400**	≤50
	Milk powder	mg·mL^{-1}	≤25	≤200**	≥400
	Putty powder	mg·mL^{-1}	≥200	≥200**	≤50
	Soil	mg·mL^{-1}	≥400	≥400**	≥400
	Sucrose	mg·mL^{-1}	≤100	≥400**	≥400
Viscera	Fresh heart	mg·mL^{-1}	≥800	≥800	≥800
	Fresh liver	mg·mL^{-1}	≤50	≤200**	≤50
	Fresh lung	mg·mL^{-1}	≤400	≥800	≤100
	Fresh spleen	mg·mL^{-1}	≤200	≥400	≤100
	Decomposed heart	mg·mL^{-1}	≤100	≥400**	≤100
	Decomposed liver	mg·mL^{-1}	≤50	≤100**	≤200
	Decomposed lung	mg·mL^{-1}	≤100	≤200**	≤200
	Decomposed spleen	mg·mL^{-1}	≤100	≤100**	≤100

** The sensitivity of UPT-LF strip improved at that tolerance limit.

samples are shown in detail in Figure 5. The properties of real samples as a complex of interference factors resulted in the complicated influence on UPT–LF strips fabricated by antibodies with various properties. Given the mild nature of gourmet powder, putty powder, soil, and sucrose as well as the close arrangement of the cardiac muscle cells of fresh heart that could not be easily combined with bacteria, the UPT–LF assay was not significantly affected by the five kinds of samples. Moreover, it showed ideal tolerance limits (reaching the highest interfering concentrations) (Table 1), except for the false-positive results of the Ban-UPT-LF strip for sucrose and those of the Ype-UPT-LF strip for gourmet powder and putty power at high interfering concentrations. By contrast, fruit juice with low pH, flour with high viscosity, fresh liver, decomposed heart, decomposed liver, decomposed lung, and decomposed spleen with thick and diversiform bio-macromolecules at high interfering concentrations caused false-positive results by increasing the signal of the negative and positive tests simultaneously and restricting the tolerance limit to a relatively lower level (Table 1). The only exceptions were the non-influence of flour and decomposed heart for Bru-UPT-LF strip and the false-negative of decomposed spleen for Ype-UPT-LF strip. The influences of milk powder, fresh lung, and fresh spleen on the UPT–LF assay were more complicated. At high interfering concentrations, fresh lung and fresh spleen samples led to false-positive results, no influence, and false-negative results for the Ban-UPT-LF, Bru-UPT-LF, and Ype-UPT-LF strips, respectively. Milk powder led to false-negative, false-positive, and no influence results for the Ban-UPT-LF, Bru-UPT-LF, and Ype-UPT-LF strips, respectively.

Similar to that in single-factor interference, the Bru-UPT-LF strip showed the strongest tolerance to real samples, with tolerance limits to flour, fresh heart, fresh lung, and fresh spleen reaching the highest interfering concentrations (Table 1). Aside from the abovementioned real samples, the remaining samples significantly increased the sensitivity of Bru-UPT-LF strip from 10^6 cfu·mL^{-1} to 10^5 cfu·mL^{-1} without any changes in specificity (Figure 5).

Operation error tolerance of the UPT–LF assay

As POCT assay is operated by non-professionals for level-one emergency response, operation error caused by apparatus and operator error cannot be avoided. Therefore, the influence of operation error on sensitivity and specificity should be evaluated. For the UPT–LF assay, the possible operation errors came from the sample measure, sample-treating buffer measure, and loading mixture measure. In Figure 6, with the results of the standard operation used as control, the T/C ratios of negative and weak positive samples under the influence of operation error are shown in a bar graph to indicate the changes in specificity and sensitivity. The volume deviation of the sample (from −5% to +200%), sample-treating buffer (from −22% to +44%), and loading mixture (from −30% to +30%) did not cause any loss of sensitivity and specificity to the UPT–LF assay. The volume change in sample and sample-treating buffer altered the concentration of detected targets in the loading mixture and led to a corresponding decrease or increase in T/C ratios. However, when the volume of the sample and the sample-treating buffer changed with the same proportion, the signals were mainly dependent on the effective volume of loading mixture flowing through the detection line,

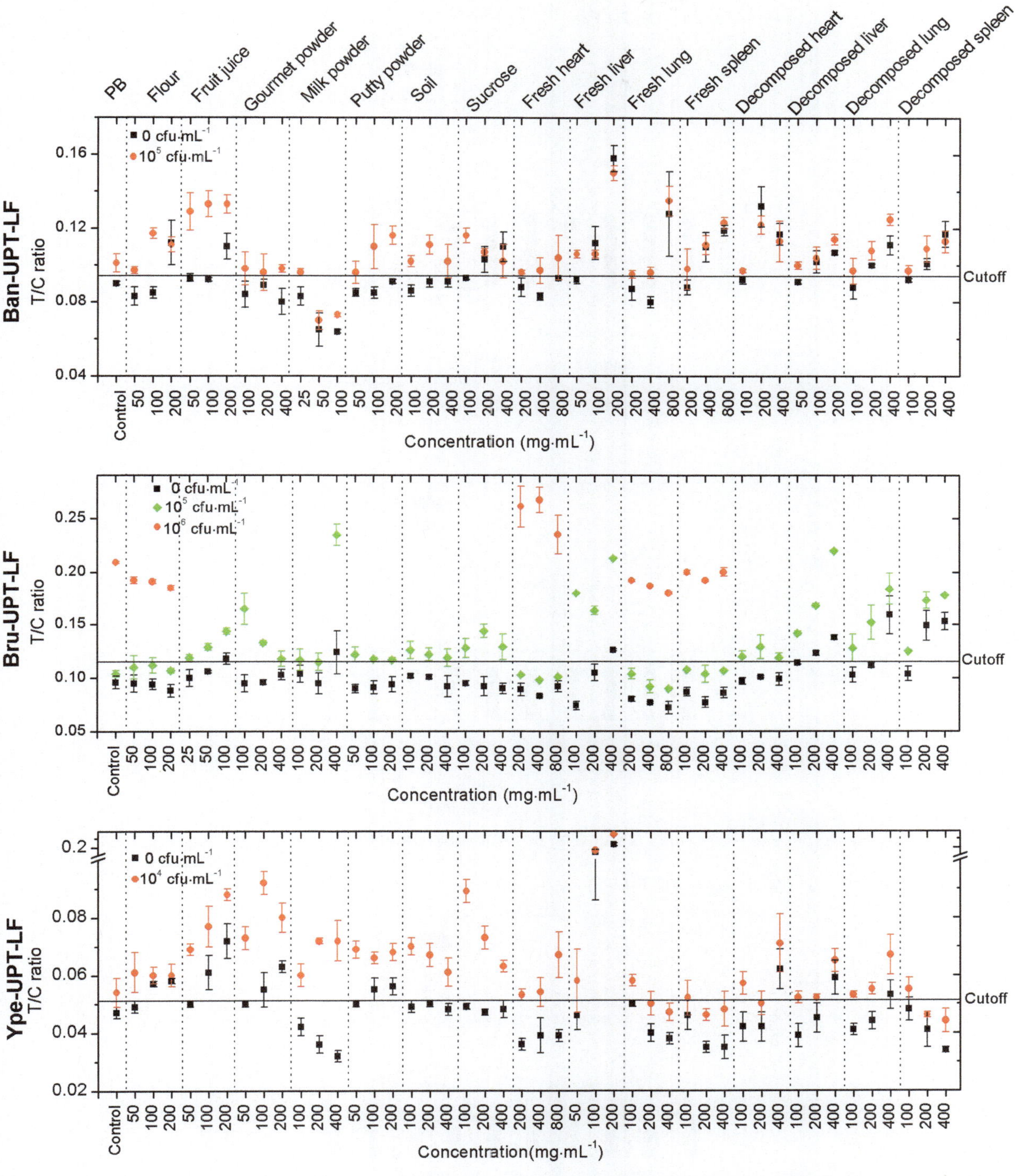

Figure 5. Tolerance of the UPT–LF strips to real sample interference. Under the influence of powders and viscera with gradient concentrations, UPT-LF strip maintained the sensitivity and specificity [T/C ratio of the positive sample (red circles) is higher than the cut-off threshold and T/C ratio of the negative sample (black squares) is lower than the cut-off threshold] and even improved the sensitivity at certain concentrations of the real samples [T/C ratio of the positive sample (green diamonds) is higher than the cut-off threshold].

which was limited by the flow rate of the nitrocellulose filter and the bed volume of absorbent pad. Thus, no significant change in T/C ratios was observed for the error of the loading mixture measure.

Conclusions

The concentration of bacteria is often high in infected animals from the nature foci and in the powders used in bioterrorism. Therefore, a rapid screening test is required for routine disease

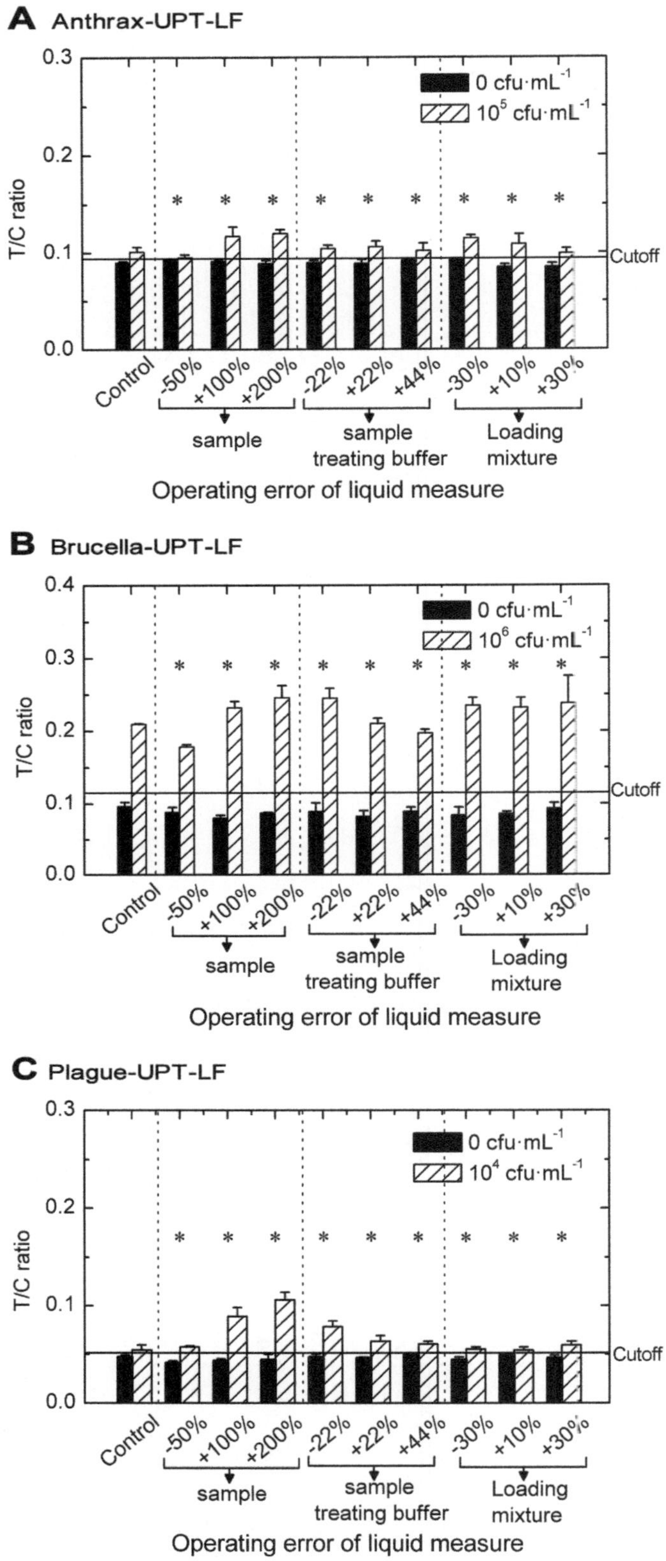

Figure 6. Tolerance of the UPT–LF assay to operation error. Ban-UPT-LF (A), Bru-UPT-LF (B), and Ype-UPT-LF (C) maintained their sensitivity and specificity (*) under the operation error of samples measure (from −50% to +200%), sample treating-buffer measure (from −22% to +44%), and loading mixture measure (from −30% to +30%).

surveillance and incident management. Pre-treatment or bacterial culture of the sample, which is needed for the detection of complex specimens, always takes a long time and entails a complicated operation process that cannot be performed by a non-professional. An improved UPT–LF assay applied for POCT was developed and evaluated comprehensively in this study. Sample tolerance and operation error tolerance were analyzed to reveal the feasibility and practical application of the newly developed technology. A complex sample could be directly detected through simple dissolution or homogenization by a non-professional. The stable optical characteristics of UCP and the covalent bond between UCP and biomolecules ensure that the UPT-LF assay is free from sample influences and non-standard operation errors. Detection using a biosensor can be realized by only pressing a key on the front panel of an upgraded UPT biosensor. Indeed, the procedure is simple and convenient for non-professional operators.

In the LF assay, some specific common antigens, which are surface exposed and important for virulence, are used in the preparation of antibodies that can recognize as many species as possible, such as F1 antigen [22] and plasminogen activator protein [16] of *Y. pestis*, anthrose tetrasaccharide of *B. anthacis* [19], outer membrane proteins and O-polysaccharide of *Brucella* [20,21,38]. However, most surface antigens are difficult to isolate and purify because they are amphipathic molecules. Even antigens obtained by bioengineering technology tend to lose their activity. To ensure the activity, specificity and recognition of as many species as possible, we used the representative inactivated bacteria to immunize the Balb/c mice prior to antibody preparation. The preliminary mAb was chosen by high-throughput screening using ELISA method coated by the representative bacteria. Specificity was also evaluated using the bacteria that shared close genetic relatedness and similar transmission routes with the target. The results demonstrated that UPT–LF assay with antibodies prepared by the methods had high specificity and yielded no false-positive results even when the concentration of nonspecific bacteria reached 10^9 cfu·mL^{-1}.

After comprehensive evaluation, we believe that the UPT–LF assay can be used as an alternative method for infectious disease surveillance in the natural foci, point-of-care screening, environmental surety of health, and incident management of biowarfare and bioterrorist attacks.

Author Contributions

Conceived and designed the experiments: LZ RY. Performed the experiments: PZ XL CW YZ FH CL. Analyzed the data: LZ PZ. Contributed reagents/materials/analysis tools: LZ RY PZ. Wrote the paper: LZ PZ RY.

References

1. Beyer W, Turnbull PC (2009) Anthrax in animals. Mol Aspects Med 30: 481–489.
2. Porche DJ (2002) Biological and chemical bioterrorism agents. J Assoc Nurses AIDS Care 13: 57–64.
3. Prentice MB, Rahalison L (2007) Plague. Lancet 369: 1196–1207.
4. Makino S, Cheun HI (2003) Application of the real-time PCR for the detection of airborne microbial pathogens in reference to the anthrax spores. J Microbiol Methods 53: 141–147.
5. Dichtwald S, Weinbroum AA (2008) Bioterrorism and the anaesthesiologist's perspective. Best Pract Res Clin Anaesthesiol 22: 477–502.
6. Inglesby TV, Dennis DT, Henderson DA, Bartlett JG, Ascher MS, et al. (2000) Plague as a biological weapon: medical and public health management. Working Group on Civilian Biodefense. JAMA 283: 2281–2290.
7. Senior K (2012) Anthrax–where does the real threat lie? Lancet Infect Dis 12: 436–437.
8. Begier EM, Asiki G, Anywaine Z, Yockey B, Schriefer ME, et al. (2006) Pneumonic plague cluster, Uganda, 2004. Emerg Infect Dis 12: 460–467.
9. Amoako KK, Thomas MC, Kong F, Janzen TW, Hahn KR, et al. (2012) Rapid detection and antimicrobial resistance gene profiling of *Yersinia pestis* using pyrosequencing technology. J Microbiol Methods 90: 228–234.
10. Derzelle S, Mendy C, Laroche S, Madani N (2011) Use of high-resolution melting and melting temperature-shift assays for specific detection and identification of *Bacillus anthracis* based on single nucleotide discrimination. J Microbiol Methods 87: 195–201.
11. Garin-Bastuji B, Blasco JM, Marín C, Albert D (2006) The diagnosis of brucellosis in sheep and goats, old and new tools. Small Rumin Res 62: 63–70.
12. Gomes-Solecki MJ, Savitt AG, Rowehl R, Glass JD, Bliska JB, et al. (2005) LcrV capture enzyme-linked immunosorbent assay for detection of *Yersinia pestis* from human samples. Clin Diagn Lab Immunol 12: 339–346.
13. Sainath Rao S, Mohan KV, Nguyen N, Abraham B, Abdouleva G, et al. (2010) Peptides panned from a phage-displayed random peptide library are useful for the detection of *Bacillus anthracis* surrogates *B. cereus* 4342 and *B. anthracis* Sterne. Biochem Biophys Res Commun 395: 93–98.
14. Shott JP, Galiwango RM, Reynolds SJ (2012) A quality management approach to implementing point-of-care technologies for HIV diagnosis and monitoring in sub-Saharan Africa. J Trop Med 2012: 651927.
15. Sokhna C, Mediannikov O, Fenollar F, Bassene H, Diatta G, et al. (2013) Point-of-care laboratory of pathogen diagnosis in rural Senegal. PLoS Negl Trop Dis 7: e1999.
16. Simon S, Demeure C, Lamourette P, Filali S, Plaisance M, et al. (2013) Fast and simple detection of *Yersinia pestis* applicable to field investigation of plague foci. PLoS One 8: e54947.
17. Wang DB, Tian B, Zhang ZP, Deng JY, Cui ZQ, et al. (2013) Rapid detection of *Bacillus anthracis* spores using a super-paramagnetic lateral-flow immunological detection system. Biosens Bioelectron 42: 661–667.
18. Tuemmers C, Luders C, Rojas C, Serri M, Castillo C, et al. (2013) Detection of *Brucella canis* by immunochromatography method in vague dogs captured in Temuco city, Chile, 2011. Rev Chilena Infectol 30: 395–401.
19. Kuehn A, Kovac P, Saksena R, Bannert N, Klee SR, et al. (2009) Development of antibodies against anthrose tetrasaccharide for specific detection of *Bacillus anthracis* spores. Clin Vaccine Immunol 16: 1728–1737.
20. Genc O, Buyuktanir O, Yurdusev N (2011) Development of an individual rapid test based on enzymatic immunofiltration assay for detection of anti-*Brucella* abortus antibody in bovine sera. J Vet Diagn Invest 23: 49–56.
21. Nielsen K, Smith P, Yu WL (2007) Detection of anti-*Brucella* antibodies in llama (Llama glama). J Immunoassay Immunochem 28: 61–66.
22. Chanteau S, Rahalison L, Ratsitorahina M, Mahafaly M, Rasolomaharo M, et al. (2000) Early diagnosis of bubonic plague using F1 antigen capture ELISA assay and rapid immunogold dipstick. Int J Med Microbiol 290: 279–283.
23. Hampl J, Hall M, Mufti NA, Yao YM, MacQueen DB, et al. (2001) Upconverting phosphor reporters in immunochromatographic assays. Anal Biochem 288: 176–187.
24. Corstjens PL, de Dood CJ, van der Ploeg-van Schip JJ, Wiesmeijer KC, Riuttamaki T, et al. (2011) Lateral flow assay for simultaneous detection of cellular- and humoral immune responses. Clin Biochem 44: 1241–1246.
25. Corstjens PL, van Lieshout L, Zuiderwijk M, Kornelis D, Tanke HJ, et al. (2008) Up-converting phosphor technology-based lateral flow assay for detection of Schistosoma circulating anodic antigen in serum. J Clin Microbiol 46: 171–176.
26. van Dam GJ, de Dood CJ, Lewis M, Deelder AM, van Lieshout L, et al. (2013) A robust dry reagent lateral flow assay for diagnosis of active schistosomiasis by detection of Schistosoma circulating anodic antigen. Exp Parasitol 135: 274–282.
27. Ongagna-Yhombi SY, Corstjens P, Geva E, Abrams WR, Barber CA, et al. (2013) Improved assay to detect Plasmodium falciparum using an uninterrupted, semi-nested PCR and quantitative lateral flow analysis. Malar J 12: 74.
28. Niedbala RS, Feindt H, Kardos K, Vail T, Burton J, et al. (2001) Detection of analytes by immunoassay using up-converting phosphor technology. Anal Biochem 293: 22–30.
29. Qu Q, Zhu Z, Wang Y, Zhong Z, Zhao J, et al. (2009) Rapid and quantitative detection of *Brucella* by up-converting phosphor technology-based lateral-flow assay. J Microbiol Methods 79: 121–123.
30. Yan Z, Zhou L, Zhao Y, Wang J, Huang L, et al. (2006) Rapid quantitative detection of *Yersinia pestis* by lateral-flow immunoassay and up-converting phosphor technology-based biosensor. Sens Actuators B Chem 119: 656–663.
31. Li L, Zhou L, Yu Y, Zhu Z, Lin C, et al. (2009) Development of up-converting phosphor technology-based lateral-flow assay for rapidly quantitative detection of hepatitis B surface antibody. Diagn Microbiol Infect Dis 63: 165–172.
32. Hong W, Huang L, Wang H, Qu J, Guo Z, et al. (2010) Development of an up-converting phosphor technology-based 10-channel lateral flow assay for profiling antibodies against *Yersinia pestis*. J Microbiol Methods 83: 133–140.
33. Ouellette AL, Li JJ, Cooper DE, Ricco AJ, Kovacs GT (2009) Evolving point-of-care diagnostics using up-converting phosphor bioanalytical systems. Anal Chem 81: 3216–3221.
34. Li W, Zhou L, Wang J, Hu K, Wang J, et al. (2006) Development of up-converting phosphor immunochromatography for fast and quantitive detection of *Bacillus anthracis* spore. Clin J Microbiol Immunol 26: 761–763.

35. Corstjens P, Zuiderwijk M, Brink A, Li S, Feindt H, et al. (2001) Use of up-converting phosphor reporters in lateral-flow assays to detect specific nucleic acid sequences: a rapid, sensitive DNA test to identify human papillomavirus type 16 infection. Clin Chem 47: 1885–1893.
36. Tamborrini M, Holzer M, Seeberger PH, Schurch N, Pluschke G (2010) Anthrax spore detection by a luminex assay based on monoclonal antibodies that recognize anthrose-containing oligosaccharides. Clin Vaccine Immunol 17: 1446–1451.
37. Koehler TM (2009) *Bacillus anthracis* physiology and genetics. Mol Aspects Med 30: 386–396.
38. Nielsen K, Smith P, Yu WL, Elmgren C, Nicoletti P, et al. (2007) Second generation competitive enzyme immunoassay for detection of bovine antibody to *Brucella abortus*. Vet Microbiol 124: 173–177.

9

Resistance Inducers Modulate *Pseudomonas syringae* pv. Tomato Strain DC3000 Response in Tomato Plants

Loredana Scalschi[1], Gemma Camañes[1], Eugenio Llorens[1], Emma Fernández-Crespo[1], María M. López[2], Pilar García-Agustín[1], Begonya Vicedo[1]*

1 Grup de Bioquímica i Biotecnologia, Area de Fisiologia Vegetal, Departament de Ciències Agràries i del Medi Natural, ESTCE, Universitat Jaume I, Castellón, Spain,
2 Departamento de Protección Vegetal y Biotecnología, Instituto Valenciano de Investigaciones Agrarias (IVIA), Apartado Oficial, Moncada, Valencia, Spain

Abstract

The efficacy of hexanoic acid (Hx) as an inducer of resistance in tomato plants against *Pseudomonas syringae* pv. tomato DC3000 was previously demonstrated, and the plant response was characterized. Because little is known about the reaction of the pathogen to this effect, the goal of the present work was to determine whether the changes in the plant defence system affect the pathogen behaviour. This work provides the first demonstration of the response of the pathogen to the changes observed in plants after Hx application in terms of not only the population size but also the transcriptional levels of genes involved in quorum sensing establishment and pathogenesis. Therefore, it is possible that Hx treatment attenuates the virulence and survival of bacteria by preventing or diminishing the appearance of symptoms and controlling the growth of the bacteria in the mesophyll. It is interesting to note that the gene transcriptional changes in the bacteria from the treated plants occur at the same time as the changes in the plants. Hx is able to alter bacteria pathogenesis and survival only when it is applied as a resistance inducer because the changes that it promotes in plants affect the bacteria.

Editor: Szabolcs Semsey, Niels Bohr Institute, Denmark

Funding: LS is the recipient of a PhD fellowship from the Ministerio de Educación (Grant AP2008-01064) (http://www.mecd.gob.es/servicios-al-ciudadano-mecd/catalogo-servicios/becas-ayudas-subvenciones/para-estudiar/universidad/doctorado.html). This work was supported by AGL2010-22300-C03-02: National R&D Plan (http://www.idi.mineco.gob.es/portal/site/MICINN/menuitem.7eeac5cd345b4f34f09dfd1001432ea0/?vgnextoid=fe5aec1eb658c310VgnVCM1000001d04140aRCRD&lang_choosen=en), and P1•1B2013-75: Projecte d'investigació científica i desenvolupament tecnologic del Pla de promoció de la investigació de la Universitat Jaume I (http://www.uji.es/ocit/pinv13/#1). The funders had no role in study design, data collection and analysis, decision to publish, or preparation of the manuscript.

Competing Interests: The authors have declared that no competing interests exist.

* Email: bvicedo@camn.uji.es

Introduction

Resistance-inducing agents that activate the natural defences of plants have been identified, and it is known that their application protects plants through the induction of defences that are effective against a wide spectrum of pathogens. Their mechanism of action has been studied in different pathosystems over the last years [1–6]. However, all of the previous studies were focused on plant defence responses, and little is known about how the pathogen reacts against the induction of plant defences.

Previous investigations by our research group have shown that the resistance-inducer hexanoic acid (Hx) protects *Solanum lycopersicum* plants against the necrotroph *Botrytis cinerea*, a pathogen that has proven difficult to chemically control [1,6]. Furthermore, its efficacy was demonstrated in *Arabidopsis thaliana* [7]. We also observed an approximately 50% reduction in disease symptoms against *Pseudomonas syringae* pv. tomato strain DC3000 (*Pst* DC3000) in treated plants compared with non-treated plants [6], which suggests that Hx induces resistance against this pathogen. The mechanisms involved in the induction of resistance have been studied in plants [4], but the bacterial response to this process remains unelucidated. Therefore, the present study was undertaken to gain insight into the mechanism of action of Hx, particularly its effect on bacterial pathogens.

We chose the plant pathogen *Pst* DC3000 because it represents an important pathogen that infects economically important crops. In nature, this pathogen causes bacterial speck disease in tomato and bell pepper and can also infect *Arabidopsis* and *Brassica* species in the laboratory [8,9]. Moreover, *Pst* DC3000 is considered a model organism for molecular studies of plant-pathogen interactions [10,11].

The infection cycle of *Pst* DC3000 is typical for a foliar pathogenic bacterium. In a successful disease cycle, *Pst* DC3000 generally shows two lifestyles that are spatially and temporally interconnected: an initial epiphytic phase and a subsequent endophytic phase in the apoplast [12]. Before initiating disease, *Pst* DC3000 strains achieve large epiphytic population sizes on healthy leaves [13]. Bacterial populations are capable of coordinating the expression of some traits in a cell-density-dependent manner through the process known as quorum-sensing (QS) [14,15]. *Pst* DC3000 and many other Gram-negative bacteria use N-acyl homoserine lactone (AHL) as the QS signal molecule [16]. In addition, *psyI* of *Pst* DC3000 functions as an AHL synthase, and *psyI* expression, as well as AHL production, are under the control of the sigma regulator PsrA. Elevated levels of AHL due to higher population densities result in the expression of virulence factors and secondary metabolites that mediate colonization of the host.

Pst DC3000 contains more than thirty type III effectors [17,18]. One of the most important effectors for the virulence of the bacteria is AvrPtoB (hopAB2), which has been found to act inside plant cells to suppress programmed cell death associated with plant immunity and increases pathogen growth and disease formation [19]. The effectors, together with the T3SS apparatus, are encoded by the hrp/hrc regulon. The *hrpL* gene encodes the alternate RNA polymerase sigma factor σL, which is one of the primary transcription factors that modulate the expression of the T3SS [20].

Pst DC3000 produces a polyketide toxin named coronatine (COR) [21]. COR is a chlorosis-inducing phytotoxin that contributes to bacterial growth and lesion formation or expansion in the host tissue. COR acts as a virulence factor in various *P. syringae* pathovars by generating increased lesion size and bacterial multiplication [22]. Several studies have shown that COR plays a critical role in pathogenesis [23,24]. Moreover, it is known that COR controls stomata opening upon bacterial infection [12]. COR consists of two distinct moieties that are synthesized by separate pathways, and an amide linkage then joins the two moieties: the polyketide coronafacic acid (CFA) and coronamic acid (CMA), an ethylcyclopropyl amino acid derived from isoleucine [25].

The role of CorA, the major magnesium transporter, in pathogenesis has been studied in a number of organisms, such as *Salmonella enterica* serovar *typhimurium*, in which it is required for full virulence [26]. A role for CorA has also been established in the human gastric pathogen *Helicobacter pylori*, where it is required for viability under limiting Mg^{2+} conditions [27]. In addition to being necessary for bacterial growth, magnesium is a cofactor in ATP-requiring enzymatic reactions [28] and is directly involved in membrane stability [29].

In this study, we investigated whether the resistance inducer Hx displays any direct antimicrobial effect on the target bacterium by performing *in vitro* assays with different Hx concentrations to determine its effect on bacterial growth, viability, and gene expression. Moreover, the effect of Hx treatment on the bacteria during plant infection was also investigated through an analysis of the expression of genes related to the different components of pathogenicity and virulence explained above. From our *in vitro* experiments, we can conclude that higher concentrations of Hx may have a temporary bacteriostatic effect against *Pst* DC3000 or a short-term bactericidal effect.

On the other hand, we previously observed that Hx treatment against *Pst* DC3000 affects the JA and SA pathways and prevents stomata opening induced by coronatine *in planta* [4]. Concurrently, in this study, we noticed a reduction in the expression of bacterial genes involved in quorum sensing establishment and pathogenesis, such as COR synthesis genes, genes associated with the type III secretion system, and genes responsible for effector synthesis in treated plants.

Materials and Methods

Bacterial strains and growth conditions

The *P. syringae* pv. tomato strain used in the present study was DC3000, rifampicin resistent. Rifampicin was added to King B medium (KB) at 50 μg mL^{-1}. Mutants used were Δ*hrpL*, (Beuzon, C. Dep. Biología Celular, Genética y Fisiología Universidad de Malaga, Spain), Δ*hrpA* (He, S. Dep. of Plant Pathology, University of Kentucky, Lexington), Δ*cma-cfa* (Brooks, D. Department of Biology, Washington University, St. Louis) and Δ*pto(ΔavrPto-ΔavrPtoB*) (Martin, G. Boyce Thompson Institute for Plant Research, Tower Rd., Ithaca,). They were grown in King's B medium (KB) [54] at 28°C with their specific antibiotics.

Bacterial growth assay

We tested Hx effect against *Pst* DC3000 and several mutants. Long time experiments were performed in LB medium to which, the resistance inducer was added at a final concentration of 0.6, 1.5, 5, 10 and 20 mM. pH of medium was adjusted to 7 after Hx was added and before adding the bacteria. The strain was precultivated in LB overnight to obtain the inoculum. The bacteria were harvested by centrifugation, washed and resuspended in sterilized $MgSO_4$ (10 mM).

The growth assay was carried in a Bioscreen C analysator (Labsystem, Finland). In a total volume of 350 μl in microtiter wells using an initial bacterial density of about 1×10^3 cfu/mL. Bacterial growth was monitored by measuring optical density every 10 min with periodic shaking for 96 h at 26°C. The results were printed out as growth curves.

Moreover Hx effect to short time points was studied with M9 medium (Sigma) to wich Hx was added at a final concentration of 10 and 20 mM. Bacteria were preincubated as described above and added to final concentration of 1×10^3 cfu/ml. *Pst* DC3000 was incubated with shaking for 16 h to 26°C. After this time, live and dead cell were quantified as described below.

Live and dead cell quantification

The proportion of living vs. dead cells was quantified using the fluorescent LIVE/DEAD BacLight Bacterial Viability Kit, L13152 (Molecular Probes, Invitrogen, Paisley, UK). For live and dead cell quantification, 50 μl of bacterial suspension were mixed with 25 μl of each of the two components of the LIVE/DEAD BacLight kit and incubated in the dark for 20 min. The viability of the cells was determined by enumerating the live and dead cells using an epifluorescence microscope Leica IRB equipped with a Leica DBC300F camera (Leica Microsystems CMS GmbH, Wetzlar, Germany). Live and dead cells were enumerated in 15 microscopic fields at 1000× magnification.Two independent experiments were performed with five repetitions for each treatment.

Plant inoculation

Four-week-old tomato plants of *Solanum lycopersicum* L. cv. Ailsa Craig were treated with Hx and inoculated as described by Scalschi *et al.* [4].Hx treatment was applied by soil drench to 4-week-old plants grown in jiffy peat pellets at a concentration of 0.6 mM, 72 h before bacterial inoculation. For inoculation, Pst DC3000 was grown in KB at 28°C for 24 h. Bacterial suspensions were adjusted to 5×10^5colony-forming units (CFU)/mL in sterile $MgSO_4$ (10 mM) containing 0.01% of the surfactant Silwet L-77 (Osi Specialties, Danbury, CT, USA), as described previously [11].Pathogen inoculation was performed by dipping the third and fourth leaves into the bacterial suspension.

The disease rate was scored at 72 hpi by determining the percentage of dark-brown spots on the leaf surface and by counting the colony forming units in KB medium. At least three samples for colony counting and 20 samples for disease rate scoring were taken for each treatment over a 3-day period. For bacteria gene expression analysis, samples were taken at 2, 6, 10, 18, 48 and 72 h after inoculation. Each experiment was independently conducted at least three times.

Long term experiments were performed in the same conditions like those described above for 30 days. The samples were taken at 4, 10, 20 and 30 days after inoculation. Each experiment was independently conducted at least three times.

Bacterial RNA extraction and cDNA synthesis

Bacterial RNA extractions were carried out from *in vitro*-cultured cells after 18 h and from cells grown *in planta* at different time intervals.

For RNA isolation from *in vitro* cultured bacteria, *Pst* DC3000 was grown on KB agar plates for 48 h at 28°C. Hx was added to 1×10^6 bacterial cells suspensions to final concentrations of 1.5 mM, 5 mM and 10 mM. Cells were incubated by shaking for 18 h before RNA extraction. Bacterial suspensions were stabilized with RNA protect bacterial reagent (Qiagen) before centrifugation. Cells pellets were immediately frozen in liquid N_2 and kept at −80°C prior to RNA extraction. We used the Qiagen RNeasy Bacteria Mini Kit for isolation of RNA samples according to manufacturer's instructions.

To extract bacterial RNA from infected plants we used a protocol for extracting RNA from *P.syringae* recovered from infected leaves as described by Yu *et al.* [55]. RNA quantification was performed with a UV-VIS UNICAM Model Heλiosβ spectrophotometer. Reverse transcription was carried out from 1 µg of total RNA for gene expression analysis, by using OMNISCRIPT (Qiagen) and random hexamer primers (Promega).

Expression analysis by qRT-PCR

Specific primers for real-time PCR were designated by using Primer3 program according to the sequence of the corresponding gene. Primers used for the assay are described in Table 1. 4 µl of diluted cDNA were used per 25 µl qPCR Quantitative real-time PCR was performed with SybrGreen Premix ExTaq (Takara) in a Cepheid Smart Cycler system Relative levels of the monitored genes were normalized with *recA* and used as an internal reference.

Statistical analysis

Statistical analysis was carried out using a one-way analysis of variance in the Statgraphics-plus software of Windows V.5 (Statistical Graphics Corp., Rockville, MD, USA). The means were expressed with standard errors and compared using a Fisher's least-significant difference test at the 95% confidence interval. All the experiments were repeated at least three times.

Results

1. Analysis of bacterial behaviour *in vitro*

1.1 Antibacterial activity of Hx. The growth of *Pst* DC3000 in the absence or presence of different concentrations of hexanoic acid (Hx) (0.6, 1.5, 5, 10, and 20 mM) was tested in LB medium. Higher concentrations were not used because damage to the tomato plants in our experimental assays was observed at concentrations higher than 20 mM. The bacterial growth curves were generated using a BioScreen system for 96 h (Fig. 1), and the results show that Hx at concentrations of 10 and 20 mM promoted slow bacterial growth and that the lag phase was markedly longer in these samples compared with those exposed to lower concentrations (20 and 8 h, respectively). Moreover, Hx at a concentration of 20 mM produced an inhibition of 15% in bacterial growth. The addition of Hx at concentrations of 0.6, 1.5, and 5 mM did not reduce the growth of the target compared with the control.

We can conclude that Hx at a concentration of 10 and 20 mM could have a temporary bacteriostatic effect against *Pst* DC3000 because at these concentrations the lag phase lasts longer. However a bactericidal effect cannot be ruled out, at these concentrations.

1.2 Analysis of Hx bactericidal effect. The bactericidal effect of Hx was investigated by labelling the bacteria with the

Table 1. Primers used for Real-Time PCR.

Gene	Funtion	Primer
cfa1	CFA synthesis	F 5′-AAAACCATCGTCGACATTCTG-3′
		R 5′-GTTGGCGTTGAGGTCGATA-3′
cmaB	CMA synthesis	F 5′-AATTCGACACCCGACAAGAC-3′
		R 5′-ACTAGGGGCTTCAGGTCCAT-3′
cfl	CMA and CFA union	F 5′-ACAGCTGAAGCAGCACTTGA-3′
		R 5′-CGAGGATCTCTCGGTAGTCG-3′
hrpL	Transcription factor σ^L	F 5′-TCTCCAGTGCGTGTTTCTTG-3′
		R 5′-AGCTTTCCTGATACGGCTGA-3′
hrpA	*hrp* pilus	F 5′-CCTCCAAACTCACCAACCTT-3′
		R 5′-CGGACTCTTTACTGGCCTTG-3′
avrPtoB	AvrPtoB type III effector	F 5′-ACCCTATCGCGTCACAATTC-3′
		R 5′-CATGAACGCCAGGTCCTTAT-3′
psyI	AHL synthase	F 5′-GGCTTGAATGGAATGTTCGT-3′
		R 5′-CAGGTGTTGATCAGCCGTAA-3′
psrA	Sigma regulator of *psyI*	F 5′-AGACGCATAACCTGCTCGAT-3′
		R 5′-ACATGACGCTGGTCAATGAA-3′
cma X	Magnesium transporter	F 5′-AGCGGGACATTTATGGTCAG-3
		R 5′-CTGTTGTTCAGCTCGTTCCA-3
recA	Internal reference	F 5′-CGGCAAGGGTATCTACCTCA-3′
		R 5′-CTTTGCAGATTTCCGGGTTA-3′

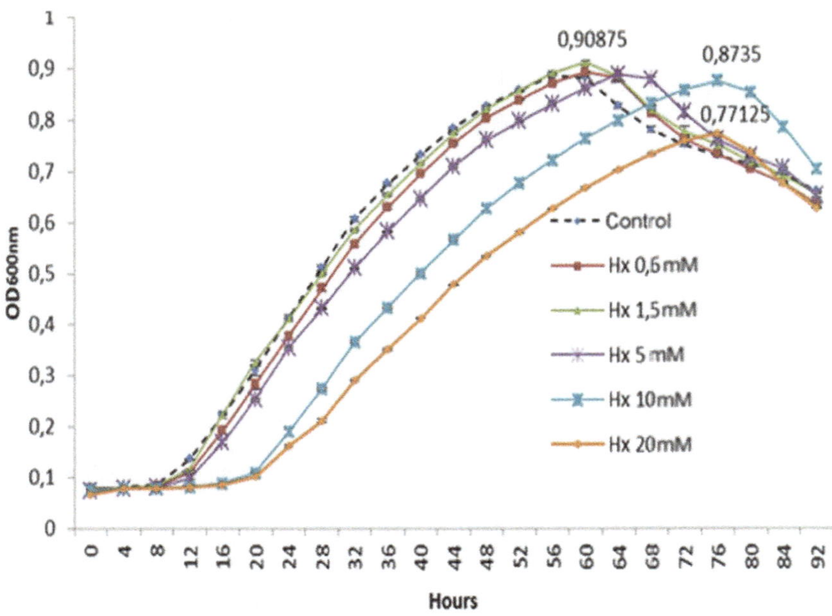

Figure 1. Influence of different hexanoic acid (Hx) concentrations on the growth rate of *Pst* DC3000. The results are presented as growth curves (in hours) of *Pst* DC3000 during its culture in the presence of 0.6, 1.5, 5, 10, and 20 mM Hx. The optical density at 600 nm was measured at 10-min intervals. The results represent the means with standard deviation.

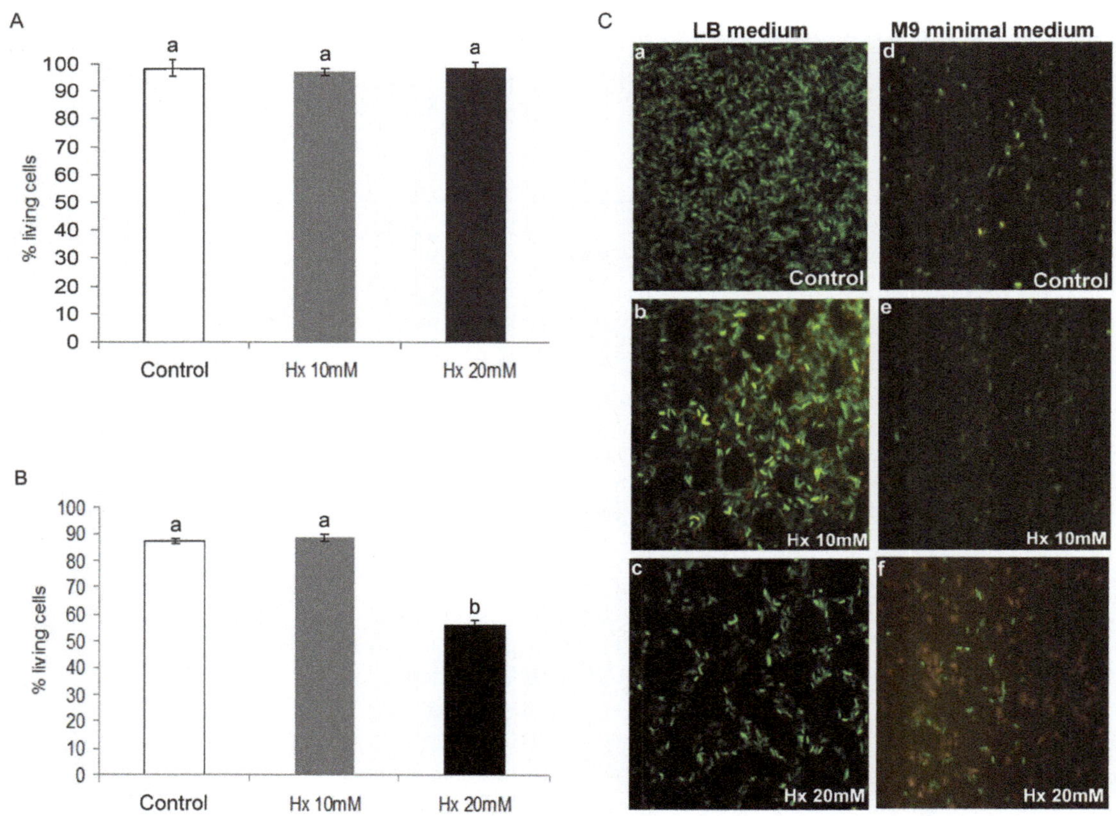

Figure 2. Bactericidal effect of Hx treatment at concentrations of 10 and 20 mM on *Pst* DC3000. Long-term effect of Hx in LB medium (A). Short-term effect of Hx in minimal medium (B). The charts illustrate the percentage of live cells, which was determined using a Live/Dead Bacterial Viability Kit. The images show the viable (green) and dead (red) cells and illustrated that aggregates are formed after exposure to Hx concentrations of 10 mM and 20 mM only in the long-term experiments (C.2, C.3).

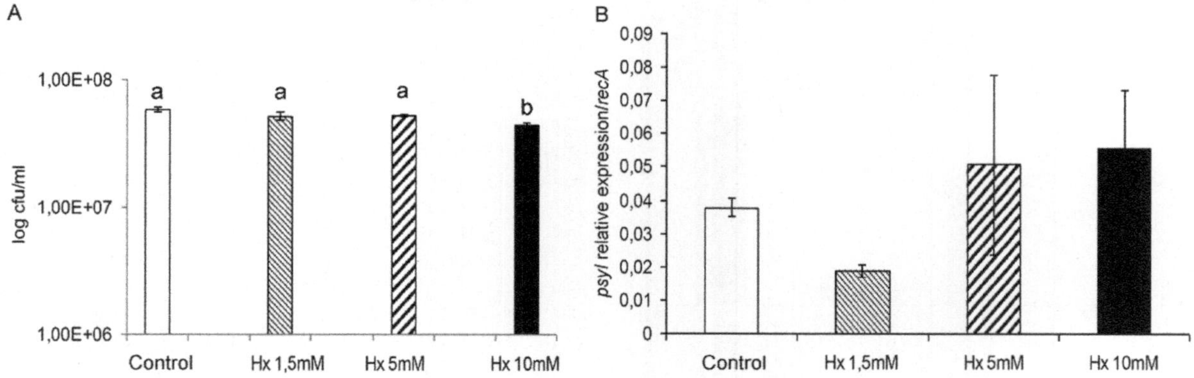

Figure 3. Effect of Hx on bacterial growth and QS-related genes. The bacteria were grown in LB medium and LB plus Hx at 1.5, 5, and 10 mM for 18 h and then plated on KB agar plates to count the colony forming units (A). The transcript levels of *psyI* were determined by qRT-PCR. The relative expression levels of *psyI* were normalized to those of *recA* (B). The results represent the means and standard deviations from three different experiments.

Live/Dead Cell Viability Assay kit, a two-colour fluorescence assay based on membrane integrity. The bactericidal effect was assessed using bacteria incubated for 96 h in LB medium with an Hx concentration of 10 and 20 mM (based on the results described above) and compared to that obtained with the untreated control. A very low number of dead bacteria (red labelled) were observed both in the untreated control and in the groups with 10 and 20 mM of Hx, and no significant differences were observed between the groups (Fig. 2A). Moreover, the non-treated controls had higher concentrations of living bacteria (Figs. 2Ca, 2Cb, and 2Cc). The presence of dead bacteria in the untreated controls appeared to be due to the experimental conditions used for growth, i.e., nutrient availability was a limiting factor due to the higher concentration that reached the control bacteria. Thus, Hx at concentrations of 10 and 20 mM caused no statistically significant changes in bacterial viability over a long period of time. This result may be because the bacteria enter the viable but non-culturable stage (VBNC) (Fig. 2Cc). In addition, at these concentrations, we microscopically observed the formation of multicellular aggregates that were not observed in the controls and the other treatment groups (Fig. 2Ca), and these aggregates were better defined in the group treated with 20 mM Hx (Fig. 2C.c).

Taking into consideration these results, the next question we addressed was whether the direct effect of Hx at 10 and 20 mM on the bacteria is a short-term bactericidal effect, which implies a reduction in the initial bacterial population followed by its recovery over the long term. To test this, the bactericidal effect was tested in bacteria that were incubated for 16 h in M9 minimal medium with Hx at concentrations of 10 and 20 mM. As shown in Figure 2B, a low number of dead bacteria were observed both in the untreated control bacteria and in the bacteria incubated with Hx at 10 mM. However, the bacteria incubated with Hx at 20 mM presented a clear short-term bactericidal effect of Hx because the number of nonviable cells represents 44% of the total number of cells (Figs. 2Cd, 2Ce, and 2Cf).

Taking together all of these results, we can conclude that the Hx effect on the target bacteria is dependent on the concentration: Hx at 10 mM has a temporary bacteriostatic effect, whereas Hx at 20 mM has a short-term bactericidal effect on half of the cells, which may explain the partial recovery of population growth over the long term.

1.3 Hx affects bacterial gene expression *in vitro*. We also performed *in vitro* assays to determine the Hx effect on bacterial gene expression. To achieve this goal, we selected genes related to the pathogenesis and survival of *Pst* DC3000, such as quorum sensing establishment-related genes (*psyI*), coronatine synthesis-related genes (*cfa1*, *cmaB* and *cfl*), genes selected as representative of the type III secretion system and the type III secretion system-associated pilus, respectively (*hrpL* and *hrpA*), and genes responsible for effector synthesis (*avrPtoB*). Hx was added at different concentrations (1.5, 5, and 10 mM) to the medium, and the bacteria were cultivated in the Hx-supplemented medium for 18 h at 28°C. After 18 h of incubation, we observed a significantly lower bacterial concentration in the group treated with 10 mM Hx (Fig. 3A).

The monitoring of the expression of *psyI* revealed very low levels in both the controls and the bacteria incubated with different concentrations of Hx. The expression of this gene was slightly higher in the bacteria incubated with Hx at 5 and 10 mM. However, at 1.5 mM Hx, the expression of this gene was lower (Fig. 3B).

The transcript levels of *cfa1* and *cfl* (Figs. 4A and 4C) were induced in bacteria grown with Hx at a concentration of 10 mM compared with those grown without Hx or with 1.5 and 5 mM Hx. However, the *cmaB* (Fig. 4B) expression level was very low, and practically equal amounts of its transcripts were detected in all of the samples or slightly lower amounts were detected in the presence of Hx.

Similar results to those obtained for *cfa1* and *cfl* were observed in the analysis of *hrpL*, *hrpA*, and *avrPtoB* (Figs. 5A, 5B, and 5C). In contrast, the analysis of the expression of *cmaX*, a CorA-like Mg^{2+} transporter, revealed similar levels of expression regardless of the concentration of Hx (Fig. 5D).

Taken together, these results demonstrate that the expression of the pathogenesis-related genes *cfl*, *cfa1*, *hrpL*, *hrpA*, and *avrPtoB* is enhanced with exposure to Hx at a concentration of 10 mM, which also affected bacterial growth. In contrast, the *cmaB* and *cmaX* expression levels were essentially equal regardless of the Hx concentration. These results indicate the involvement of the pathogenesis system in bacterial survival under abiotic stress conditions.

To confirm this hypothesis, the behaviour of mutants defective in these genes (*ΔhrpL*, *ΔhrpA*, *Δcma-cfa*, and *Δpto* (*ΔAvrPto-ΔAvrPtoB*)) was analysed under the same conditions to test the antimicrobial activity of Hx. Each mutant showed a similar level of

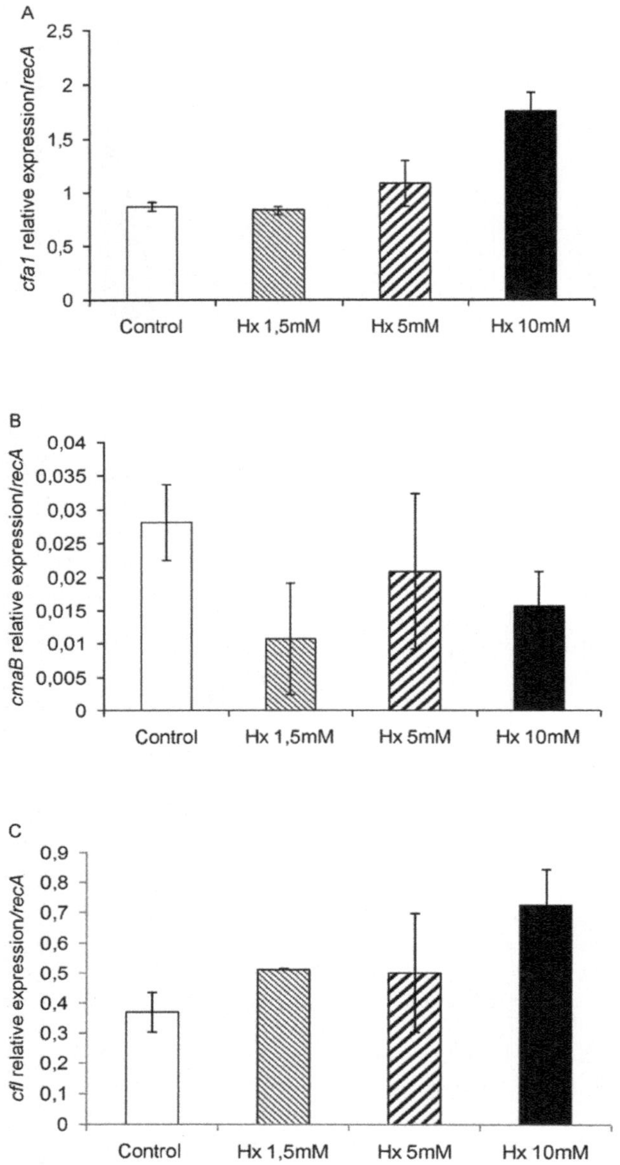

Figure 4. Relative expression of COR synthesis genes (*cfa1* (A), *cmaB* (B), and *cfl* (C)) during bacterial growth in the presence of different Hx concentrations. Template cDNAs were generated from the total RNA extracted from bacteria grown as described in Figure 3. The *recA* gene was used as an endogenous reference gene. The results represent the means and standard deviation from three different experiments.

survival to *Pst* DC3000, regardless of the Hx concentration (data not shown).

2. Analysis of bacterial behaviour *in planta*

Assays were conducted under the conditions described by Scalschi *et al.* [4]. The efficacy of Hx treatment was confirmed at both 48 hpi and 72 hpi by the reduction in disease symptoms and the reduction in the number of colony-forming units observed in the treated plants compared with the untreated plants, and major differences were observed at 72 hpi (Figs. 6A and 6B). Furthermore, we examined the expression of the genes analysed above after the infection of tomato plants with *Pst* DC3000 to determine whether their expression was altered by the treatment.

2.1 QS is affected by Hx treatment. The gene expression was evaluated 2, 4, 10, 18, 48, and 72 h after infection. At the early stages, due to the low bacterial concentration, the obtained values exhibited a broad range, and we thus selected the 48- and 72-h time points for further analyses. At these time points, the greatest changes in the defence system of the treated plants were induced against *Pst* DC3000 [4].

The epiphytic phase of *Pst* DC3000 is considered one of the first steps in the infection process [30,31]. Thus, the transcriptional activity of the *psyI* and *psrA* genes was monitored during the infection of treated and untreated tomato plants. As illustrated in Figure 7A, the expression of *psyI* was significantly lower in the bacteria from the treated plants at 48 hpi. Although the *psyI* expression level was decreased in bacteria from both untreated and treated plants at 72 hpi, the difference between these bacteria was maintained.

In contrast, the monitoring of the relative expression of *psr*A revealed very low levels of expression at both 48 and 72 hpi. However, its expression was higher in the bacteria from untreated plants at 48 hpi (Fig. 7B).

2.2 Expression of COR synthesis genes is reduced by Hx treatment. To investigate the effect of Hx treatment on *Pst* DC3000, the expression levels of the coronatine synthesis genes *cfa1*, *cmaB*, and *cfl* were measured in bacteria extracted from both treated and untreated tomato plants and were compared with those obtained in culture media (Fig. 8). As expected, the transcription levels of *cfa1* and *cfl* were increased at 48 hpi and were higher in the bacteria from untreated plants compared with those from the treated plants. Although the expression of *cfl* was significantly decreased in the bacteria from the untreated plants at 72 hpi, its transcription levels exceeded those obtained in the bacteria from the treated plants. In contrast, *cfa1* expression was hardly detected at this time point (Figs. 8A and 8C). A similar increase was observed with *cmaB* at 48 hpi, but elevated levels in its expression were recorded in the bacteria from the treated plants at 72 hpi (Fig. 8B).

2.3 Type III secretion system is affected by Hx treatment. We assayed the expression of the *hrpL* and *hrpA* genes, which were selected as representative of the type III secretion system and the type III secretion system-associated pilus, respectively, and the expression of the *avrPtoB* gene, which is responsible for effector synthesis (Fig. 9).

To our surprise, we did not observe a strong induction of *hrpL* upon infection; on the contrary, we found that *hrpL* was induced to higher levels *in vitro* than *in planta*. Although Hx treatment inhibits the expression of this gene at 48 h, its expression was enhanced in the treated plants compared to the untreated ones at 72 hpi (Fig. 9A).

In contrast, the expression of *hrpA*, the major component of the type III secretion system-associated pilus, was stronger in the infected plants than *in vitro*, with the exception of the bacteria incubated with Hx at 10 mM, which exhibited similar expression levels to those observed in the infected plants. *In planta*, this gene has a similar behaviour to that observed with *hrpL* (Fig. 9B).

Furthermore, the expression of the gene involved in the synthesis of the specific effector AvrPtoB was monitored (Fig. 9C). At both 48 and 72 hpi, substantial increases in the *avrPtoB* transcription levels were recorded in the untreated plants compared with the treated plants.

2.4 Other effects of Hx acid treatment. The relative expression levels of the genes related with the CorA-like Mg^{2+} transporter, namely *cmaX* and *corA*, were measured in bacteria

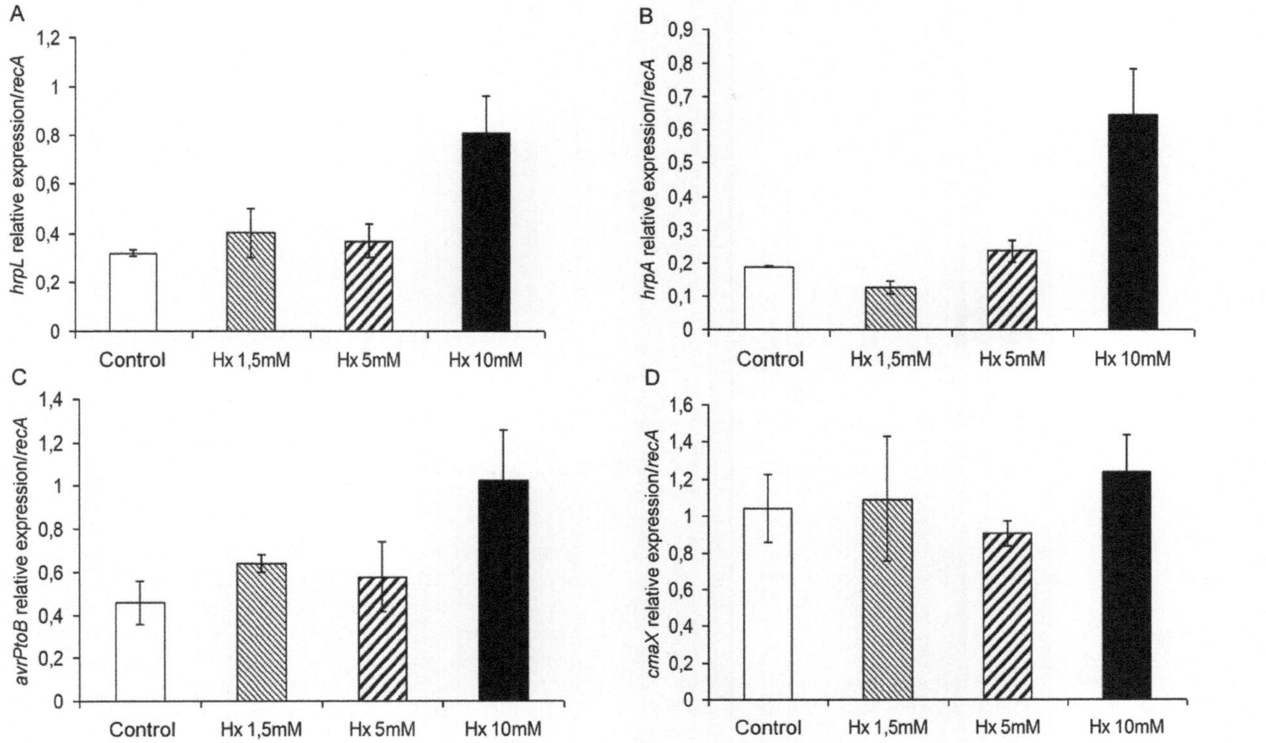

Figure 5. Expression of type III secretion system- and other virulence factor-related genes in response to different hexanoic acid concentrations. The transcript levels of *hrpL* (A), *hrpA* (B), *avrPtoB* (C) and *cmaX* (D) were examined as described in Figure 3, and the *recA* gene was used as an internal reference. The results represent the means and standard deviation from three different experiments.

recovered from treated and untreated plants (Fig. 9D). At 48 hpi, the transcription level of *cmaX* was decreased by more than 10-fold in the bacteria recovered from treated plants.

2.5 Is *Pseudomonas* able to overcome Hx-IR over time?. The findings that the bacterial populations in treated plants increase even though fewer symptoms are observed and that some of the pathogenesis-related genes are expressed at 72 h led us to ask whether the bacteria can overcome the induced defence mechanisms in plants. To study the long-term bacterial behaviour, similar experiments were conducted over a 30-day period.

After 20 days, an increase in disease severity followed by abscission of the infected leaves was observed in the untreated plants. The treated plants presented a better physical condition than the untreated ones, as determined by the lack of extension of infection to the younger uninoculated leaves and the absence of leaf abscission (Figs. 10B and 10C).

The bacterial growth was higher in the untreated plants than in the treated ones. Although the bacterial populations were increased in both the treated and non-treated plants after four days, at longer time points, the number of pathogens was markedly

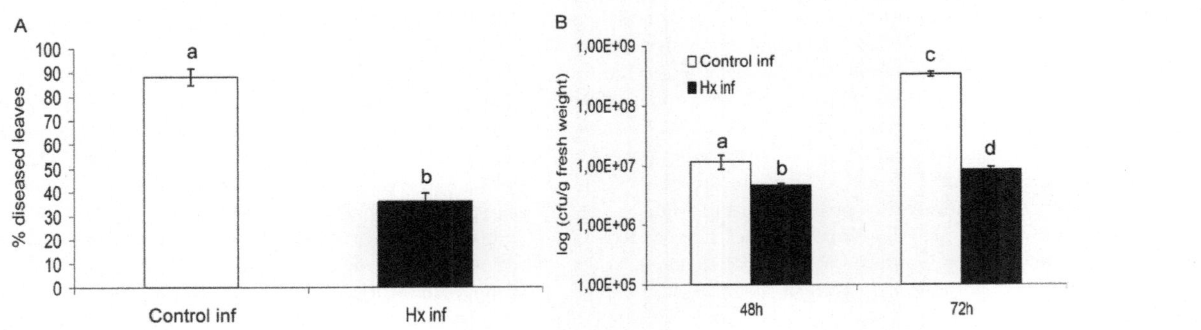

Figure 6. Hexanoic acid-induced resistance against *Pst* DC3000. Four-week-old tomato plants were treated with Hx through soil drench and dip inoculated with *Pst* DC3000. Seventy-two hours after inoculation, the disease rating was scored by measuring the percentage of infected leaves in relation to the total number of analysed leaves (A) and by recounting the bacterial populations through plating in agar-King's B medium (B). The data show the average values ± standard error ($n = 20$). The different letters represent statistically significant differences ($P < 0.05$; least-significant difference test).

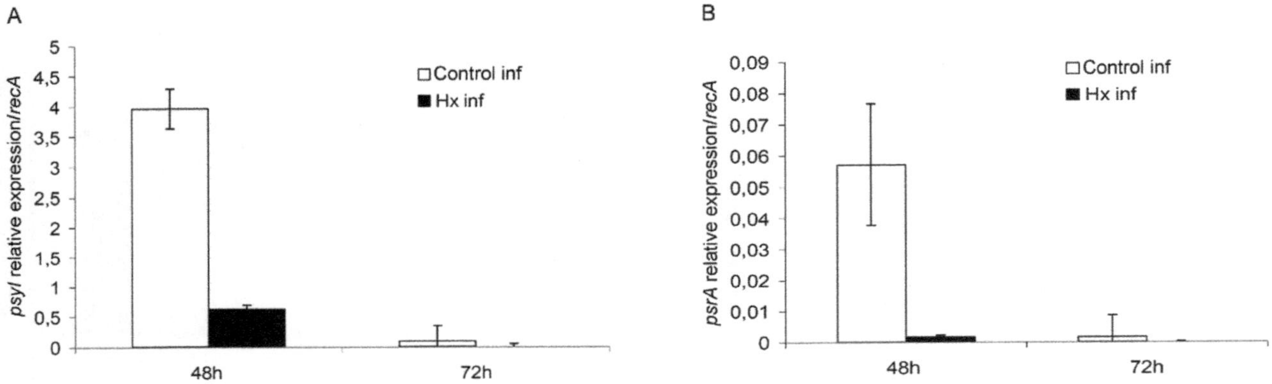

Figure 7. Effect of hexanoic acid treatment on quorum sensing establishment. The relative expression of *psyI* (A) and *psrA* (B) was determined by qRT-PCR during the infection of treated and untreated tomato plants with *Pst* DC3000. The relative expression levels of *psyI* and *psrA* were normalized to those of *recA*. The results represent the means and standard deviation from three different experiments.

decreased in the treated plants. In fact, in most of the samples, the bacterial population was under the limit of detection (Fig. 10A).

Discussion

In previous works, we demonstrated that Hx presents antifungal activity against the phytopathogen *B. cinerea in vitro* at nonphytotoxic concentrations [1]. Similar studies were performed by other researchers who evaluated the *in vitro* effect of several inducers on the fungus *Mycosphaerella fijiensis* [32], but little is known about the effect of resistance inducers on bacteria. Previously, we demonstrated the efficacy of Hx as a plant defence inducer against *Pst* DC3000, and we studied the molecular and physiological mechanisms behind Hx-IR [4]. To elucidate the details of the mechanism of action of Hx, the present study was focused on bacterial response both to the direct effect of Hx, which was applied *in vitro* to the culture medium, as to the changes observed in the plant defence mechanisms in response to Hx, which was applied as an inducer agent.

The main findings of the *in vitro* experiments were the following: Hx at 10 mM has a bacteriostatic but not bactericidal effect against *Pst* DC3000, whereas 20 mM Hx exerts an initial and partial bactericidal effect, which may explain the lower final concentrations of bacteria observed in the long-term experiments.

Furthermore, the formation of bacterial aggregates at these concentrations may be correlated to the adaptation of the bacteria to the stress conditions generated by the action of Hx at high concentrations, may explain the slow growth or the entry into the VBNC stage of the cells, and may be considered a mechanism through which the bacteria protect themselves under stress conditions, as reported for *Erwinia amylovora* [33] and *Ralstonia solanacearum* [34]. The bacterial counts obtained using the Live/Dead Kit are directly related to the real state of the bacteria and not only to their ability to form colonies in solid medium.

These results are further supported by the gene expression analysis of virulence- and quorum sensing-related genes, which revealed that there was no effect of Hx at concentrations lower than 10 mM. The COR synthesis genes, with the exception of *cmaB*, were relatively over expressed when Hx was added at a concentration of 10 mM to the medium. Another aspect that must be taken into consideration is that the analysis of these genes was performed in the exponential growth phase. The higher expression of *cfa1* compared to that of *cmaB* is in agreement with the results of other researchers, who observed that CFA is released with coronatine in the culture medium [35]. Moreover, previous studies performed by other researchers revealed that COR production is regulated by temperature and that the expression of *cma* genes increases during the exponential growth phase while their expression is temperature independent during the stationary phase [36]. The same researchers also reported that the activity of *cma* expression is dependent on the genetic background of the host strain.

In contrast, direct contact with the host did not appear to be required for *hrpA*, *hrpL*, and *avrPtoB* expression because their level of expression in the *in vitro* controls was similar or even exceeded those observed *in planta*. These results are supported by the findings reported by Haapalainen *et al.* [37]. Moreover, the increase in *hrpA*, *hrpL*, and *avrPtoB* expression due to exposure to Hx at 10 mM may be attributed to the abiotic stress caused by higher concentrations of Hx and indicates an important role for T3SS in bacterial adaptation and survival. Rahme *et al.* [38] showed that the *hrp* gene expression in *P. syringae* pv. phaseolicola is under the control of multiple physiological and environmental factors, specifically pH, osmotic strength, and catabolite repression. According to these authors, the optimal expression of all *hrp* genes *in vitro* is obtained under stress conditions that stimulate the environment of the leaf apoplast with respect to a low osmolyte content and pH 5.5. Similar results were obtained by van Dijk *et al.* [39], who demonstrated that T3SS is environmentally regulated at both the transcriptional and post-transcriptional levels and that environmental conditions, such as temperature, and pH play an important role in both gene expression and protein secretion. In contrast, the increase in T3SS-associated gene expression may be related to the growth pattern. Taking into consideration that at the lag phase lasts longer at this Hx concentration, the expression of these genes at 18 h may be required for the bacteria to pass from the lag phase to the exponential phase. Recently, Lee *et al.* [40] revealed the importance of the T3SS and specific effectors for the development of aggregates on the leaf surface. Thus, the increase in the transcription levels of genes that encode the structural and regulatory proteins associated with T3SS may be correlated with the aggregate formation observed after exposure to 10 mM Hx. Additionally, the elevation of the relative expression of *psyI* at this concentration supports our results because elevated levels of the quorum-sensing signal molecule AHL result in the expression of virulence factors that may be necessary for bacterial adaptation and survival under stress conditions.

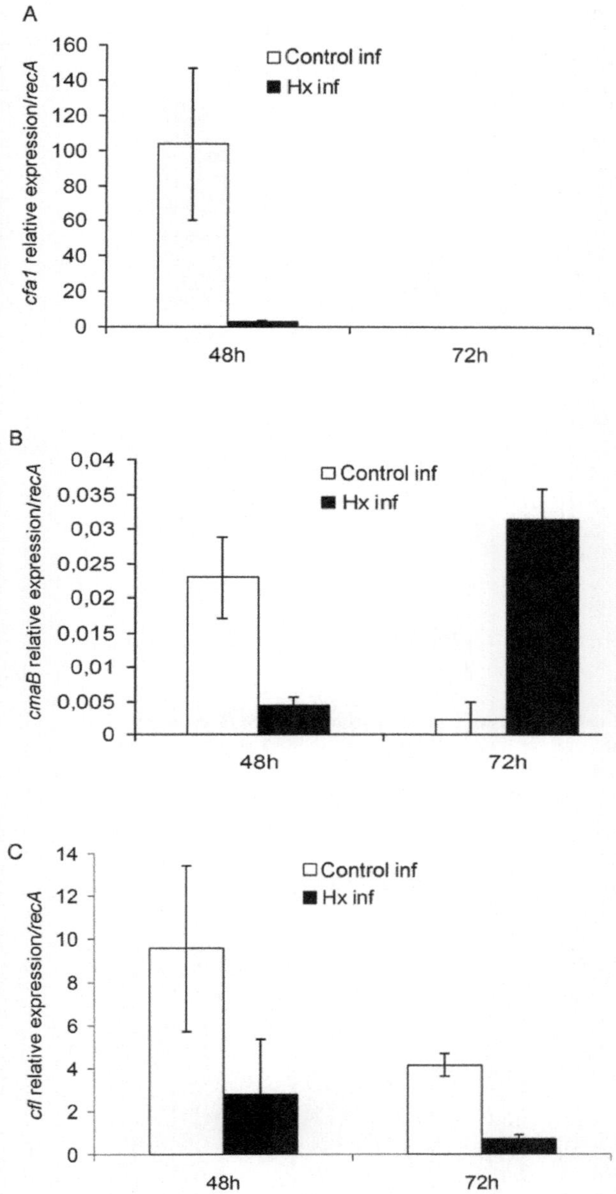

Figure 8. Hx acid treatment reduces expression of COR synthesis genes. The relative expression level of *cfa1* (A), *cmaB* (B) and *cfl* (C) were measured at 48 and 72 h postinoculation, using qRT-PCR with specific primers. The gene expression was normalized with *recA* that was used as endogenous references. The results represent the means of three different experiments with standard deviation.

To investigate the involvement of these factors in bacterial adaptation to stress conditions and survival, we used different mutants with impairments in the analysed genes. Although the mutants did not show reduced survival, the activation of these genes may contribute but not be essential for survival, and their involvement in bacterial adaptation cannot be ruled out. Because *P. syringae* has numerous other effectors, one or more of these may fulfil this role.

Regarding the changes that *Pst* DC3000 undergoes *in planta*, the study of gene expression patterns related to virulence or quorum-sensing establishmen *in planta* may help illuminate how the bacteria can survive and overcome plant defences and, in our case, how the bacteria responds to the induced plant defences since previous works described the main signalling pathways that are involved in basal and induced resistance to this pathogen. The study of hormonal patterns and the expression of marker genes revealed that the activation of these pathways occurs starting 48 hpi [4], which is correlated with transcriptional changes in bacterial genes and disease onset.

The dip-inoculation method used in our studies forces the bacteria to enter plants through natural openings present on the plant surface to establish infection sites within the apoplast. The fact that the treatment lost its effectiveness when the plants were infected by vacuum infiltration [4] indicates the possibility that the treatment may act in the early stages of bacterial infection, such as entry into host tissue and bacterial colonization of the intercellular spaces between plant cells.

The results obtained from our study of the factors involved in the early stages of infection, such as *psyI* and *psrA*, point to the fact that QS is reached earlier in the untreated plants compared with the treated ones, thereby delaying its establishment. It is known that QS in bacteria regulates the gene expression of virulence factors. Thus, this delay can be added to an activation of plant defences that may prevent or at least mitigate pathogen attack. Moreover, Chatterjee *et al.* [41] demonstrated that *psrA* has a positive effect on disease because the mutation of this gene affects the time of the onset of symptoms and their severity. These phenotypes correlate with the observed reduced expression of *hrpL*, which encodes an alternate sigma factor. Thus, in addition to delaying the establishment of QS, Hx treatment may diminish *Pst* DC3000 virulence by reducing the *PsrA* expression levels.

In previous works, we found that Hx treatment affects the action of COR by preventing stomatal reopening after bacterial infection [4]. One of the purposes of this study was to establish whether the changes in the infected plants due to Hx treatment could produce an alteration in the expression of genes involved in the synthesis of this molecule. Interestingly, we observed that Hx treatment reduces the expression of the COR synthesis genes *cfa1*, *cmaB*, and *cfl* at 48 hpi. This finding has an important implication in the enhanced resistance of treated plants, which presented markedly reduced disease severity compared with the untreated plants, because COR has been shown to suppress plant defence responses [23]. In contrast, at 72 hpi, elevated *cmaB* expression levels were recorded in the treated plants. Surprisingly, we also observed an increment in *hrpL* expression at this time point. According to Sreedharan *et al.* [42], HrpL modulates *cfa1* and *cmaB* expression. It is not clear why this increase occurs only in the case of *cmaB*, although it is possible that the bacteria have an excess of CFA due to the higher expression levels of *cfa1* at 48 hpi, which could result in the synthesis of COR. This would therefore involve a delay in symptom development, which was not observed in the long-term experiments. Thus, this discrepancy in the synthesis of the two components of COR in treated plants leads to a delay of disease onset and may influence the survival of the bacteria. On the other hand, the transcription of these genes in infected plants was extensively induced compared with that recorded in medium, which suggests that host signals are involved in their full activation [43].

Another factor that is important in the pathogenicity of *P. syringae* is the type III secretion system, which is encoded by the *hrp/hrc* regulon. In general, *hrp* genes are induced in plants and minimal nutrient media but repressed in complex media. Their expression is influenced by carbon and nitrogen sources, pH, osmolarity, temperature, and plant signal molecules [44]. In the bacteria from treated plants, a delay in *hrpL* and *hrpA* expression

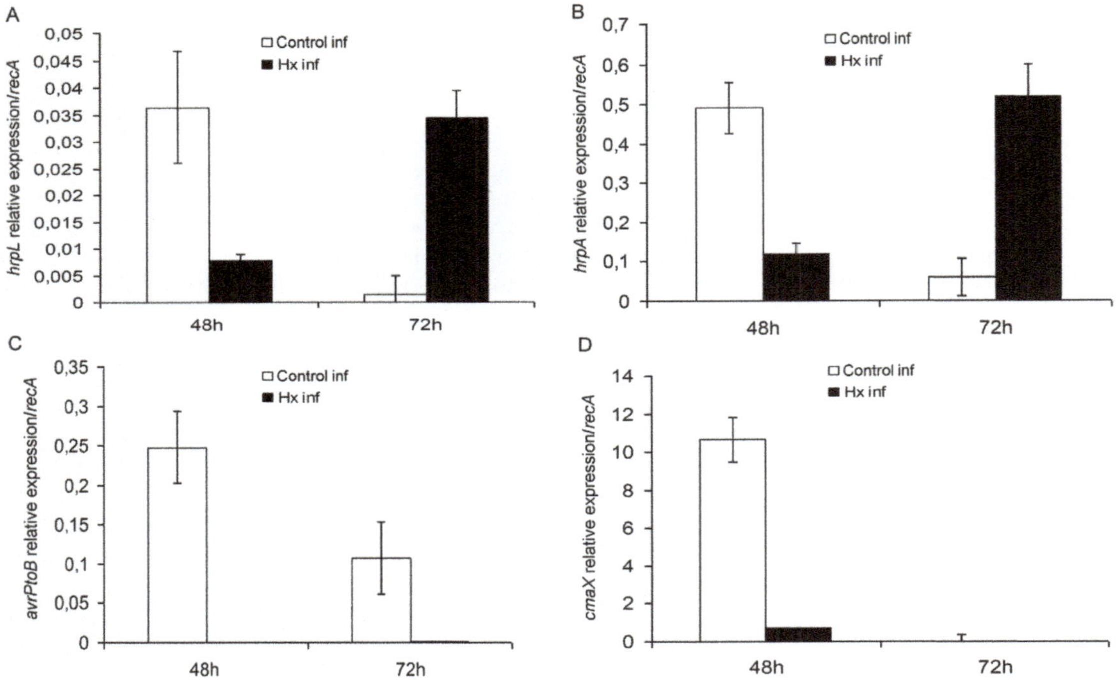

Figure 9. Effect of Hx acid treatment on Type III secretion system and other virulence factors. The transcript levels of type III secretion system associated genes (*hrpL* (A), *hrpA* (B), *avrPtoB* (C)) and CorA-like Mg^{2+} transporter (*cmaX* (D)) was determined at various time points after infection of tomato plants with *P.syringae* DC3000.The *recA* gene was used as endogenous references. The results represent the means of three different experiments with standard deviation.

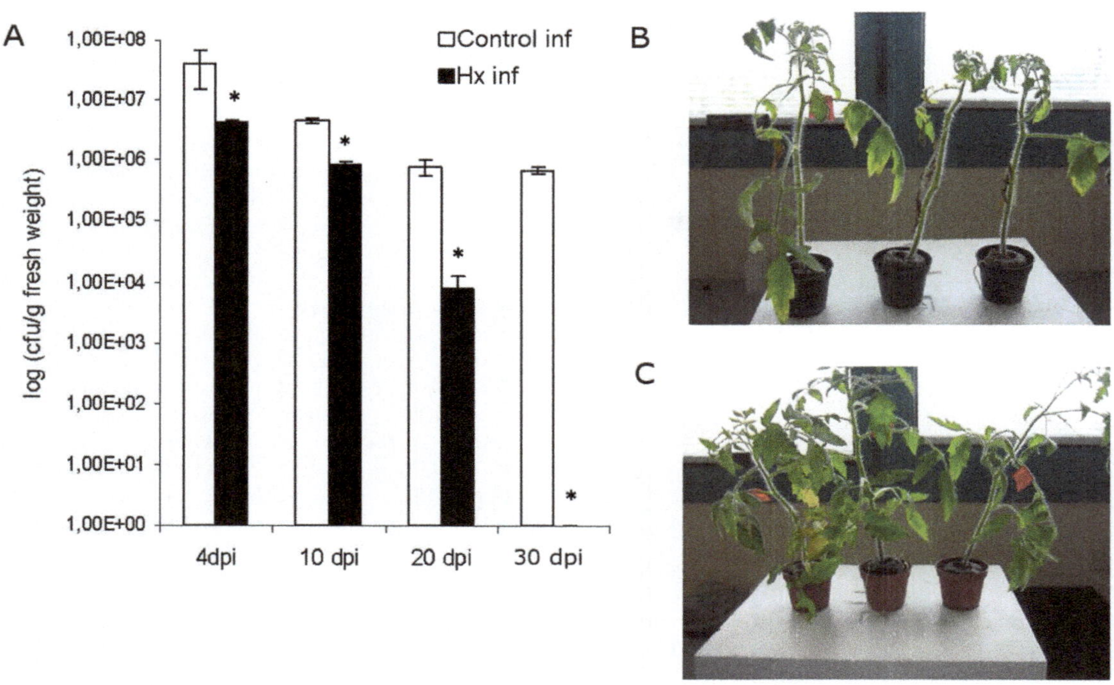

Figure 10. Hx treatment has a long lasting effect. Plants were treated with 0.6 mM Hx and dip inoculated. The bacterial growth was evaluated at 4, 10, 20 and 30 days after inoculation with *P.syringae* DC3000 (10^5 cfu/ml) (A). The pictures of representative untreated (B) and Hx treated (C) plants were taken 20 days after inoculation.

was observed. Although the expression of *hrpL* was 10-fold higher *in vitro* than *in planta*, other researchers have clearly shown that T3SS requires HrpL for its function [45]; however, the low expression levels observed *in planta* may be an indication of the existence of additional levels of regulation that have not been previously described. These results may also indicate that HrpL does not accumulate to high intracellular levels, suggesting that some type of feedback regulation could take place between the secretion and production of HrpL, perhaps through the degradation of intracellular HrpL [37].

In *Pst* DC3000, the T3SS is induced when the bacteria enter the leaf mesophyll through stomata. This bacterium can control the volume of type-three protein secretion by the expression level of HrpA pilin and the number of pili formed on the cell surface. The amount of HrpA protein needed per individual T3SS machinery depends on the distance between the bacterium and the plant cell [46]. It is possible that Hx treatment delays the contact of the bacterium with the plant cells and that this delay is responsible for the elevated levels of *hrpA* transcripts observed later in the treated plants.

Given that the current view of molecular plant-pathogen interactions is that pathogens secrete effectors into the apoplast and/or cytoplasm of host plants to suppress host defences, facilitate colonization, and trigger symptom production [47,48], several effectors have been studied. AvrPtoB has been found to act inside plant cells to suppress programmed cell death associated with plant immunity and increases pathogen growth, dissemination, and disease formation [19,49]. Once infection is established, this effector induces host programmed cell death to gain access to nutrients or aid pathogen dissemination. The fact that Hx treatment diminishes *avrPtoB* expression *in planta* is supported by the results reported by Abramovitch and Martin [49] because the symptoms are greatly reduced, even though the size of the bacterial population is increased to a lesser extent compared with untreated plants.

The virulence of pathogenic bacteria is also affected by divalent cations, of which Mg^{2+} plays a central role. CorA or CmaX is the major Mg^{2+} transporter. The importance of CorA or CmaX in virulence was previously reported [26,50]. We observed that Hx treatment clearly reduces *cmaX* expression, and this decrease may affect the production of several plant cell wall-degrading exoenzymes, such as pectate lyase, polygalacturonase, cellulose, and protease, which are responsible for the maceration of plant tissue. This reduction could also affect bacterial survival and multiplication *in planta*. Our data are supported by the results obtained by Kersey *et al.* [51] *in planta*. These researchers noted a significant reduction in the cfu of the $CorA^-$ mutant recovered from infected tissue compared with the $CorA^+$ parent.

Because several genes were found to be expressed in the bacteria from the treated plants at 72 hpi, we then investigated whether the bacterial population could increase to a point where the bacteria might overcome the plant defence responses. The results obtained in the long-term assays discard this hypothesis and point to the induction of defence as a control strategy that reduces the pathogenicity by limiting the rate of pathogen multiplication [52] and by changing the expression pattern of virulence-related genes.

The conclusions we can draw for the effects of Hx on bacteria are that Hx has a direct *in vitro* effect on bacteria at high concentrations and that its effect on treated plants demonstrates that it acts as an inducer of resistance. In the latter case, the Hx effect is not conditioned by its presence in the plant, as demonstrated by our previous finding that Hx does not accumulate in treated plants because a chromatographic analysis revealed no detectable trace of either Hx or 13C-Hx above the detection limits (10 $\mu g\ L^{-1}$) [6]. Thus, one of the most interesting outcomes of this study is that the two effects are opposite: the direct effect of Hx consists of the activation of the bacterial genes involved in virulence, and its effect as an inducer of resistance is based on the inactivation of these genes, a delay in their expression, or a discrepancy in their activation that leads to the development of less symptoms and the decreased ability of the bacteria to grow and may even compromise its long-term survival. It is interesting to note that the transcriptional changes in the bacterial genes in the treated plants occur at the same time point (48 hpi) as the changes in the plants. Additionally, the differences in the bacterial population size are more pronounced between the treated and non-treated plants at 72 h. Because most of the bacterial genes studied are repressed at 48 h, this repression apparently prevents bacterial growth. This finding suggests that the pathogenesis and survival of the bacteria is affected by plant response when Hx is applied as a resistance inducer and that Hx affects both aspects of the molecular "arms race" (i.e., the plant immune system and pathogen virulence effectors) [53] to benefit the plant immune system.

Supporting Information

File S1 File S1 includes the following: Data S1. Table presenting data underlying the results described in figure 1. **Data S2.** Table presenting data underlying the results described in figure 2. **Data S3.** Table presenting data underlying the results described in figure 3. **Data S4.** Table presenting data underlying the results described in figure 4. **Data S5.** Table presenting data underlying the results described in figure 5. **Data S6.** Table presenting data underlying the results described in figure 6. **Data S7.** Table presenting data underlying the results described in figure 7. **Data S8.** Table presenting data underlying the results described in figure 8. **Data S9.** Table presenting data underlying the results described in figure 9. **Data S10.** Table presenting data underlying the results described in figure 10.

Acknowledgments

The authors are grateful to Instituto Valenciano de Investigaciones Agrarias (IVIA). We thank Esther Nogales (IVIA) for her help in Bioscreen and VBNC analysis.

Author Contributions

Conceived and designed the experiments: LS BV. Performed the experiments: LS BV GC EL EFC. Analyzed the data: LS BV GC. Contributed reagents/materials/analysis tools: PG BV ML. Contributed to the writing of the manuscript: LS BV EL ML.

References

1. Leyva MO, Vicedo B, Finiti I, Flors V, Del Amo G, et al. (2008) Preventive and post-infection control of Botrytis cinerea in tomato plants by hexanoic acid. Plant Pathology 57: 1038–1046.
2. Llorens E, Fernandez-Crespo E, Vicedo B, Lapena L, Garcia-Agustin P (2013) Enhancement of the citrus immune system provides effective resistance against Alternaria brown spot disease. Journal of Plant Physiology 170: 146–154.
3. Pastor V, Luna E, Mauch-Mani B, Ton J, Flors V (2012) Primed plants do not forget. Environmental and Experimental Botany 10.1016/j.envexpbot.2012.02.013.
4. Scalschi L, Vicedo B, Camanes G, Fernandez-Crespo E, Lapena L, et al. (2013) Hexanoic acid is a resistance inducer that protects tomato plants against Pseudomonas syringae by priming the jasmonic acid and salicylic acid pathways. Molecular Plant Pathology 14: 342–355.

5. Ton J, Mauch-Mani B (2004) beta-amino-butyric acid-induced resistance against necrotrophic pathogens is based on ABA-dependent priming for callose. Plant Journal 38: 119–130.
6. Vicedo B, Flors V, Leyva MD, Finiti I, Kravchuk Z, et al. (2009) Hexanoic Acid-Induced Resistance Against Botrytis cinerea in Tomato Plants. Molecular Plant-Microbe Interactions 22: 1455–1465.
7. Kravchuk Z, Vicedo B, Flors V, Camanes G, Gonzalez-Bosch C, et al. (2011) Priming for JA-dependent defenses using hexanoic acid is an effective mechanism to protect Arabidopsis against B. cinerea. Journal of Plant Physiology 168: 359–366.
8. Xin X-F, He SY (2013) Pseudomonas syringae pv. tomato DC3000: A Model Pathogen for Probing Disease Susceptibility and Hormone Signaling in Plants. Annual Review of Phytopathology, Vol 51 51: 473–498.
9. Zhao YF, Thilmony R, Bender CL, Schaller A, He SY, et al. (2003) Virulence systems of Pseudomonas syringae pv. tomato promote bacterial speck disease in tomato by targeting the jasmonate signaling pathway. Plant Journal 36: 485–499.
10. Buell CR, Joardar V, Lindeberg M, Selengut J, Paulsen IT, et al. (2003) The complete genome sequence of the Arabidopsis and tomato pathogen Pseudomonas syringae pv. tomato DC3000. Proceedings of the National Academy of Sciences of the United States of America 100: 10181–10186.
11. Katagiri F, Thilmony R, He SY (2002) The Arabidopsis thaliana-pseudomonas syringae interaction. The Arabidopsis book/American Society of Plant Biologists 1: e0039–e0039.
12. Melotto M, Underwood W, He SY (2008) Role of stomata in plant innate immunity and foliar bacterial diseases. Annual Review of Phytopathology 46: 101–122.
13. Rouse DI, Nordheim EV, Hirano SS, Upper CD (1985) A model relating the probability of foliar disease incidence to the population frequencies of bacterial plant-pathogens. Phytopathology 75: 505–509.
14. Loh J, Pierson EA, Pierson LS, Stacey G, Chatterjee A (2002) Quorum sensing in plant-associated bacteria. Current Opinion in Plant Biology 5: 285–290.
15. von Bodman SB, Bauer WD, Coplin DL (2003) Quorum sensing in plant-pathogenic bacteria. Annual Review of Phytopathology 41: 455–482.
16. Bertani I, Sevo M, Kojic M, Venturi V (2003) Role of GacA, LasI, RhlI, ppk, PsrA, vfr and ClpXP in the regulation of the stationary-phase sigma factor rpoS/RpoS in Pseudomonas. Archives of Microbiology 180: 264–271.
17. Cui H, Xiang T, Zhou J-M (2009) Plant immunity: a lesson from pathogenic bacterial effector proteins. Cellular Microbiology 11: 1453–1461.
18. Lindeberg M, Stavrinides J, Chang JH, Alfano JR, Collmer A, et al. (2005) Proposed guidelines for a unified nomenclature and phylogenetic analysis of type III Hop effector proteins in the plant pathogen Pseudomonas syringae. Molecular Plant-Microbe Interactions 18: 275–282.
19. Lin N-C, Martin GB (2005) An avrPto/avrPtoB Mutant of Pseudomonas syringae pv. tomato DC3000 Does Not Elicit Pto-Mediated Resistance and Is Less Virulent on Tomato. Molecular Plant-Microbe Interactions 18: 43–51.
20. Xiao YX, Heu SG, Yi JS, Lu Y, Hutcheson SW (1994) Identification of a putative alternate sigma factor and characterization of a multicomponent regulatory cascade controlling the expression of *Pseudomonas syringae* pv. *syringae* Pss61 hrp and hrmA genes. Journal of Bacteriology 176: 1025–1036.
21. Bender CL, Alarcon-Chaidez F, Gross DC (1999) Pseudomonas syringae phytotoxins: Mode of action, regulation, and biosynthesis by peptide and polyketide synthetases. Microbiology and Molecular Biology Reviews 63: 266–+.
22. Bender CL, Stone HE, Sims JJ, Cooksey DA (1987) Reduced pathogen fitness of Pseudomonas syringae pv. tomato Tn5 insertions defective in coronatine production. Physiol Mol Plant Pathol 30: 273–283.
23. Brooks DM, Bender CL, Kunkel BN (2005) The Pseudomonas syringae phytotoxin coronatine promotes virulence by overcoming salicylic acid-dependent defences in Arabidopsis thaliana. Molecular Plant Pathology 6: 629–639.
24. Yao N, Imai S, Tada Y, Nakayashiki H, Tosa Y, et al. (2002) Apoptotic cell death is a common response to pathogen attack in oats. Molecular Plant-Microbe Interactions 15: 1000–1007.
25. Ichihara A, Shiraishi K, Sato H, Sakamura S, Nishiyama K, et al. (1977) The structure of coronatine [22]. Journal of the American Chemical Society 99: 636–637.
26. Papp-Wallace KM, Nartea M, Kehres DG, Porwollik S, McClelland M, et al. (2008) The CorA Mg2+ channel is required for the virulence of Salmonella enterica serovar Typhimurium. Journal of Bacteriology 190: 6517–6523.
27. Pfeiffer J, Guhl J, Waidner B, Kist M, Bereswill S (2002) Magnesium uptake by CorA is essential for viability of the gastric pathogen Helicobacter pylori. Infection and Immunity 70: 3930–3934.
28. Reinhart RA (1988) Magnesium-metabolism - a review with special reference to the relationship between intracellular content and serum levels. Archives of Internal Medicine 148: 2415–2420.
29. Nikaido H, Vaara M (1985) MOLECULAR-BASIS OF BACTERIAL OUTER-MEMBRANE PERMEABILITY. Microbiological Reviews 49: 1–32.
30. Beattie GA, Lindow SE (1999) Bacterial colonization of leaves: A spectrum of strategies. Phytopathology 89: 353–359.
31. Hirano SS, Upper CD (2000) Bacteria in the leaf ecosystem with emphasis on Pseudomonas syringae - A pathogen, ice nucleus, and epiphyte. Microbiology and Molecular Biology Reviews 64: 624–653.
32. Mogollón Ortiz ÁM, Castaño Zapata J (2012) Evaluación in vitro de Inductores de Resistencia sobre Mycosphaerella fijiensis Morelet. Revista Facultad Nacional de Agronomía, Medellín 65: 6327–6336.
33. Ordax M, Marco-Noales E, Lopez MM, Biosca EG (2006) Survival strategy of Erwinia amylovora against copper: Induction of the viable-but-nonculturable state. Applied and Environmental Microbiology 72: 3482–3488.
34. Alvarez B, Lopez MM, Biosca EG (2008) Survival strategies and pathogenicity of Ralstonia solanacearum phylotype II subjected to prolonged starvation in environmental water microcosms. Microbiology-Sgm 154: 3590–3598.
35. Ferguson IB, Mitchell RE (1985) STIMULATION OF ETHYLENE PRODUCTION IN BEAN LEAF-DISKS BY THE PSEUDOMONAD PHYTOTOXIN CORONATINE. Plant Physiology 77: 969–973.
36. Weingart H, Stubner S, Schenk A, Ullrich MS (2004) Impact of temperature on in planta expression of genes involved in synthesis of the Pseudomonas syringae phytotoxin coronatine. Molecular Plant-Microbe Interactions 17: 1095–1102.
37. Haapalainen M, van Gestel K, Pirhonen M, Taira S (2009) Soluble Plant Cell Signals Induce the Expression of the Type III Secretion System of Pseudomonas syringae and Upregulate the Production of Pilus Protein HrpA. Molecular Plant-Microbe Interactions 22: 282–290.
38. Rahme LG, Mindrinos MN, Panopoulos NJ (1992) Plant and environmental sensory signals control the expression of hrp genes in *Pseudomonas syringae* pv *phaseolicola*. Journal of Bacteriology 174: 3499–3507.
39. van Dijk K, Fouts DE, Rehm AH, Hill AR, Collmer A, et al. (1999) The Avr (effector) proteins HrmA (HopPsyA) and AvrPto are secreted in culture from Pseudomonas syringae pathovars via the Hrp (type III) protein secretion system in a temperature- and pH-sensitive manner. Journal of Bacteriology 181: 4790–4797.
40. Lee J, Teitzel GM, Munkvold K, del Pozo O, Martin GB, et al. (2012) Type III Secretion and Effectors Shape the Survival and Growth Pattern of Pseudomonas syringae on Leaf Surfaces. Plant Physiology 158: 1803–1818.
41. Chatterjee A, Cui Y, Hasegawa H, Chatterjee AK (2007) PsrA, the Pseudomonas sigma regulator, controls regulators of epiphytic fitness, quorum-sensing signals, and plant interactions in Pseudomonas syringae pv. tomato strain DC3000. Applied and Environmental Microbiology 73: 3684–3694.
42. Sreedharan A, Penaloza-Vazquez A, Kunkel BN, Bender CL (2006) CorR regulates multiple components of virulence in Pseudomonas syringae pv. tomato DC3000. Molecular Plant-Microbe Interactions 19: 768–779.
43. Nurnberger T, Brunner F, Kemmerling B, Piater L (2004) Innate immunity in plants and animals: striking similarities and obvious differences. Immunological Reviews 198: 249–266.
44. Lindgren PB (1997) The role of hrp genes during plant-bacterial interactions. Annual Review of Phytopathology 35: 129–152.
45. Xiao YX, Hutcheson SW (1994) A single promoter sequence recognized by a newly identified alternate sigma factor directs expression of pathogenicity and host range determinants in *Pseudomonas syringae*. Journal of Bacteriology 176: 3089–3091.
46. Boureau T, Routtu J, Roine E, Taira S, Romantschuk M (2002) Localization of hrpA-induced Pseudomonas syringae pv. tomato DC3000 in infected tomato leaves. Molecular Plant Pathology 3: 451–460.
47. Chisholm ST, Coaker G, Day B, Staskawicz BJ (2006) Host-microbe interactions: Shaping the evolution of the plant immune response. Cell 124: 803–814.
48. Jones JDG, Dangl JL (2006) The plant immune system. Nature 444: 323–329.
49. Abramovitch RB, Martin GB (2005) AvrPtoB: A bacterial type III effector that both elicits and suppresses programmed cell death associated with plant immunity. Fems Microbiology Letters 245: 1–8.
50. Haque MM, Tsuyumu S (2005) Virulence, resistance to magainin II, and expression of pectate lyase are controlled by the PhoP-PhoQ two-component regulatory system responding to pH and magnesium in Erwinia chrysanthemi 3937. Journal of General Plant Pathology 71: 47–53.
51. Kersey CM, Agyemang PA, Dumenyo CK (2012) CorA, the magnesium/nickel/cobalt transporter, affects virulence and extracellular enzyme production in the soft rot pathogen Pectobacterium carotovorum. Molecular Plant Pathology 13: 58–71.
52. Miller MR, White A, Boots M (2005) The evolution of host resistance: Tolerance and control as distinct strategies. Journal of Theoretical Biology 236: 198–207.
53. Nomura K, Mecey C, Lee Y-N, Imboden LA, Chang JH, et al. (2011) Effector-triggered immunity blocks pathogen degradation of an immunity-associated vesicle traffic regulator in Arabidopsis. Proceedings of the National Academy of Sciences of the United States of America 108: 10774–10779.
54. King EO, Ward MK, Raney DE (1954) Two simple media for the demonstration of pyocyanin and fluorescin. J Lab Clin Med 44: 301–307.
55. Yu X, Lund SP, Scott RA, Greenwald JW, Records AH, et al. (2013) Transcriptional responses of Pseudomonas syringae to growth in epiphytic versus apoplastic leaf sites. Proceedings of the National Academy of Sciences of the United States of America 110: E425–E434.

Characterization of Lignocellulolytic Activities from a Moderate Halophile Strain of *Aspergillus caesiellus* Isolated from a Sugarcane Bagasse Fermentation

Ramón Alberto Batista-García[1,2]☯, Edgar Balcázar-López[1,2]☯, Estefan Miranda-Miranda[3], Ayixón Sánchez-Reyes[1,2], Laura Cuervo-Soto[1,2], Denise Aceves-Zamudio[2], Karina Atriztán-Hernández[4], Catalina Morales-Herrera[2], Rocío Rodríguez-Hernández[2], Jorge Folch-Mallol[2]*

1 Facultad de Ciencias, Universidad Autónoma del Estado de Morelos, Cuernavaca, Morelos, Mexico, **2** Centro de Investigación en Biotecnología, Universidad Autónoma del Estado de Morelos, Cuernavaca, Morelos, México, **3** Centro Nacional de Investigación Disciplinaria en Parasitología Veterinaria, Instituto Nacional de Investigaciones Forestales Agrícolas y Pecuarias, Cuernavaca, Morelos, Mexico, **4** Facultad de Ciencias Biológicas, Universidad Autónoma del Estado de Morelos, Cuernavaca, Morelos, Mexico

Abstract

A moderate halophile and thermotolerant fungal strain was isolated from a sugarcane bagasse fermentation in the presence of 2 M NaCl that was set in the laboratory. This strain was identified by polyphasic criteria as *Aspergillus caesiellus*. The fungus showed an optimal growth rate in media containing 1 M NaCl at 28°C and could grow in media added with up to 2 M NaCl. This strain was able to grow at 37 and 42°C, with or without NaCl. *A. caesiellus* H1 produced cellulases, xylanases, manganese peroxidase (MnP) and esterases. No laccase activity was detected in the conditions we tested. The cellulase activity was thermostable, halostable, and no differential expression of cellulases was observed in media with different salt concentrations. However, differential band patterns for cellulase and xylanase activities were detected in zymograms when the fungus was grown in different lignocellulosic substrates such as wheat straw, maize stover, agave fibres, sugarcane bagasse and sawdust. Optimal temperature and pH were similar to other cellulases previously described. These results support the potential of this fungus to degrade lignocellulosic materials and its possible use in biotechnological applications.

Editor: Melanie R. Mormile, Missouri University of Science and Technology, United States of America

Funding: The work was partially funded by grant CB-153789-Q from the National Council for Science and Technology (CONACyT) from the Mexican government. The funders had no role in study design, data collection and analysis, decision to publish, or preparation of the manuscript.

Competing Interests: Jorge L. Folch-Mallol is currently an Academic Editor for PLOS ONE but this does not alter the authors' adherence to PLOS ONE Editorial policies and criteria.

* Email: jordi@uaem.mx

☯ These authors contributed equally to this work.

Introduction

For many years, extremophile microorganisms were described exclusively in the Archae and Eubacteria domains [1]. These peculiar organisms have been found and studied in environments with very extreme conditions such as temperature, pressure, pH, salinity and radiation, among others. Physiological and metabolic studies on extremophile microorganisms derived in the development of biotechnological applications with incidence in the agricultural, pharmacological and environmental industries [2–6]. The description of microbial diversity in extremophile ecosystems has been of great interest for a long time until our days. However, the report of eukaryotic extremophiles took many years.

In particular, environments with high concentrations of salt, called saline and hypersaline environments, have been attractive ecosystems for studying the structures of microbial communities that inhabit them [7–9], since sodium chloride has been considered an effective microbicide. For example, treating food with salt is an effective method of preserving meat since ancient times [10]. The isolation and characterization of microorganisms in habitats with over 1 M NaCl is important for the identification of metabolites and/or robust proteins with potential industrial applications, and to understand the cellular physiology, molecular biology and biochemistry that support the survival of these organisms under extreme conditions [11–13].

Numerous studies have described halophilic bacterial genera in hypersaline ecosystems [5], [8], [14–21]. However the first report of a fungus isolated from a solar saltern appeared in the year 2000 [22]. Since then, many studies on biodiversity and physiology have reported the characterization of halophilic fungi present in saline and hypersaline ecosystems. Many species of Ascomycetes, and some Basidiomycetes, have been described with the ability to grow in these environments [22], [23]. Currently, the presence of fungi in the structure of microbial communities that compose some ecosystems with extreme salinity conditions has been described [24]. Until 2009, only 10 orders of fungi known to tolerate low water activity (a_w) due to high salt concentrations have been reported [22].

Among the most studied halophilic fungal genera are *Cladosporium* [22], [23], [25], *Wallemia*, *Scopulariopsis*, *Alternaria* [24], [26] and some species of *Aspergillus* and *Penicillium* [23]. Particularly in the *Aspergillus* genera, *A. niger, A. sydowwi, A. flavus, A. tubingensis* and *A. versicolor* have been isolated and described as halotorelant and halophile fungi as part of the hypersaline environments [22]. The study of these fungi has allowed the characterization of halophilic enzymes with interesting properties [27]. For example, cellulases and xylanases have been described from halophilic microorganisms, filamentous fungi included, with interesting biochemical properties such as activity in the presence of high salt concentrations, at acidic pH, or in ionic liquids [11], [13], [28–33]. In general, halophilic microorganisms can be isolated from the sea, saline or hypersaline lakes, solar salterns, and salted foods, among other habitats with high concentrations of salt.

It was assumed for a long time that extremophile microorganisms were strictly growing in extreme conditions; however this assumption was proved to be false because there are many examples of extremophiles isolated from non-extreme environments [34–36]. In this way, some reports demonstrate that halophilic and/or halotolerant microorganisms are not restricted to saline or hypersaline habitats and can be found almost everywhere in non-saline environments [37–46]. The general principles of microbial ecology, the physiological plasticity and metabolic versatility of the microorganisms do not impede the isolation of halophile and/or halotolerant organisms from non-saline environments.

The isolation of halophile and/or halotolerant fungi able to grow in lignocellulosic materials is a very interesting source for the search of industrially useful enzymes, potentially capable of producing fermentable sugars from agricultural wastes. One of the major limitations of using enzymes in industry is obtaining robust biocatalysts capable of operating under rigorous industrial conditions. Because of this demand, the study of halophilic and halotolerant microorganisms is an appropriate strategy for the characterization of enzymes such as cellulases, esterases, lipases and xylanases with potential application in biorefineries. Sugarcane bagasse is considered a recalcitrant, high cellulose content agroindustrial waste. The microbial communities growing on sugarcane bagasse, especially filamentous fungi must have efficient enzymes to obtain energy and carbon from lignocellulose degradation. The aims of this work were to characterize a filamentous fungus isolated from sugarcane bagasse with the potential to grow in concentrations of sodium chloride higher than 1 M, and to analyse the lignocellulolytic activities it produces.

Materials and Methods

Isolation and preservation of microorganisms

Three grams of non-sterile sugarcane bagasse (as the main carbon source and from where the microorganisms were isolated) were mixed with 2.5 g of sterile soil (as a source of organic matter) into 1000 mL Erlenmeyer flasks containing 250 mL of Vogels medium as described in [47] and a final concentration of 2 M NaCl. Fermentation was performed at 25°C and 150 rpm for 30 days. Primary isolation of cellulolytic microorganisms was done taking one mL of the sugar bagasse culture and serial dilutions were performed up to 10^{-10}. Two hundred μL of each dilution were inoculated in Petri plates containing Vogels medium supplemented with 2% carboxymethylcellulose (CMC, Sigma Catalogue No. C5678) and 0.5 M NaCl. The cultures were incubated for 10 days at 30°C and were observed daily for secondary isolation of cellulolytic bacteria and/or fungi. A fungus was isolated in Potato Dextrose Agar (PDA) and Saboraud agar media (both from DIFCO). This fungus was stored at 4°C in saline solution (0.5% NaCl) supplemented with glycerol (20%). Isolates were performed in triplicate to achieve representativeness of the obtained microbial populations.

Identification of a fungal moderate halophilic strain

Mycelium from a fungus growing on PDA plates was collected after 10-days for genomic DNA isolation according to a previously reported method [48]. For identification, we analysed molecular markers previously described to be distinctive to filamentous fungi. A fragment of the 18S ribosomal DNA was amplified by PCR using primers nu-SSU-0817 and nu-SSU-1536 as described by Bornenman and Hartin [49]. Also, regions of the 28S large subunit RNA gene (D1–D2) and internal transcribed spacers 1 (ITS1 region) were amplified. These regions have been particularly useful for the molecular identification of fungi. The primers and conditions used for these PCR reactions have been previously described by Peterson [50] and Hinrikson *et al.* [51], respectively.

Amplicons were analysed in 1% agarose gel electrophoresis in 1x TBE buffer [52] and purified from the agarose gels using a commercial gel extraction kit (Fermentas Catalogue No. K0513) and sequenced in both directions using the same primers used for the amplification. Sanger sequencing was performed at the Sequencing Unit of the Instituto de Biotecnología of the Universidad Nacional Autónoma de México.

The sequences were analysed using the website of the National Centre for Biotechnology Information (NCBI) (www.ncbi.nih.gov). BLAST search was performed to determine similar sequences. Phylogeny studies allowed defining the relationship of these sequences with those obtained in the BLAST analysis. Phylogenetic analysis was performed online with the server Phylogeny.fr (www.phylogeny.fr). This platform considers various bioinformatics programs to reconstruct a robust phylogenetic tree from a set of sequences. These tools allowed identification of the fungal strain considering molecular criteria.

Micromorphological identification was performed to complete a polyphasic vision for the microbial identification. Size, shape and grouping of conidiophores, morphological aspects of the colony, and the fungal hyphae were analysed. The aspect of the colony on different media (PDA, Saboraud agar and Malt extract agar (MEA, from DIFCO) was also considered.

Growth Rate Determination

Specific growth rate of the fungal colony (expressed as mm/day) was determined as follows. Plugs of 7 mm of diameter obtained from a fungal pre-culture in Vogel's medium supplemented with 2% CMC were inoculated in the same medium at different temperatures (28, 37, and 42°C) or supplemented with NaCl (0, 0.5, 1, 1.5, 2 or 3 M, final concentration). The diameter of the colony was measured every 24 hours for 10 days. Experiments were performed in triplicate for subsequent statistical analysis of data.

Solid-state Fermentation

Solid-state fermentations were performed using the following autoclaved substrates: agave fibre (*Agave fourcroydes*), sugarcane bagasse (*Saccharum officinarum*), maize stover (*Zea mays*), wheat straw (*Triticum aestivum*), and pine sawdust (*Pinus sylvestris*).

Erlenmeyer flasks of 500 mL including 2 g of each substrate were inoculated with two plugs of 7 mm of diameter of the fungal strain previously grown on PDA plates. Humidity in the system was maintained by adding 20% (w/v) of distilled water to the solids. Fermentation was allowed to take place at 28°C for 10 days. Subsequently, soluble fermentation products were collected in 10 mL in 60 mM citrate buffer pH 5.

Enzymatic activity determinations

Cellulase activity was determined qualitatively and quantitatively. For qualitative determinations, seven-day cultures of the fungus grown on agar Vogel's medium supplemented with 2% CMC and NaCl (0, 0.5, 1, 1.5, 2 M, final concentration) at 28°C were set up. Petri dishes with the colonies were then inundated with approximately 15 mL of Congo red (1% diluted in distilled water) for 10–15 minutes. Subsequently, these dishes were washed three times with approximately 15 mL of a 1 M NaCl solution. Discoloration halos around the colony indicated the degradation of cellulose due to the production of cellulases [53], [54]. Determinations were performed in triplicate.

For quantitative determination of cellulase activity, plugs of 7 mm of diameter from pre-cultures of the fungus grown in Vogels medium with 2% CMC were inoculated in 500 mL Erlenmeyer flasks with 100 mL of the same medium plus NaCl (0, 0.5, 1, 1.5, 2 M). The flasks were incubated for 9 days at 28°C and 150 rpm. Enzymatic activity and protein concentration were determined every 24 h from 2 mL of the supernatants of these liquid cultures. Enzymatic activity was assessed by the production of reducing sugars from polymeric substrates using the 3,5-dinitrosalysilic acid (DNS) assay described by Miller [55]. Protein concentration was determined by the Lowry method [56]. Enzymatic specific activity is expressed as IU/mg protein.

For cellulases activity measurements, CMC (2%) dissolved in 50 mM citrate buffer pH 5 was used as a substrate. The enzymatic reaction contained 200 μL of supernatant, 300 μL of 50 mM citrate buffer pH 5 and 500 μL of substrate solution. The reaction mixtures were incubated at 50°C for 30 minutes. Briefly, 50 μL aliquots were taken every 5 minutes (after adding the supernatant to the reaction mixture) up to 45 minutes and then mixed with 50 μL of a DNS solution, boiled for 5 minutes and immediately cooled on ice for 5 minutes. Finally 500 μL of water were added and absorbance measured at λ 540 nm in a spectrophotometer (BioMate, ThermoSpectronic). Reducing sugars concentration was extrapolated from a glucose standard curve ranging from 0.1 to 2 mg/mL; concentration values were plotted against time, and the slope was calculated to determine the velocity of the reaction. Concentration of released reducing sugars *vs.* time was used to calculate enzymatic activities, where 1 IU is defined as 1 μmol of glucose equivalent released per minute, under the assayed conditions. For specific activity calculation, protein concentration in mg/mL was determined against a bovine serum albumin (BSA) standard curve [57]. Xylanases activity was measured in a similar way using 2% oat xylan (Sigma Catalogue No. X0627) as substrate. The absorbance readings were compared against a standard curve of xylose (0.1 to 2 mg/mL).

MnP activity was determined spectrophotometrically (λ 270 nm) by following the formation of Mn^{3+}–malonate complex at pH 4.5 in 50 mM sodium malonate buffer with 0.5 mM $MnSO_4$ [58]. To start the reaction, H_2O_2 was added to a final concentration of 0.1 mM [59]. The reaction was followed for 30 seconds at room temperature. ΔAbs min^{-1} was converted to UL^{-1} using the malonate extinction coefficient of 11 590 $M^{-1}cm^{-1}$ [58]. MnP specific activity is expressed as IU/mg protein.

Laccase activity was monitored by oxidation of 1 mM of 2,2′-azino-bis(3-ethylbenzthiazoline-6-sulfonic acid) (ABTS) in acetate buffer (pH 4.5), measuring formation of the cation radical ($\varepsilon436 = 2.9 \times 104\ M^{-1}\ cm^{-1}$). The reaction mixture volume was 1 mL and was incubated for 30 minutes at room temperature. One unit of laccase activity is defined as the amount of enzyme catalysing the oxidation of 1 μmol of substrate per minute. ΔAbs min^{-1} was determined at λ 420 nm using spectrophotometer (BioMate, ThermoSpectronic).

Lipase/esterase and esterase activities were assessed using 2-naphthyl acetate (Sigma Catalogue No. N6875) and 4-nitrophenyl acetate (Sigma Catalogue No. N8130) as substrates, respectively. A stock solution (250 μM) of 2-naphthyl acetate was prepared in phosphate buffer saline (PBS) pH 6.5, while a stock solution of 4-nitrophenyl acetate (300 μM) was prepared in potassium phosphate buffer pH 6. The reaction volume was 1 mL and it was followed for 30 minutes at room temperature. The determination of these activities was measured spectrophotometrically at λ 538 nm for detection of 2-naphtol and at λ 410 nm for 4-nitrophenol products of enzymatic hydrolysis of lipase/esterase and esterase respectively, using the molar extinction coefficient for each substrate (23 598 $M^{-1}cm^{-1}$ for 2-naphthol and 17 700 $M^{-1}cm^{-1}$ for 4-nitrophenol). Lipase/esterase and esterase specific activities are expressed as IU/mg protein.

To determine all activities we plotted the absorbance values *vs* time, and then calculate the value of the slope of the best fitted line to the experimental points. Previously, we determined the protein concentration as described above. Finally we calculated the volumetric enzymatic and specific enzymatic activities (express as IU/mg protein).

Triplicate independent assays were performed and three readings for each sample were taken in all cases. A BioMate, ThermoSpectronic spectrophotometer was used for all the measurements.

Zymograms

Zymograms were performed to identify cellulase and xylanase activities from the supernatants of the fungal culture on Vogel's medium supplemented with 2% CMC and NaCl (0, 0.5, 1, 1.5, 2 M), or from soluble products recovered from the solid-state fermentations. Zymograms were performed as described in Quiroz-Castañeda *et al.* [57]. Briefly, 50 μg of protein were loaded per lane in the gels for the experiments in media containing NaCl, or 30 μg from the samples of the solid-state fermentation experiments. For cellulases, the gels were embedded with 2% CMC, while 2% oat xylan was used in the gels for xylanases. *Trichoderma viride* cellulase (Sigma Catalogue No. 1794), and xylanase (Sigma Catalogue No. 3876) were included as controls. Once the native-PAGE electrophoresis was carried out, gels were incubated with a 1% Congo red solution (in water) for 30 minutes at room temperature and then washed 3 times with a 1 M NaCl solution. The cellulase activity developed as clear bands. The molecular weight of the bands was estimated against a protein marker (Fermentas Catalogue No. 26612).

Optimal temperature and pH of cellulase activity

Enzyme reactions were performed as described earlier at different incubation temperatures (20, 30, 40, 50, 60, 70 and 80°C) in 50 mM sodium citrate buffer, pH 5. Different pH conditions ranging from 3 to 8 were tested at 50°C in citrate or phosphate buffer depending of the pH tested. All measurements were determined in triplicate.

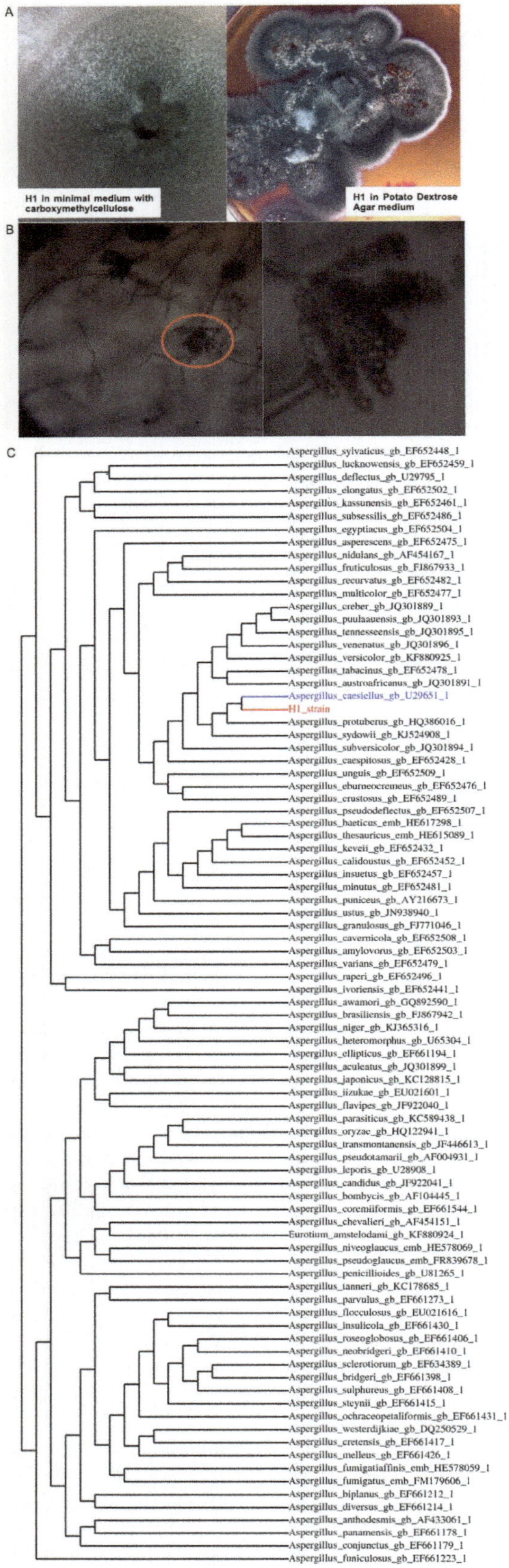

Figure 1. (A) *A. caesiellus* grown in Vogel's medium supplemented with CMC (2%) with 0.5 M NaCl and PDA medium. (B) Microcultures of *A. caesiellus* in Saboraud Agar medium. (C) Molecular phylogeny for the D1–D2 domain of 28S rDNA gene.

Thermal-stability of the cellulase activity

Cellulases thermostability was determined by incubating 200 μL of supernatant for 60 minutes at the following temperatures: 30, 40, 50, 60 and 70°C. Subsequently, the samples were cooled on ice for 5 minutes and the residual activity was determined as described before (at 50°C and pH 5). All measurements were performed in triplicate.

Statistical calculations

For statistical treatment of experimental data, the arithmetic mean and the standard deviation were calculated. Simple classification ANOVA tests were applied to determine significant differences between the different cases. Firstly, the assumptions of ANOVA were revised: analysis of homogeneity of variance (Hartley-Cochran-Bartlett test) and normal distribution (Kolmogorov-Smirnov and Lilliefors tests) were performed [60]. Subsequently ANOVAs were conducted to demonstrate the similarities or differences between the data of the population of samples. Finally, a post hoc analysis that defines the order of the differences found in the ANOVAs was developed. The Fisher LSD, Tukey HSD and Duncan tests were considered for the post hoc analyses [60], [61]. The use of these three tests ensures greater statistical robustness of the proposed analysis. All statistical calculations were performed in STATISTICA (last version for computers).

Results and Discussion

Isolation of microorganisms

No bacterial or yeast growth was detected in the primary isolation of microorganisms, whereas a well-represented filamentous fungus was isolated in all serial dilutions up to 10^{-5}. This fungal strain was named H1 and showed the capacity to grow in 0.5 M NaCl and CMC as a carbon resource. During the secondary isolation in selective culture media (Sabouraud Agar or PDA added with 0.5 M NaCl) the fungus grew as a dark green colony on the aerial mycelium and brownish-grey in the vegetative mycelium (Figure 1). The production of metabolites secreted as dark brown and bright yellow pigments was evident. The selection forces we used to isolate microorganisms able to degrade lignocellulosic material under high salinity conditions were a complex source of carbon and energy (sugarcane bagasse) and 2 M NaCl. High NaCl concentrations inhibit the growth of most microorganisms; this ratio may explain the very low representation of the microbial population obtained during primary isolation taking in account that it came from a non-saline environment [1], [62–64].

Identification of H1

Macroscopic observations of the colony and micromorphological characteristics (hyphae, conidiophores and conidial shape) suggested that the isolate belonged to the genus *Aspergillus* (Figure 1).

Light green grainy colonies were observed in Vogel's medium with 2% CMC and 0.5 M NaCl, while medium dark green colonies were observed in PDA (Figure 1A). Microscopic examination of the culture mycelia showed columnar conidiophores of approximately 350 μm in length with vesicles 20–35 μm in diameter and pale conidia with globose and subglobose and smooth spores. Spinose or roughened conidia were never observed. The conidia length was between 5 and 6 μm while the conidia width was approximately 3 μm (Figure 1B). These observations coincide with the characteristics described for *Aspergillus caesiellus* in the CBS-KNAW Fungal Biodiversity Centre (www.cbs.knaw.nl).

To confirm the identity of this strain, molecular markers for H1 strain were analysed. We chose to amplify a fragment of the 18S rDNA, the D1–D2 domain (28S rDNA) and the ITS1 region; the determined sequences were annotated in the GenBank with the accession numbers KJ476140, KJ476141 and KJ476142, respectively. BLAST DNA sequences analyses showed similarity with various species of *Aspergilli* demonstrating that H1 belongs to the *Aspergilli* genus.

Many molecular markers with taxonomic value have been described in the genus *Aspergillus*. These include the mitochondrial cytochrome *b* gene [65–67], a putative aflatoxin pathway regulatory gen (*aflR*) [68], the DNA topoisomerase II gene (*TOP2*) [69], the β-tubulin gene [70] and different rDNA gene regions [71]. Among the regions of the rDNA genes the most representative are the 5′end of the large-subunit 28S rDNA gene (D1–D2 region) [50] and the internal transcribed spacers 1 and 2 (ITS1 and ITS2) regions between the small- and large-subunit rDNA genes [51], [71], [72]. It has been reported that for the *Aspergilli*, the ITS1 and D1–D2 regions are the most variable and therefore the most useful for molecular identification of this genus species [51].

To define the species related to H1 strain, phylogenetic reconstructions based on the amplified sequences for each of the molecular markers were studied. Sequences with more than 80% identity and 70% coverage obtained through Blastn analysis were considered to reconstruct the phylogenies. The phylogenetic analysis of the 18S rDNA fragment was inconclusive for species identification. In this case, the H1 sequence does not group with any particular species of *Aspergillus,* although the closest relative to H1 was *Aspergillus versicolor* (Figure S1). On the other hand the analysis of molecular phylogenies for sequences of the ITS regions were also not conclusive because H1 strain was grouped individually on a single branch of the phylogenetic tree (Figure S1). In this case, the most closely related species was *Aspergillus niger* (Figure S1).

Phylogeny analysis obtained for the sequences of the D1–D2 domain allowed us to propose the identity of strain H1 as *Aspergillus caesiellus*. Numerous studies show this region as very useful for identification filamentous fungi, not only for *Aspergillus* species [73–79]. In this case, our sequence was grouped only with a reference strain of *A. caesiellus* (*A. caesiellus* NRRL a-14879 annotated in the CBS-KNAW Fungal Biodiversity Centre). In this phylogeny it was found that *Aspergillus sydowii* and *Aspergillus protuberus* were the closest species to H1 (Figure 1C).

A. sydowii and *A. protuberus* are classified in the *Aspergillus* section *Versicolores* which was established as the *A. versicolor* group by Thom and Churche [80] and it was after reconsidered by Thom and Raper [81]. *A. protuberus* does not produce soluble pigments or exudates on MEA and its colonies show reverse light pinkish yellow to pinkish yellow colour. Moreover it presents conidia with finely roughened walls that can also be ellipsoidal to pyriform [82]. *A. sydowii* either produce soluble pigments or exudates on MEA and exhibit colonies with unpigmented reverse to brownish pink colour [82]. *A. sydowii* isolates growing specifically on Saboraud media appear as dark green colonies with a white fringe (www.thunderhouse4-yuri.blogspot.ie) and excrete purple pigments [83]. Its colonies on PDA are blue-green colour, reverse reddish and often with reddish exudates. Red-brown colour in PDA and Czapeks agar media is a very distinctive characteristic of *A. sydowii* [83], [84]. Besides, the micromorphological characterization of this species confirms spherical, and very echinulate or spinose (rough, jagged texture) conidia [83], [84].

Strain H1 shows significant differences with all the above descriptions because the presence of abundant exudates and dark soluble pigments (non-reddish and non-purple) on Saboraud agar, MEA and PDA media was detected and microculture conidia with the previously described characteristics for *A. sydowii* and *A. protuberus* were not observed. Additionally the aspect of the H1′s colony does not coincide with those described for the previous two species. Cultural and morphological criteria support the molecular identification proposal for strain H1 because they coincide phylogenetically with H1 and the rooted closest species. Systematics and taxonomy of fungi, especially for the genus *Aspergillus* spp., continue up to day using morpho-culture characters for the distinction between related species [82], [85–88]. On the other hand, *A. protuberus* cannot grow at 37°C [82], while H1 showed thermotolerant behaviour (see next section). In our case, it was critical to evaluate the colony texture, colour on the nutrient media, excretions of exudates and soluble pigments and the characteristics of the conidia to conclude the taxonomic identification. Thus, attending molecular and micromorphological criteria H1 strain was identified as *Aspergillus caesiellus*. We consider that the identification of the strain H1 is robust and conclusive according to the current criteria for identification of filamentous fungi. The polyphasic approach for the identification of microorganisms ensures better taxonomic conclusion according to several studies [89], [90].

A. caesiellus has been poorly studied. There are very few reports about the biology and physiology of this filamentous fungus. Some of these studies have reported the potential of this fungus to produce keratinase [91], amylases [92], and invertases [93]. However, little is known about the biology of this species; besides the original report, only few studies describe the isolation of *A. caesiellus* from different sources including air and dust, [94], a marine sponge [95], and chicken litter [91]. *A. caesiellus* has also been reported as a pathogen fungus or opportunistic pathogen [96], [97].

To the best of our knowledge, this is the first report describing the isolation of *A. caesiellus* from samples of lignocellulosic material. This strain is able to grow in different NaCl concentrations and tolerant up to 2 M NaCl. Phylogenetically related species such as *A. versicolor* and *A. sydowii* have been described as halotolerant and halophilic by some authors [22], [98–101]. These species were related to H1 strain in the constructed molecular phylogenies, a finding that also supports its molecular identification. This study demonstrates the flexibility and plasticity of microbial physiology and the possibility to isolate halotolerant and/or halophilic microorganisms from non-hypersaline environments. The assumptions of microbial ecology are general and valid but not absolute in these terms.

A. caesiellus H1 is a moderate halophile and thermotolerant fungus

The growth curves of *A. caesiellus* H1 showed statistically significant greater growth rates in the presence of NaCl as compared to those without NaCl for the three temperatures tested (28, 37 and 42°C) (Figure 2). H1 growth was inhibited only in 3 M NaCl. The higher specific growth rate was obtained in cultures of H1 at 37°C and 1 M NaCl. Similar specific growth rates were observed under conditions of 0.5 and 1 M NaCl in experiments at 28°C, while at 42°C, the higher specific growth rate is achieved in the culture conditions of 1.5 M NaCl (Table 1). These results

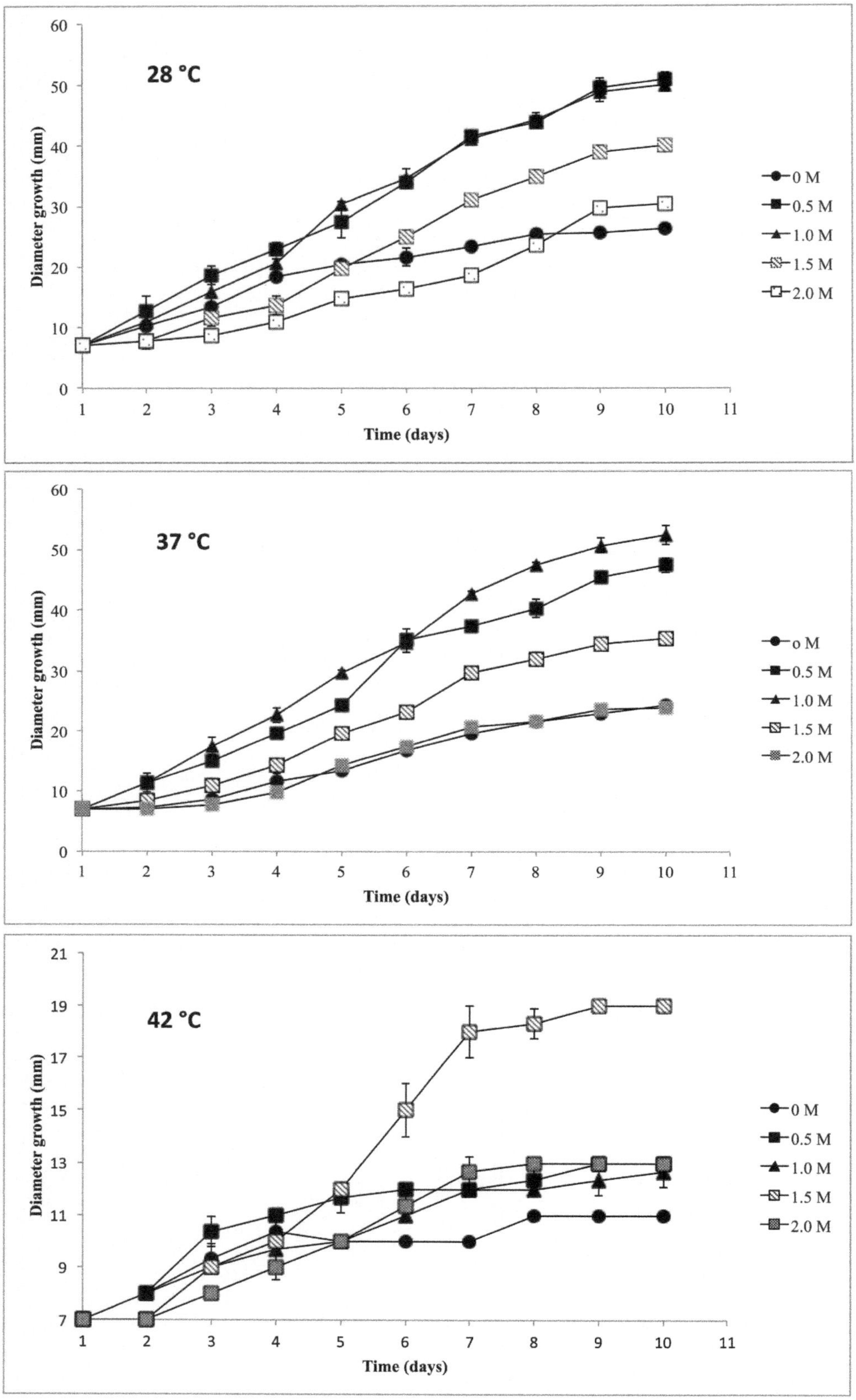

Figure 2. Growth curves of H1 at different temperatures and NaCl concentrations.

Table 1. Specific growth rate (mm/day) of the strain H1 at different temperatures and NaCl concentrations.

NaCl (M)	Growth rate 28°C	Growth rate 37°C	Growth rate 42°C
0	2.17±0.03[d]	2.17±0.07[e]	0.39±0.03[d]
0.5	5.12±0.22[a]	4.80±0.02[b]	0.62±0.03[c]
1.0	5.22±0.02[a]	5.40±0.10[a]	0.63±0.07[c]
1.5	4.18±0.14[b]	3.59±0.07[c]	1.62±0.04[a]
2.0	2.81±0.05[c]	2.28±0.10[d e]	0.80±0.01[b]

Different letters indicate different statistical orders.

show that the H1 is also a thermotolerant strain. Thermotolerant organisms can grow in a range of temperatures above 40°C [102], [103].

There is still much controversy about the definition of extremophile microorganisms. However, one of the most accepted, defines extremophile organisms as those showing optimal growth parameters in different environmental conditions than those normal for humans [104]. In the particular case of halophile microorganisms, there are also disputes about which is the best definition. Certainly it is even difficult to define the boundaries between the concepts: halotolerant and halophile. The establishment of the difference is harder in fungi, even up to date the limits between halotolerant and halophilic strains is not outspoken [46]. However, these definitions are more than fifty years old. Larsen [105] and Kushner and Kamekura [106] introduced three important definitions: moderate halophiles, extreme halophiles and halotolerant.

Halotolerance and halophilia are very clearly defined for Bacteria and Archae. Prokaryotes are classified as obligated halophiles when they require NaCl for optimal growth. Bacteria can be grouped in two categories according with the NaCl concentration required: extremely halophilic or moderately halophilic. In general, definitions mark off extremely halophilic bacteria when they require NaCl concentrations of 2.5–5.2 M (15–30%) for their optimal growth rate, while moderate halophiles grow optimally in media containing 0.5–2.5 M (3–15%) NaCl [107–110]. Some authors have defined for convenience halophilic bacteria as microorganisms that form colonies on agar plates with 20% added NaCl [44]. General definitions as that of Madigan *et al.* [104] classify halophile organisms as those requiring high concentrations of NaCl (1.5 M) to growth optimally. In contrast, halotolerant microorganisms are defined as those showing optimal growth parameters in the absence of salt, but can survive in not-common NaCl concentrations [111], [112].

For fungi, the term halophile was introduced for first time in 1975 to describe xerophilic species inhabiting food that showed superior growth in media with NaCl as the controlling solute [113]. Gunde-Cimerman *et al.* [22], [24] proposed as halophilic or xerophilic fungi those growing well at a_w of ≤0.85, corresponding to 17% NaCl (3 M) or 50% glucose added to their growth medium. In the same review [22] the authors noted that these fungi have different halophilic characteristics when they are compared with the majority of halophilic prokaryotes. Plemenitaš *et al.* [114] described that most halophilic fungi do not require salt for viability because they can show adaptive properties to grow in any salinity range, from freshwater to saturated NaCl concentrations.

Thus, there are several criteria to classify microorganisms attending their growth and survival in different NaCl concentrations. We consider that *A. caesiellus* H1 can be classified as a moderate halophile fungus because, although it does not exhibit the best growth rate at NaCl concentration higher than 17%, it has greater specific growth rates in the presence of high salt concentrations than those in the absence of NaCl. However, the H1 strain was not isolated from a hypersaline environment and is capable of growing without NaCl in the medium. Our proposed classification attends the halotolerance and halophilic concepts reviewed by different authors. This shows that moderate extremophile microorganisms can live in environments that are not extreme (see Introduction).

This study is the first report of a moderate halophile strain of *A. caesiellus*. Other species have been described in the genus as halophile/halotolerant, for example, *A. versicolor, A. sydowii, A. flavus*, etc. Some species, such as *A. versicolor and A. sydowii* have been found as part of the fungal microbiota of hypersaline environments [22], [99], [100], [115–117].

The physiology and metabolism of microorganisms with the ability to grow in high concentrations of NaCl ensure a number of applications of biotechnological interest [118–121]. In particular, the study of the ability of *A. caesiellus* H1 for degradation of lignocellulosic material became an attraction for our research group. H1 is a strain isolated from sugarcane bagasse, able to grow at a range of NaCl between 0.5 and 2 M and uses cellulose as the sole source of carbon and energy. These qualities point to *A. caesiellus* H1 as a good candidate to study some of its lignocellulolytic enzymes.

Cellulases activity

Cellulase enzymatic activity was observed at its highest level within the sixth day of culture in 1.5 M NaCl, using CMC as only carbon source at 28°C (Figure 3). From the fifth day on, cultures of H1 in the presence of varying salt concentrations showed higher cellulase specific activity compared to cultures in the absence of NaCl with exception of the 2 M condition, which from day 7 on anyway showed similar activities to the control culture (without NaCl) (Figure 3). To the best of our knowledge this is the first report of cellulase activity for *A. caesiellus*.

Zymograms of supernatants of *A. caesiellus* H1 cultures in CMC as a carbon source showed two isoforms for cellulases (of around 50 and 35 kDa), regardless of the NaCl concentration (Figure 4). The 50 kDa activity band diminishes at high salt concentrations (1.5 and 2 M), since the amount of protein loaded was the same for all the lanes. We were not able to detect any bands in glucose grown cultures in this experiment suggesting that cellulase expression is repressed by glucose (Figure 4, lane 1). These results indicate that cellulases from *A. caesiellus* H1 are capable of functioning at concentrations up to 2 M NaCl, since growth is supported in this condition with CMC as a sole carbon source.

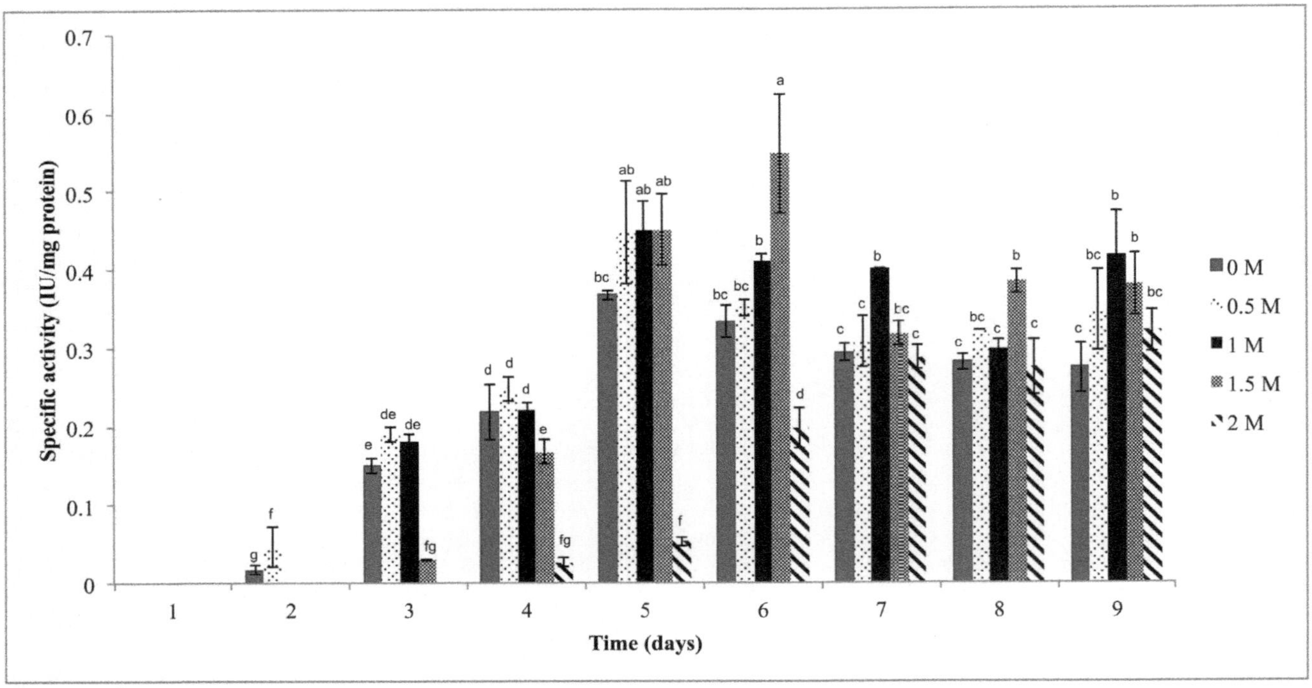

Figure 3. Cellulase activity of H1 cultures on CMC (2%) as the sole carbon source at different NaCl concentrations.

Solid-state fermentation of natural substrates by H1

We were interested in testing this fungal strain for enzyme production of lignocellulosic substrates in similar conditions to those found in the field, so solid state fermentations were set up in the absence of NaCl. Wheat straw, maize stover and agave fibres (in that order) were the substrates where the best cellulase and xylanase activities were obtained during solid-state fermentation. Xylanase production was favoured over cellulase production when these substrates were used, and it was higher in wheat straw and maize stover (Figure 5). When a PAGE native gel with embedded xylan was stained with Congo red, four bands of approximately 10, 12, 15 and 20 kDa with xylanase activity were observed for the maize stover fermentation (Figure 6A, lane 3). The 12, 15 and 20 kDa bands also are induced, although in different proportions, when wheat straw and agave fivers were used as substrates, being the 15 kDa band very abundant in the agave fibres fermentation (Figure 6A, lanes 2 and 5). In the sugar cane bagasse fermentation only the 15 kDa band was observed (Figure 6A, lane 4). Again, no xylanase activity was observed in the zymogram when glucose was used as the carbon source Figure 6A, lane 6), indicating a repression by this sugar for xylanases expression. Distinct band patterns were observed for different substrates suggesting a differential expression of xylanases according to the substrate used.

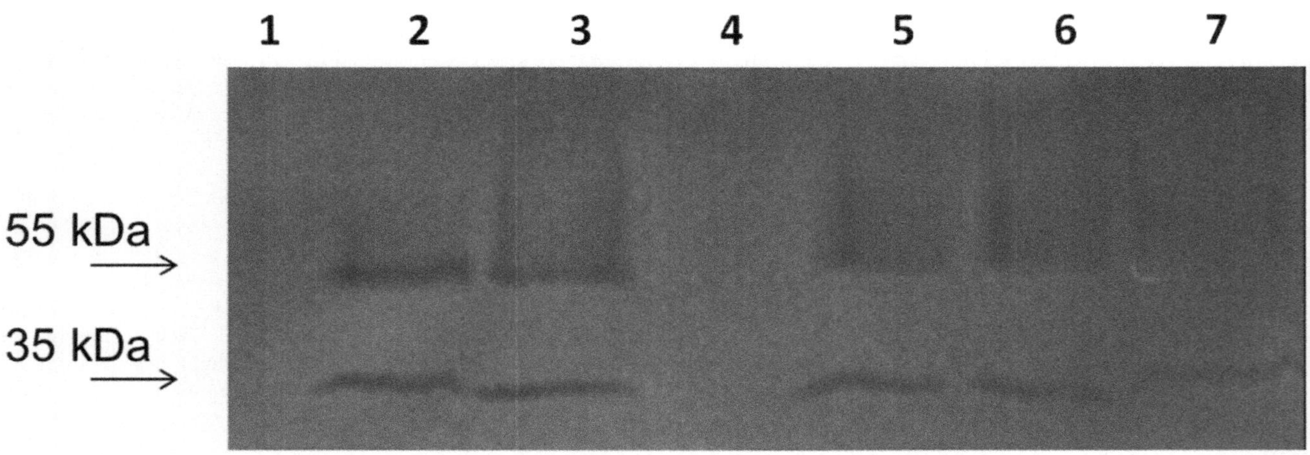

Figure 4. Zymogram for detection of extracellular cellulases from liquid cultures of strain H1 at different NaCl concentrations in **CMC (2%) as the sole carbon source.** Lane 1: Control culture in 2% glucose without NaCl. Lane 2: Culture without NaCl. Lane 3: Culture with 0.5 M NaCl. Lane 4: No sample charge. Lane 5: Culture with 1 M NaCl. Lane 6: Culture with 1.5 M NaCl. Lane 7: Culture with 2 M NaCl. Arrows indicate the position were the molecular weight markers migrated.

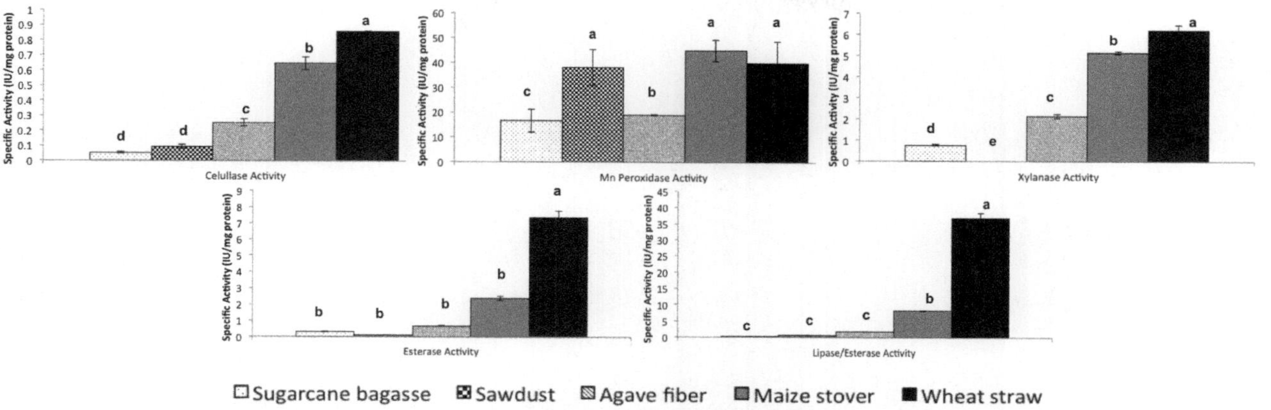

Figure 5. Enzymatic activities from solid-state fermentation of H1 in different substrates. Different letters indicate different statistical orders.

In a native gel with cellulose stained with Congo red, we observed three bands with cellulase activity for wheat straw and maize stover cultures, while only two bands were evident in sugarcane bagasse and agave fibres (Figure 6B). Note that in cultures with glucose as a carbon source, only a low basal expression of cellulases is observed (Figure 6B), suggesting again that glucose represses the expression of most of the isoforms of these enzymes. Sawdust proved the worst substrate for induction of these enzymes (Figure 5).

A. caesiellus H1 also showed strong Mn peroxidase activity (Figure 5). No statistically significant differences in the Mn peroxidase activity measured in the cultures grown on wheat straw, maize stover and sawdust were observed. In the latter substrates the best Mn peroxidase activity was obtained (Figure 5). Cultures of the fungus in agave fibres and sugarcane bagasse produced lower Mn peroxidase activity with statistically significant differences between them (Figure 5).

Fungal laccases have been widely studied for their industrial applications. However, no laccase activity for H1 strain was detected in any of the conditions we tested.

Wheat straw, maize stover and agave fibres were the substrates where the higher esterase and lipase/esterase activity were determined (Figure 5). This fungus showed a very high lipase/esterase activity with 2-naphthyl acetate as a substrate in wheat straw and maize stover (Figure 5). In general, the substrates where the best enzymatic activities were observed were wheat straw and maize stover. These results support the potential of this fungus to degrade lignocellulosic material and its potential biotechnological applications. Experiments are under way to test H1 for lignocellulosic substrates degradation in the presence of NaCl.

Optimum temperature and thermo-stability of cellulases activity

No statistically significant differences in the cellulase activity of the fungus at 50, 60 and 70°C were observed (Figure 7A). When fungal supernatant was incubated for one hour at 60°C, it kept 40% of the cellulases activity (residual activity). This result suggests that *A. caesiellus* cellulases are thermostable. One of the important factors for the industrial application of cellulolytic enzymes is its

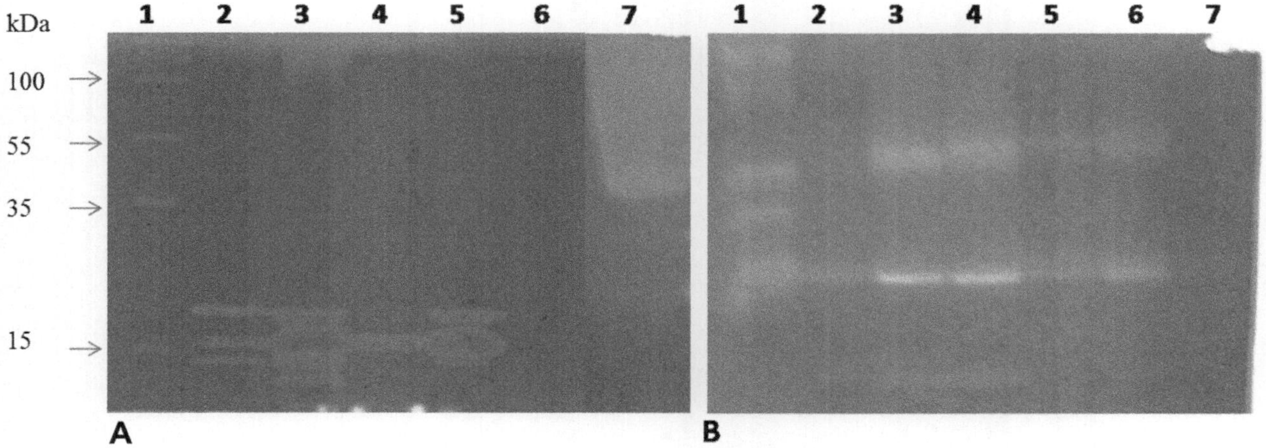

Figure 6. Zymograms for detection of extracellular xylanases and cellulases from solid-state cultures of H1 in different natural substrates. (A) Native gel for xylanases. Line 1: Molecular Weight marker. Line 2: Culture in wheat straw. Line 3: Culture in. maize stover Line 4: Culture in. sugarcane bagasse Line 5: Culture in. agave fibre Line 6: Culture in glucose. Line 7: Positive control. Xylanasas from *Trichoderma viridae.* (B) Native gel for cellulases. Line 1: Positive control. Cellulases from *Trichoderma viridae.* Line 2: Culture in glucose. Line 3: Culture in wheat straw. Line 4: Culture in maize stover. Line 5: Culture in sugarcane bagasse. Line 6: Culture in agave fibre. Line 7: Culture in sawdust. The arrows indicate bands with activity.

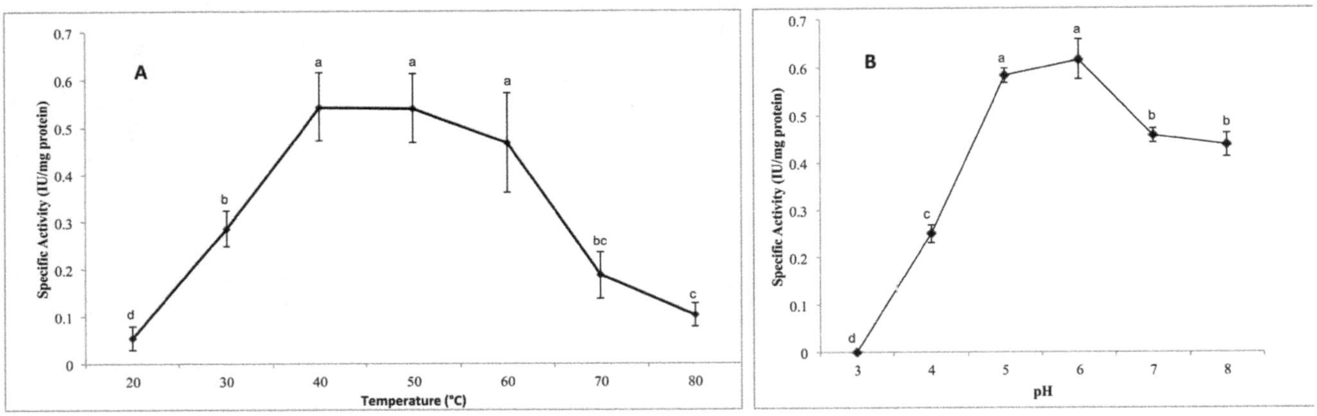

Figure 7. (A) Optimal temperature of celullase activity of H1. (B) Optimal pH of celullase activity of H1. Different letters indicate different statistical orders.

sturdiness [57], [122–125]. Our results are comparable with those obtained by Liu *et al.* [95] and Raddadi *et al.* [126], and different than those reported by Pang *et al.* [127], since they lost the cellulase activity when supernatant was heated for one hour at 50°C. In our case, the residual activity decreases to 22% (residual activity) when the supernatants were exposed for one hour at 70°C. A similar result was reported by Narra *et al.* [128], who found a dramatic decrease of the cellulolytic activity at this temperature.

Cellulase Ph Optimum

The best cellulase activity was observed at pH values between 5 and 6 (Figure 7B). This result is consistent for other cellulases [95], [128]. However, some cellulases show optimal pH in the neutral or basic range [127], [129].

In this work we have described the isolation and characterization of a moderate halophile *A. caesiellus* strain that produces thermostable cellulases and other biotechnological interesting activities. Our results show that this strain has a great potential for lignocellulose degradation and could be used for biorefinery applications.

Supporting Information

Figure S1 Phylogenies for molecular taxonomic identification of moderate halophile strain H1. (A) Molecular phylogeny considering the sequence of the fragment of the 18S ribosomal DNA. (B) Molecular phylogeny considering the sequence of the regions of the ITS1 region.

Acknowledgments

We are grateful to Dr. Claudia Martínez-Anaya for her critical review of this manuscript and Mss. Aspasia Giannika for her help in the image editing.

Author Contributions

Conceived and designed the experiments: JF-M EM-M RB-G EB-L. Performed the experiments: RB-G EB-L LC-S DA-Z KA-H CM-H AS-R. Analyzed the data: JF-M AS-R RB-G EB-L. Contributed reagents/materials/analysis tools: RR-H. Contributed to the writing of the manuscript: RB-G EB-L JF-M.

References

1. Oren A (2002) Molecular ecology of extremely halophilic Archaea and Bacteria. FEMS Microbiology Ecology 39, 1–7.
2. Nicolaus B, Lama L, Manca MC, Gambacorta R (1999) Extremophiles: Polysaccharides and enzymes degrading polysaccharides (INIST-CNRS, Cote INIST: L 27265, 35400008510647.0040: Research signpost, Trivandrum, INDE).
3. Conte F (2008) Engineering of glycoside hydrolases of biotechnological interest from extremophilic organisms. Doctoral thesis. Universitá di Napoli Federico II, Italy.
4. Marques E, Uesugi C (2013) Avaliação de bactérias extremófilas facultativas na produção de fitomassa do híbrido "urograndis" de eucalipto, a partir de sementes. Revista Árvore 37, 41–47.
5. Bhattacharya A, Pletscheke B (2014) Review of the enzymatic machinery of *Halothermothrix orenii* with special reference to industrial applications. Enzyme and Microbial Technology 55, 159–169.
6. Díaz-Tena E, Rodríguez-Ezquerro A, López de Lacalle Marcaide LN, Gurtubay Bustinduy L, Elías-Sáenz A (2014) A sustainable process for material removal on pure copper by use of extremophile bacteria. Journal of Cleaner Production. In Press, Corrected Proof: dx.doi.org/10.1016/j.jclepro.2014.01.061.
7. Gomes J, Steiner W (2004) The Biocatalytic Potential of Extremophiles and Extremozymes. Food Technol. Biotechnol. 42, 223–235.
8. Mesbah NM, Wiegel J (2009) *Natronovirga wadinatrunensis* gen. nov., sp. nov. and *Natranaerobius trueperi* sp. nov., halophilic, alkalithermophilic microorganisms from soda lakes of the Wadi An Natrun, Egypt. Int. J. Syst. Evol. Microbiol. 59, 2042–2048.
9. Hamamura N, Itai T, Liu Y, Reysenbach AL, Damdinsuren N, et al. (2014) Identification of anaerobic arsenite-oxidizing and arsenate-reducing bacteria associated with an alkaline saline lake in Khovsgol, Mongolia: Arsenic transformation in a Mongolian saline lake. Environmental Microbiology Reports. DOI:10.1111/1758-2229.12144.
10. Gunde-Cimerman N, Zalar P, de Hoog S, Plemenitaš A (2000) Hypersaline waters in salterns – natural ecological niches for halophilic black yeasts. FEMS Microbiology Ecology 32, 235–240.
11. Zhang T, Datta S, Eichler J, Ivanova N, Axen SD, et al. (2011) Identification of a haloalkaliphilic and thermostable cellulase with improved ionic liquid tolerance. Green Chem. 13, 2083–2090.
12. Zhang G, Li S, Xue Y, Mao L, Ma Y (2012) Effects of salts on activity of halophilic cellulase with glucomannanase activity isolated from alkaliphilic and halophilic Bacillus sp. BG-CS10. Extremophiles 16, 35–43.
13. Gunny AN, Arbain D, Edwin Gumba R, Jong BC, Jamal P (2014) Potential halophilic cellulases for in situ enzymatic saccharification of ionic liquids pretreated lignocelluloses. Bioresource Technology 155, 177–181.
14. Antón J, Oren A, Benlloch S, Rodríguez-Valera F, Amann R, et al. (2002) *Salinibacter ruber* gen. nov., sp. nov., a novel, extremely halophilic member of the Bacteria from saltern crystallizer ponds. Int. J. Syst. Evol. Microbiol. 52, 485–491.
15. Simmons CW, Claypool JT, Marshall MN, Jabusch LK, Reddy AP, et al. (2014) Characterization of bacterial communities in solarized soil amended with lignocellulosic organic matter. Applied Soil Ecology 73, 97–104.
16. Sorokin DY, Kolganova TV (2013) Bacterial chitin utilization at halophilic conditions. Extremophiles 18, 243–248.

17. Muruga B, Anyango B (2013) A Survey of Extremophilic Bacteria in Lake Magadi, Kenya. Columbia International Publishing American Journal of Molecular and Cellular Biology 2, 14–26.
18. Challacombe JF, Majid S, Deole R, Brettin TS, Bruce D, et al. (2013) Complete genome sequence of *Halorhodospira halophila* SL1. Stand Genomic Sci 8, 206–214.
19. Mesbah NM, Hedrick DB, Peacock AD, Rohde M, Wiegel J (2007) *Natranaerobius thermophilus* gen. nov., sp. nov., a halophilic, alkalithermophilic bacterium from soda lakes of the Wadi An Natrun, Egypt, and proposal of Natranaerobiaceae fam. nov. and Natranaerobiales ord. nov. Int. J. Syst. Evol. Microbiol. 57, 2507–2512.
20. Leela A (2013) Catalytic properties and industrial potential of halophilic extracellular protease from salt – pan isolates. PhD Thesis. Bharathidassan University.
21. Roldán-Carrillo T, Castorena-Cortés G, Reyes-Avila J, Zapata-Peñasco I, Olguín-Lora P (2013) Effect of porous media types on oil recovery by indigenous microorganisms from a Mexican oil field. Journal of Chemical Technology & Biotechnology 88, 1023–1029.
22. Gunde-Cimerman N, Ramos J, Plemenitaš A (2009) Halotolerant and halophilic fungi. Mycological Research 113, 1231–1241.
23. Butinar L, Zalar P, Frisvad JC, Gunde-Cimerman N (2005) The genus *Eurotium* - members of indigenous fungal community in hypersaline waters of salterns. FEMS Microbiol. Ecol. 51, 155–166.
24. Gunde-Cimerman N, Butinar L, Sonjak S, Turk M, Uršič V, et al. (2005) Halotolerant and Halophilic Fungi from Coastal Environments in the Arctics. In Adaptation to Life at High Salt Concentrations in Archaea, Bacteria, and Eukarya, N. Gunde-Cimerman, A. Oren, and A. Plemenitaš, eds. (Springer Netherlands), 397–423.
25. Zalarc P, Hoog GS de, Schroers HJ, Crous J, Groenewald JZ, et al. (2007) Phylogeny and ecology of the ubiquitous saprobe *Cladosporium shpaerospermum*, with descriptions of seven new species from hypersaline environments. Studies in Mycology 58, 157–183.
26. Zalar P, Sybren de Hoog G, Schroers HJ, Frank JM, Gunde-Cimerman N (2005) Taxonomy and phylogeny of the xerophilic genus *Wallemia* (Wallemiomycetes and Wallemiales, cl. et ord. nov.). Antonie Van Leeuwenhoek 87, 311–328.
27. DasSarma S, Aurora P (2012) Halophiles. In eLS, John Wiley & Sons, Ltd, Chichester. United Kindom. DOI:10(9780470015902), a0000394.
28. Molitoris HP, Buchalo AS, Kurchenko I, Nevo E, Rawal BS, et al (2000) Physiological diversity of the first filamentous fungi isolated from the hypersaline dead sea. Fungal Diversity 5, 55–70.
29. Kutateladze NZ (2009) Selection of Microscopic Fungi - Pectinase Producers. Bulletin of the Georgian National Academy of Science 3.
30. Li X, Wang HL, Li T, Yu HY (2012) Purification and characterization of an organic solvent-tolerant alkaline cellulase from a halophilic isolate of *Thalassobacillus*. Biotechnol. Lett. 34, 1531–1536.
31. Xue DS, Chen HY, Lin DQ, Guan YX, Yao SJ (2012) Optimization of a natural medium for cellulase by a marine *Aspergillus niger* using response surface methodology. Appl. Biochem. Biotechnol. 167, 1963–1972.
32. Ren W, Zhang F, Yang X, Tang S, Ming H, et al. (2013) Purification and properties of a SDS-resistant xylanase from halophilic *Streptomonospora* sp. YIM 90494. Cellulose 20, 1947–1955.
33. Liu X, Huang Z, Zhang X, Shao Z, Liu Z (2014) Cloning, expression and characterization of a novel cold-active and halophilic xylanase from *Zunongwangia profunda*. Extremophiles 18, 441–450.
34. Onishi H, Fuchi H, Konomi K, Hidaka O, Kamekura M (1980) Isolation and distribution of a variety of halophilic bacteria and their clas- sification by salt-response. Agric Biol Chem 44, 1253–1258.
35. Horikoshi K (1999) Alkaliphiles: some applications of their products for biotechnology. Microbiol Mol Biol Rev 63, 735–750.
36. Saiz-Jimenez C, Laiz L (2000) Occurrence of halotolerant/halophilic bacterial communities in deteriorated monuments. Int Biodeterior Biodegrad 46, 319–326.
37. Abdel-Fafez S (1981) Halophilic fungi of desert soils in Saudi Arabia. Mycopathologia 75, 75–80.
38. Moubasher A, Abdel-Hafez S, Bagy M, Abdel-Satar M (1990) Halophilic and halotolerant fungi in cultivated desert and salt marsh soils from Egypt. Acta Mycologica 26, 65–81.
39. Nielsen P, Fritze D, Priest FG (1995) Phenetic diversity of alkaliphilic *Bacillus* strains: proposal for nine new species. Microbiology 141, 1745–1761.
40. Castillo G, Demoulin V (1997) NaCl salinity and temperature ef- fects on growth of three wood-rotting basidiomycetes from a Papua New Guinea coastal forest. Mycol Res 101, 341–344.
41. Arahal DR, Marquez MC, Volcani BE, Schleifer KH, Ventosa A (1999) *Bacillus marismortui* sp. nov., a new moderately halophilic species from the Dead Sea. Int J Syst Bacteriol 49, 521–530.
42. Piñar G, Ramos C, Rölleke S, Schabereiter-Gurtner C, Vybiral D, et al. (2001) Detection of indigenous *Halobacillus* populations in damaged ancient wall paintings and building materials: molecular monitoring and cultivation. Appl Environ Microbiol 67, 4891–4895.
43. Santos SX, Carvalho CC, Bonfá MR, Silva R, Gomes E (2004) Screening for pectinolytic activity of wood-rotting Basidio- mycetes and characterization of the enzymes. Folia Microbiol 49, 46–52.
44. Echigo A, Hino M, Fukushima T, Mizuki T, Kamekura M, et al. (2005) Endospore of halophilic bacteria of the family Bacillaceae isolated from non-saline Japanese soil may be transported by Kosa event (Asian dust storm). Saline Systems 1, 1–13.
45. Abrahão MC, Gugliotta AM, Da Silva R, Fujieda RJ, Boscolo M, et al. (2008) Ligninolytic activity from newly iso-lated basidiomycete strains and effect of these enzymes on the azo dye orange II decolourisation. Annals Microbiol 58, 427–432.
46. Arakaki R, Monteiro D, Boscolo R, Gomes E (2013) Halotolerance, ligninase production and herbicide degradation ability of basidiomicetes strains. Brazilian Journal of Microbiology 44, 1207–1214.
47. Vogel HJ (1956) A convenient growth medium for *Neurospora* (Medium N). Microbial Genet Bull 13, 42–43.
48. Kuhad RC, Kapoor RK, Lal R (2004) Improving the yield and quality of DNA isolated from white-rot fungi. Folia Microbiol 49, 112–116.
49. Borneman J, Hartin RJ (2000) PCR Primers that amplify fungal rRNA genes from environmental samples. Appl Environ Microbiol 66, 4356–4360.
50. Peterson SW (2000) Integration of modern taxonomic methods for Penicillium and Aspergillus. Phylogenetic relationships in Aspergillus based on rDNA sequence analysis. (Amsterdam: Harwood Academic Publishers.), 323–355.
51. Hinrikson HP, Hurst SF, Lott TJ, Warnock DW, Morrison CJ (2005) Assessment of ribosomal large-subunit D1–D2, internal transcribed spacer 1, and internal transcribed spacer 2 regions as targets for molecular identification of medically important *Aspergillus* species. J. Clin. Microbiol. 43, 2092–2103.
52. Sambrook J, Russell DW (2001) Molecular Cloning: A Laboratory Manual (Cold Spring Harbor Laboratory Press).
53. Teather RM, Wood PJ (1982) Use of Congo red-polysaccharide interactions in enumeration and characterization of cellulolytic bacteria from the bovine rumen. Appl. Environ. Microbiol. 43, 777–780.
54. Onsori H, Zamani MR, Motallebi M, Zarghami N (2005) Identification of over producer strain of endo-β-1,4-glucanase in *Aspergillus* Species: Characterization of crude carboxymethyl cellulase. African Journal of Bacteriology 4, 26–30.
55. Miller GL (1959) Use of Dinitrosalicylic Acid Reagent for Determination of Reducing Sugar. Anal. Chem. 31, 426–428.
56. Lowry OH, Rosebrough NJ, Farr AL, Randall RJ (1951) Protein Measurement with the Folin Phenol Reagent. J. Biol. Chem. 193, 265–275.
57. Quiroz-Castañeda RE, Balcázar-López E, Dantán-González E, Martinez A, Folch-Mallol J, et al. (2009) Characterization of cellulolytic activities of *Bjerkandera adusta* and *Pycnoporus sanguineus* on solid wheat straw medium. Electronic Journal of Biotechnology 12.
58. Järvinen J, Taskila S, Isomäki R, Ojamo H (2012) Screening of white-rot fungi manganese peroxidases: a comparison between the specific activities of the enzyme from different native producers. AMB Express 2, 62.
59. Wariishi H, Valli K, Gold MH (1992) Manganese(II) oxidation by manganese peroxidase from the basidiomycete *Phanerochaete chrysosporium*. Kinetic mechanism and role of chelators. J. Biol. Chem. 267, 23688–23695.
60. Hoaglin DC, Welsch RE (1978) The Hat Matrix in Regression and ANOVA. The American Statistician 32, 17–22.
61. Fisher RA, Yates F (1949) Statistical tables for biological, agricultural and medical research. Edinburgh, Oliver and Boyd Ltd. Ed. 3, 112.
62. Evans S, Hansen RW, Stone HM, Schneegurt MA (2013) Isolation and Characterization of Halotolerant Soil Fungi from the Great Salt Plains of Oklahoma (USA). Cryptogamie, Mycologie 34, 329–341.
63. Konte T, Plemenitaš A (2013) The HOG signal transduction pathway in the halophilic fungus *Wallemia ichthyophaga*: identification and characterization of MAP kinases WiHog1A and WiHog1B. Extremophiles 17, 623–636.
64. Shtarkman YM, Koçer ZA, Edgar R, Veerapaneni RS, D'Elia T, et al. (2013) Subglacial Lake Vostok (Antarctica) Accretion Ice Contains a Diverse Set of Sequences from Aquatic, Marine and Sediment-Inhabiting Bacteria and Eukarya. PLoS ONE 8, e67221.
65. Wang L, Yokoyama K, Takahasi H, Kase N, Hanya Y, et al. (2001) Identification of species in *Aspergillus* section Flavi based on sequencing of the mitochondrial cytochrome b gene. International Journal of Food Microbiology 71, 75–86.
66. Yokoyama K, Wang L, Miyaji M, Nishimura K (2001) Identification, classification and phylogeny of the *Aspergillus* section Nigri inferred from mitochondrial cytochrome b gene. FEMS Microbiology Letters 200, 241–246.
67. Spiess B, Buchheidt D (2013) Protocol of a LightCycler PCR Assay for Detection and Quantification of Aspergillus fumigatus DNA in Clinical Samples of Neutropenic Patients. In Laboratory Protocols in Fungal Biology, V.K. Gupta, M.G. Tuohy, M. Ayyachamy, K.M. Turner, and A. O'Donovan, eds. (Springer New York), 315–320.
68. Chang PK, Bhatnagar D, Cleveland TE, Bennett JW (1995) Sequence variability in homologs of the aflatoxin pathway gene *aflR* distinguishes species in *Aspergillus* section Flavi. Appl. Environ. Microbiol. 61, 40–43.
69. Kanbe T, Yamaki K, Kikuchi A (2002) Identification of the Pathogenic *Aspergillus* Species by Nested PCR Using a Mixture of Specific Primers to DNA Topoisomerase II Gene. Microbiology and Immunology 46, 841–848.
70. Geiser DM, Pitt JI, Taylor JW (1998) Cryptic speciation and recombination in the aflatoxin-producing fungus *Aspergillus flavus*. Proc. Natl. Acad. Sci. USA 95, 388–393.

71. Iwen PC, Hinrischs SH, Rupp ME (2002) Utilization of the internal transcribed spacer regions as molecular targets to detect and identify human fungal pathogens. Med. Mycol. 40, 87–109.
72. White TJ, Bruns T, Lee S, Taylor J (1990) Amplification and direct sequencing of fungal ribosomal RNA genes for phylogenetics (In M. A. Innis, D. H. Gelfand, J. J. Sninsky, and T. J. White (ed.), PCR protocols-a guide to methods and applications. Academic Press, San Diego, Calif).
73. Guarro J (2012) Taxonomía y biología de los hongos causantes de infección en humanos. Enfermedades Infecciosas y Microbiología Clínica 30, 33–39.
74. Kwiatkowski NP, Babiker WM, Merz WG, Carroll KC, Zhang SX (2012) Evaluation of Nucleic Acid Sequencing of the D1/D2 Region of the Large Subunit of the 28S rDNA and the Internal Transcribed Spacer Region Using SmartGene IDNS Software for Identification of Filamentous Fungi in a Clinical Laboratory. The Journal of Molecular Diagnostics 14, 393–401.
75. Massire C, Buelow DR, Zhang SX, Lovari R, Matthews HE, et al. (2013) PCR Followed by Electrospray Ionization Mass Spectrometry for Broad-Range Identification of Fungal Pathogens. J. Clin. Microbiol. 51, 959–966.
76. Odero V, Galán-Sánchez F, García-Agudo L, García-Tapia AM, Guerrero-Lozano I, et al. (2013) Fungemia por *Trichosporon asahii* en paciente con neoplasia hematológica. Revista Iberoamericana de Micología. http://dx.doi.org/10.1016/j.riam.2013.09.001.
77. Santos DW, Padovan AC, Melo AS, Gonçalves SS, Azevedo VR, et al. (2013) Molecular Identification of Melanised Non-Sporulating Moulds: A Useful Tool for Studying the Epidemiology of Phaeohyphomycosis. Mycopathologia 175, 445–454.
78. Alam MZ, Alam Q, Jiman-Fatani A, Kamal MA, Abuzenadah AM, et al. (2014) *Candida* identification: a journey from conventional to molecular methods in medical mycology. World Journal Microbiol Biotechnol 1–15.
79. Kim MJ, Park PW, Ahn JY, Kim KH, Seo JY, et al. (2014) Fatal Pulmonary Mucormycosis Caused by *Rhizopus microsporus* in a Patient with Diabetes. Annals of Laboratory Medicine 34, 76.
80. Thom G, Church M (1926) The *Aspergilli*. Baltimore, MD: Williams and Wilkins Co.
81. Thom G, Raper K (1945) Manual of the *Aspergilli*. Boltimore, MD: Williams and Willkins Co.
82. Jurjevic Z, Stephen W, Bruce W (2012) *Aspergillus* section Versiolores: nine new species and multilocus DNA sequence based phylogeny. IMA Fungus: The Global Mycological Journal 3, 59–79.
83. Panchagnula B, Terlir R (2013) Morphological and molecular characterization of a lipolytic marine fungal isolate *Aspergillus sydowii* strain BTSS1005. International Journal of Biology, Pharmacy and Allied Sciences 2, 1488–1500.
84. Bindiya P, Ramana T (2012) Optimization of lipase production from an indigenously isolated marine *Aspergillus sydowii* of Bay of Bengal. Journal of Biochemical Technology 3, 203–211.
85. Klich M (2002) Identification of common *Aspergillus* species. Centraalbureau voor Schimmelcultures, Utrecht, The Netherlands.
86. Sugui J, Peterson S, Clark L, Nardone G, Folio L, et al. (2012) *Aspergillus tanneri* sp. nov., a new pathogen that causes invasive disease refractory to antifungal therapy. J. Clin. Microbiol 50, 3309–3317.
87. Krimitzas A, Pyrri I, Kouvelis V, Kapsanaki-Gotsi E, Typas M (2013) A phylogenetic analysis of Greek isolates of *Aspergillus* species based on morphology and nuclear and mitocondrial gene sequences. BioMed Research International. Doi:10.1155/2013/260395.
88. Barton R (2013) Laboratory diagnosis of invasive aspergillosis: from diagnosis to prediction of outcome. Scientifica. Doi:10.1155/2013/459405.
89. Passarini MR, Santos C, Lima N, Berlinck RS, Sette LD (2013) Filamentous fungi from the Atlantic marine sponge *Dragmacidon reticulatum*. Arch Microbiol 195, 99–111.
90. Krishnamurthi S (2014) The Study of Prokaryotic diversity of a landfull environment using a polyphasic taxonomic approach. INFLIBNET. http://hdl.handle.net/10603/14893.
91. Marcondes NR, Taira CL, Vandresen DC, Svidzinski TI, Kadowaki MK, et al. (2008) New Feather-Degrading Filamentous Fungi. Microb Ecol 56, 13–17.
92. De Paris LD, Scheufele FB, Teixeira T, Guerreiro TL, Hasan SD (2010) Estudo do crescimento de *A. casiellus* em farelo de soja convencional para produção de enzimas. Estudos Tecnológicos Em Engenharia 6.
93. Novaki L, Hasan S, Kadowaki M, Andrade D (2010) Produção de invertase por fermentação em estado sólido a partir de farelo de soja. ENGEVISTA 12.
94. Lugauskas A, Krikštaponis A (2004) Filamentous Fungi Isolated in Hospitals and Some Medical Institutions in Lithuania. Indoor and Built Environment 13, 101–108.
95. Liu D, Zhang R, Yang X, Wu H, Xu D, et al. (2011) Thermostable cellulase production of *Aspergillus fumigatus* Z5 under solid-state fermentation and its application in degradation of agricultural wastes. International Biodeterioration & Biodegradation 65, 717–725.
96. Denning DW (1998) Invasive aspergillosis. Clinical Infectious Diseases 26, 584–589.
97. Browman J, Panizza B, Gandhi M (2007) Sphenoid sinus fungal balls. Annals of Otology, Rhinology and Laryngology 116, 514–519.
98. Abdel-Hafez SI (1981) Halophilic fungi of desert soils in Saudi Arabia. Mycopathologia 75, 75–80.
99. Butinar L, Frisvad JC, Gunde-Cimerman N (2011) Hypersaline waters – a potential source of foodborne toxigenic aspergilli and penicillia. FEMS Microbiology Ecology 77, 186–199.
100. Nayak S, Gonsalves V, Nazareth S (2012) Isolation and salt tolerance of halophilic fungi from mangroves and solar salterns in Goa - India. IJMS 41, 164–172.
101. Oren A, Gunde-Cimerman N (2012) Fungal Life in the Dead Sea. In Biology of Marine Fungi, C. Raghukumar, ed. (Springer Berlin Heidelberg), 115–132.
102. Kumar M, Yadav AN, Tiwari R, Prasanna R, Saxena AK (2014) Deciphering the diversity of culturable thermotolerant bacteria from Manikaran hot springs. Ann Microbiol 1–11.
103. Maller A, Silva TM, Damásio AR, Hirata IY, Jorge JA, et al. (2013) Functional properties of a manganese-activated exo-polygalacturonase produced by a thermotolerant fungus *Aspergillus niveus*. Folia Microbiol 58, 615–621.
104. Madigan M, Martinko J, Stahl D, Clark D (2012) Brock Biology of Microorganisms (San Francisco, USA: (13 Eth) Pearson Education).
105. Larsen H (1962) Halophilism. The Bacteria 4, 297–342.
106. Kushner DJ, Kamekura M (1988) Physiology of halophilic eubacteria. In *Halophilic bacteria Volume I*. Boca Raton, CRC Press 109–140.
107. González C, Gutíerrez C, Ramírez C (1978) *Halobacterium vallismortis* sp. nov. an amylolytic and carbohydrate-metabolizing, extremely halophilic bacterium. Can J Microbiol 24, 710–715.
108. Hao MV, Kocur M, Komagata K (1984) *Marinococcus* gen nov., a new genus for motile cocci with meso-diaminopimelic acid in the cell wall; and *Marinococcus albus* sp. nov. and *Marinococcus halophilus* (Novitsky and Kushner) comb. nov. J Gen Appl Microbiol 30, 449–459.
109. Zvyagintseva IS, Tarasov AL (1987) Extreme halophilic bacteria from saline soils. Mikrobiologiya 56, 839–844.
110. Bouchotroch S, Quesada E, del Moral A, Llamas I, Bejar V (2001) *Halomonas maura* sp. nov., a novel moderately halophilic, exopolysaccharide-producing bacterium. Int J Syst Evol Microbiol 51, 1625–32.
111. Kivisto A, Karp M (2011) Halophilic anaerobic fermentative bacteria. Review. J Biotechnol 152, 114–124.
112. Yoon JH, Kang KH, Park YH (2003) *Halobacillus salinus* sp. nov., iso-lated from a salt lake on the coast of the East Sea in Korea. Int J Syst Evol Microbiol 53, 687–93.
113. Pitt J, Hocking A (1985) Fungi and food spoilage, First edition. Academic Press, Sydney.
114. Plemenitaš A, Vaupctic T, Lenassi M, Kogej T, Gunde-Cimerman N (2008) Adaptation of extremely yeast Horteae werneckii to increased osmolarity: a molecular perspective at a glance. Studies in Mycology 61, 67–75.
115. Dave G, Modi H (2013) Phytase producing microbial species associated with rhizosphere of mangroves in an Arid Coastal Region of Dholara. Academia Journal of Biotechnology 1, 027–035.
116. Dhanasekaran P (2014) Anti-infective potential of halophilic actinomycetes isolated from saltpan. Malaya Journal of Biosciences 1, 37–40.
117. Trincone A (2013) Biocatalytic Processes Using Marine Biocatalysts: Ten Cases in Point. Current Organic Chemistry 17, 1058–1066.
118. Margesin R, Schinner F (2001) Potential of halotolerant and halophilic microorganisms for biotechnology. Extremophiles 5, 73–83.
119. Babavalian H, Amoozegar MA, Pourbabaee AA, Moghaddam MM, Shakeri F (2013) Isolation and identification of moderately halophilic bacteria producing hydrolytic enzymes from the largest hypersaline playa in Iran. Microbiology 82, 466–474.
120. Bonfá MR, Grossman MJ, Piubeli F, Mellado E, Durrant LR (2013) Phenol degradation by halophilic bacteria isolated from hypersaline environments. Biodegradation 24, 699–709.
121. Oren A, Bardavid RE, Mana L (2014) Perchlorate and halophilic prokaryotes: implications for possible halophilic life on Mars. Extremophiles 18, 75–80.
122. Parawira W (2012) Enzyme research and applications in biotechnological intensification of biogas production. Critical Reviews in Biotechnology 32, 172–186.
123. Bhat MK, Bhat S (1997) Cellulose degrading enzymes and their potential industrial applications. Biotechnology Advances 15, 583–620.
124. Valderrama B, Ayala M, Vazquez-Duhalt R (2002) Suicide Inactivation of Peroxidases and the Challenge of Engineering More Robust Enzymes. Chemistry & Biology 9, 555–565.
125. Jiang Y, Liu X, Chen Y, Zhou L, He Y, et al. (2014) Pickering emulsion stabilized by lipase-containing periodic mesoporous organosilica particles: A robust biocatalyst system for biodiesel production. Bioresource Technology 153, 278–283.
126. Raddadi N, Cherif A, Daffonchio D, Fava F (2013) Halo-alkalitolerant and thermostable cellulases with improved tolerance to ionic liquids and organic solvents from *Paenibacillus tarimensis* isolated from the Chott El Fejej, Sahara desert, Tunisia. Bioresource Technology 150, 121–128.
127. Pang H, Zhang P, Duan CJ, Mo XC, Tang JL, et al. (2009) Identification of Cellulase Genes from the Metagenomes of Compost Soils and Functional Characterization of One Novel Endoglucanase. Curr Microbiol 58, 404–408.
128. Narra M, Dixit G, Divecha J, Kumar K, Madamwar D, et al. (2014) Production, purification and characterization of a novel GH 12 family endoglucanase from *Aspergillus terreus* and its application in enzymatic degradation of delignified rice straw. International Biodeterioration & Biodegradation 88, 150–161.
129. Li X, Yu HY (2013) Characterization of a halostable endoglucanase with organic solvent-tolerant property from *Haloarcula* sp. G10. International Journal of Biological Macromolecules 62, 101–106.

An Antifungal Role of Hydrogen Sulfide on the Postharvest Pathogens *Aspergillus niger* and *Penicillium italicum*

Liu-Hui Fu[1]◔, Kang-Di Hu[1]◔, Lan-Ying Hu[1]◔, Yan-Hong Li[1], Liang-Bin Hu[2], Hong Yan[3], Yong-Sheng Liu[1], Hua Zhang[1]*

1 School of Biotechnology and Food Engineering, Hefei University of Technology, Hefei, China, **2** School of Food Science, Henan Institute of Science and Technology, Xinxiang, China, **3** College of Chemical and Environmental Engineering, Harbin University of Science and Technology, Key Laboratory of Green Chemical Technology of College of Heilongjiang Province, Harbin, China

Abstract

In this research, the antifungal role of hydrogen sulfide (H_2S) on the postharvest pathogens *Aspergillus niger* and *Penicillium italicum* growing on fruits and under culture conditions on defined media was investigated. Our results show that H_2S, released by sodium hydrosulfide (NaHS) effectively reduced the postharvest decay of fruits induced by *A. niger* and *P. italicum*. Furthermore, H_2S inhibited spore germination, germ tube elongation, mycelial growth, and produced abnormal mycelial contractions when the fungi were grown on defined media in Petri plates. Further studies showed that H_2S could cause an increase in intracellular reactive oxygen species (ROS) in *A. niger*. In accordance with this observation we show that enzyme activities and the expression of superoxide dismutase (SOD) and catalase (CAT) genes in *A. niger* treated with H_2S were lower than those in control. Moreover, H_2S also significantly inhibited the growth of *Saccharomyces cerevisiae*, *Rhizopus oryzae*, the human pathogen *Candida albicans*, and several food-borne bacteria. We also found that short time exposure of H_2S showed a microbicidal role rather than just inhibiting the growth of microbes. Taken together, this study suggests the potential value of H_2S in reducing postharvest loss and food spoilage caused by microbe propagation.

Editor: John Calvert, Emory University, United States of America

Funding: This work was supported by the Natural Science Foundation of China (31271803 to LYH, 31301820 to HZ, 31300133 to KDH) http://www.nsfc.gov.cn/, the Scientific Research Foundation for the Returned Overseas Chinese Scholars (SRF for ROCS, SEM to HZ), the Natural Science Foundations of Anhui Province (11040606M85 to HZ) and the Anhui Provincial Education Department (2012AJZR0028 to HZ). The funders had no role in study design, data collection and analysis, decision to publish, or preparation of the manuscript.

Competing Interests: The authors have declared that no competing interests exist.

* Email: hzhanglab@gmail.com

◔ These authors contributed equally to this work.

Introduction

Generally, about 20% of harvested fruits and vegetables undergo decay during postharvest storage [1]. Considerable postharvest decay is caused by plant fungal pathogens [2]. It has been reported that *Aspergillus niger* can induce the spoilage of fruits such as cherry tomatoes and grapes, and *Penicillium italicum* can cause postharvest blue mold of citrus fruit [3,4,5]. Decay caused by food-borne bacterial pathogens AR also a major concern due to the increasing demands for food safety [6]. For instance, both *Salmonella typhimurium* and *Staphylococcus aureus* in contaminated food are leading causes of gastroenteritis [7,8]. The application of synthetic chemical as germicides is a primary method to control postharvest decay [9]. However, chemical control faces two intractable problems: first, the inevitable development of pathogen resistance; and second, a range of generally used germicides are under review in many countries due to health safety issues [10]. Thus, there is a growing need to develop alternative treatments of postharvest disease that are more enduring and safe.

Hydrogen sulfide (H_2S), traditionally thought as a toxic gas, has proved to be a gaseous signaling molecule after nitric oxide and carbon monoxide in animals [11]. Accumulating evidence shows multiple roles of H_2S in plant development, abiotic stresses, and postharvest senescence [12,13,14,15,16]. Nitric oxide has also been shown to extend postharvest storage of fruits and to inhibit the growth of postharvest pathogens [17,18]. Lai et al. [19] found that the inhibitory effect of NO on the spores of *Penicillium expansum* was associated with oxidative damage. Similarly, it has been found that exogenous H_2S application can prolong postharvest storage of strawberry, fresh-cut kiwifruit, broccoli and mulberry fruit by modulating the antioxidant system [15,20,21,22]. The concentration of the applied H_2S required to delay senescence in strawberry is quite low, indicating that fumigation of fruits with H_2S gas could be safe and practical [15]. However, there is limited data on the relations between H_2S and postharvest pathogens. The earliest related research on this topic was reported by Marsh [23] who found that H_2S was toxic to germinating spores of *Sclerotinia fructicola*. Recently, Hu et al. [24] showed that H_2S could prolong postharvest storage of fresh-cut pears and inhibit fungal growth, although the underlining mechanism of the antifungal role of H_2S is unknown.

In this work, we investigated the antifungal effect of H_2S on the postharvest pathogens *A. niger* and *P. italicum* inoculated on fruits, as well as on the growth of these fungi on Petri dishes with defined media. We also examined the effect of H_2S on baker's yeast (*Saccharomyces cerevisiae*), *Rhizopus oryzae*, the human pathogen *Candida albicans*, and several food-borne bacteria, including *Staphyloccocus aureus*, *Salmonella typhimurium*, *Listeria monocytogenes*, *Bacillus subtilis*, *Bacillus thuringiensis*, *Escherichia coli* and *Enterobacter aerogenes*.

Materials and methods

Materials

Six different fruits, apple (*Malus domestica*), kiwifruit (*Actinidia deliciosa*), pear (*Pyrus bretschneideri* Rehd.), sweet orange (*Citrus sinensis*), mandarin (*Citrus reticulata*) and tomato (*Lycopersicon esculentum*), used in this work were supplied by a fruit market in Hefei, Anhui province, China. Unwounded and healthy fruits, all of a similar size and maturity, were selected for experimentation. Pure fungal and bacterial isolates used in this research were kindly supplied by School of Biotechnology and Food Engineering, Hefei University of Technology, Anhui, People's Republic of China, except *Candida albicans* (SC5314) which was kindly bestowed by Prof. Jianli Sang at College of Life Science, Beijing Normal University. Three molds (*Aspergillus niger*, *Penicillium italicum*, *Rhizopus oryzae*), two yeasts (*Saccharomyces cerevisiae*, *Candida albicans*) and seven bacteria (*Staphyloccocus aureus*, *Salmonella typhimurium*, *Listeria monocytogenes*, *Bacillus subtilis*, *Bacillus thuringiensis*, *Escherichia coli* and *Enterobacter aerogenes*) were used in this study. Molds, yeasts and bacteria were cultured on potato dextrose agar (PDA), yeast peptone dextrose (YPD) agar and nutrient agar in Petri dishes, respectively. Sodium hydrosulfide (NaHS) was purchased from Sigma and used as exogenous H_2S donor [24]. Aqueous NaHS solutions (150 mL) at different concentrations were placed in the bottom of sealed containers (volume 3 L) to release H_2S to fumigate fruits or pathogens. Aqueous NaHS solutions could steadily release H_2S gas since 30 min, and almost keep at the same level till 24 h. Thus NaHS solutions are renewed daily.

Antifungal Assay of H_2S on Postharvest Fruits Infected with *A. niger* and *P. italicum*

A. niger and *P. italicum* spores were harvested from isolates after 6 d growth at 25°C, and suspended in sterile physiological saline. The suspensions were then filtered through sterile gauze to remove mycelium and adjusted to 1×10^6 spores per ml. Apples, kiwifruits, pears, sweet oranges, mandarins and tomatoes were washed with tap water and sterilized with 75% ethanol. The surfaces of these fruits were wounded at five different sites and each of the wounds (2 mm diameter and 4 mm deep) was injected with 5 μL of spore suspension [25]. After air-drying, three replicates of plant samples were fumigated with 0.5 mM NaHS solution (150 mL) or H_2O (as control) in sealed containers at 25°C. NaHS solutions were changed daily.

Antifungal Effects of H_2S on *A. niger* and *P. italicum*

A 2 μL spore suspension (1×10^6 spores per mL) was inoculated in the centre of a 9-cm diameter Petri dish and NaHS solutions at different concentrations (0.00, 0.01, 0.05, 0.10, 0.50 and 2.5 mM) were added in the bottom of sealed containers to fumigate the pathogens at 25°C and the NaHS solutions were renewed daily. The colony diameters (mm) of *A. niger* and *P. italicum* were recorded for 8 d. The lowest NaHS concentration that resulted in failure of colony formation after 4 d incubation was regarded as the minimal inhibitory concentration (MIC) [26]. To investigate whether H_2S treatment could kill the pathogens or just inhibit their growth, *A. niger* and *P. italicum* were first treated with 2.5 mM NaHS solution for 1 day and allowed to recover with water fumigation for another 8 d at 25°C.

Determination of Spore Germination and Germ Tube Elongation

Spore germination and germ tube elongation was measured according to Lazar et al. [18] and Liu et al. [26] with minor modifications. 20 μL of spore suspension (1×10^6 spores per ml) was placed on a 7-mm diameter plug of PDA which was then placed on a glass slide. Each plug-slide was held in a Petri dish with moistened filter paper, and then placed in sealed container at 25°C. Spore germination (germ tube longer than two times the diameter of the corresponding spore) was assessed under a microscope on three occasions (6, 12 and 24 h) and the length of germ tube was recorded at 12 h for *A. niger* and 24 h for *P. italicum*.

Examination of Mycelial Growth and Micro-Morphology

To test the effect of H_2S on mycelial growth and hyphal micromorphology, three different sites on a Petri plate were spot-inoculated with 2 μL spore suspension (1×10^6 spores per ml). After 3 d of water fumigation, plates were treated with NaHS solutions at MIC i.e. 0.5 mM, 2.5 mM or water (as control) for 1 d at 25°C and mycelial growth and morphological change were recorded as described by Liu et al [26].

Detection of ROS with Fluorescence Probe in *A. niger*

A redox-sensitive fluorescent probe 2′,7′-dichlorofluorescein diacetate (DCHF-DA) was used for the analysis of intracellular ROS level according to Savi et al [27]. *A. niger* mycelia grown on PDA plates were incubated with 10 μM DCHF-DA at 37°C for 20 min in the dark. After washing the fungi with double-distilled water for three times, the fluorescence of dichlorofluorescein DCF (the oxidation product of DCHF-DA; excitation at 485 nm, emission at 530 nm) were observed using a Nikon Eclipse 80i fluorescence microscope (Nikon, Japan). Non-stained *A. niger* was used as negative control.

Determination of SOD and CAT Activity in *A. niger*

A. niger mycelium grown on PDA media were suspended in 50 mM ice-cold phosphate buffer (pH 7.8, 1 mM EDTA and 1 mM PMSF) and an equal volume acid-washed glass beads (0.4–0.6 mm diameter, Sigma) were added to 1 mL of *A. niger* suspension in an Eppendorf tube. The cells were broken in a FastPrep-24 Tissue Homogenizer (MP Biomedicals, USA) for 40 s at 5500 rpm, repeated four times with 2 min on ice between intervals. The mixture was centrifuged at 12,000 rpm at 4°C for 10 min to obtain the supernatant. Activities of SOD and CAT were determined by procedures described by Giannopolitis and Ries [28] and Beers and Sizer [29] respectively. One SOD unit was defined as the amount of enzyme that inhibits the rate of nitroblue tetrazolium (NBT) reduction by 50% and one unit of CAT was defined as a decrease of 0.1 OD at 240 nm. Activity was expressed as $U\cdot mg^{-1}$ protein. The protein content was measured according to Bradford [30] using bovine serum albumin as standard.

Reverse Transcriptase-Polymerase Chain Reaction (RT-PCR)

A. niger mycelium grown on PDA medium and an equal weight of acid-washed glass beads were added in RNAiso plus (TaKaRa,

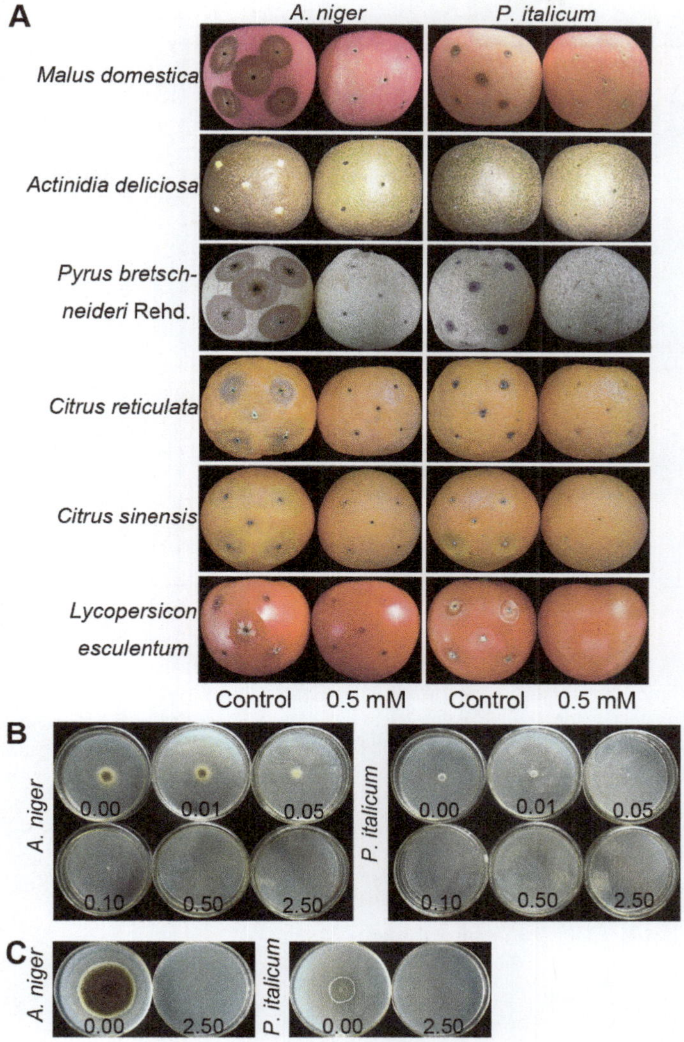

Figure 1. Effect of H_2S on *A. niger* and *P. italicum* growth in inoculated fruits and on defined media. A, photographs of six kinds of postharvest fruits inoculated with *A. niger* and *P. italicum* when fumigated with 0.0 mM (control) or 0.5 mM NaHS solution for 9 d at 25°C; B, photographs of *A. niger* and *P. italicum* grown on Petri plates on defined media after fumigation with 0.00, 0.01, 0.05, 0.10, 0.50 and 2.50 mM NaHS solutions for 2 d at 25°C; C, photographs of *A. niger* and *P. italicum* after 1 d of 2.5 mM NaHS treatment and recovered in water fumigation for another 8 d at 25°C.

Japan) solution. The cells were bead-beaten (5500 rpm) for four rounds of 40 s with 2 min on ice between intervals. Total RNA and cDNA were obtained according to the manufacturer's instructions of RNAiso plus (TaKaRa, Japan) and PrimeScript RT Master Mix (TaKaRa, Japan), respectively. PCR conditions were as follows: initial denaturation at 94°C for 5 min, followed by cycles of 94°C for 30 s, 50°C for 30 s, and 72°C for 30 s. Actin-encoding gene (ANI_1_106134) expression was used as a control for SOD-encoding genes (ANI_1_840184, ANI_1_470064 and ANI_1_1170064) and CAT-encoding gene (ANI_1_2390104) [31]. Primers used for RT-PCR and corresponding PCR product sizes are shown as following: ANI_1_106134-For/Rev, 5′-AAAATGCTGCTGCGAATG-3′/5′-GCGGATAGCTGAACGAGAT-3′, 238 bp; ANI_1_840184-For/Rev, 5′-CGCTGCTCTTGCTGCTAT-3′/5′-GTCCTGGTTCTGTGAAATCG-3′, 153 bp; ANI_1_470064-For/Rev, 5′-CACGGATAACAGCGACTAG-3′/5′-TGAGCAGATTTGAGCACCTT-3′, 269 bp; ANI_1_1170064-For/Rev, 5′-GCTAATGCTGGACGGAACT-3′/5′-TGGTCTGTGAAGGGAAAGG-3′, 164 bp; ANI_1_2390104-For/Rev, 5′-GAATCGGCATCAACCTCC-3′/5′-CACGCTCATCCATCACTT-3′, 285 bp. The band densities of RT-PCR results were quantified with Image J software (NIH, USA).

Antimicrobial Activity of H_2S on Baker's Yeast, *C. albicans*, *R. oryzae*, and some Food-Borne Bacteria

S. cerevisiae, *C. albicans*, *S. aureus*, *S. typhimurium*, *L. monocytogenes*, *B. subtilis*, *B. thuringiensis*, *E. coli* and *E. aerogenes* suspensions were harvested from medium cultured for 24 h, and diluted to about 2×10^3 cells per mL. Suspensions (100 μL) were evenly spread on YPD or nutrient agar in a Petri dish and each plate was placed in a sealed container and fumigated with H_2S released from NaHS solutions (150 mL) at 25°C. Colony formation (CFU: Colony-Forming Units) of three replicates was recorded daily for 4 d. The lowest NaHS concentration that resulted in failure to form colonies after 2 d

Table 1. Colony diameter of *A. niger* and *P. italicum* after exposure to H_2S at 25°C.

Strains	Treatment (mM NaHS)	Treatment time			
		2 d	4 d	6 d	8 d
		Diameter (mm)	Diameter (mm)	Diameter (mm)	Diameter (mm)
A. niger	0.00	27.1±0.9 a	41.5±3.1 a	51.4±2.8 a	58.6±2.8 a
	0.01	24.2±1.0 b	39.6±1.7 b	47.0±0.7 b	54.7±0.7 b
	0.05	17.8±0.3 c	39.5±1.5 b	48.3±0.5 b	54.9±2.0 b
	0.10	4.2±0.5 d	32.8±1.0 c	43.3±1.5 c	52.9±1.3 b
	0.50	0 e	0 d	0 d	0 c
	2.50	0 e	0 d	0 d	0 c
P. italicum	0.00	9.2±0.3 a	17.3±1.1 a	22.2±1.3 a	25.3±1.3 a
	0.01	8.5±0.7 b	16.7±0.5 a	21.2±1.0 a	24.0±1.4 a
	0.05	3.8±0.5 c	12.4±0.1 b	18.3±0.7 b	21.0±1.0 b
	0.10	0 d	10.0±0.2 c	16.5±0.6 c	20.4±1.3 b
	0.50	0 d	0 d	0 d	0 c
	2.50	0 d	0 d	0 d	0 c

Different letters mean significance of difference between the treatments ($P<0.05$, ANOVA, LSD). Diameter: colony diameter.

incubation was regarded as the minimal inhibitory concentration (MIC). For *R. oryzae*, a 2 μL spore suspension (1×10^6 spores per mL) harvested from a 7-d culture on plate was inoculated 4 sites on a 9-cm diameter Petri dish and NaHS solutions at different concentrations (0.00, 0.05, 0.10, 0.50, 3.00 and 10.00 mM) were added in the bottom of sealed containers to fumigate *R. oryzae* at 25°C and the NaHS solutions were renewed daily. To investigate whether H_2S treatment could kill these strains or just inhibit their growth, strains were first treated with 2.5 mM NaHS solution for 1 d and then recovered in a water atmosphere for another 6 d at 25°C.

Statistical Analysis

Each experiment was repeated three times. Statistical significance was tested by one-way analysis of variance (ANOVA) using IBM SPSS Statistics (SPSS version 20.0), and the results were expressed as the means ± standard deviation (SD). Least significant difference test (LSD) was performed on all data following ANOVA tests to test for significant ($P<0.05$ or $P<0.01$) differences between treatments.

Table 2. Spore germination and germ tube elongation of *A. niger* and *P. italicum* after exposure to H_2S at 25°C.

Strains	Treatment (mM NaHS)	Treatment time				
		6 h	12 h		24 h	
		Germ (%)	Germ (%)	El (μm)	Germ (%)	El (μm)
A. niger	0.00	0 a	73.3±1.8 a	37.3±5.0 a	100±0.0 a	–
	0.01	0 a	62.9±2.9 b	12.6±1.5 b	100±0.0 a	–
	0.05	0 a	3.3±0.5 c	2.7±0.9 c	100±0.0 a	–
	0.10	0 a	0 d	0 c	77.8±2.5 b	–
	0.50	0 a	0 d	0 c	0 c	–
	2.50	0 a	0 d	0 c	0 c	–
P. italicum	0.00	0 a	53.0±2.1 a	–	100±0.0 a	155.6±14.1 a
	0.01	0 a	0 b	–	100±0.0 a	116.4±12.5 b
	0.05	0 a	0 b	–	84.5±2.2 b	10.7±2.1 c
	0.10	0 a	0 b	–	10.1±2.6 c	4.5±0.7 cd
	0.50	0 a	0 b	–	0 d	0 d
	2.50	0 a	0 b	–	0 d	0 d

Different letters mean significance of difference between the treatments ($P < 0.05$, ANOVA, LSD). Germ: spore germination percentage; El: germ tube elongation. The symbol "–" stands for not determined at this time point.

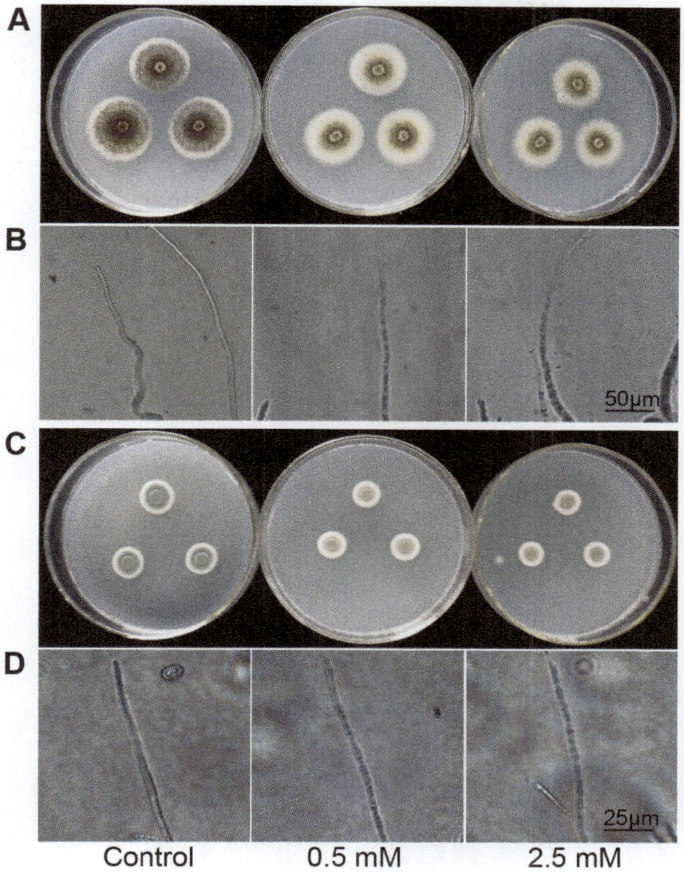

Figure 2. Effect of H_2S fumigation on mycelial growth and micro-morphology of *A. niger* and *P. italicum* grown on defined media. Colonies were cultured for 3 d of water fumigation at 25°C, and then exposed to H_2S released from 0.5 mM and 2.5 mM NaHS for 1 d. A and B, mycelial growth and micro-morphology of *A. niger*; C and D, mycelial growth and micro-morphology of *P. italicum*.

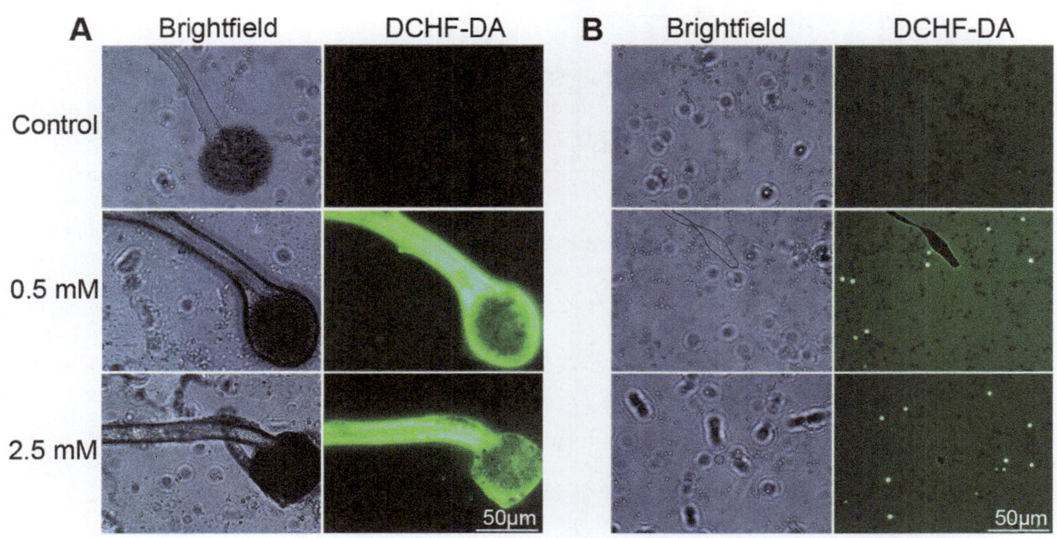

Figure 3. Effect of H_2S fumigation on ROS level in *A. niger*. *A. niger* cells in Figure 2 were used for ROS detection. A, ROS staining in sporangia and sporangiophores; B, ROS staining in *A. niger* spores. Left parts of A and B shows the bright field images, and right parts fluorescence images.

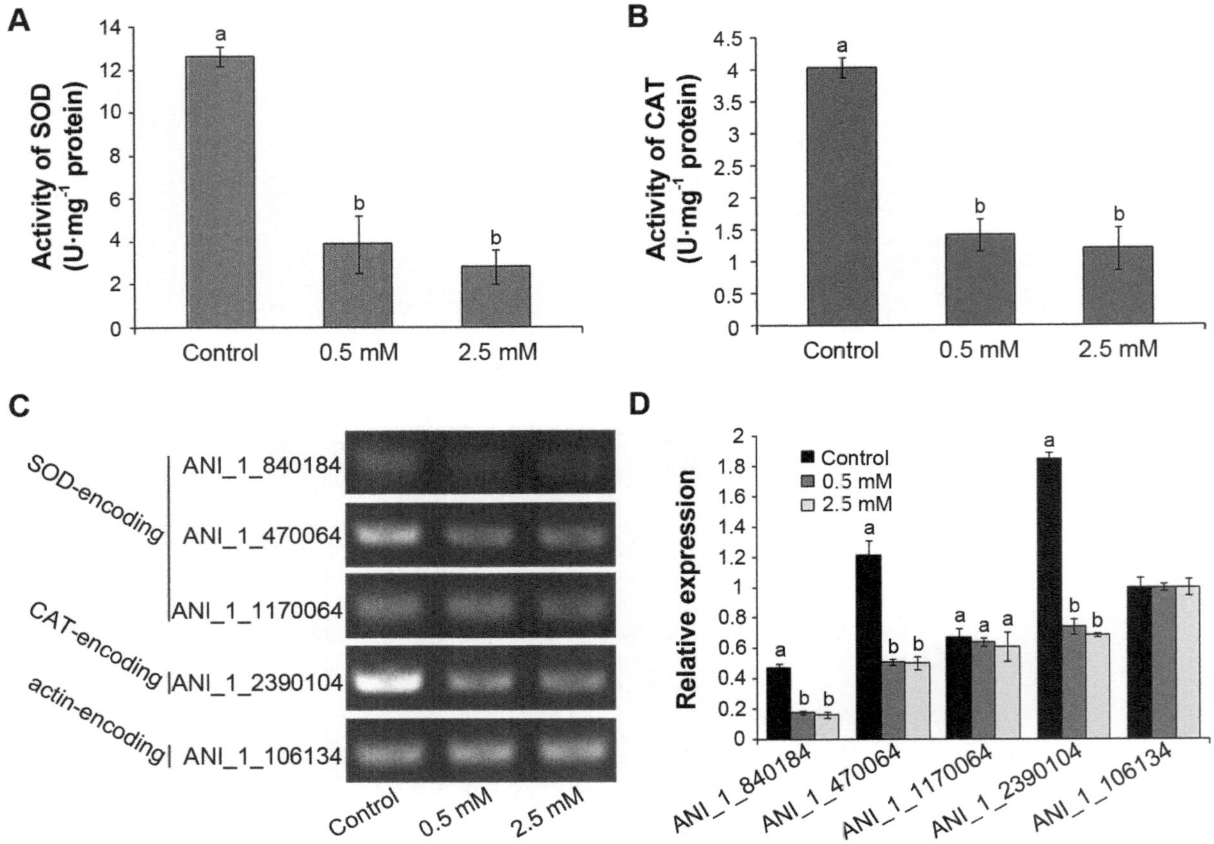

Figure 4. Effect of H_2S fumigation on enzyme activities and gene expression of SOD and CAT in *A. niger*. Colonies were cultured for 3 d in water atmosphere at 25°C, and then exposed to H_2S released from 0.5 mM and 2.5 mM NaHS for 1 d. A, SOD activity in *A. niger*; B, CAT activity in *A. niger*; C, RT-PCR results to analyze gene expressions of SOD and CAT in *A. niger*. D, relative expression levels by quantifying RT-PCR results in C (with the actin-encoding gene value [ANI_1_106134] set as 1.0) with Image J software. Different letters mean significance of difference between the treatments (P<0.01, ANOVA, LSD).

Results

Effect of H_2S on Postharvest Condition of Fruits Inoculated with *A. niger* and *P. italicum*

As shown in Figure 1A, H_2S released by 0.5 mM NaHS solution effectively controlled fruit decay caused by *A. niger* and *P. italicum*. After infection with *A. niger* or *P. italicum*, apples, pears and tomatoes from control condition began to decay at day 3, and kiwifruits, sweet oranges and mandarins at day 5, while fruits from H_2S treatment remained decay-free (data not shown). When stored for 9 d, obvious signs of decay were found in control fruits, whereas lesion diameters of fruits from H_2S treatment remained basically unchanged (Figure 1A).

Effect of H_2S on Growth of Cultured *A. niger* and *P. italicum*

As shown in Figure 1B and Table 1, H_2S significantly ($P<0.05$) inhibited colony growth of *A. niger* or *P. italicum* when grown on YPD plates in a dose-dependent manner. For *A. niger* at day 2, colony diameter with 0.1 mM NaHS was approximately 16% of that in control. In contrast, colony formation of *P. italicum* was not observed when NaHS solution was increased to 0.1 mM at day 2 (Figure 1B and Table 1). The data in Table 1 show that MIC, the minimum concentration of NaHS that resulted in failure of colony formation after 4 d of incubation, for both *A. niger* or *P. italicum* was 0.5 mM. In addition, when *A. niger* or *P. italicum* were treated with 2.5 mM NaHS for 1 day and recovered in a water atmosphere for another 8 d, growth of both fungi was completely suppressed (Figure 1C).

Effect of H_2S on Spore Germination and Germ Tube Elongation

The data in Table 2 show that exposure to H_2S significantly ($P<0.05$) inhibited spore germination and germ tube elongation in *A. niger* and *P. italicum* in a dose-dependent manner. At 12 h, germination percentages of *A. niger* spores in 0.01 mM and 0.05 mM NaHS were approximately 86% and 4.5% of control, respectively, while no germination of *P. italicum* spores occurred in 0.1 mM NaHS. At 24 h, more spores of both fungi germinated at H_2S concentrations below 0.1 mM NaHS. However, 0.5 mM and 2.5 mM NaHS totally inhibited spore germination in *A. niger* and *P. italicum*. The length of germ tubes decreased by 93% at 12 h for both *A. niger* and *P. italicum* at 24 h in 0.05 mM NaHS treatment compared with that of water control. Complete inhibition of spore germination and germ tube elongation occurred in the presence of NaHS concentrations higher than 0.5 mM.

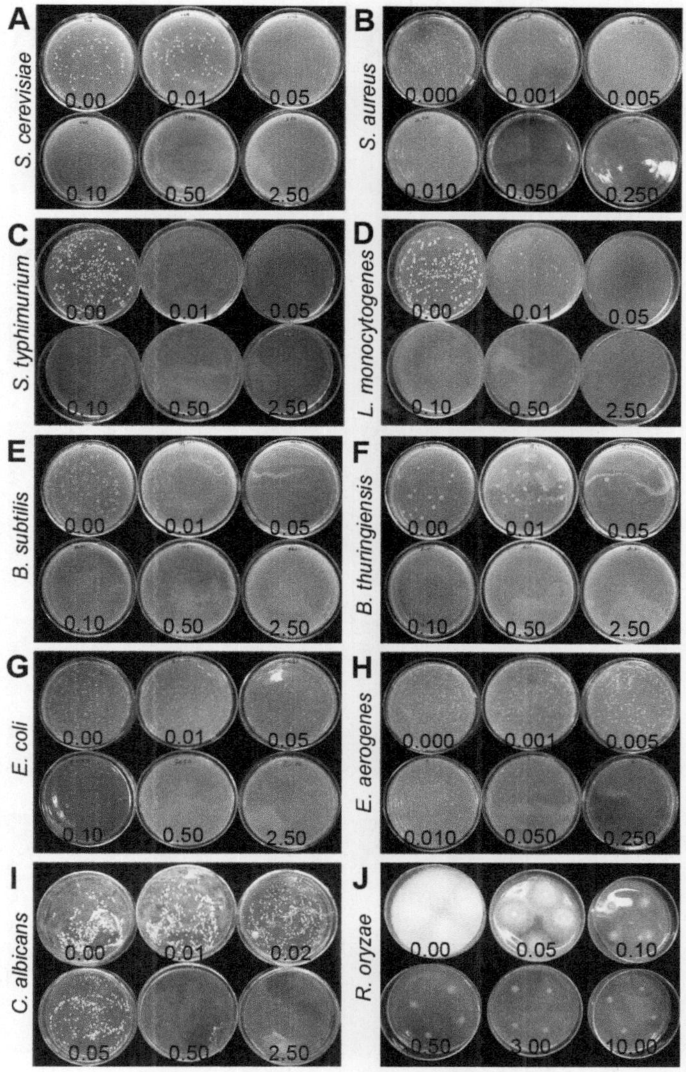

Figure 5. Effect of H_2S fumigation on grown of baker's yeast, *C. albicans*, *R. oryzae*, and some food-borne bacteria growth on defined media. A–F, photographs of *S. cerevisiae*, *S. aureus*, *S. typhimurium*, *L. monocytogenes*, *B. subtilis* and *B. thuringiensis* after 3 d exposure to H_2S released from different concentrations of NaHS at 25°C; G–H, photographs of *E. coli* and *E. aerogenes* after exposure to H_2S for 1 d; I–J, photographs of *C. albicans* and *R. oryzae* after exposure to H_2S for 4 d and 7 d, respectively.

Effect of H_2S on Mycelial Growth and Hyphal Morphology

Mycelial growth and hyphal morphology of *A. niger* and *P. italicum* are presented in Figure 2. As shown in Figure 2A and C, H_2S treatment reduced mycelial diameter compared with control. To further test the effect of H_2S on mycelial micro-morphology, mycelia were examined by light microscopy and morphological deformities (Figure 2B and D) were found. Abnormal contraction of mycelial cytoplasm appeared in *A. niger* and *P. italicum* when fumigated with 0.5 mM or 2.5 mM NaHS for as short as 1 d. For *A. niger*, 2.5 mM treatment even led to the fragmentation of mycelial cytoplasm (Figure 2B).

Effect of H_2S on ROS Level in *A. niger*

The oxidant sensitive probe 2′,7′-dichlorofluorescein diacetate (DCHF-DA) was used to assess the ROS level in *A. niger*. Treatment with 0.5 mM or 2.5 mM NaHS solution significantly increased ROS level in the sporangia and sporangiophores (Figure 3A) compared with water control. However, not all spores from H_2S treatments showed a higher level of ROS (Figure 3B), which might be due to the fact that only peripheral spores in sporangia were exposed to H_2S.

Effect of H_2S on Enzyme Activities and Expression of SOD and CAT Genes in *A. niger*

As shown in Figure 4A and B, H_2S treatments significantly ($P< 0.01$) decreased both SOD and CAT activities in *A. niger* compared with the control. SOD activity in H_2S treatment was reduced by 69% (0.5 mM NaHS) and 78% (2.5 mM NaHS) compared with control, respectively. Similarly, CAT activity in H_2S treatments was only 35% (0.5 mM NaHS) and 30% (2.5 mM NaHS) of that in control, respectively. The expression of SOD and CAT genes was examined by RT-PCR and the data shown in Figure 4C and D. Compared with controls, exposure of *A. niger* to H_2S significantly decreased the expression of the SOD-

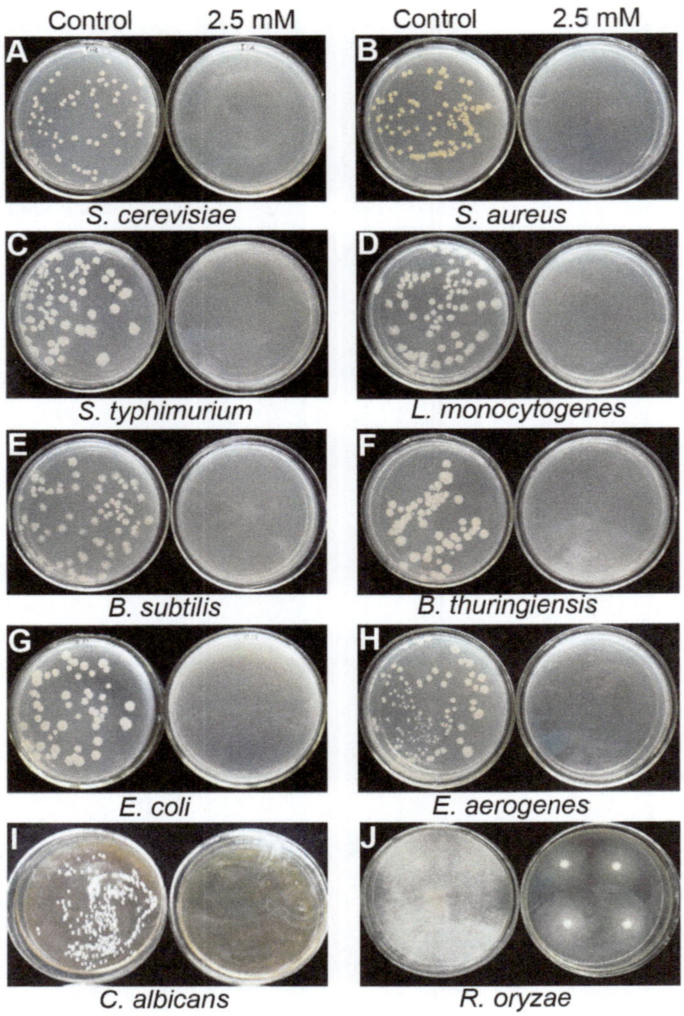

Figure 6. Effect of short term H_2S exposure to baker's yeast, *C. albicans*, *R. oryzae*, and some food-borne bacteria grown on defined media. After 1 d of 2.5 mM NaHS treatment and another 6 d in water atmosphere at 25°C, the photographs were taken. A–H show photographs of *S. cerevisiae*, *S. aureus*, *S. typhimurium*, *L. monocytogenes*, *B. subtilis*, *B. thuringiensis*, *E. coli* and *E. aerogenes*, respectively; I–J show photographs of *C. albicans* and *R. oryzae* recovered in water atmosphere for 2 and 4 d after 2.5 mM NaHS treatment for 1 d.

encoding genes ANI_1_840184 and ANI_1_470064, and the CAT-encoding gene ANI_1_2390104, while no obvious change appeared in the transcript of the SOD-encoding gene ANI_1_1170064. The enzyme activity and gene expression assays of SOD and CAT indicated that H_2S negatively regulated antioxidant system in *A. niger*.

Effect of H_2S on Baker's Yeast, *C. albicans*, *R. oryzae* and some Food-Borne Bacteria

To study possible effect of H_2S on baker's yeast, *C. albicans*, *R. oryzae* and some food-borne bacteria, we examined the growth of *S. cerevisiae*, *C. albicans*, *R. oryzae*, *S. aureus*, *S. typhimurium*, *L. monocytogenes*, *B. subtilis*, *B. thuringiensis*, *E. coli* and *E. aerogenes* grown on culture media. The data in Figure 5 and Table 3 show that H_2S had a significant ($P<0.05$) antimicrobial effect on *S. cerevisiae*, *C. albicans* and on bacteria in a dose-dependent manner. *S. aureus* was the most sensitive to H_2S treatment. The MIC was determined to be 0.01 mM for *S. aureus*, 0.05 mM for *S. cerevisiae*, *S. typhimurium*, *L. monocytogenes* and *B. subtilis*, 0.1 mM for *B. thuringiensis*, 0.25 mM for *E. aerogenes*, 0.5 mM for *C. albicans* and 2.5 mM for *E. coli*. In contrast, *R. oryzae* showed high resistance to H_2S fumigation though H_2S did inhibit the colony growth of *R. oryzae*, and the MIC might be very high (Figure 5). Moreover, after 1 d exposure to 2.5 mM NaHS, no colony formation was observed in any strains after recovery in a water atmosphere for a further 6 d (Figure 6) except *R. oryzae*, suggesting that high concentrations of H_2S exhibited a microbicidal role rather than just inhibiting the growth of microbes.

Discussion

The results reported in this work show the possibility of an alternative strategy for postharvest fruit storage, based on exposure of infected fruit to H_2S released by NaHS. Our previous work found that exogenous H_2S could extend the postharvest life of strawberry, fresh-cut kiwifruit, broccoli and pear [15,20,21,24], while there are fewer data on an antifungal role of H_2S against postharvest pathogens. Haneklaus et al. [32] reported that uptake

Table 3. Colony formation of yeasts and bacteria after exposure to H_2S at 25°C.

Strains	Treatment (mM NaHS)	Treatment time			
		1 d	2 d	3 d	4 d
		Number (CFU)	Number (CFU)	Number (CFU)	Number (CFU)
S. cerevisiae	0.00	0 a	81.0±7.2 a	81.0±7.2 a	81.0±7.2 a
	0.01	0 a	85.7±9.0 a	85.7±9.0 a	85.7±9.0 a
	0.05	0 a	0 b	20.5±2.1 b	47.5±4.9 b
	0.10	0 a	0 b	0 c	0 c
	0.50	0 a	0 b	0 c	0 c
	2.50	0 a	0 b	0 c	0 c
S. aureus	0.000	0 a	73.0±2.8 a	351.5±3.5 a	351.5±3.5 a
	0.001	0 a	20.0±1.4 b	114.0±2.8 b	114.0±.8 b
	0.005	0 a	0.5±0.7 c	18.5±0.7 c	30.0±1.4 c
	0.010	0 a	0 c	11.5±0.7 d	20.5±2.1 d
	0.050	0 a	0 c	0 e	0 e
	0.250	0 a	0 c	0 e	0 e
S. typhimurium	0.00	357.0±24.0 a	368.5±19.1 a	368.5±19.1 a	368.5±19.1 a
	0.01	0 b	8.0±2.8 b	26.5±2.1 b	42.5±0.7 b
	0.05	0 b	0 b	0.5±0.7 c	3±4.2 c
	0.10	0 b	0 b	0 c	0 c
	0.50	0 b	0 b	0 c	0 c
	2.50	0 b	0 b	0 c	0 c
L. monocytogenes	0.00	0 a	316.5±20.5 a	316.5±20.5 a	316.5±20.5 a
	0.01	0 a	40.0±9.9 b	62.5±14.8 b	72.5±12.0 b
	0.05	0 a	0 c	2.0±1.4 c	2.5±2.1 c
	0.10	0 a	0 c	0 c	0 c
	0.50	0 a	0 c	0 c	0 c
	2.50	0 a	0 c	0 c	0 c
B. subtilis	0.00	0 a	78.3±8.7 a	78.3±8.7 a	78.3±8.7 a
	0.01	0 a	47.0±18.3 b	48.7±18.0 b	56.3±9.7 b
	0.05	0 a	0 c	0.7±0.6 c	0.7±0.6 c
	0.10	0 a	0 c	0 c	0 c
	0.50	0 a	0 c	0 c	0 c
	2.50	0 a	0 c	0 c	0 c
B. thuringiensis	0.00	8.3±3.5 a	21.0±2.0 a	21.7±1.2 a	23.0±2.0 a
	0.01	6.0±1.0 a	16.7±4.2 b	19.3±3.1 a	21.0±1.0 b
	0.05	0.7±0.6 b	0.7±0.6 c	0.7±0.6 b	0.7±0.6 c
	0.10	0 b	0 c	0 b	0 c
	0.50	0 b	0 c	0 b	0 c
	2.50	0 b	0 c	0 b	0 c
E. coli	0.00	143.7±11.6 a	193.0±24.8 a	193.0±24.8 a	193.0±24.8 a
	0.01	102.7±16.0 b	157.0±14.7 b	157.0±14.7 b	157.0±14.7 b
	0.05	61.3±0.6 c	128.3±8.3 c	128.3±8.3 c	128.3±8.3 c
	0.10	38.0±8.5 d	94.0±11.3 d	127.7±8.1 c	127.7±8.1 c
	0.50	0 e	43.0±23.9 e	83.0±22.9 d	97.7±10.1 d
	2.50	0 e	0 f	0 e	0 e
E. aerogenes	0.000	225.5±4.9 a	236.5±4.9 a	236.5±4.9 a	236.5±4.9 a
	0.001	205.5±7.8 b	218.0±7.1 a	218.0±7.1 b	218.0±7.1 b
	0.005	114.0±4.2 c	141.5±2.1 b	141.5±2.1 c	141.5±2.1 c

Table 3. Cont.

Strains	Treatment (mM NaHS)	Treatment time			
		1 d	2 d	3 d	4 d
		Number (CFU)	Number (CFU)	Number (CFU)	Number (CFU)
	0.010	71.0±1.4 d	137.5±3.5 b	137.5±3.5 c	137.5±3.5 c
	0.050	37.5±2.1 e	89.0±1.4 c	120.0±2.8 d	120.0±2.8 d
	0.250	0 f	0 d	0 e	0 e
C. albicans	0.00	–	–	–	401.0±14.8 a
	0.01	–	–	–	378.7±10.2 a
	0.02	–	–	–	325.3±8.3 b
	0.05	–	–	–	135.6±11.5 c
	0.50	–	–	–	0 d
	2.50	–	–	–	0 d

Different letters mean significance of difference between the treatments ($P<0.05$, ANOVA, LSD). CFU: Colony-Forming Units. The symbol "–" stands for not determined at this time point.

of 10 μM/h H_2S by the pathogen would produce a fungicidal effect. It has also been shown that endogenous H_2S release showed a significant rise when agricultural crops suffered from fungal infection [33,34], suggesting an important role of H_2S in plant defense against fungal attack. In this study, we found an effective antifungal effect of H_2S to the postharvest pathogens *A. niger* and *P. italicum* inoculated on fruits and grown on defined medium (Figure 1 and Table 1). Consistent with these observations, we previously reported that H_2S plays as a role as a fungicide to the pear pathogens *P. expansum* and *A. niger* and prolongs postharvest storage of fresh-cut pears [24].

We also studied the toxicity of H_2S to baker's yeast, the human pathogen *C. albicans* and several bacteria, and find that H_2S released by 2.5 mM NaHS plays a microbicidal role rather than just inhibiting the growth of molds, yeasts and bacteria (Figure 1C, Figure 6 and Table 3), while *R. oryzae* is more resistant to H_2S. Our work also shows that H_2S exerts its antifungal effect by affecting various aspects of fungal growth, including: 1) inhibiting spore germination and germ tube elongation (Table 2), 2) retarding mycelial growth (Figure 2A and C), and 3) causing abnormal contraction even fragmentation of mycelial cytoplasm (Figure 2B and D). Similar to the effect of H_2S towards mycelial cytoplasm that we observed, Senthilkumar et al. [35] also observed such deformities in the mycelia of *Paenibacillus* sp. HKA-15 when exposed to antibiotics.

Recently, Lai et al. [19] reported that NO could inhibit the germination of *P. expansum* spores and induce an increase of intracellular ROS which might result from increased ROS formation and decreased ROS detoxification. In the present study, we found that H_2S could increase intracellular ROS in sporangia and sporangiophores as well as in spores of *A. niger* (Figure 3A and B). It is unclear that whether H_2S directly induces the formation of ROS in *A. niger*. However, Eghbal et al. [36] have reported that H_2S exerts its cytotoxic effect against hepatocytes via inducing ROS formation and the overall reaction could be written as: $nS^{2-} + 2nO_2 \rightarrow Sn + 2nO_2^{-}$. In addition, H_2S has also been found to induce oxidative damage in *Glycera dibranchiate* and mammalian cells [37,38]. Therefore, we speculate that H_2S might directly induce ROS generation in *A. niger* and excessive ROS subsequently causes oxidative damage to molecules crucial for mycelial growth and spore germination.

Antioxidant enzymes such as SOD and CAT also play important roles in ROS elimination in response to oxidative stress [39]. Our results also show that H_2S could inhibit gene expression (Figure 4C and D) and decrease enzyme activities of SOD and CAT (Figure 4A and B), which might contribute to the increased ROS level observed in H_2S-treated *A. niger*. In contrast, the studies in postharvest storage of fruits and vegetables reveal that H_2S can eliminate ROS accumulation by improving the endogenous antioxidant system [15,20,21,22,24]. Thus H_2S exhibits different effects on microbes and plants which might be due to different tolerance to H_2S. Besides, the phenomenon that endogenous H_2S released when agricultural crops suffered from fungal infection [33,34] also supports our hypothesis. Endogenous H_2S in Gram-negative and Gram-positive bacteria is also required for antibiotic tolerance by elevating their antioxidant capacity [40]. In our study, we find that exogenous H_2S application effectively inhibits the growth of a series of bacteria (Figure 5), which can be attributed to the relatively higher level of applied H_2S than lower level in endogenous production.

The evaluation of postharvest fungi *A. niger*, *P. italicum* and *R. oryzae*, yeasts and several food-borne bacteria strongly suggests the possible commercial value of H_2S fumigation to reduce postharvest spoilage and food storage. In this work, we also show that a fungicidal role of H_2S might be associated with increased ROS accumulation in H_2S-treated fungi. Further studies involving oxidative damage will help to better understand the ROS-related mechanism by which H_2S inhibits the growth of postharvest pathogens.

Acknowledgments

We are grateful to Russell Jones, University of California at Berkeley for critical reading the manuscript and Jianli Sang at Beijing Normal University for providing the strain *Candida albicans* (SC5314).

Author Contributions

Conceived and designed the experiments: LHF KDH HZ. Performed the experiments: LHF KDH LYH. Analyzed the data: LHF LBH HY YSL HZ. Contributed reagents/materials/analysis tools: YHL LBH HY YSL. Contributed to the writing of the manuscript: LHF KDH LYH HZ.

References

1. Hodges RJ, Buzby JC, Bennett B (2011) Postharvest losses and waste in developed and less developed countries: opportunities to improve resource use. J Agric Sci 149: 37–45.
2. Heydari A, Pessarakli M (2010) A review on biological control of fungal plant pathogens using microbial antagonists. J Biol Sci 10: 273–290.
3. Tian J, Ban X, Zeng H, Huang B, He J, et al. (2011) *In vitro* and *in vivo* activity of essential oil from dill (*Anethum graveolens* L.) against fungal spoilage of cherry tomatoes. Food Control 22: 1992–1999.
4. García-Cela E, Crespo-Sempere A, Ramos AJ, Sanchis V, Marin S (2014) Ecophysiological characterization of *Aspergillus carbonarius*, *Aspergillus tubingensis* and *Aspergillus niger* isolated from grapes in Spanish vineyards. Int J Food Microbiol 173: 89–98.
5. Hernández-Montiel LG, Ochoa JL, Troyo-Diéguez E, Larralde-Corona CP (2010) Biocontrol of postharvest blue mold (*Penicillium italicum* Wehmer) on Mexican lime by marine and citrus *Debaryomyces hansenii* isolates. Postharvest Biol Technol 56: 181–187.
6. Strawn LK, Fortes ED, Bihn EA, Nightingale KK, Gröhn YT, et al. (2013) Landscape and meteorological factors affecting prevalence of three food-borne pathogens in fruit and vegetable farms. Appl Environ Microbiol 79: 588–600.
7. Reynolds A, Moffatt CR, Dyda A, Hundy R, Kaye AL, et al. (2010) An outbreak of gastroenteritis due to *Salmonella typhimurium* phage type 170 associated with consumption of a dessert containing raw egg. Commun Dis Intell Q Rep 34: 329–333.
8. Majumdar T, Chakraborty R, Raychaudhuri U (2013) Rapid electrochemical quantification of food borne pathogen *Staphylococcus aureus* based on hydrogen peroxide degradation by catalase. J Electrochem Soc 160: G75–G78.
9. Tripathi P, Dubey NK (2004) Exploitation of natural products as an alternative strategy to control postharvest fungal rotting of fruit and vegetables. Postharvest Biol Technol 32: 235–245.
10. Deng J, Li W, Peng XL, Hao XH (2013) Study on the potential of antifungal activity of essential oils against fungal pathogens of fruits and vegetables. J Chem Pharm Res 5: 443–446.
11. Wang R (2012) Physiological implications of hydrogen sulfide: a whiff exploration that blossomed. Physiol Rev 92: 791–896.
12. Zhang H, Hu LY, Hu KD, He YD, Wang SH, et al. (2008) Hydrogen sulfide promotes wheat seed germination and alleviates oxidative damage against copper stress. J Integr Plant Biol 50: 1518–1529.
13. Jin Z, Shen J, Qiao Z, Yang G, Wang R, et al. (2011) Hydrogen sulfide improves drought resistance in *Arabidopsis thaliana*. Biochem Biophys Res Commun 414: 481–486.
14. Jin Z, Xue S, Luo Y, Tian B, Fang H, et al. (2013) Hydrogen sulfide interacting with abscisic acid in stomatal regulation responses to drought stress in *Arabidopsis*. Plant Physiol Biochem 62: 41–46.
15. Hu LY, Hu SL, Wu J, Li YH, Zheng JL, et al. (2012) Hydrogen sulfide prolongs postharvest shelf life of strawberry and plays an antioxidative role in fruits. J Agric Food Chem 60: 8684–8693.
16. Lisjak M, Teklic T, Wilson ID, Whiteman M, Hancock JT (2013) Hydrogen sulfide: environmental factor or signalling molecule? Plant Cell Environ 36: 1607–1616.
17. Manjunatha G, Lokesh V, Neelwarne B (2010) Nitric oxide in fruit ripening: trends and opportunities. Biotechnol Adv 28: 489–499.
18. Lazar EE, Wills RBH, Ho BT, Harris AM, Spohr LJ (2008) Antifungal effect of gaseous nitric oxide on mycelium growth, sporulation and spore germination of the postharvest horticulture pathogens, *Aspergillus niger*, *Monilinia fructicola* and *Penicillium italicum*. Lett Appl Microbiol 46: 688–692.
19. Lai T, Li B, Qin G, Tian S (2011) Oxidative damage involves in the inhibitory effect of nitric oxide on spore germination of *Penicillium expansum*. Curr Microbiol 62: 229–234.
20. Gao SP, Hu KD, Hu LY, Li YH, Han Y, et al. (2013) Hydrogen sulfide delays postharvest senescence and plays an antioxidative role in fresh-cut kiwifruit. HortScience 48: 1385–1392.
21. Li SP, Hu KD, Hu LY, Li YH, Jiang AM, et al. (2014) Hydrogen sulfide alleviates postharvest senescence of broccoli by modulating antioxidant defense and senescence-related gene expression. J Agric Food Chem 62: 1119–1129.
22. Hu H, Shen W, Li P (2014) Effects of hydrogen sulphide on quality and antioxidant capacity of mulberry fruit. Int J Food Sci Technol 49: 399–409.
23. Marsh RW (1929) Investigations on the fungicidal action of sulphur. III. Studies on the toxicity of sulphuretted hydrogen and on the interaction of sulphur with fungi. J Hortic Sci 7: 237–250.
24. Hu KD, Wang Q, Hu LY, Gao SP, Wu J, et al. (2014) Hydrogen sulfide prolongs postharvest storage of fresh-cut pears (*Pyrus pyrifolia*) by alleviation of oxidative damage and inhibition of fungal growth. PLoS One 9: e85524.
25. Yu T, Chen J, Lu H, Zheng X (2009) Indole-3-acetic acid improves postharvest biological control of blue mold rot of apple by *Cryptococcus laurentii*. Phytopathology 99: 258–264.
26. Liu X, Wang J, Gou P, Mao C, Zhu ZR, et al. (2007) *In vitro* inhibition of postharvest pathogens of fruit and control of gray mold of strawberry and green mold of citrus by aureobasidin A. Int J Food Microbiol 119: 223–229.
27. Savi GD, Bortoluzzi AJ, Scussel VM (2013) Antifungal properties of Zinc-compounds against toxigenic fungi and mycotoxin. Int J Food Sci Technol 48: 1834–1840.
28. Giannopolitis CN, Ries SK (1977) Superoxide dismutases I. Occurrence in higher plants. Plant Physiol 59: 309–314.
29. Beers RF, Sizer IW (1952) A spectrophotometric method for measuring the breakdown of hydrogen peroxide by catalase. J Biol Chem 195: 133–140.
30. Bradford MM (1976) A rapid and sensitive method for the quantitation of microgram quantities of protein utilizing the principle of protein-dye binding. Anal Biochem 72: 248–254.
31. Thellin O, Zorzi W, Lakaye B, De-Borman B, Coumans B, et al. (1999) Housekeeping genes as internal standards: use and limits. J Bacteriol 75: 291–295.
32. Haneklaus S, Bloem E, Schnug E (2007) Sulfur and plant disease. In: Datnoff L, Elmer W, Huber D, editors. Mineral nutrition and plant diseases. St Paul, MN: APS Press. pp.101–118.
33. Bloem E, Riemenschneider A, Volker J, Papenbrock J, Schmidt A, et al. (2004) Sulphur supply and infection with *Pyrenopeziza brassicae* influence L-cysteine desulphydrase activity in *Brassica napus* L. J Exp Bot 55: 2305–2312.
34. Bloem E, Haneklaus S, Kesselmeier J, Schnug E (2012) Sulfur fertilization and fungal infections affect the exchange of H_2S and COS from agricultural crops. J Agric Food Chem 60: 7588–7596.
35. Senthilkumar M, Govindasamy V, Annapurna K (2007) Role of antibiosis in suppression of charcoal rot disease by soybean endophyte *Paenibacillus* sp. HKA-15. Curr Microbiol 55: 25–29.
36. Eghbal MA, Pennefather PS, O'Brien PJ (2004) H_2S cytotoxicity mechanism involves reactive oxygen species formation and mitochondrial depolarisation. Toxicology 203: 69–76.
37. Joyner-Matos J, Predmore BL, Stein JR, Leeuwenburgh C, Julian D (2010) Hydrogen sulfide induces oxidative damage to RNA and DNA in a sulfide-tolerant marine invertebrate. Physiol Biochem Zool 83: 356–365.
38. Attene-Ramos MS, Wagner ED, Gaskins HR, Plewa MJ (2007) Hydrogen sulfide induces direct radical-associated DNA damage. Mol Cancer Res 5: 455–459.
39. Kreiner M, Harvey LM, McNeil B (2002) Oxidative stress response of a recombinant *Aspergillus niger* to exogenous menadione and H_2O_2 addition. Enzyme Microb Technol 30: 346–353.
40. Shatalin K, Shatalina E, Mironov A, Nudler E (2011) H_2S: a universal defense against antibiotics in bacteria. Science 334: 986–990.

12

Identification of Novel Factors Involved in Modulating Motility of *Salmonella enterica* Serotype Typhimurium

Lydia M. Bogomolnaya[1,2], Lindsay Aldrich[1], Yuri Ragoza[1], Marissa Talamantes[1], Katharine D. Andrews[1], Michael McClelland[3], Helene L. Andrews-Polymenis[1]*

1 Department of Microbial Pathogenesis and Immunology, College of Medicine, Texas A&M University, Bryan, Texas, United States of America, **2** Institute of Fundamental Medicine and Biology, Kazan Federal University, Kazan, Russia, **3** Department of Microbiology and Molecular Genetics, University of California Irvine, Irvine, California, United States of America

Abstract

Salmonella enterica serotype Typhimurium can move through liquid using swimming motility, and across a surface by swarming motility. We generated a library of targeted deletion mutants in *Salmonella* Typhimurium strain ATCC14028, primarily in genes specific to *Salmonella*, that we have previously described. In the work presented here, we screened each individual mutant from this library for the ability to move away from the site of inoculation on swimming and swarming motility agar. Mutants in genes previously described as important for motility, such as *flgF, motA, cheY* are do not move away from the site of inoculation on plates in our screens, validating our approach. Mutants in 130 genes, not previously known to be involved in motility, had altered movement of at least one type, 9 mutants were severely impaired for both types of motility, while 33 mutants appeared defective on swimming motility plates but not swarming motility plates, and 49 mutants had reduced ability to move on swarming agar but not swimming agar. Finally, 39 mutants were determined to be hypermotile in at least one of the types of motility tested. Both mutants that appeared non-motile and hypermotile on plates were assayed for expression levels of FliC and FljB on the bacterial surface and many of them had altered levels of these proteins. The phenotypes we report are the first phenotypes ever assigned to 74 of these open reading frames, as they are annotated as 'hypothetical genes' in the Typhimurium genome.

Editor: Michael Hensel, University of Osnabrueck, Germany

Funding: This work was supported in part by NIH/NIAID grants R01AI083646, R56AI077645 awarded to HAP, R01 AI075093 awarded to MM. LMB was supported in part by a subsidy to support the Program of competitive growth of Kazan Federal University. The funders had no role in study design, data collection and analysis, decision to publish, or preparation of the manuscript.

Competing Interests: The authors have declared that no competing interests exist.

* Email: handrews@medicine.tamhsc.edu

Introduction

Infection with *Salmonella enterica* serotypes remains a serious human and animal health problem worldwide. *Salmonellae* cause an array of diseases ranging from gastroenteric disease to systemic disease including Typhoid fever and bacteremia [1]. While gastroenteritis as a result of *Salmonella* infection is common worldwide, systemic diseases caused by this organism are relatively rare in the developed world. Serotype Typhimurium is one of the two most common serotypes identified from cases of clinical disease in the United States [2]. After gaining access to a susceptible host by the oro-fecal route, *Salmonella* utilizes multiple strategies to colonize and persist. *Salmonellae* have many well studied virulence factors including Type 3 Secretion Systems (T3SS-1 and -2), lipopolysaccharides (LPS), fimbria and others, and are capable of multiple types of motility including swimming and swarming. Both types of motility require the presence of functional flagella that are composed of many proteins and consist of the following structures: the basal body, the hook and the filament. Flagellar biosynthesis is a complex, tightly regulated process where gene products are produced in the order of flagellar apparatus assembly. The current understanding of flagella structure and regulation in *Salmonella* is nicely summarized in recent review by Chevance and Hughes [3]. Motility is linked to virulence in many pathogenic bacteria [4,5].

Swimming motility is directed movement through liquid, which is assayed using semi-solid "swimming" media, containing a low concentration of agar. Under these conditions individual bacteria swim through medium-filled spaces between agar [6]. Swimming motility is closely linked to chemotaxis [7], the ability to orient bacteria along certain chemical gradients, and is thought to allow bacteria to detect and pursue nutrients or avoid unwanted repellents [7]. Ultimately, this kind of motility allows these organisms to avoid unfavorable environments for colonization and to reach and maintain preferred niches for colonization.

Swarming motility is a multi-cellular phenomenon involving the coordinated and rapid movement of a bacterial population across a semisolid surface [8]. Factors on which swarming is known to depend include bacterial cell density, media composition and surface moistness [9–12]. Swarming is not just another form of motility but rather a part of alternative growth state and is characterized by change in gene expression of nearly one third of genome in *Salmonella* Typhimurium [4].

Recent large-scale studies have been performed to identify genetic determinants required for each type of motility in *E. coli*, an organism closely related to *Salmonella*. Both transposon mutants in *E.coli* K-12 [13], and the Keio collection, a collection of targeted deletion mutants in each non-essential open reading frame in *E. coli* K-12 [14], were screened to identify mutants with reduced motility. These comprehensive studies demonstrated that numerous genes are involved in regulation of motility in bacteria.

We have generated a library of 1023 targeted single gene deletion mutants (SGD) in virulent *Salmonella enterica* serotype Typhimurium ATCC14028s [15]. This library contains mutants in nearly all *Salmonella*-specific open reading frames as well as an additional 100 genes shared between *Salmonella* and other *Enterobacteriaceae*. We screened this collection to systematically identify those mutants in our collection that had either reduced or enhanced ability to move away from the site of inoculation on swimming and/or swarming motility agar. We identified 160 mutants with altered motility in at least one condition. Mutants with previously known motility defects (flagellar, LPS biosynthesis, chemotaxis genes) were correctly identified in our screen. We identified mutants in nine additional genes that are unable to move away from the site of inoculation on both types of agar. Furthermore, we found that the ability to swim and swarm could be uncoupled in some mutants. We also identified a significant number of mutants that had enhanced motility of one or both types ('hypermotility'). Finally, many of the motility phenotypes we identified belong to mutants with mutations in genes annotated as of unknown function (FUN, or 'orphan' genes), and are thus the first phenotypes of any kind described for these genes.

Materials and Methods

Bacterial Strains and Media

All *Salmonella enterica* serotype Typhimurium strains used in this study were derived from ATCC14028, including HA420, a spontaneous Nalidixic acid resistant isolate [16]. Targeted deletion mutants screened in this study have been described previously [15].

Salmonella strains used in this study were routinely grown on LB agar or broth, or on M9 minimal media. The growth of all mutant strains was tested on minimal M9 media [17], prior to use of these mutants in motility assays. Antibiotics were added in the following concentrations as appropriate: 50 mg/L Kanamycin sulfate, 50 mg/L of Nalidixic acid [15,16]. Media for assaying swimming and swarming motility have been described previously [11]. Swimming was assayed on plates containing 0.3% Difco Bacto Agar (LB Miller base 25 g/L), while swarming motility was assayed on 0.6% Difco Bacto Agar (LB Miller base 25 g/L, and 0.5% glucose).

Screening Individual Mutants for Swimming and Swarming Motility

Our collection of targeted deletion mutants was assayed in 96-well format for both swimming and swarming on large agar plates (15 cm diameter). Strains were inoculated into the appropriate agar with 96-pin replicator, incubated at 37°C, and closely monitored for the duration of the assay. Wild type ATCC14028s and ATCC14028r (smooth and rough LPS) were included as positive and negative controls on each motility plate. Several hours post-inoculation (3.5 hours for swimming, 5 hours for swarming), the swimming and swarming ability of each strain was evaluated by estimating the diameter of the spread of the bacteria and assigning a motility score. Individuals scoring motility were blinded to the identity of the mutants being scored, but were aware of the location of positive and negative controls inoculated on every plate. Motility was scored on a scale of 0 (completely non-motile) to 10 (extremely hypermotile) relative to wild type ATCC14028s, which was always assigned a motility score of 5. Our scoring system allowed us to identify mutants with a range of hypo- and hyper- motility. Large-scale screening assays were performed in triplicate and were repeated on at least three separate occasions with the entire collection of mutants. The mean of the motility scores for a given assay were determined for each mutant and these are shown in Table S1. Mutants used in our large scale screening were not transduced into a new genetic background, thus motility phenotypes associated with gene loss will be confirmed in future studies.

Mutants with the most severe phenotypes (less than 25% motility of wild type or hypermotile mutants) were further evaluated by measuring the diameter of the swimming or swarming colony as compared to positive and negative controls. This step was required for identification and removal from the future study a few false positive motility candidates that during primary screening were scored as non-motile or hypermotile due to uneven transfer of cells with 96-pin replicator. Overnight cultures of each mutant were grown and normalized by OD_{600}. 3 µl was spotted on motility plates, incubated and scored as described above. The diameter of each swimming or swarming colony was measured, and compared to positive (WT ATCC14028s, smooth) and negative (ATCC14028r, rough mutant or *motA*) control isolates. Experiments were conducted in triplicate, and were repeated on three separate occasions. The identity of each mutant that displayed a statistically significant phenotype in our motility assays was verified by PCR using primers flanking the site of deletion if not previously verified. Statistical significance was determined using a Student's *t*-test and a p-value of <0.05.

Evaluation of Flagellin Expression

Each mutant with reduced or enhanced motility was evaluated to determine the level of FljB and FliC produced in the bacterial cell, and to determine the amount of each of these proteins that reach the bacterial surface as compared to the isogenic wild type organism. Bacterial cultures grown to stationary phase were normalized by OD_{600} and bacteria were collected by centrifugation (Eppendorf 5415D). In order to shear flagella from the bacterial surface, pelleted bacteria were resuspended in 1 ml PBS and subjected to 5 minutes of vortexing (Vortex-genie, Scientific Industries) [18]. Sheared protein was precipitated using Trichloroacetic acid (TCA) (6% final concentration) overnight at 4°C, washed twice with 300 µl of acetone and resuspended in SDS sample buffer. Precipitated protein from sheared fractions was evaluated by SDS-PAGE and Western Analysis using antibodies against FliC and FljB (Difco). The remaining bacterial pellet (without sheared flagella) was solubilized in SDS sample buffer and examined by 12% SDS-PAGE and Coomassie staining to ensure equivalent loading. Wild type organisms and *ΔfljB* and *ΔfliC* mutants were used as positive and negative controls in these assays.

Results and Discussion

Screening of a collection of targeted deletion mutants in *Salmonella* to identify determinants of motility

We have screened a library of targeted single-gene deletion mutants in many genes that are specific to *Salmonellae* and not shared with close relatives, in addition to approximately 100 mutants in shared genes that served as controls, for motility phenotypes on plates [15] (Figure 1). We identified 160 mutants with altered motility. Of these mutants, only 29 were previously connected to motility either in *Salmonella* or *E. coli*. We divided the resulting set of mutants with motility phenotypes into four

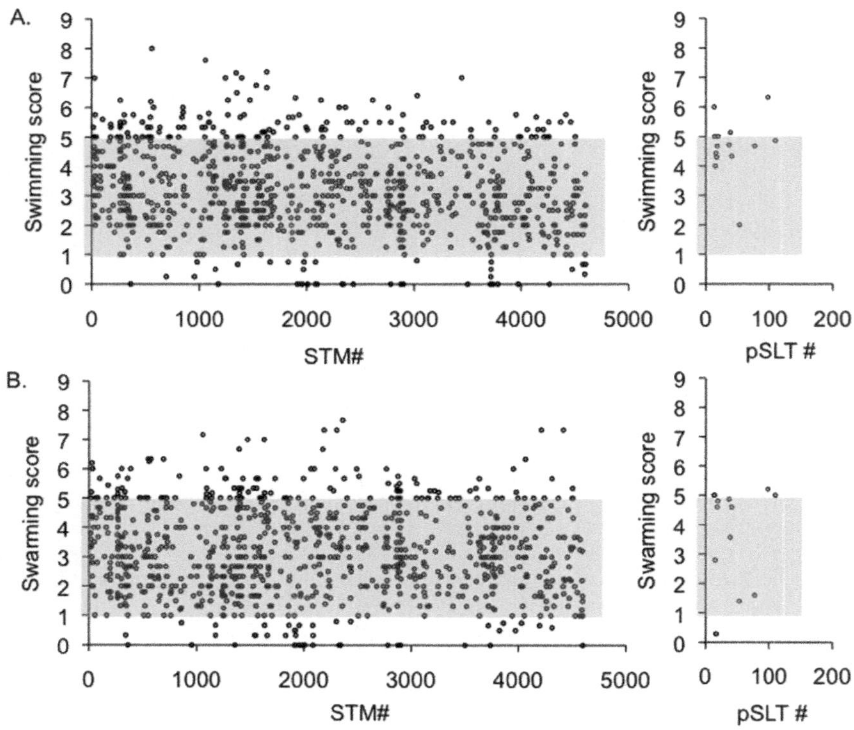

Figure 1. Motility scores of single deletion mutants of *Salmonella* Typhimurium ATCC14028 on swimming (A) and swarming (B) agar. Stationary phase cultures were transferred to motility agar, and motility was scored after 3.5 and 5 hours of incubation at 37°C on swimming or swarming plates, respectively. Motility was scored on the scale from 0 to 10 with wild type motility equal to 5. Data are presented as average swimming or swarming score from experiments with triplicate samples, performed on three independent occasions. Dots located outside of shaded area indicate scores greater than 5.25 and lower than 1.25 to define mutants with increased or reduced (<25 % of wild type) motility.

categories: (a) Mutants with defects in both types of motility; (b) Mutants with reduced swimming motility, swarming motility is unaffected; (c) Mutants reduced swarming motility only, swimming is unaffected; (d) Mutants that were hypermotile (Figure 2).

Mutants with reduced swimming and swarming motility

We identified 21 mutants that moved less than or equal to 25% of wild type from the initial site of inoculation (i.e. impaired swimming and swarming). This group contains mutations in 12 genes previously reported to be needed for swimming and swarming motility, including flagellar genes *flhA, flhB, flgF, flgG, motA, motB*, LPS biosynthesis genes *rfbN, rfbK, rfaI*, chemotaxis gene *cheY* and others (Table 1), validating our approach. Our targeted deletion collection assayed in this study contained only eight mutants in known flagellar genes, and each of the mutants in these genes had the expected non-motile phenotype.

In addition, we identified general motility phenotypes for mutants in 9 genes not previously described to be involved in motility (Table 1). Four of these mutants had deletions in genes with putative functions in transport (*aroD*), in transcription (*tctD*) and in signal transduction (*STM0343, STM0551*). The remaining five mutants with reduced movement from the site of inoculation on swimming and swarming agar had deletion mutations in fimbrial genes *sthE, STM4595* [19], *STM0699, STM2010* and *STM2880*, a gene with unknown function encoded within SPI-1 [4,20]. None of the mutants in this group had a growth defect in LB-broth (Figure S1). We note that our screening assay does not allow us to determine why mutants do not appear to spread from the site of inoculation. As such, classification of these mutants into categories such as defects in flagellar motility, motor activity, or chemotaxis is a fascinating area of future investigation.

We hypothesized that some of the phenotypes we observed could be due to an inability to export flagellin to the bacterial surface. The external portion of the flagellum is a helical filament composed of flagellin proteins, FliC or FljB [21]. In our wild type (ATCC14028) population FliC appeared to be the major flagellin expressed (accounts for>70 % of total flagellin) and FljB expression was weak (data not shown). We tested all 22 mutants with reduced swimming and swarming motility for expression of flagellins on the bacterial surface by evaluating the amount of FliC and FljB in the sheared fraction of surface proteins by Western analysis.

Mutants in *flgF, flgG*, and *rfbK*, genes known to be important for motility [14,22], were both non-motile and displayed less flagellins on the bacterial surface in our screen. For other mutants however, (*motB, cheY*) the presence of filaments is known not to be sufficient for the ability of these mutants to move from the site of inoculation on motility plates. Our mutants in *motB* and *cheY* displayed wild type levels of flagellins on the cell surface yet were unable to spread from the site of inoculation as previously reported in the literature [23,24] (Figure 3).

Of the nine mutants with reduced ability to move from the site of inoculation on swimming and swarming agar plates in our screen that were not previously known to be needed for this ability, all had levels of FliC on the bacterial surface comparable to wild type (Figure 3). Several of the mutants we identified in this category (*STM0343, STM0551, STM0699)* expressed less of the minor flagellin FljB on the bacterial surface compared to wild type

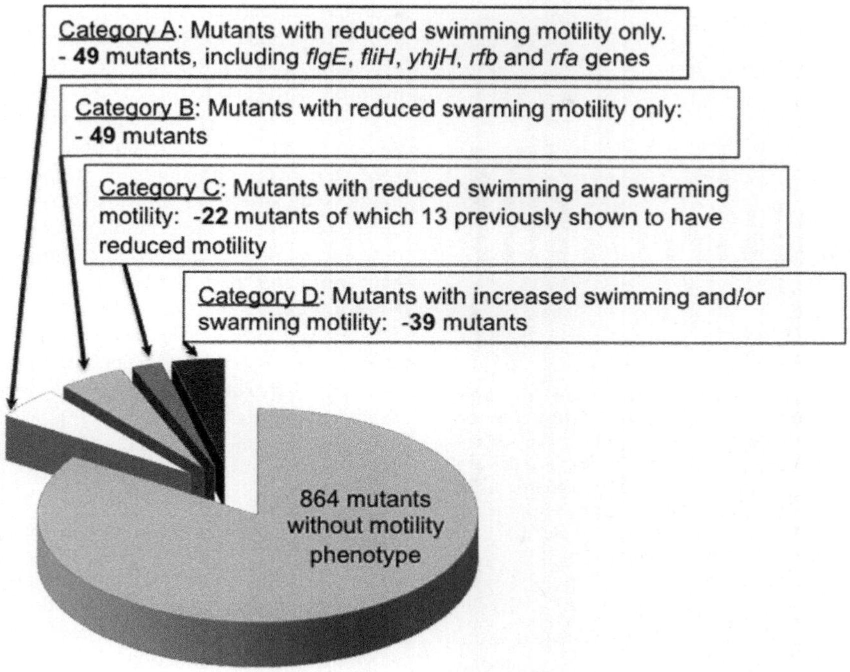

Figure 2. Distribution of motility phenotypes in single gene knock out collection.

Table 1. Pathway clustering of mutants with altered motility.

Functional categories*	Swimming and swarming defect	Reduced swimming only	Reduced swarming only
Motility	***flgF**, flgG, flhA, flhB, motA, motB, fliD, fliM***	***flgE***, *ssaV*, ***fliH***, *stjC****	*ssaU, pefD, pefC*
Cell envelope biogenesis	***rfbN, rfaI***	***rfbP, rfbM, rfbJ, rfbC, rfbD, rfaJ, rfaG, rfaQ***	
Signal transduction	*STM0343, STM0551,* ***cheY,***	*phoQ,* ***yhjH****, yjcC*	*yciR*
Carbohydrate transport & metabolism	***rfbK***	*STM0722, STM4424*	*STM0860, STM3780*
Transport and metabolism	*aroD*	*aroA, fur, yfeJ, pdxK,* ***tatC***	*yliB, STM1635, sfbA, mgtB, sodA, STM0163, STM1546, STM0857*
Transcription	*tctD*	*invF, STM2912, STM3696, STM4417,* ***arcA***	*STM0859, ydiP, torR, STM4315*
Replication, recombination and repair		*STM1005*	*STM1861*
Translation, ribosomal structure and biogenesis		*valS*	*STM1552*
Posttranslational modification, protein turnover, chaperones		*STM2743, sspA*	
Energy production and conversion			*STM0762, STM0858*
Cell cycle control, cell division, chromosome partitioning			*STM2594*
Defense mechanism			*STM4262*
Not in COGs	*STM0699, STM2010, STM2880, sthE, STM4595*	*STM0289, STM0295, STM0660, STM0971, STM1040, STM1331, ssaG, STM2374, sptP, sipA, STM3026,* ***rfaL****, yibR,* ***rfaP****, STM3783, STM4216,STM4219, sRNA candidate C1023*	*STM0056, STM0362, STM1131, pagC, STM1254, STM1258, STM1543, srfC, STM1632, STM1856, STM1926, STM1958, STM2303, STM2508, ygaU, STM3125, lpfE, STM3944, STM4030.S, STM4197, STM4204, STM4529, STM4574, STM4599, invR*

*Based on COGs (Clusters of Orthologous Groups of protein).
**Mutants with previously known motility phenotype are shown in bold.
***Mutants in fimbrial genes are underlined.

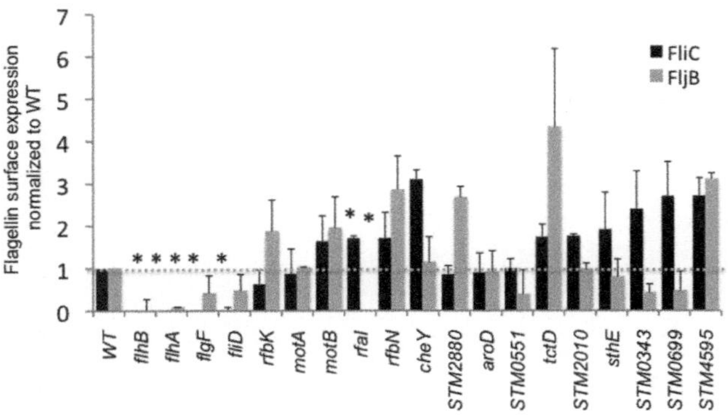

Figure 3. Reduced motility on semi-solid media in some mutants correlated with reduced expression of flagellin on the bacterial surface. Flagellins were sheared from bacterial surface, analyzed by Western blotting with antibodies to FliC (black bars) and FljB (grey bars) and blots were quantified by densitometry using Quantity One software. Surface expression of flagellins was normalized to the level on the surface of the wild type strain. Samples for each mutant were prepared in three independent experiments.

ATCC14028 (Figure 3, Figure S2). Two of these mutants (*STM0343* and *STM0551*) have deletions in genes encoding EAL-domains containing proteins, predicted to be involved in the metabolism of cyclic diguanosine monophosphate (c-di-GMP) [25,26]. Purified STM0551 possesses phosphodiesterase activity *in vitro* that is abolished by a point mutation in the EAL domain [25]. Based on the current model of c-di-GMP metabolism in bacteria, mutation in c-di-GMP phosphodiesterases results in accumulation of c-di-GMP leading to decreased expression of flagellins, and loss of motility [27]. Furthermore *STM0551* appears to be a negative regulator of type 1 fimbria. Activation of *fim* genes includes the activation of a negative regulator of motility *fimZ* [25]. Over-expression of FimZ is known to repress swimming motility in *Salmonella* [28]. In agreement with these findings, deletion of *STM0551* abrogates outward movement on both swimming and swarming agar in *Salmonella*. Thus, data from our screen is consistent with previously published work, and shows that the amount of flagellin sheared from the bacterial surface can be correlated with motility for some but not all mutants.

Mutants with reduced swimming motility

We identified 49 mutants that had reduced movement from the site of inoculation on swimming motility plates (less than 25% of wild type swimming) but appeared to behave similarly to the wild type organism on swarming motility plates (Table 1, Table S1). We were surprised that the ability to perform these two different types of motility could be uncoupled, and we note that further microscopic study of these mutants during swarming will provide clues regarding these seemingly paradoxical phenotypes. Sixteen mutants identified in this group were known to have reduced swimming motility including *flgE*, *fliH*, *arcA*, *yhjH* and *tatC* [4,13,14,29–33]. In addition to mutants mentioned above, we identified 33 mutants that had reduced motility on swimming agar but normal movement on swarming agar that were not previously known to be involved in swimming motility (Table S3). Some mutants in this category had deletions in genes encoding proteins involved in intracellular transport and metabolism, signal transduction and regulation of transcription based on COG predictions (Clusters of Orthologous Groups of proteins) [34] (Table 1). For example, our *Δfur* mutant was severely impaired on swimming agar, but not on swarming agar. We found that deletion of *phoQ*, also resulted in reduced swimming but not swarming motility.

Mutants in a number of genes necessary for LPS biosynthesis/assembly (*rfaG, rfaI, rfaL, rfaQ, rfaP, rfbC, rfbD, rfbJ, rfbM, rfbP, yibR*) had severely impaired movement on swimming motility agar. In *E. coli*, lipopolysaccharide (LPS) biosynthesis genes are not required for swimming motility [14]. Our work shows that *Salmonella* appears to require LPS biosynthesis for movement on swimming motility plates (Table 1). This finding is not unprecedented however, as others have shown that in *Salmonella rfaP and rfaJ* and in *Pseudomonas rfaL*, are required for swimming motility [22,35,36].

Mutants that had reduced outward movement from the site of inoculation on swimming agar but normal swarming movement, also included deletion mutants in genes encoding important virulence factors such as fimbria (*stjC, stdD*), Type Three Secretion System components *invF, sipA, sptP* (T3SS 1), and *ssaV* (T3SS 2) as well as Type Six Secretion System (T6SS, SPI-6 encoded) components *STM0289* and *STM0295* [37,38]. STM0289 is a member of Vrg family of proteins required for effectors delivery via T6SS. Ours is the first description of T6SS encoded on SPI-6 potential involvement in motility. Finally, fourteen mutants, the largest group of mutants that swam poorly but swarmed normally, are not annotated or do not have a previously described function or phenotype. Thus, we describe the first functional data and potential phenotypes for these genes.

Mutants with reduced swarming motility

We identified 49 mutants with reduced ability to move outward from the site of inoculation on swarming agar (less than 25% of wild type) but with normal outward migration on swimming agar (Table S4). None of the genes that we identified in this category were directly implicated previously in the ability to swarm. A limited number of reports describe genes required for swarming but not swimming motility. Those include *E.coli recA*, *S.*Typhimurium *flhE* and *Proteus mirabilis waaL* [39–42] and were not present in the library used for the current study. Deletion mutants in a number of transcriptional regulators, including STM1355 and *torR* (STM3824) appear to be involved in the regulation of swarming motility in *Salmonella*.

Several of these non-swarming mutants had deletions in genes involved in energy production/conversion and transport/metabolism as predicted by COG assignments (12 out of 49, Table 1). For example, mutants in *STM0762*, a gene that encodes fumarase, a TCA cycle enzyme involved in energy production, cannot swarm. Several additional mutants, *ΔSTM0849, ΔSTM0056 and ΔSTM1258,* have putative functions in transport or energy production. *STM0849* is a homolog of *yliB*, a ppGpp-dependent glutathione importer encoded on *yliABCD* operon [43–45] and *STM0056* is annotated as a putative oxalacetate decarboxylase gamma subunit [46]. Our results are consistent with previous proteomic approaches demonstrating that genes associated with energy production and *de novo* synthesis are required for swarming [47] and with the current thinking that swarming is an energetically costly process.

We identified a mutant in *srfC as* poor swarmer but normal swimmer. This gene is clustered with the flagellar class 2 genes and was determined to be under FlhDC control [48]. A polar mutation in an upstream gene in the *srfABC* operon, *srfB*, or deletion of the whole operon affected only swarming motility [49]. Similar to recent reports on swarming motility of *E. coli*, *Pseudomonas aeruginosa* and *Xenorhabdus nematophila* [14,50,51], mutations in fimbrial genes, *pefC* and *pefD* (encoding for usher and chaperone proteins, respectively) and in *lpfE* (long polar fimbrial minor

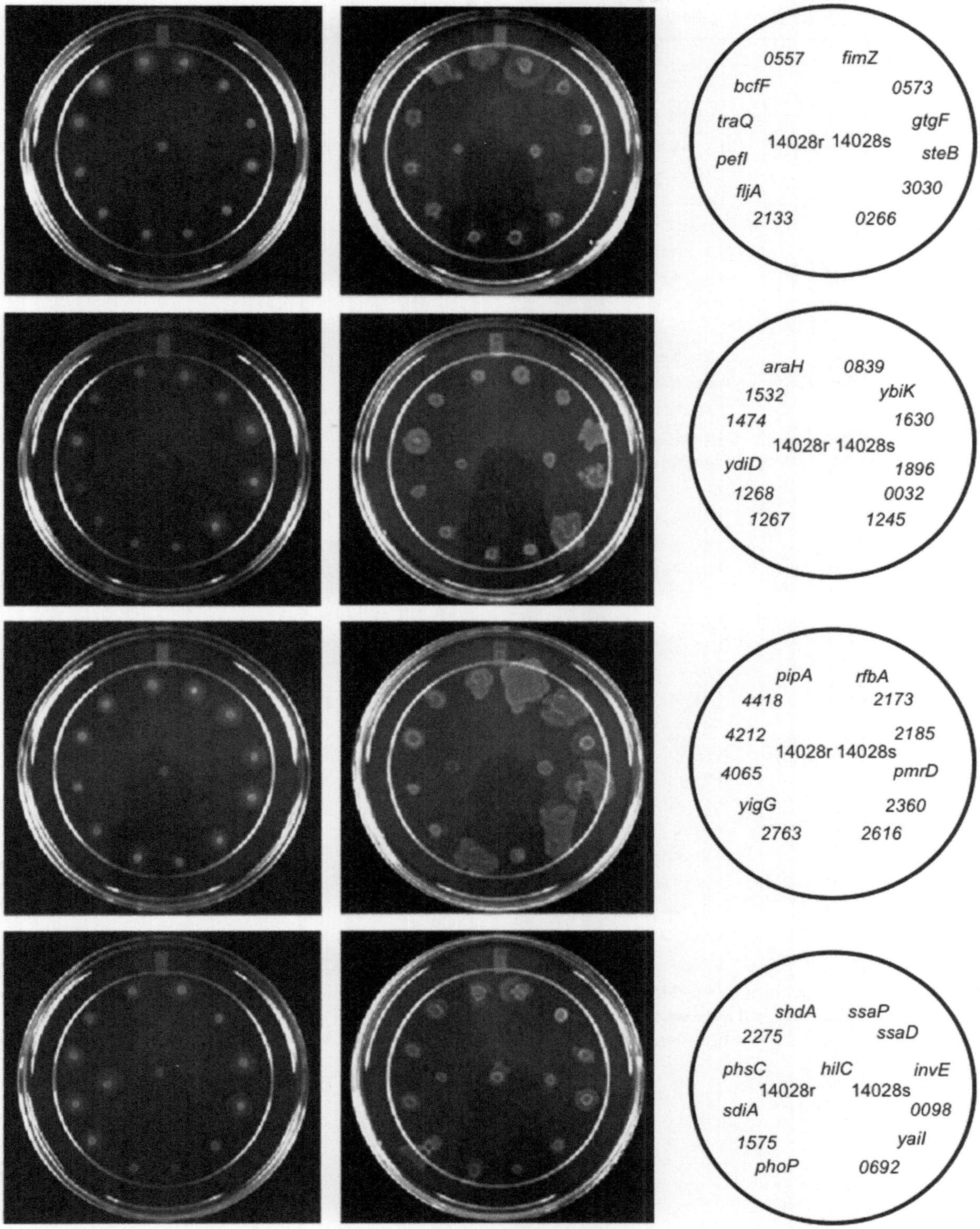

Figure 4. Confirmation of hypermotility phenotypes. Stationary cultures normalized by OD_{600} were spotted on swimming or swarming agar along with wild type strain, ATCC14028s. Rough strain, ATCC 14028r, was used as a negative control on swarming agar. After 3.5 and 5 hours of incubation at 37°C on swimming or swarming plates, respectively, the diameter of each bacterial growth area was measured in three independent experiments, with each experiment performed in triplicate.

Table 2. Mutants with enhanced motility.

STM	Gene	Motility compared to wild type*	
		Swimming Mean ± SD	**Swarming Mean ± SD**
STM0026	*bcfF*	**1.32±0.26****	**2.55±0.64**
STM0032		**1.49±0.24**	1.42±0.75
STM0098		**1.65±0.37**	**1.46±0.29**
STM0266		1.13±0.35	**1.59±0.45**
STM0387	*yaiI*	**1.42±0.23**	1.15±0.25
STM0549	*fimZ*	**1.32±0.20**	1.89±1.13
STM0557	*gtrC*	**1.38±0.24**	**2.13±0.52**
STM0573		1.14±0.17	**1.64±0.32**
STM0839		**1.20±0.20**	**1.57±0.67**
STM0847	*ybiK*	**1.53±0.49**	**1.36±0.29**
STM1087	*pipA*	**1.30±0.22**	**1.50±0.35**
STM1231	*phoP*	**1.39±0.34**	**1.69±0.56**
STM1344	*ydiV*	**1.29±0.16**	0.93±0.18
STM1350	*ydiD*	**1.61±0.79**	1.20±0.63
STM1395	*ssaD*	**1.41±0.30**	0.93±0.24
STM1417	*ssaP*	**1.47±0.27**	1.25±0.64
STM1575		**1.53±0.19**	1.35±0.40
STM1629	*steB*	0.98±0.26	**1.71±0.33**
STM1630		**1.72±0.37**	**1.62±0.50**
STM1896		**1.50±0.21**	**1.94±1.01**
STM1950	*sdiA*	**1.60±0.26**	**1.49±0.45**
STM2063	*phsC*	**1.46±0.27**	1.23±0.30
STM2095	*rfbA*	**1.41±0.26**	2.08±1.11
STM2133		1.08±0.10	**1.55±0.46**
STM2173		**1.41±0.22**	**1.62±0.70**
STM2185		**1.62±0.22**	**1.49±0.56**
STM2304	*pmrD*	**1.49±0.23**	**1.71±0.44**
STM2360		**1.35±0.16**	1.43±0.74
STM2513	*shdA*	**1.37±0.23**	1.14±0.26
STM2616		**1.31±0.28**	0.90±0.09
STM2763		**1.68±0.35**	1.52±0.79
STM2770	*fljA*	1.05±0.08	**1.39±0.26**
STM2867	*hilC*	**1.67±0.26**	**1.58±0.34**
STM2897	*invE*	**1.56±0.30**	1.41±0.51
STM3954	*yigG*	**1.36±0.22**	1.19±0.33
STM4212		1.14±0.24	**1.47±0.45**
STM4418		**1.26±0.21**	1.40±0.68
PSLT013	*pefI*	0.98±0.19	**1.18±0.11**
PSLT098	*traQ*	1.04±0.22	**1.33±0.23**

* - Diameter of swimming and swarming rings were measured and compared to wild type. Results are shown as the mean of six independent experiments.
** - Bold indicates statistical significance, $p<0.05$.

protein) strongly affected swarming motility with no significant effect on swimming motility.

Interestingly, we found that a *fliB* mutant was defective in swarming, but not in swimming. FliB methylates the lysine residues of flagellin [52]. Previous studies showed that the loss of *fliB* did not affect swimming motility and it is thought that flagellin methylation by FliB is required for *Salmonella* virulence but not for flagellin function [48]. Our data suggest that flagellin methylation is required for swarming motility.

Swarming and virulence are linked in several bacteria [53]. 21 out of 49 of our mutants with defects in swarming are deleted for genes associated with virulence. *STM1131, STM2303,*

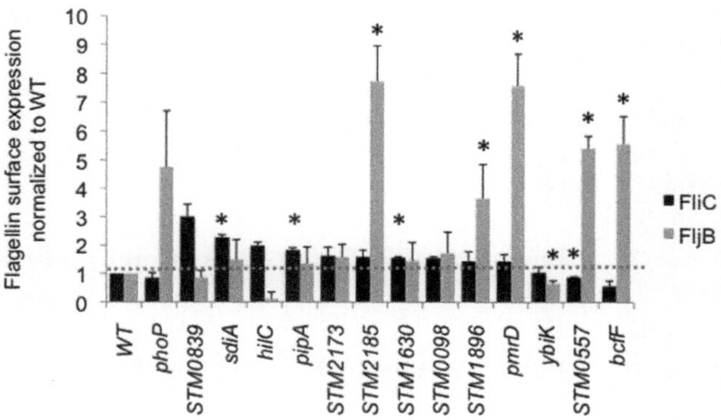

Figure 5. Elevated levels of flagellin were present on the bacterial surface of some mutants with improved swimming and swarming motility. Flagellins were sheared from bacterial surface, analyzed by Western blotting with antibodies to FliC and FljB and blots were quantified by densitometry using Quantity One software. Surface expression of flagellins was normalized to the level on the surface of the wild type strain. Samples for each mutant were prepared in three independent experiments.

STM4030.S, *STM4262* (*siiF*) are reported to be needed for full virulence in mice and calves [15,54,55] and mutants in these genes have reduced swarming in our assays. *pagC*, *mgtB* and *STM0859* also had reduced motility in our assays and are part of the *phoPQ* regulon in *Salmonella* [56–58]. Finally, swarming motility was also compromised in mutants that had deletions of genes encoded on SPI-14 (*STM0859*), a region important for virulence in chickens [59]. STM0859 was recently reported to be co-regulated with the type three secretion system encoded on SPI-2 [60].

We identified a group of genes that when deleted reduce the ability to swarm to less than 25% of the ability of the wild type organism without significant reductions in the ability to move away from the site of inoculation on swimming motility plates. Some of the products of these genes may be involved in generating movement, signaling when conditions exist for swarming, surface properties or other qualities that are directly involved in the ability to swarm, while others may affect swarming indirectly. Careful quantification and microscopic examination of each of these mutants will be useful to fit the corresponding genes into an overall framework with respect to their roles in swarming motility.

The identification of hypermotile mutants

As we were using 96-well format to screen our deletion mutant library for motility phenotypes, multiple observations of each screening plate were required in order to identify phenotypes before colonies intersected obscuring individual phenotypes. Early and frequent observation allowed us to identify mutants that had increased ability to move away from the site of inoculation relative to the wild type organism. We observed thirty-nine mutants to have a larger swimming and/or swarming ring diameter as compared to the wild type observed on the same plate at the same time point post inoculation (Figure 4). Of this group, we found 14 mutants that had generalized improvement in motility, 17 mutants with increased swimming motility, and 8 mutants with increased swarming motility compared to wild type (Table 2). Movies S1, S2, and S3 show examples of mutants with increased swarming motility, able to move across swarming agar faster than wild type cells (See Supporting Information Legend).

Hypermotility phenotypes have not been well described in *Salmonella*, and there are only a limited number of reports describing hypermotility as a phenotype in bacteria, primarily in *Pseudomonas aeruginosa* and *Proteus mirabilis* [61–69] (Table S5). Thus, there is little existing data to easily validate and benchmark the hypermotility phenotypes we observed, yet a few mutations are known to promote hypermotility. First, over-expression of FimZ, a positive regulator of Type I fimbriae and negative regulator of flagellar motility, represses swimming motility in *Salmonella* [28]. A *fimZ* deletion mutant is expected to be hypermotile, and consistent with this prediction our deletion mutant in *fimZ* was indeed hypermotile (Figure 4 and Table 2).

Second, deletion of *STM1344* (*ydiV*), annotated by COG database as a gene involved in signal transduction, resulted in improved swimming motility in our assays (Table 2). YdiV negatively regulates motility via binding to the $FlhD_4C_2$ complex to prevent interaction of this complex with DNA [70]. YdiV also functions as an adaptor protein that binds FlhD and delivers $FlhD_4C_2$ to ClpXP protease for proteolytic degradation [71] and connects flagellar gene expression to nutrient starvation [72]. In agreement with previously reported role of EAL-containing protein YdiV in negative regulation of motility in uropathogenic *E. coli* [73] and *S.* Typhimurium [74], we found that *ydiV* deletion mutant in *S.* Typhimurium was more motile on swimming agar.

Third, in our screen a *ΔphoP* mutant displayed increased motility both on swimming and swarming plates (Table 2, Figure 4). PhoP, the response regulator DNA-binding protein of PhoP/PhoQ two component system [75,76], negatively regulates motility in various microorganisms including *Photorhabdus luminescens, Pseudomonas aeruginosa*, *Proteus mirabilis* and in uropathogenic *E. coli*, where a *phoP* nulls are hypermotile [66,77–79]. Moreover, null allele in *Proteus mirabilis phoP* homolog exhibits an elevated level of *flhDC* [66]. Finally, in uropathogenic *E. coli* inactivation of *phoP* results in the increased expression of flagellin on the bacterial surface [78]. Similar to uropathogenic *E. coli phoP* mutant in *S.* Typhimurium increased motility correlated with increased flagellin expression on the bacterial surface (Figure 5).

We hypothesized that enhanced motility of at least some mutants could be due to increased expression of flagellins on the bacterial cell surface as was previously shown for *phoP* mutant in uropathogenic *E. coli* [78] and for *ydiV* mutant in *Salmonella* and uropathogenic *E. coli* [73,80]. Therefore, we examined the amount of FliC and FljB on the bacterial surface for each mutant that displayed hypermotility both on swimming and swarming agar (Figure 5, Figure S2). We found that while our *ΔphoP* mutant had wild type level of expression for FliC, it also had more FljB on the bacterial surface than the isogenic wild type organism. Furthermore, our examination of flagellar proteins sheared from the bacterial surface showed that thirteen out of fourteen of our hypermotile mutants had more FliC or FljB on the bacterial surface (Figure 5, Figure S2) than the wild type. It seems plausible that the hypermotility phenotype we observed for this highly diverse group of mutants could be the result of up-regulation of flagella.

Interestingly, mutations in several fimbrial operons also result in hypermotility phenotypes, *ΔSTM0026* (*bcfF*) mutants were hypermotile and the *bcfABCDEFG* (bovine colonization factor) operon has previously been implicated in virulence in mice [81]. Inactivation of *bcfF* improves biofilm formation on HEp-2 tissue culture cells and chicken intestinal epithelium in comparison to wild type [82]. Mutants in FimZ (STM0549) were also hypermotile. Inactivation of FimZ reduces expression from the P_{fimA} promoter and prevents serovar Typhimurium from making type I fimbriae [83,84]. FimZ also binds to the P_{flhDC} promoter and represses the expression of *flhDC* operon.

Concluding remarks

In this study we evaluated the ability of 1023 defined non-polar single gene deletions of *Salmonella* Typhimurium to move away from the site of inoculation on swimming and swarming agar plates. Over 90% of deletions in our collection were introduced in genes present exclusively in *Salmonella* or closely related pathogenic bacteria from the *Enterobacteriaceae*. Many of the mutants used in this study did not have any functions previously assigned. We confirmed motility phenotypes associated with loss of genes involved in flagellar regulon, LPS biosynthesis and chemotaxis. We identified a number of novel contributors to bacterial motility including both known and uncharacterized genes in our pathogenicity-biased collection. Furthermore, we found that there are genes needed for both types motility, and there are genes that make a unique contribution to different kinds of motility. Finally, we identified mutations in a number of genes result in increased motility. Determination of the molecular mechanisms of the improved motility is a fascinating area of the future work.

Supporting Information

Figure S1 Mutants with severe defects in swimming and swarming motility grow indistinguishably from wild type. Overnight cultures were subcultured at 1/100 ration in LB-broth and incubated at 37°C with shaking. Bacterial growth was monitored by OD600 in three independent experiments.

Figure S2 Flagellin expression on the cell surface correlates with the ability to move on swimming and swarming agar for some, but not all mutants. Flagellins sheared from the bacterial surface from strains with decreased (A) or increased (B, C) motility grown in LB-broth were analyzed by Western blotting with μ-FliC and μ-FljB sera. The whole cell lysates for each sample were also analyzed by SDS-PAGE and stained with Coomassie as a loading control.

Table S1 Swimming and swarming motility scores from high-throughput screening.

Table S2 Confirmation of motility defects (loss of>75% motility compared to wild type) in mutants identified in primary screening.

Table S3 Mutants with defect in swimming motility only.

Table S4 Mutants with defect in swarming motility.

Table S5 Previously described hypermotility in bacteria.

Table S6 Densitometry data underlying Figures 3 and 5, and representative images of Western blots used to generate this data.

Movie S1 WT_400x_6h.mov shows examples of the swarming motility of the wild type organism.

Movie S2 fimZ_400x_6h.mov shows the swarming motility of a deletion mutant in fimZ that has increased swarming motility relative to the wild type.

Movie S3 STM1630_400x_6h.mov shows the swarming motility of a deletion mutant in STM1630 that has increased swariming motility relative to the wild type.

Author Contributions

Conceived and designed the experiments: LMB HLAP MM. Performed the experiments: LMB LA YR MT KDA. Analyzed the data: LMB KDA MM HLAP. Contributed reagents/materials/analysis tools: LMB MM HLAP. Wrote the paper: LMB MM HLAP.

References

1. Haraga A, Ohlson MB, Miller SI (2008) Salmonellae interplay with host cells. Nat Rev Microbiol 6: 53–66.
2. Rabsch W, Tschape H, Baumler AJ (2001) Non-typhoidal salmonellosis: emerging problems. Microbes Infect 3: 237–247.
3. Chevance FF, Hughes KT (2008) Coordinating assembly of a bacterial macromolecular machine. Nat Rev Microbiol 6: 455–465.
4. Wang Q, Frye JG, McClelland M, Harshey RM (2004) Gene expression patterns during swarming in Salmonella typhimurium: genes specific to surface growth and putative new motility and pathogenicity genes. Mol Microbiol 52: 169–187.
5. Ottemann KM, Miller JF (1997) Roles for motility in bacterial-host interactions. Mol Microbiol 24: 1109–1117.
6. Harshey RM (2003) Bacterial motility on a surface: many ways to a common goal. Annu Rev Microbiol 57: 249–273.
7. Adler J (1966) Chemotaxis in bacteria. Science 153: 708–716.
8. Fraser GM, Hughes C (1999) Swarming motility. Curr Opin Microbiol 2: 630–635.
9. Berg HC (2005) Swarming motility: it better be wet. Curr Biol 15: R599–600.
10. Chen BG, Turner L, Berg HC (2007) The wetting agent required for swarming in Salmonella enterica serovar typhimurium is not a surfactant. J Bacteriol 189: 8750–8753.
11. Toguchi A, Siano M, Burkart M, Harshey RM (2000) Genetics of swarming motility in Salmonella enterica serovar typhimurium: critical role for lipopolysaccharide. J Bacteriol 182: 6308–6321.
12. Wang Q, Suzuki A, Mariconda S, Porwollik S, Harshey RM (2005) Sensing wetness: a new role for the bacterial flagellum. EMBO J 24: 2034–2042.
13. Girgis HS, Liu Y, Ryu WS, Tavazoie S (2007) A comprehensive genetic characterization of bacterial motility. PLoS Genet 3: 1644–1660.
14. Inoue T, Shingaki R, Hirose S, Waki K, Mori H, et al. (2007) Genome-wide screening of genes required for swarming motility in Escherichia coli K-12. J Bacteriol 189: 950–957
15. Santiviago CA, Reynolds MM, Porwollik S, Choi SH, Long F, et al. (2009) Analysis of pools of targeted Salmonella deletion mutants identifies novel genes affecting fitness during competitive infection in mice. PLoS Pathog 5: e1000477.
16. Bogomolnaya LM, Santiviago CA, Yang HJ, Baumler AJ, Andrews-Polymenis HL (2008) 'Form variation' of the O12 antigen is critical for persistence of Salmonella Typhimurium in the murine intestine. Mol Microbiol 70: 1105–1119.
17. Sambrook J, Fritsch EF, Maniatis T (1989) Molecular cloning: a laboratory manual. Cold Spring Harbor, N.Y.: Cold Spring Harbor Laboratory.
18. Guard-Petter J (1997) Induction of flagellation and a novel agar-penetrating flagellar structure in Salmonella enterica grown on solid media: possible consequences for serological identification. FEMS Microbiol Lett 149: 173–180.
19. McClelland M, Sanderson KE, Spieth J, Clifton SW, Latreille P, et al. (2001) Complete genome sequence of Salmonella enterica serovar Typhimurium LT2. Nature 413: 852–856.
20. Lawley TD, Chan K, Thompson LJ, Kim CC, Govoni GR, et al. (2006) Genome-wide screen for Salmonella genes required for long-term systemic infection of the mouse. PLoS Pathog 2: e11.
21. Chilcott GS, Hughes KT (2000) Coupling of flagellar gene expression to flagellar assembly in Salmonella enterica serovar typhimurium and Escherichia coli. Microbiol Mol Biol Rev 64: 694–708.
22. Nevola JJ, Stocker BA, Laux DC, Cohen PS (1985) Colonization of the mouse intestine by an avirulent Salmonella typhimurium strain and its lipopolysaccharide-defective mutants. Infect Immun 50: 152–159.
23. Mariconda S, Wang Q, Harshey RM (2006) A mechanical role for the chemotaxis system in swarming motility. Mol Microbiol 60: 1590–1602.

24. Blair DF, Kim DY, Berg HC (1991) Mutant MotB proteins in Escherichia coli. J Bacteriol 173: 4049–4055.
25. Wang KC, Hsu YH, Huang YN, Yeh KS (2012) A previously uncharacterized gene stm0551 plays a repressive role in the regulation of type 1 fimbriae in Salmonella enterica serotype Typhimurium. BMC Microbiol 12: 111.
26. Ahmad I, Lamprokostopoulou A, Le Guyon S, Streck E, Barthel M, et al. (2011) Complex c-di-GMP signaling networks mediate transition between virulence properties and biofilm formation in Salmonella enterica serovar Typhimurium. PLoS One 6: e28351.
27. Wolfe AJ, Visick KL (2008) Get the message out: cyclic-Di-GMP regulates multiple levels of flagellum-based motility. J Bacteriol 190: 463–475.
28. Clegg S, Hughes KT (2002) FimZ is a molecular link between sticking and swimming in Salmonella enterica serovar Typhimurium. J Bacteriol 184: 1209–1213.
29. Reynolds MM, Bogomolnaya L, Guo J, Aldrich L, Bokhari D, et al. (2011) Abrogation of the twin arginine transport system in Salmonella enterica serovar Typhimurium leads to colonization defects during infection. PLoS One 6: e15800.
30. Paul K, Nieto V, Carlquist WC, Blair DF, Harshey RM (2010) The c-di-GMP binding protein YcgR controls flagellar motor direction and speed to affect chemotaxis by a "backstop brake" mechanism. Mol Cell 38: 128–139.
31. Kato Y, Sugiura M, Mizuno T, Aiba H (2007) Effect of the arcA mutation on the expression of flagella genes in Escherichia coli. Biosci Biotechnol Biochem 71: 77–83.
32. Simm R, Morr M, Kader A, Nimtz M, Romling U (2004) GGDEF and EAL domains inversely regulate cyclic di-GMP levels and transition from sessility to motility. Mol Microbiol 53: 1123–1134.
33. Zhao R, Amsler CD, Matsumura P, Khan S (1996) FliG and FliM distribution in the Salmonella typhimurium cell and flagellar basal bodies. J Bacteriol 178: 258–265.
34. Tatusov RL, Natale DA, Garkavtsev IV, Tatusova TA, Shankavaram UT, et al. (2001) The COG database: new developments in phylogenetic classification of proteins from complete genomes. Nucleic Acids Res 29: 22–28.
35. Ilg K, Endt K, Misselwitz B, Stecher B, Aebi M, et al. (2009) O-antigen-negative Salmonella enterica serovar Typhimurium is attenuated in intestinal colonization but elicits colitis in streptomycin-treated mice. Infect Immun 77: 2568–2575.
36. Abeyrathne PD, Daniels C, Poon KK, Matewish MJ, Lam JS (2005) Functional characterization of WaaL, a ligase associated with linking O-antigen polysaccharide to the core of Pseudomonas aeruginosa lipopolysaccharide. J Bacteriol 187: 3002–3012.
37. Blondel CJ, Jimenez JC, Contreras I, Santiviago CA (2009) Comparative genomic analysis uncovers 3 novel loci encoding type six secretion systems differentially distributed in Salmonella serotypes. BMC Genomics 10: 354.
38. Folkesson A, Lofdahl S, Normark S (2002) The Salmonella enterica subspecies I specific centisome 7 genomic island encodes novel protein families present in bacteria living in close contact with eukaryotic cells. Res Microbiol 153: 537–545.
39. Gomez-Gomez JM, Manfredi C, Alonso JC, Blazquez J (2007) A novel role for RecA under non-stress: promotion of swarming motility in Escherichia coli K-12. BMC Biol 5: 14.
40. Stafford GP, Hughes C (2007) Salmonella typhimurium flhE, a conserved flagellar regulon gene required for swarming. Microbiology 153: 541–547.
41. Lee J, Harshey RM (2012) Loss of FlhE in the flagellar Type III secretion system allows proton influx into Salmonella and Escherichia coli. Mol Microbiol 84: 550–565.
42. Morgenstein RM, Clemmer KM, Rather PN (2010) Loss of the waaL O-antigen ligase prevents surface activation of the flagellar gene cascade in Proteus mirabilis. J Bacteriol 192: 3213–3221.
43. Haneda T, Sugimoto M, Yoshida-Ohta Y, Kodera Y, Oh-Ishi M, et al. (2010) Comparative proteomic analysis of Salmonella enterica serovar Typhimurium ppGpp-deficient mutant to identify a novel virulence protein required for intracellular survival in macrophages. BMC Microbiol 10: 324.
44. Traxler MF, Summers SM, Nguyen H-T, Zacharia VM, Hightower GA, et al. (2008) The global, ppGpp-mediated stringent response to amino acid starvation in Escherichia coli. Molecular microbiology 68: 1128–1148.
45. Suzuki H, Koyanagi T, Izuka S, Onishi A, Kumagai H (2005) The yliA, -B, -C, and -D genes of Escherichia coli K-12 encode a novel glutathione importer with an ATP-binding cassette. J Bacteriol 187: 5861–5867.
46. Gillespie JJ, Wattam AR, Cammer SA, Gabbard JL, Shukla MP, et al. (2011) PATRIC: the comprehensive bacterial bioinformatics resource with a focus on human pathogenic species. Infect Immun 79: 4286–4298.
47. Kim W, Surette MG (2004) Metabolic differentiation in actively swarming Salmonella. Mol Microbiol 54: 702–714.
48. Frye J, Karlinsey JE, Felise HR, Marzolf B, Dowidar N, et al. (2006) Identification of new flagellar genes of Salmonella enterica serovar Typhimurium. J Bacteriol 188: 2233–2243.
49. Wang Q, Mariconda S, Suzuki A, McClelland M, Harshey RM (2006) Uncovering a large set of genes that affect surface motility in Salmonella enterica serovar Typhimurium. J Bacteriol 188: 7981–7984.
50. Kohler T, Curty LK, Barja F, van Delden C, Pechere JC (2000) Swarming of Pseudomonas aeruginosa is dependent on cell-to-cell signaling and requires flagella and pili. J Bacteriol 182: 5990–5996.
51. Chandra H, Khandelwal P, Khattri A, Banerjee N (2008) Type 1 fimbriae of insecticidal bacterium Xenorhabdus nematophila is necessary for growth and colonization of its symbiotic host nematode Steinernema carpocapsiae. Environ Microbiol 10: 1285–1295.
52. Stocker BAD, McDonough MW, Ambler RP (1961) A gene determining presence or absence of epsilon-N-methyl-lysine in Salmonella flagellar protein. Nature 189: 556–558.
53. Verstraeten N, Braeken K, Debkumari B, Fauvart M, Fransaer J, et al. (2008) Living on a surface: swarming and biofilm formation. Trends Microbiol 16: 496–506.
54. Chaudhuri RR, Peters SE, Pleasance SJ, Northen H, Willers C, et al. (2009) Comprehensive identification of Salmonella enterica serovar typhimurium genes required for infection of BALB/c mice. PLoS Pathog 5: e1000529.
55. Morgan E, Campbell JD, Rowe SC, Bispham J, Stevens MP, et al. (2004) Identification of host-specific colonization factors of Salmonella enterica serovar Typhimurium. Mol Microbiol 54: 994–1010.
56. Groisman EA, Mouslim C (2006) Sensing by bacterial regulatory systems in host and non-host environments. Nat Rev Microbiol 4: 705–709.
57. Navarre WW, Halsey TA, Walthers D, Frye J, McClelland M, et al. (2005) Co-regulation of Salmonella enterica genes required for virulence and resistance to antimicrobial peptides by SlyA and PhoP/PhoQ. Mol Microbiol 56: 492–508.
58. Monsieurs P, De Keersmaecker S, Navarre WW, Bader MW, De Smet F, et al. (2005) Comparison of the PhoPQ regulon in Escherichia coli and Salmonella typhimurium. J Mol Evol 60: 462–474.
59. Shah DH, Lee MJ, Park JH, Lee JH, Eo SK, et al. (2005) Identification of Salmonella gallinarum virulence genes in a chicken infection model using PCR-based signature-tagged mutagenesis. Microbiology 151: 3957–3968.
60. Yoon H, McDermott JE, Porwollik S, McClelland M, Heffron F (2009) Coordinated regulation of virulence during systemic infection of Salmonella enterica serovar Typhimurium. PLoS Pathog 5: e1000306.
61. Waite RD, Rose RS, Rangarajan M, Aduse-Opoku J, Hashim A, et al. (2012) Pseudomonas aeruginosa possesses two putative type I signal peptidases, LepB and PA1303, each with distinct roles in physiology and virulence. J Bacteriol 194: 4521–4536.
62. Caiazza NC, Merritt JH, Brothers KM, O'Toole GA (2007) Inverse regulation of biofilm formation and swarming motility by Pseudomonas aeruginosa PA14. J Bacteriol 189: 3603–3612.
63. Kuchma SL, Ballok AE, Merritt JH, Hammond JH, Lu W, et al. (2010) Cyclic-di-GMP-mediated repression of swarming motility by Pseudomonas aeruginosa: the pilY1 gene and its impact on surface-associated behaviors. J Bacteriol 192: 2950–2964.
64. Merritt JH, Brothers KM, Kuchma SL, O'Toole GA (2007) SadC reciprocally influences biofilm formation and swarming motility via modulation of exopolysaccharide production and flagellar function. J Bacteriol 189: 8154–8164.
65. Merritt JH, Ha DG, Cowles KN, Lu W, Morales DK, et al. (2010) Specific control of Pseudomonas aeruginosa surface-associated behaviors by two c-di-GMP diguanylate cyclases. MBio 1.
66. Wang WB, Chen IC, Jiang SS, Chen HR, Hsu CY, et al. (2008) Role of RppA in the regulation of polymyxin b susceptibility, swarming, and virulence factor expression in Proteus mirabilis. Infect Immun 76: 2051–2062.
67. Clemmer KM, Rather PN (2008) The Lon protease regulates swarming motility and virulence gene expression in Proteus mirabilis. J Med Microbiol 57: 931–937.
68. Liaw SJ, Lai HC, Ho SW, Luh KT, Wang WB (2001) Characterisation of p-nitrophenylglycerol-resistant Proteus mirabilis super-swarming mutants. J Med Microbiol 50: 1039–1048.
69. Belas R, Schneider R, Melch M (1998) Characterization of Proteus mirabilis precocious swarming mutants: identification of rsbA, encoding a regulator of swarming behavior. J Bacteriol 180: 6126–6139.
70. Li B, Li N, Wang F, Guo L, Huang Y, et al. (2012) Structural insight of a concentration-dependent mechanism by which YdiV inhibits Escherichia coli flagellum biogenesis and motility. Nucleic Acids Res 40: 11073–11085.
71. Takaya A, Erhardt M, Karata K, Winterberg K, Yamamoto T, et al. (2012) YdiV: a dual function protein that targets FlhDC for ClpXP-dependent degradation by promoting release of DNA-bound FlhDC complex. Mol Microbiol 83: 1268–1284.
72. Wada T, Morizane T, Abo T, Tominaga A, Inoue-Tanaka K, et al. (2011) EAL domain protein YdiV acts as an anti-FlhD4C2 factor responsible for nutritional control of the flagellar regulon in Salmonella enterica Serovar Typhimurium. J Bacteriol 193: 1600–1611.
73. Simms AN, Mobley HL (2008) Multiple genes repress motility in uropathogenic Escherichia coli constitutively expressing type 1 fimbriae. J Bacteriol 190: 3747–3756.
74. Simm R, Remminghorst U, Ahmad I, Zakikhany K, Romling U (2009) A role for the EAL-like protein STM1344 in regulation of CsgD expression and motility in Salmonella enterica serovar Typhimurium. J Bacteriol 191: 3928–3937.
75. Miller SI, Kukral AM, Mekalanos JJ (1989) A two-component regulatory system (phoP phoQ) controls Salmonella typhimurium virulence. Proc Natl Acad Sci U S A 86: 5054–5058.
76. Groisman EA, Chiao E, Lipps CJ, Heffron F (1989) Salmonella typhimurium phoP virulence gene is a transcriptional regulator. Proc Natl Acad Sci U S A 86: 7077–7081.

77. Brinkman FS, Macfarlane EL, Warrener P, Hancock RE (2001) Evolutionary relationships among virulence-associated histidine kinases. Infect Immun 69: 5207–5211.
78. Alteri CJ, Lindner JR, Reiss DJ, Smith SN, Mobley HL (2011) The broadly conserved regulator PhoP links pathogen virulence and membrane potential in Escherichia coli. Mol Microbiol 82: 145–163.
79. Derzelle S, Turlin E, Duchaud E, Pages S, Kunst F, et al. (2004) The PhoP-PhoQ two-component regulatory system of Photorhabdus luminescens is essential for virulence in insects. J Bacteriol 186: 1270–1279.
80. Stewart MK, Cummings LA, Johnson ML, Berezow AB, Cookson BT (2011) Regulation of phenotypic heterogeneity permits Salmonella evasion of the host caspase-1 inflammatory response. Proc Natl Acad Sci U S A 108: 20742–20747.
81. Weening EH, Barker JD, Laarakker MC, Humphries AD, Tsolis RM, et al. (2005) The Salmonella enterica serotype Typhimurium lpf, bcf, stb, stc, std, and sth fimbrial operons are required for intestinal persistence in mice. Infect Immun 73: 3358–3366.
82. Ledeboer NA, Frye JG, McClelland M, Jones BD (2006) Salmonella enterica serovar Typhimurium requires the Lpf, Pef, and Tafi fimbriae for biofilm formation on HEp-2 tissue culture cells and chicken intestinal epithelium. Infect Immun 74: 3156–3169.
83. Yeh KS, Hancox LS, Clegg S (1995) Construction and characterization of a fimZ mutant of Salmonella typhimurium. J Bacteriol 177: 6861–6865.
84. Yeh KS, Tinker JK, Clegg S (2002) FimZ binds the Salmonella typhimurium fimA promoter region and may regulate its own expression with FimY. Microbiol Immunol 46: 1–10.

13

Role of ARF6, Rab11 and External Hsp90 in the Trafficking and Recycling of Recombinant-Soluble *Neisseria meningitidis* Adhesin A (rNadA) in Human Epithelial Cells

Giuseppe Bozza[1☯¤a], Mirco Capitani[2☯], Paolo Montanari[1], Barbara Benucci[1], Marco Biancucci[1¤b], Vincenzo Nardi-Dei[1], Elena Caproni[1], Riccardo Barrile[1¤c], Benedetta Picciani[2¤d], Silvana Savino[1], Beatrice Aricò[1], Rino Rappuoli[1], Mariagrazia Pizza[1], Alberto Luini[3], Michele Sallese[2*], Marcello Merola[1,4*]

1 Novartis Vaccines, Siena, Italy, **2** Unit of Genomic Approaches to Membrane Traffic, Fondazione Mario Negri Sud, S. Maria Imbaro (CH), Italy, **3** Institute of Protein Biochemistry, CNR, Naples, Italy, **4** Department of Biology, University of Naples "Federico II", Naples, Italy

Abstract

Neisseria meningitidis adhesin A (NadA) is a meningococcus surface protein thought to assist in the adhesion of the bacterium to host cells. We have previously shown that NadA also promotes bacterial internalization in a heterologous expression system. Here we have used the soluble recombinant NadA (rNadA) lacking the membrane anchor region to characterize its internalization route in Chang epithelial cells. Added to the culture medium, rNadA internalizes through a PI3K-dependent endocytosis process not mediated by the canonical clathrin or caveolin scaffolds, but instead follows an ARF6-regulated recycling pathway previously described for MHC-I. The intracellular pool of rNadA reaches a steady state level within one hour of incubation and colocalizes in endocytic vesicles with MHC-I and with the extracellularly labeled chaperone Hsp90. Treatment with membrane permeated and impermeable Hsp90 inhibitors 17-AAG and FITC-GA respectively, lead to intracellular accumulation of rNadA, strongly suggesting that the extracellular secreted pool of the chaperone is involved in rNadA intracellular trafficking. A significant number of intracellular vesicles containing rNadA recruit Rab11, a small GTPase associated to recycling endosomes, but do not contain transferrin receptor (TfR). Interestingly, cell treatment with Hsp90 inhibitors, including the membrane-impermeable FITC-GA, abolished Rab11-rNadA colocalization but do not interfere with Rab11-TfR colocalization. Collectively, these results are consistent with a model whereby rNadA internalizes into human epithelial cells hijacking the recycling endosome pathway and recycle back to the surface of the cell via an ARF6-dependent, Rab11 associated and Hsp90-regulated mechanism. The present study addresses for the first time a meningoccoccal adhesin mechanism of endocytosis and suggests a possible entry pathway engaged by *N. meningitidis* in primary infection of human epithelial cells.

Editor: Matthew Seaman, Cambridge University, United Kingdom

Funding: This work is entirely funded by Novartis Vaccines; no external entities have contributed to cover project expenses. The funders had no role in study design, data collection and analysis, decision to publish, or preparation of the manuscript.

Competing Interests: PM, VND, SS, BA, RR, and MP are employees of Novartis Vaccines, whose company funded this study. GB, RB, and MB were students working at Novartis and BB is a student still working at Novartis. EC was a post-doctoral fellow working at Novartis. MM is a Novartis Scientific collaborator and was responsible for this project. The antigen of this study is one of the components of Bexsero, a commercially available vaccine against meningococcus B. There are no further patents, products in development, or marketed products to declare. This does not alter the authors' adherence to all the PLOS ONE policies on sharing data and materials.

* Email: sallese@negrisud.it (MS); marcello.merola@novartis.com (MM)

☯ These authors contributed equally to this work.

¤a Current address: European Medicines Agency, London, United Kingdom
¤b Current address: Northwestern University of Chicago, Evanston, IL, United States of America
¤c Current address: Wyss Institute for Biologically Inspired Engineering, Harvard University, Boston, MA, United States of America
¤d Current address: Department of Diagnostic and Laboratory Pathology, IRCCS Foundation, "Istituto Nazionale Tumori", Milan, Italy

Introduction

Neisseria meningitidis (meningococcus) is a Gram-negative diplococcus that causes severe invasive disease and represents one of the most devastating bacterial infections. Although fatal if not treated on time, meningococcus invasion appears to be more an undesirable event of a usually commensal bacterium, probably due to a combination of host susceptibility and strain specific propensity to invasiveness [1,2]. The pathophysiology of the bacterium *N. meningitidis* is a process that requires several steps: penetration of the epithelial or mucosal barrier, reaching and surviving the bloodstream, crossing the blood-brain barrier, and eventually causing meningitis through extracellular proliferation [3]. Epithelia are the primary target of bacterial colonization, an event basically asymptomatic and common to both non-virulent and virulent strains. Disease is a rare event compared to the extent of meningococcal nasopharynx colonization. [2,4,5]. Experimental data support attachment of the bacterium to nonciliated cells of the respiratory epithelium [6] and transcellular route of passage through this barrier [6–9]. A recent report shows that bacterial capsule and type 4 pili are important for epithelial cell transcytosis [9] but host and pathogen players involved in this process are far from being defined.

Neisseria meningitidis adhesin A (NadA) was identified through genome wide analysis and afterward proved to be expressed on the bacterial surface [10–12]. NadA is a trimeric outer membrane protein whose gene is present in three out of four known hypervirulent lineages of serogroup B strains [13,14]. NadA is classified as a trimeric autotransporter adhesion (TAA), a family of outer membrane adhesin, which are present only in Gram negative bacteria. TAAs differ from classical autotrasporter protein by the homotrimeric structure hanging on the bacterial surface and the Type Vc secretion system [10,11,15–17]. Like other TAA proteins, NadA has a common modular organization, composed of i) a conserved C-terminal membrane anchor through which the protein is translocated to the cell surface, ii) a long central alpha helical domain (stalk) with high propensity to form coiled-coil structures, iii) an N-terminal globular head that has been associated with binding to specific cellular receptors and iiii) a cleavable signal sequence [18,19]. Generally, TAAs (previously classified as Oligomeric coiled-coil Adhesins (Oca) family [20]) mediate bacterial interaction with host cells or extracellular matrix (ECM) proteins or induce invasion into target cells [21–29]. The specificity of the biological function is thought to reside in the head, while the stalk is thought to ensure adequate exposure of the head from the outer membrane [19].

In previous reports, we have used a NadA heterologous expression system to demonstrate the invasive ability acquired by an *Escherichia coli* strain exposing surface NadA. Our data support the role of NadA in the uptake of bacteria by Chang cells, a human epithelial cell line [10,30]. A recombinant NadA (rNadA), expressed in *E. coli* and purified in a soluble form in absence of the anchor (translocator) domain, preserves its immunogenic properties and is included in a multicomponent vaccine against meningococcus B (Bexsero) [31,32]. A peculiar feature of rNadA, perhaps unique among all members of the TAA family, is the ability to preserve a stable trimeric structure in solution [10,13,33]. This recombinant soluble homo-trimer still binds eukaryotic cells [10,13,33–35]. The gain-of-function phenotype acquired by heterologous bacteria expressing NadA and the conserved binding characteristics shown by the recombinant protein provide an opportunity to dissect the function of this adhesin in host-pathogen interaction(s).

When expressed on the surface of *E. coli*, NadA promotes bacterial internalization [10], by an undefined mechanism. Eukaryotic cells internalize a variety of extracellular substances as well as plasma membrane proteins via a remarkable diversity of endocytic pathways generally classified as clathrin-dependent or independent according to the involvement of clathrin [36–38]. Endocytic vesicles converge into early endosomes (EEs), the main sorting compartment from which most of the cargos are rapidly recycled back to the plasmamembrane (PM), while some are sorted into late endosomes and destined either to lysosomes or to the trans Golgi network (TGN) [39,40]. While lysosome targeting is achieved by clathrin-coated endosomes, mechanisms of clathrin-independent endocytosis have been identified more recently and are emerging as complex, multi-pattern processes of internalization still poorly characterized [41,42]. In particular, ADP ribosylation Factor 6 (ARF6) has been involved in internalization and sorting of cargos targeted to recycling endosome [43]. Initially described for the major histocompatibility complex (MHC) class I trafficking [44], the list of membrane proteins requiring this factor for internalization and recycling is growing [42,43]. While ARFs coordinate vesicle formation [45], key regulators of endosome trafficking include the Rabs, a family of small GTPase proteins associated to all kind of vesicles [46]. The more than 60 Rabs found in eukaryotic cells communicate with each other through recruitment of effectors that regulate the correct destination of vesicles [39,47].

In our attempt to identify eukaryotic cell proteins involved in the NadA-mediated interaction of bacteria, we found that such processes are sensitive to inhibition of Hsp90 activity, and the host chaperone itself was found to bind NadA *in vitro* [30]. Hsp90 is a ubiquitous molecular chaperone crucial for maintenance of cellular homeostasis [48]. The Hsp90 family of chaperones includes organelle specialized isoforms and two cytoplasm/nucleus localized factors, the inducible Hsp90α and the constitutive Hsp90β. More recently, several lines of evidence suggest that Hsp90α is also secreted by normal cells in response to stress or insults, and by several tumor cell lines (reviewed in [49]). The major functions of this extracellular pool have been identified as protective response to several kinds of stress, assistance to antigen presentation and promoting tissue invasion by cancer cells [49–52]. Results from our previous study also indicated a protective role of Hsp90 toward NadA mediated invasion, although we did not differentiate the possible role of extracellular and intracellular pools of the chaperone [30].

In this report we have used the soluble recombinant $NadA_{\Delta 351-405}$ (rNadA) and demonstrated that it was able to internalize upon binding to the Chang human epithelial cell line. Evidence is provided on the stability of the rNadA trimer at 37°C and on the temperature dependence of NadA binding to Chang cells. In our analysis of the internalization pathway exploited by NadA, we found that the process triggered by NadA was clathrin independent. As reported for MHC class I internalization and recycling to cell surface, ARF6-dependence and recruitment of Rab 11 on rNadA vesicles suggested that the adhesin is sorted in a recycling endosomes pathway. The effect of two Hsp90 inhibitors on the internalization and recycling of the adhesin indicated that extracellular Hsp90 is involved in the trafficking of rNadA.

Materials and Methods

Cell culture and transfection

Chang epithelial cells (Wong-Kilbourne derivative, clone 1-5c-4, human conjunctiva, ATCC CCL-20.2) were grown in Dulbecco's Modified Eagle Medium (DMEM) supplemented with 15 mM glutamine, 100 U/ml penicillin, 100 μg/ml streptomycin, 10% FCS and maintained at 37°C in a controlled humidified atmosphere containing 5% CO_2. Transfection of Chang cells was performed using Hiperfect (Qiagen), jetPEI (Polyplus transfection) or lipofectamine (Invitrogen) according the manufacturer's instructions.

Antibodies, plasmid vectors and reagents

Anti-MHC-I (Biolegends); anti-EEA1 (Novus Biologicals); anti-ARF6 (Santa Cruz Biotechnology); anti-Rab5 (BD transduction Laboratories); anti clathrin (Sigma); anti-M6PR kindly provided by J. Gruenberg University of Geneva, Switzerland; anti-HSP90 (SPA-830, Stressgene); Alexa-488-conjugated Phalloidin, Cy3-conjugated transferrin, Alexa488- and 543-conjugated secondary antibodies (Invitrogen); Alexa488-conjugated anti-MHC-I (Santa cruz Biotechnology). Antibodies against rNadA were previously described in [53]. Alexa-488-conjugated and Alexa-633 conjugated rNadA were obtained by labelling with Alexa Fluor Microscale Protein Labeling Kit (Molecular Probes-Invitrogen) following manufacturer instructions. cDNA expressing the wild type and ARF6-Q67L mutant were kindly provided by J. Donaldson (to M.S. and A.L.), National Institutes of Health, Bethesda, USA. 17-AAG (17-N-allylamino-17-demethoxygeldanamycin) and Wortmannin were from Sigma. FITC-GA was synthesized by labogen (Milan - Italy). The myc-AP180C plasmid containing the C-terminal portion of the AP180 gene fused in frame with N-terminal myc tag sequence was generated as described elsewhere [54]. Generation of plasmids coding for C-terminal EGFP fusion Rabs was performed as described previously [55].

Binding of rNadA to human Chang epithelial cells

Chang cells were non-enzymatically detached from the support by using cell dissociation solution (Sigma), harvested and resuspended in DMEM medium supplemented with 1% FBS.

Then 3×10^5 cells were mixed with recombinant rNadA (200 µg/ml) for 30 minutes at four different temperatures (RT, 4°C, 30°C and 37°C). After two washes with 1% FBS in phosphate-buffered saline, cells were incubated with the mouse monoclonal anti-NadA antibody 9F11 (1.25 µg/ml to 160 µg/ml) or IgG mouse isotype control (Sigma) for 1 hr at 4°C. Samples were washed twice and then incubated for 30 minutes at 4°C with Allophycocyanin-conjugated goat $F(ab)_2$ antibody to mouse IgG (1:100, Jackson ImmunoResearch) and finally the cells were analyzed with the FACS-Scan flow cytometer Canto II. The mean fluorescence intensity for each sample was calculated.

To analyze the time dependence of NadA binding to Chang, 3×10^5 cells were incubated with recombinant rNadA (200 µg/ml) at 37°C for different period of time (1 minute to 2 hours), washed and incubated with the mouse monoclonal anti-NadA antibody 9F11 (20 µg/ml) or IgG mouse isotype control for 1 hr at 4°C. FACS analysis was performed as described above.

Size Esclusion-HPLC–multiangle laser light-scattering (MALLS) analysis

Four identical aliquots (100 µl) of rNadA (1,7 mg/ml in 10 mM NaH_2P0_4 150 mM NaCl pH 7.2) were heated at 37°C for 30 min, 1, 2 and 4 hrs respectively. For each time point, sample was loaded on an analytical size exclusion TSK Super SW3000 (4.6×300-mm column of 4 µm and 300 Å) with a separation range on globular proteins from 10 to 500 kDa (Tosoh Bioscience, Tokyo, Japan). Samples were eluted isocratically in 0.1 M NaH_2P0_4, 0,1 M $NaHSO_4$ buffer at pH 7.0. MALLS analyses were performed in real time using a multi-angle light-scattering detector Dawn TREOS (Wyatt Corp., Santa Barbara, CA, USA), with an incident laser of 658 nm; intensity of the scattered light was measured at 3 angles simultaneously. Data elaboration was performed using Astra V software (Wyatt). Zimm formalism was used to determine the weight-average molecular mass (MW) in daltons and polydispersity index (MW/Mn) for each oligomer present in solution.

Full length trimeric NadA purification on size exclusion chromatography

A 300 µl aliquot of rNadA (1,7 mg/ml Buffer 10 mM NaH_2P0_4 150 mM NaCl pH 7.2) was heated at 37°C for 4 hours. At end of this time period, sample was loaded on the gel filtration column. Size exclusion was performed using a Superdex 75 10–300 GL (GE-Healtcare) and elution was performed isocratically in 0.1 M NaH_2P0_4 pH 7.2, 0.15 M NaCl. Protein containing fractions were identified by OD at 280 nm and collected.

Immunofluorescence staining and confocal microscopy

Chang cells were cultured on coverslips to reach 70–80% of confluence, washed three times in medium without serum and then incubated with rNadA (200 µg/ml) in the same medium supplemented with 1% FCS. Incubation was performed at 37°C for the indicated times. Cells were then fixed in paraformaldehyde for 8–10 min at 37°C, permeabilized and incubated with primary antibodies for 1 hr at room temperature. Afterwards, cells were washed three times with PBS, incubated with fluorophore-conjugated secondary antibody for 30–45 min at room temperature, washed again three times in PBS and mounted on slides.

In vivo staining of MHC-I was performed as follows: cells were washed three times in medium without serum and incubated with a mouse monoclonal antibody against MHC-I (10–30 µg/ml) for 1 hr at 4°C. Afterwards, cells were washed 3 times and incubated with recombinant rNadA (200 µg/ml) for 1 hr at 37°C in a medium supplemented with 1% FCS, then fixed and permeabilized. rNadA was stained following the standard procedure while MHC-I was revealed using a fluorescence-conjugated secondary antibody directed against the mouse monoclonal primary antibody.

In vivo staining of HSP90 was performed as follows: cells were washed three times in medium without serum and incubated with a rabbit polyclonal HSP90 antibody (50 µg/ml) for 2–4 hrs at 4°C. Afterward, cells were washed 3 times and incubated with recombinant rNadA (200 µg/ml) for 1 h at 37°C in a medium supplemented with 1% FCS, then fixed and permeabilized. rNadA was stained following the standard procedure while HSP90 was detected using a Alexa543-conjugated secondary antibody directed against the rabbit polyclonal primary antibody.

Samples were analyzed by confocal microscopy (LSM 510, Zeiss) using a 60× oil-immersion objective, maintaining the pinhole of the objective at 1 airy unit. Images were scanned using an Argon 488 laser, a HeNe 543 laser and a HeNe 633 laser, under non-saturating conditions (pixel fluorescence below 255 arbitrary units).

The colocalization analysis and the quantification of immunofluorescence (IF) intensity of rNadA in the cells was performed with LSM510-3.2 software (Zeiss). To assess the colocalization we removed the background immunofluorescence by adjusting the threshold levels and used the histo and colocalization functions of the above software. This software provides two colocalization coefficients that ranges from 0 (no colocalization) to 1 (complete colocalization). The colocalization coefficients indicate the amount of pixels of the channel A that colocalizes with pixels from channel B and viceversa. Finally, we expressed the colocalization extent as a percentage over the total immunofluorescence per channel. The immunofluorescence (IF) intensity was calculated as total immunofluorescence of rNadA in the cell divided by the area of the cell and expressed as arbitrary units (A.U.).

rNadA uptake in the presence of Hsp90 inhibitors

Internalization was performed by adding rNadA to the culture medium at a final concentration of 200 µg/ml and incubating at 37°C for the indicated period of time. Chang cells grown at about 50% confluence were pre-treated overnight with 0.5 µM 17-AAG and the then incubated with recombinant NadA (200 µg/ml) at 37°C for 1, 4 or 16 hrs in presence of the same concentration of 17-AAG. When 10 µM 17-AAG or FITC-GA were used, cells were grown to 70–80% confluence and pretreated for 1 hr with the inhibitors before adding rNadA (200 µg/ml) and subsequently incubated at 37°C for 1,4 or 16 hrs in the continued presence of the inhibitors. To temporarily permeabilize cells, growth medium was substituted with 0.01% saponin in PBS for 30 seconds at room temperature. Cells were then washed 3 times with PBS before adding fresh medium and incubate at 37°C for the indicated time. Afterwards, cells were fixed and labeled as detailed above.

AKT detection in Chang cell lysates

Chang cells were seeded on 24-well tissue culture plates (6–8×10^4 cells per well) and incubated at 37°C overnight. The following day, cells were treated with HSP90 inhibitors as described above. Total cell extracts were prepared in RIPA buffer (Sigma) supplemented with complete protease inhibitor (Roche).

Equal amounts of proteins were prepared in NuPAGE SDS Buffer under reducing conditions and separated on NuPAGE polyacrylamide gels. Proteins were transferred to nitrocellulose membranes for Western blot analysis. Membranes were then blocked with PBS containing 0.1% Tween 20 (PBST)+10% dried skim milk at room temperature for 1 h. After extensive washings in PBST, proteins bound on nitrocellulose membranes were detected with specific anti-Akt rabbit antibody (Cell Signaling Technology) followed by HRP-conjugated goat anti-rabbit secondary antibody.

Time-lapse Microscopy

Chang cells at 70–80% of confluence on glass-bottom Petri dishes (MatTek Corporation-USA) were incubated overnight at 37°C. The day after, cells were incubated with Alexa488-conjugated anti-MHC-I antibody and 2.5 µg/ml Alexa633-conjugated rNadA, or 5 µg/ml of Cy3-conjugated transferrin and 2.5 µg/ml Alexa-488-conjugated rNadA, or pretreated overnight with 0.5 µM of 17AAG and then with Alexa-488-conjugated rNadA for 60 min at 37°C. Cells were washes three times and supplemented with the medium buffered with 20 mM Hepes pH 7.2. The Petri dish was then placed on a stage of LSM-510 confocal microscope (Zeiss, Germany) equipped with a thermoregulation device at 37°C. Images were acquired every 2 (for MHC-I and NadA; movie S1) or 4 seconds (for transferrin and rNadA, movie S2; rNadA in 17AAG treated or untreated cells, movie S3 and S4 respectively) at 25% laser power intensity.

Results

Purified rNadA binds Chang cells in a time and temperature dependent manner

In previous reports, rNadA binds Chang epithelial cells more than other human cell lines [10]. A temperature dependence has also been suggested, as binding to human monocytes/macrophages was detectable at 37°C but not at 0°C [35]. We sought to investigate the temperature dependence and kinetics of NadA binding to Chang epithelial cells. Non permeabilized Chang cells were incubated with rNadA and, after washing, the bound fraction was revealed by FACS analysis with anti-NadA antibody.

At first this analysis was performed after 30 minutes of incubation at different temperatures. The result is shown in figure 1A. The rNadA binding to Chang cells was temperature dependent, giving the highest value at 37°C while dropping to about 60% at 30°C (figure 1A). Lowering the incubation temperatures at 20°C caused a decrease of the binding ability at less than 20% of the value obtained at 37°C. However, even at 4°C the amount of cell-bound rNadA was measurably higher than the background (independent IgG used as control) Setting incubation temperature at 37°C, we then analyzed the kinetics of rNadA binding to Chang cells (figure 1B). The maximum level of surface bound rNadA was observed at 30 minutes of incubation and decreased at longer incubation times. The time-dependent decrease of surface localized rNadA was obtained with various monoclonal and polyclonal antibodies used for detection (data not shown).

To exclude a destabilization of the trimeric rNadA structure at 37°C over time, we incubated the recombinant adhesin for 30 min, 1, 2 and 4 hrs and analyzed the products by analytical size exclusion chromatography and MALLS (Multi Angle Laser Light Scattering). The results of such analysis, presented in figure S1 and table S1, showed that the native trimer is the predominant species (about 95%) although the rNadA preparation include C-terminus deleted forms previously characterized [33]. The 4 hour treatment at 37°C completely removed these truncated species and the full length homogenous trimeric population can be purified as shown in figure 1, panels C–E.

The size exclusion chromatography analysis of the sample before and after thermal treatment is shown in figure 1C and 1D, respectively. The SDS-PAGE of the eluted samples confirmed that all contaminants were removed from the fractions containing the majority of the trimeric rNadA following 4 hour incubation at 37°C (fractions 12 and 13 of figure 1C and 1D). The purified full length trimeric rNadA eluted in peak II (lane B of figure 1E) was used at least once in all our experiments and challenged with the crude rNadA preparations. Both samples gave identical results. Thus, trimers of rNadA form a stable structure in solution which is resistant to the temperature conditions used in our studies.

rNadA internalization into Chang epithelial cells

Having excluded that the time-dependent reduction of rNadA binding to Chang cells was caused by changes in the trimeric structure or stability of the protein, we investigated a possible internalization process by confocal immunofluorescence microscopy (figure 2).

Following 15 minutes of incubation with rNadA, cells showed minor amounts of detectable rNadA apparently distributed on the cell surface (figure 2). Upon 30 minutes of incubation, rNadA appeared mostly as dot-like structures. At a longer time of incubation the fluorescence of the dot-like structures slightly increased (figure 2A and B). Hypothesizing that the dot-like structures are indicative of rNadA internalization, this phenomenon could explain, at least partially, the observed loss of rNadA fluorescence observed by FACS analysis at prolonged incubation times. However, the confocal microscopy also indicated that the intracellular pool of rNadA reaches a steady state level at about 30–60 minutes incubation time, possibly because it recycles towards the cell exterior or because of intracellular degradation.

Phosphoinositide 3-kinase (PI3K) control of rNadA internalization

Src phosphorylation of several components of the clathrin dependent endocytosis, including cortactin, arrestin, dynamin and clathrin itself, is essential for internalization [56]. In addition, Src activity is also required for the alternative caveolar endocytosis [57]. Protein kinase C is involved in the endocytosis of several G protein coupled receptor through clathrin or caveolae pathways [58,59]. The phosphoinositide 3-kinase (PI3K) is important in the early steps of endocytosis and and in the regulation of fusion and maturation dynamics of endocytic vesicles [60]. Here we used inhibitors of the different kinases to better understand the internalization pathway(s) exploited by rNadA.

Chang cells were incubated with or without kinase inhibitorsfor 1 hour with vehicle, 100 nM wortmannin (PI3K inhibitor more active on class I enzymes), 100 µM genistein (wide spectrum tyrosine kinase inhibitor known to prevent the caveolae mediated endocytosis [61]), 10 µM Bisindolylmaleimide I (BSMI, a general PKC inhibitor), and 10 µM SU6656 (Src family kinase inhibitor)

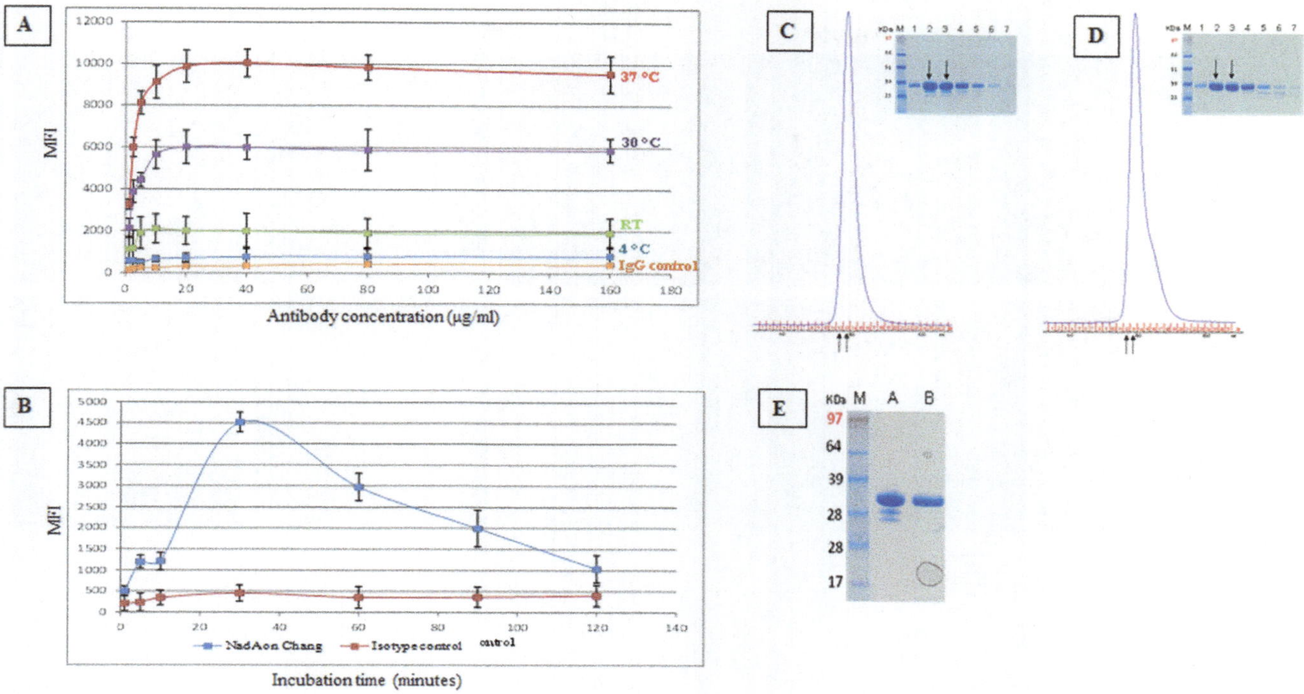

Figure 1. Purified rNadA binds Chang cells in a time- and temperature dependent manner. A and B. Chang cells were incubated with 200 µg/ml rNadA and then washed with PBS – 1% FBS. Binding was detected using anti-NadA 9F11 mouse mAb and Allophycocyanin-conjugated goat anti-mouse secondary antibody. Analysis was performed with Canto II instrument, and mean fluorescence intensity is reported (MFI). ***A***: Binding of rNadA to cells for 30 minutes at the indicated temperatures, and using varying concentrations of the primary antibody ***B***: Binding of rNadA was performed at 37°C for a period of time ranging from 1 minute to 2 hours as indicated. ***C***: SE-HPLC profile of rNadA at room temperature. ***D***: SE-HPLC profile of rNadA after heating period of 4 hrs at 37°C. ***E*** SDS-Page of collected fractions. Black arrows in **D** indicate the fractions pooled.

and then for an additional hour with rNadA. The internalization of rNadA was unaffected by the action of the tyrosine kinase inhibitors (data not shown) whereas wortmannin treatment reduced the internalization of rNadA (figure 3A and B). The cellular localization of MHC-I was unaffected by exposure to rNadA, genistein, BSMI and SU6656 (not shown), whereas wortmannin shifted the localization of MHC-I towards the PM fraction (figure 3A and C).

In these experiments, rNadA showed a certain extent of endosomal colocalization with MHC class I and we decided to further explore this observation. Antibody labelling of MHC-I following cell permeabilization would show the total pool of MHC-I (PM plus endosomally localized MHC-I). In order to visualize only the recently internalized MHC-I we decided to pre-label the PM-localized MHC-I. Chang cells were treated with an antibody against MHC-I for 1 hr at 4°C. This incubation allowed binding of the antibody to MHC-I at a temperature that prevents MHC-I internalization. The cells were then treated with rNadA for 60 minutes at 37°C, a temperature at which internalization is restored. As shown in figure 3D and E, rNadA appeared to colocalize with MHC-I in endosomes.

Confocal microscopy analysis of rNadA internalization pathway

In an attempt to elucidate the cellular entry mechanism of rNadA, we performed a confocal microscopy analysis in search of rNadA colocalization with different endocytic markers belonging to the clathrin-dependent and -independent pathways.

Chang cells were incubated with rNadA for 60 minutes at 37°C to allow internalization of the adhesin and then stained for rNadA and the endocytic markers, early endosome antigen 1 (EEA1), small GTPase Rab5, M6PR. As shown in figure 4A–C and F, rNadA showed a weak colocalization with early endosomal markers (EEA1, Rab5), as well as with the endo-lysosomal marker M6PR. We also analyzed the colocalization of internalized rNadA with clathrin itself and the transferrin receptor, a prototype cargo of the clathrin-dependent pathway. Remarkably, rNadA containing carriers did not associate with the scaffold protein clathrin, and the colocalization with transferrin receptor was negligible (figure 4D–F). As a control, we showed that transferrin receptor colocalized with clathrin in Chang cells (figure S2). Since these data suggested that the main rNadA internalization pathway does not rely on the clathrin- pathway, we sought to verify this aspect by the analysis of rNadA internalization in presence of an inhibitor of clathrin vesicles formation. The AP180 protein is a crucial adaptor involved in the early stage of chlatrin-dependent endocytosis [54].Expression of the C-terminus moiety of this adaptor, known as AP180C, inhibits transferrin and EGF receptors clathrin-dependent endocytosis by preventing coat formation and leading to clathrin redistribution [54,62]. AP180C transfected Chang cells were incubated with rNadA for 60 minutes at 37°C before to be fixed and treated for confocal analysis. As control, we followed TfR internalization both in absence and presence of the truncated AP180. Results are shown in figure 4G and quantification of rNadA, TfR and AP180C colocalization reported in graph. As expected, TfR internalization was severely impaired by AP180C expression while the intracellular vesicular pattern of rNadA was not affected (figure 4G). Furthermore, the absence of rNadA

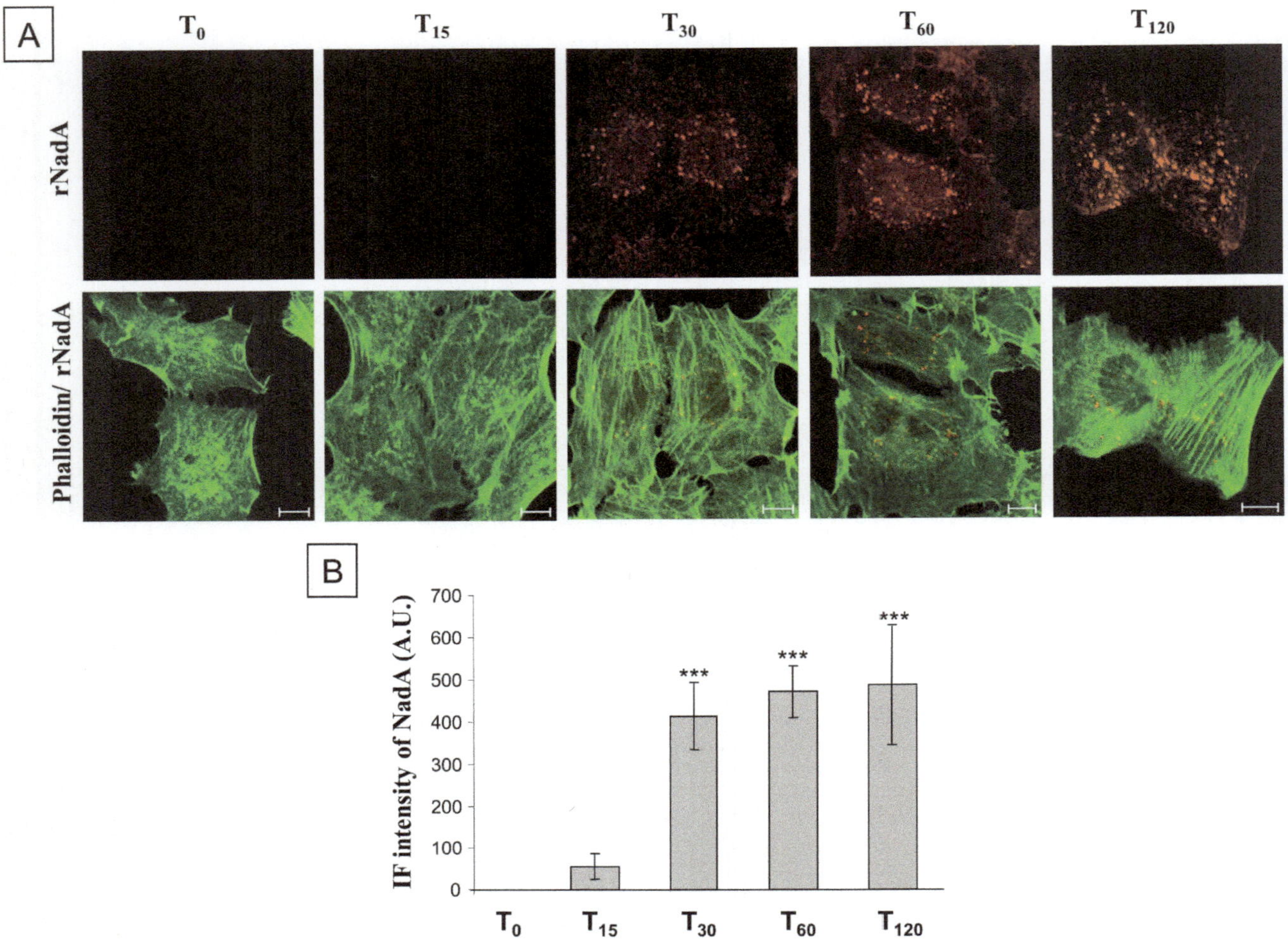

Figure 2. Time course of rNadA internalization. ***A:*** Chang cells were incubated with 200 μg/ml rNadA at 37°C for 0–120 min and then fixed, permeabilized and double stained for rNadA (upper panels; red) and Alexa488-conjugated phalloidin (green). Merged images of rNadA and phalloidin are shown in the lower panel. Scale bar is 10 μm. ***B:*** Quantification of IF intensity of rNadA in the experiment illustrated in panel A. Data are mean ± s.e.m representative of two independent experiments, each assessing 10–15 cells, and expressed as Arbitrary Units (A.U.). ***$p<0.001$, compared to T_0 treated cells (t-test).

colocalization with either TfR or AP180C confirmed that the two proteins follow different internalization pathways.

To further explore the internalization pathway used by rNadA, we analyzed the presence of rNadA in the lysosomal compartment. rNadA showed a very minor colocalization with the lysosomal marker LAMP1 upon 60 minute of incubation as well as at longer incubation times up to 24 hours (not shown). This observation is consistent with a small presence of the adhesin in the early endosomal compartment and indicates that internalized rNadA is not directed towards the degradative lysosomal pathway.

Finally, we analyzed the involvement of the clathrin-independent ARF6-regulated pathway. To verify whether endosomal sorting of internalized rNadA involves the small GTPase ARF6 [60] we monitored the internalization/recycling of rNadA in Chang cells expressing wild-type, and GTP-locked ARF6 (ARF6-Q67L). Previous studies demonstrated that ARF6-Q67L transfection leads to the intracellular accumulation of MHC-I [60]. Indeed, when mutant cells overexpressing ARF6-Q67L were incubated with rNadA overnight there was an accumulation of intracellular rNadA, while in cells overexpressing wild-type ARF6 the classical spotted distribution of rNadA was observed (figure 5).

In conclusion, this indicates that the rNadA endosomal trafficking involves the clathrin independent ARF6-regulated pathway. The carriers generated along this pathway can fuse with the early endosome, or take a more direct route towards the PM [60,63,64]. Indeed, our data suggest that rNadA may use a more direct recycling pathway without involvement of the early endosomal compartment. This hypothesis would be consistent with the negligible colocalization with the lysosomal compartment [64].

rNadA, MHC-I and transferrin internalization dynamics in live cells

In order to validate that rNadA uses the MHC-I internalization/recycling route we investigated their co-internalization in live cells. Chang cells were incubated for 1 hour with fluorescent Alexa633-conjugated rNadA and fluorescent Alexa488-conjugated anti-MHC-I antibody. Cells were rapidly washed with PBS and imaged for 5 minutes by confocal microscope. The movies showed the presence of rapidly moving carriers, containing both rNadA and MHC-I (figure 6A and movie S1). A minor fraction of carriers

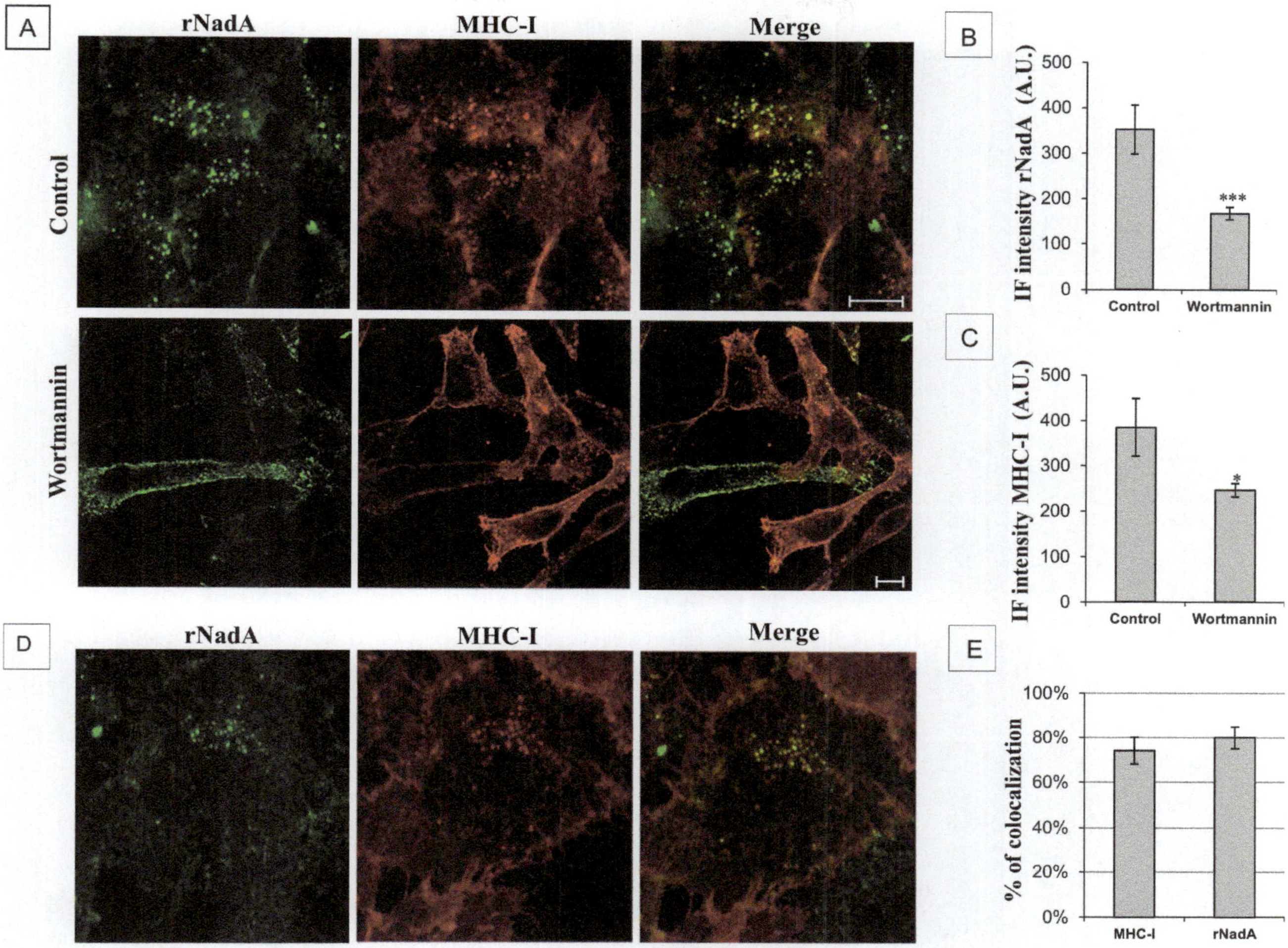

Figure 3. PI3K inhibition impairs rNadA internalization and colocalization of rNadA with MHC-I. ***A:*** Chang cells were treated with either vehicle (upper panel) or 100 nM Wortmannin (lower panel) for 1 h at 37°C, and then incubated with 200 µg/ml rNadA for an additional hour at 37°C in presence of the inhibitor or vehicle. Cells were then fixed, permeabilized and double stained for rNadA (green) and MHC-1 (red). Merged images are also shown. Scale bar 10 µm. ***B:*** Quantification of IF intensity of rNadA in the experiment illustrated in panel A. Data are mean ± s.e.m representative of two independent experiments, each assessing 20–25 cells, and expressed as Arbitrary Units (A.U.). ***$p<0.001$, compared to control cells (t-test). ***C:*** Quantification of IF intensity of MHC-I in the experiment illustrated in panel A. Data are mean ± s.e.m representative of two independent experiments, each assessing 20–25 cells, and expressed as Arbitrary Units (A.U.). *$p<0.05$, compared to control cells (t-test). ***D:*** Chang cells were treated with anti-MHC-I antibody for 1 h at 4°C and then incubated at 37°C for 1 h with 200 µg/ml rNadA. Cells were fixed and double stained for rNadA (green) and MHC-1 (red). Scale bar is 10 µm. ***E:*** Quantification of MHC-I and rNadA colocalization. MHC-I column indicate the percentage of MHC-I immunofluorescent pixels colocalizing with rNadA immunofluorescent pixels. Conversely, rNadA column indicate the percentage of rNadA immunofluorescent pixels colocalizing with MHC-I immunofluorescent pixels. Data are mean ± s.e.m representative of two independent experiments, each assessing 20–25 cells.

contained either rNadA or MHC-I. This could be due to the previously described dual targeting of surface MHC class I [65].

To cross-check that the rNadA internalization route is clathrin-independent, we tested the overlap between endocytosed rNadA and transferrin, a cargo internalized via the clathrin-dependent pathway. Chang cells were incubated for 1 hour with fluorescent Alexa488-conjugated rNadA and fluorescent Cy3-conjugated transferrin. Cells were rapidly washed with PBS and imaged as above. The movies showed the presence of rapidly moving rNadA or transferrin containing carriers indicating that the intracellular trafficking of rNadA is essentially different from that of transferrin (figure 6B and movie S2). As expected, only a minor fraction of carriers showed the presence of both rNadA and transferrin, perhaps as a result of a sorting/recycling compartment where the clathrin-dependent and independent pathways converge. Taken together, these observations confirm that the intracellular trafficking route of rNadA is clathrin independent and mostly overlaps with that of MHC-I.

Functional involvement of intracellular and external Hsp90 in internalization and trafficking of rNadA

In our recent report [30], we identified Hsp90 as a cellular factor interfering with rNadA-mediated bacterial infection. Although Hsp90 is an intracellular chaperone, its presence in the extracellular space has also been reported [50]. Here, we sought to verify the possible relationship between extracellular Hsp90 and rNadA internalization.

To this end, we selectively labeled the external Hsp90 in live cells by incubating them with an antibody against Hsp90 for 1 hr

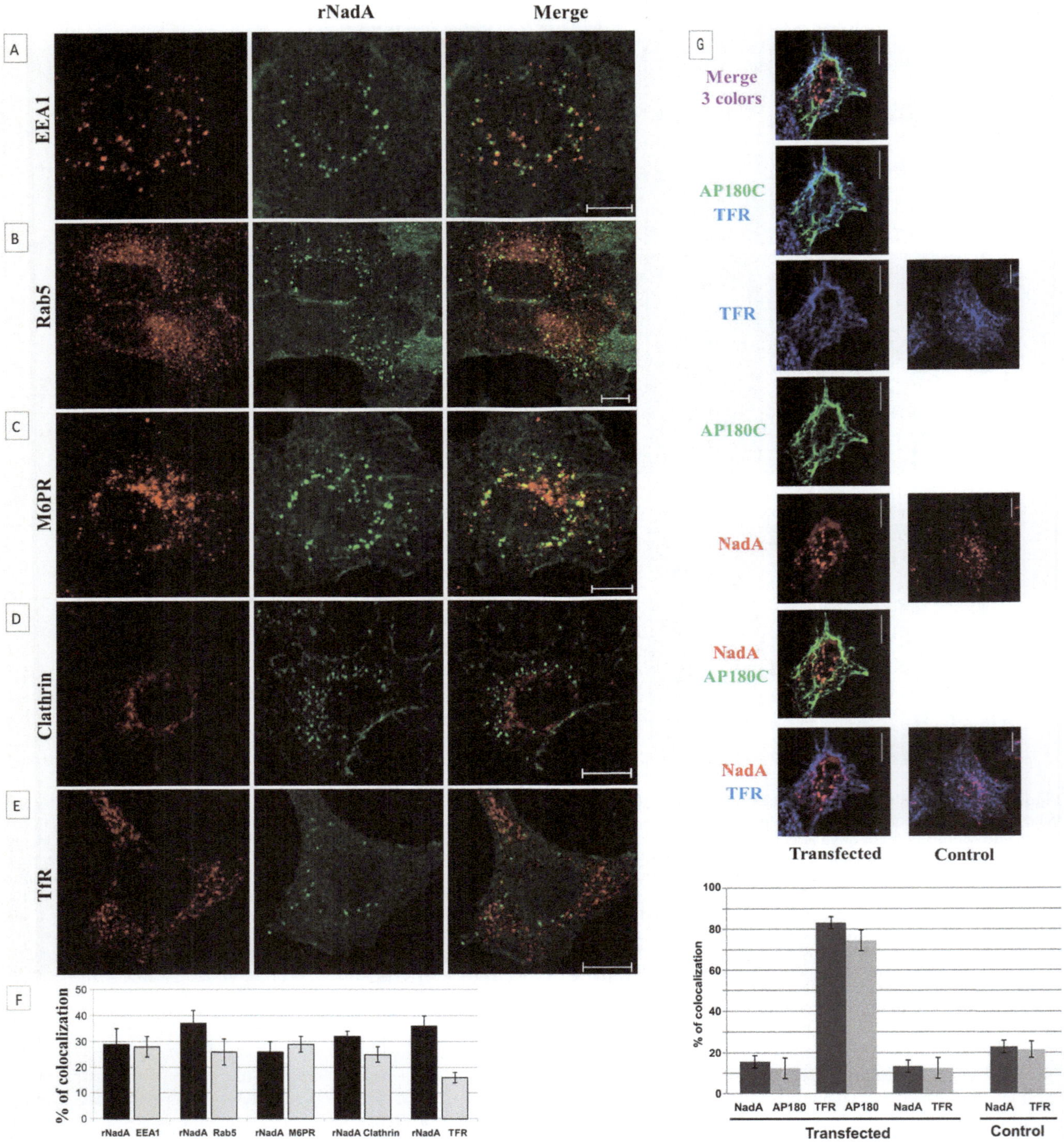

Figure 4. Colocalization of rNadA with endosome markers. *A:* Chang cells were incubated with 200 µg/ml rNadA for 1 h at 37°C, then fixed, permeabilized and double stained for rNadA (green) and EEA1 (*A*), Rab5 (*B*), M6PR (*C*), Clathrin (*D*) and TfR (*E*) (red). Merged images of the red and green signals are also shown. Images are representative of two independent experiments. Scale bar is 10 µm. *F:* Quantification of rNadA colocalization with endosome markers. The black columns indicate the percentage of rNadA immunofluorescent pixels colocalizing with the endosomal marker (EEA1, RAb5, M6PR, Clathrin, TfR as indicated) immunofluorescent pixels. Viceversa the grey columns indicate the percentage of the endosomal marker immunofluorescent pixels colocalizing with rNadA immunofluorescent pixels. Data are mean ± s.e.m representative of two independent experiments, each assessing 20–25 cells. *G:* Chang cells were transfected with an EGFR-AP180C expressing plasmid or with empty vector and incubated 24 hours at 37°C to recover. Cells were then incubated with rNadA, fixed and stained as indicated in point A. EGFR-AP190C is shown in green, rNadA in red and TfR in blu. Quantification was performed as indicated in point F.

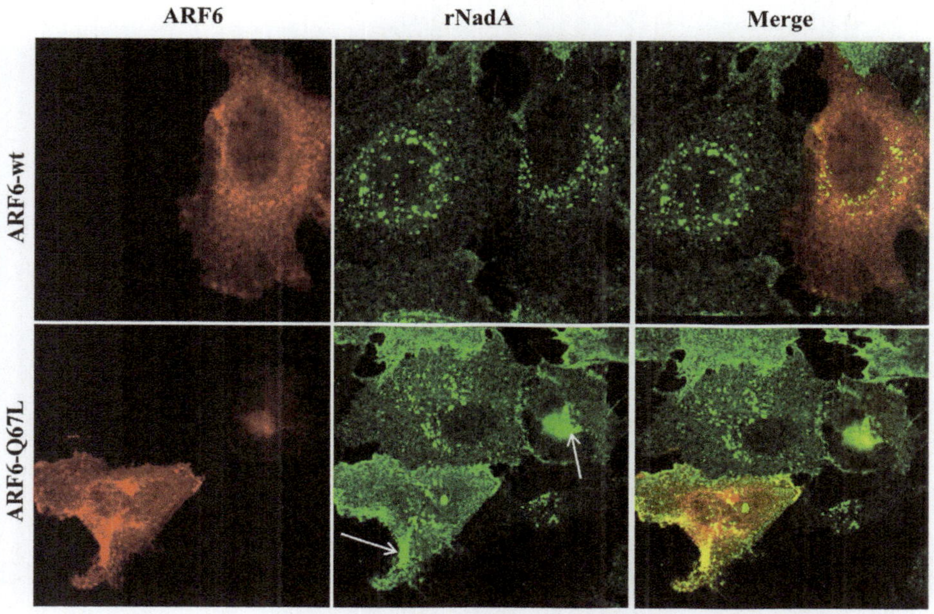

Figure 5. rNadA intracellular distribution is affected by ARF6. Chang cells were transfected in order to overexpress wild type ARF6 (ARF6-wt) or dominant-negative ARF6 (ARF6-Q67L) and then incubated overnight with 200 μg/ml rNadA at 37°C. Afterwards, cells were fixed, permeabilized and double stained for rNadA (green) and ARF6 (red). Merged images are also shown. Scale bar 10 μm. Images are representative of two independent experiments.

at 4°C. This approach allows the binding of the antibody to Hsp90 present on the external side of the PM, while the low temperature prevents internalization. Excess unbound anti-hsp90 antibodies were removed and the cells were treated with rNadA for 60 minutes at 37°C. Finally, the cells were fixed and rNadA was labeled with the appropriate primary and secondary antibodies. The anti-Hsp90 antibodies were revealed using fluorescence-conjugated secondary antibodies. As shown in figure 7A, externally labelled Hsp90

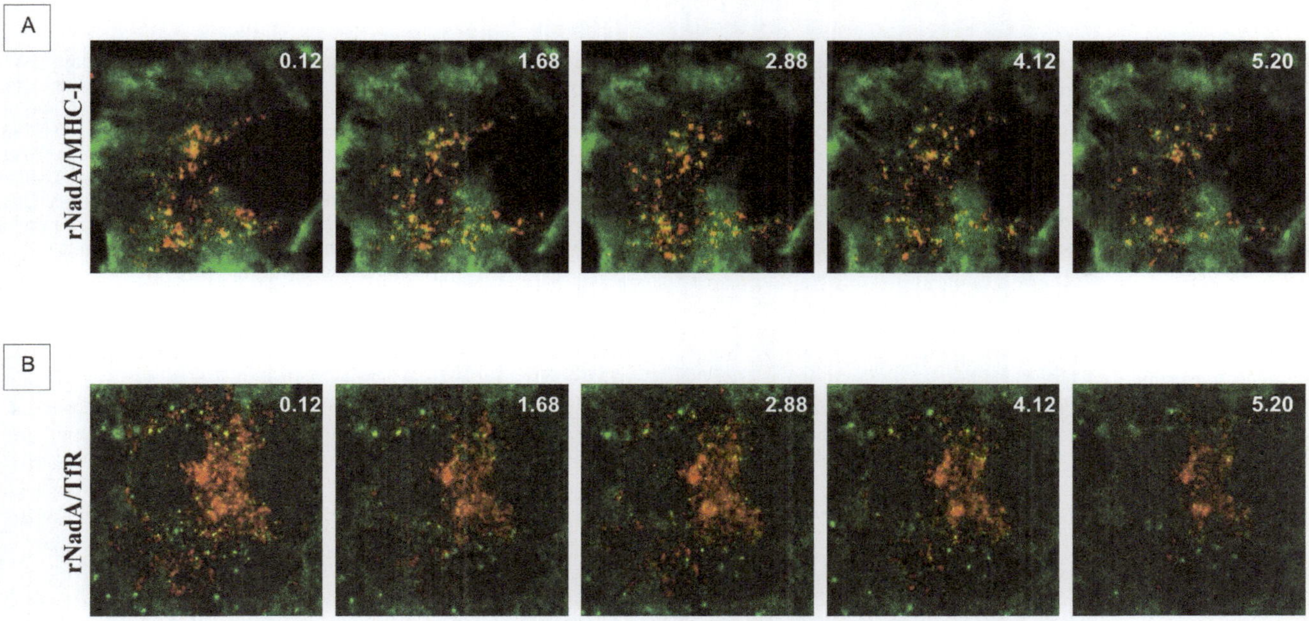

Figure 6. Time lapse rNadA internalization. ***A:*** Chang cells were incubated for 1 h with Alexa633-conjugated rNadA (shown in red) and Alexa488-conjugated anti-MHC-I antibody (green). Cells were then washed and live images were recorded every 2 seconds by confocal microscopy. Five frames from the Movie S1 are shown. Time (seconds) is indicated in the top right of each panel. ***B:*** Chang cells were incubated for 1 h with Alexa488-conjugated rNadA (green) and Cy3-conjugated transferrin (red). Cells were then washed and live images were recorded every 4 seconds by confocal microscope. Five frames from the Supplementary Movie S2 are shown. Time (seconds) is indicated in the top right of each panel.

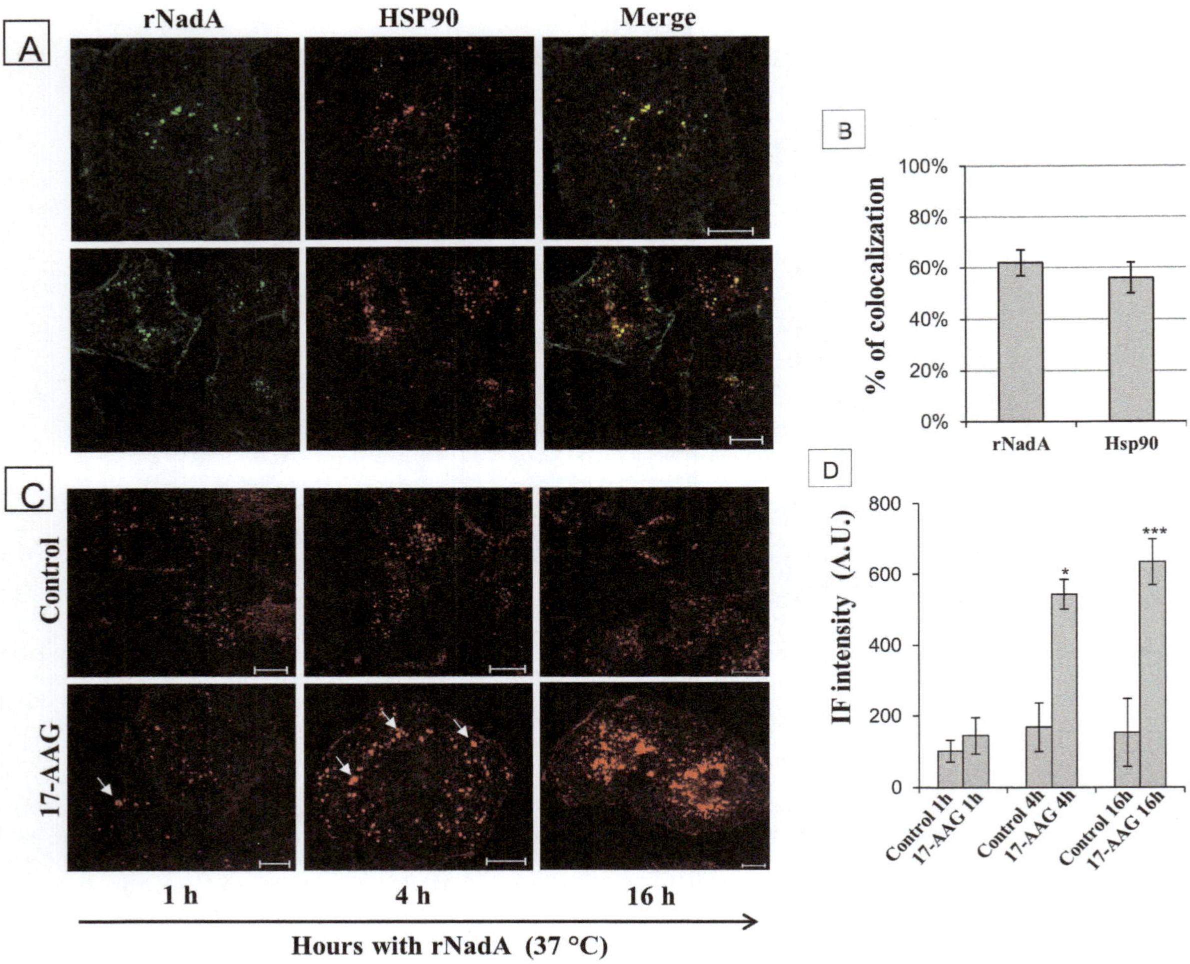

Figure 7. Colocalization of rNadA with HSP90. ***A:*** Chang cells, stained *in vivo* with a rabbit polyclonal anti-HSP90, were incubated with rNadA for 1 h at 37°C, then fixed, permeabilized and double stained for rNadA (green) and HSP90 (red). Panels are taken from two different experiments. Merged images are also shown. Scale bar 10 μm. ***B:*** Quantification of rNadA and Hsp90 colocalization. rNadA column indicate the percentage of rNadA immunofluorescent pixels colocalizing with HSP90 immunofluorescent pixels. Conversely, HSP90 columns indicate the percentage of HSP90 immunofluorescent pixels colocalizing with rNadA immunofluorescent pixels. Data are mean ± s.e.m representative of two independent experiments, each assessing 20–25 cells. ***C:*** Chang cells pre-treated overnight with either vehicle (upper panel) or 0.5 μM 17-AAG (lower panel), were incubated with 200 μg/ml rNadA for 1 h, 4 h and 16 hrs as indicated, then fixed, permeabilized and stained for rNadA (red). Intracellular accumulation of rNadA clusters are indicated by arrows. Scale bar 10 μm. ***D:*** Quantification of IF intensity of rNadA in the experiment illustrated in panel C. Data are mean ± s.e.m representative of two independent experiments, each assessing 20–25 cells, and expressed as Arbitrary Units (A.U.). ***p<0.001, compared to 1 h of 17-AAG treated cells (t-test). *p<0.05, compared to 1 h of 17-AAG treated cells (t-test).

accumulated intracellularly in compartments containing rNadA, strongly suggesting that Hsp90 present on the external side of the PM is internalized and colocalizes with the rNadA (figure 7B).

The functional role of Hsp90 in rNadA internalization was further investigated using the specific Hsp90 inhibitor 17-AAG. Chang cells pre-treated overnight with 0.5 μM of 17-AAG were then incubated with rNadA for 1, 4 and 16 hours. After 1 hour of rNadA incubation in 17-AAG-treated cells, clusters of rNadA appeared as dot-like structures as compared to control inhibitor-untreated cells (figure 7C, left panels and D). At longer times, clusters of rNadA-containing structures increased in size and number resulting in a remarkable accumulation of rNadA within the cells (figure 7C, central panels and D). At 16 hours the intracellular rNadA amount was roughly double in 17-AAG treated cells compared to control cells (figure 7C, right panels and D). Finally, we investigated the rNadA dynamics in 17-AAG treated cells. Chang cells were pre-treated overnight with 0.5 μM 17-AAG, incubated for 1 hour with fluorescent Alexa488-conjugated rNadA, rapidly washed with PBS and a movie recorded as above. In contrast to the rapidly moving rNadA containing carriers of the control cells, in 17-AAG treated cells, carriers containing rNadA appeared less mobile. In addition, their fusion to form larger structures or the emergence of small rNadA carriers from these aggregate appeared less frequent in 17-AAG than in controls (movie S3 and S4).

The functional role of external Hsp90 in rNadA internalization was further explored by means of a synthetic membrane-impermeable inhibitor, the FITC-GA, that specifically target the extracellular population of the chaperone. To verify that FITC-GA did not enter the cells we used the cellular AKT levels as a marker for the functionality of the intracellular chaperone [66]. Hsp90 inhibitors (e.g. 17-AAG) promote the proteasomal degradation of this intracellular kinase. Cells were incubated with medium containing 2.5 and 10 µM of either 17-AAG or FITC-GA for 1, 4, 24 and 48 hrs. At the end of each period, cell lysates were prepared and equal amount of proteins were analyzed by electrophoresis and Western blot to reveal the amount of intracellular AKT. As shown in figure 8A, the levels of AKT did not change until 4 hrs of treatment with the inhibitors (figure 8A, lanes 2–7). At 24 and 48 hrs the intracellular kinase was degraded in cells incubated in presence of 17-AAG at both concentrations (figure 8A, lanes 9–10, 13–14) while its amount was not modified in FITC-GA treated cells (figure 8A, lanes 11–12, 15–16). This result strongly suggested that the Hsp90 inhibitor FITC-GA do not exert any activity on the intracellular population of the chaperone.

To verify that the membrane-impermeable inhibitor was an efficient Hsp90 inhibitor, we transiently permeabilized the cells. Chang cells were treated with saponin for 30 seconds in presence of FITC-GA and, once cell integrity was restored, incubated for 1, 4, 24 and 48 hrs. The amount of intracellular AKT was then detected as previously described. As shown in figure 8B, intracellular FITC-GA was revealed to be an efficient Hsp90 inhibitor leading to AKT degradation within 24 hours (figure 8B, lanes 11–18).

Chang cells were preincubated with either 10 µM 17-AAG or 10 µM FITC-GA at 37°C for 1 hr before addition of rNadA, and were further incubated for 1 and 4 hrs in presence of the inhibitors. Cells were then fixed, stained and subjected to confocal microscopy analysis (figure 8C). Compared to the control (upper panel of figure 8C), both inhibitors led to a similar accumulation of intracellular rNadA.

Taken together, the use of these inhibitors confirmed that the rNadA intracellular trafficking is somehow dependent upon functionally active Hsp90 and emphasized the role of the extracellular pool of this chaperone.

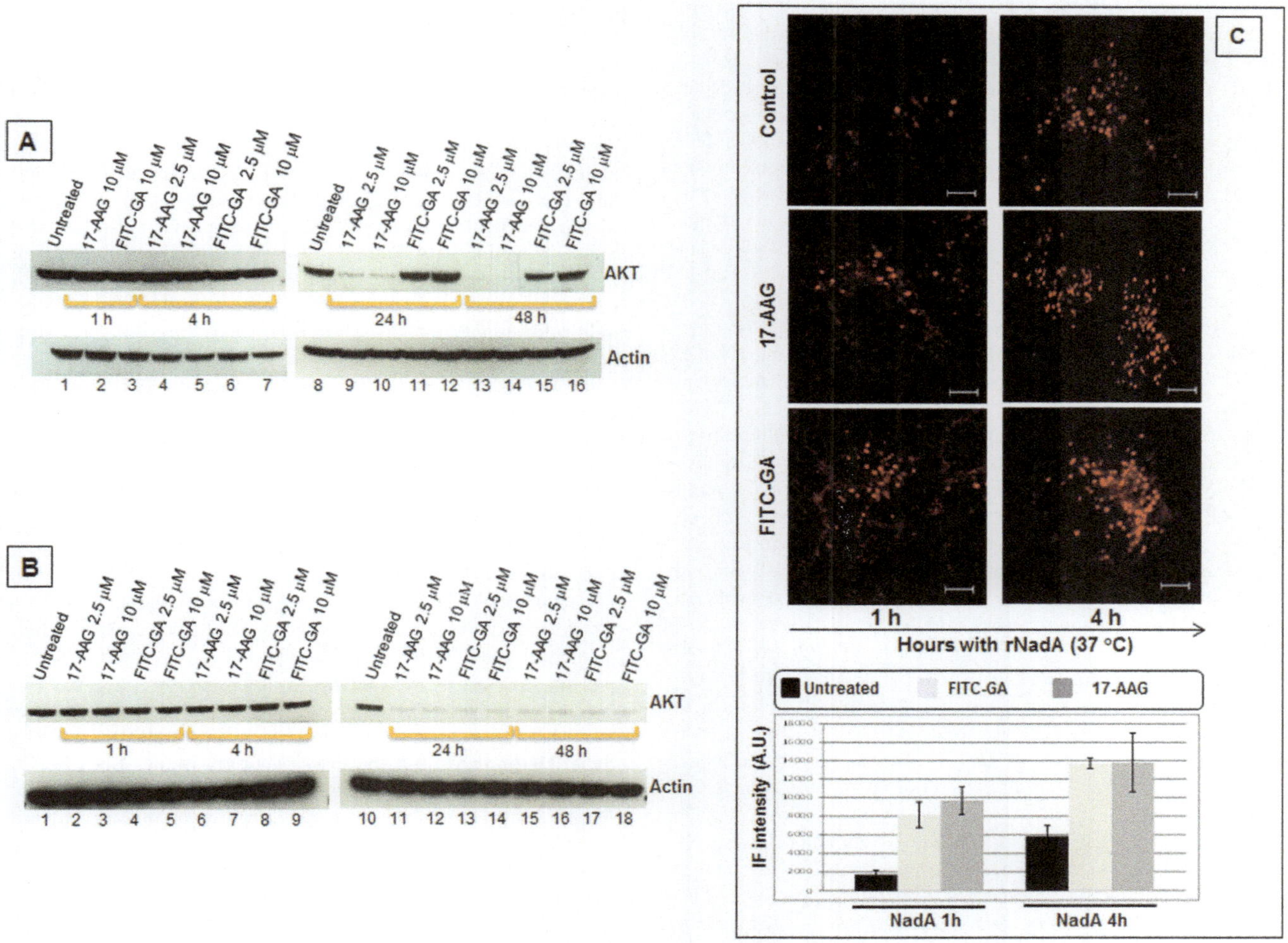

Figure 8. Inhibition of Hsp90 influences AKT degradation and intracellular accumulation of rNadA. *A:* Chang cells were incubated for 1, 4, 24 and 48 hours with either 17-AAG or FITC-GA at two different concentrations. At the end of each period cells were washed, trypsinized and RIPA-buffer total lysate prepared. Proteins from each extract were separated on NuSieve gel, blotted, and AKT or Actin (loading control) were detected using the respective primary antibodies. *B:* Chang cells were pre-treated with 0.01% saponin in PBS for 30 seconds at room temperature, washed three times with PBS and then handled as described in A. *C:* Chang cells were pre-treated for 1 hour with either vehicle (upper panels), 10 µM 17-AAG (middle panel) or 10 µM FITC-GA (lower panel), and then incubated with 200 µg/ml rNadA at 37°C for 1 (left panels) or 4 hours (right panels). Cells were then fixed, permeabilized and stained for rNadA. The drugs were present during the entire incubation period. IF intensity was calculated as mean ± s.e.m in two independent experiments, each assessing 10–15 cells and expressed as Arbitrary Units (A.U.). Scale bar 10 µm.

Colocalization of rNadA with Rab11 and effect of inhibitors

Our data strongly suggested that rNadA is reclycled back to the plasma membrane and possibly released back to the cell culture medium. Although several Rabs are known to regulate cellular membrane traffic [43,67], Rab11 has been particularly associated with recycling endosomes and vesicular exocytosis at the plasma membrane [68]. We hypothesized that rNadA was colocalizing with Rab 11. Chang cells were incubated with rNadA for 1 hr and then stained for rNadA and Rab11. Confocal microscopy revealed that vesicles containing rNadA were also associated with Rab11 (figure 9A, upper panel) strongly suggesting that the adhesin moved into a recycling pathway.

We also investigated the Rab11-rNadA colocalization in presence of the two Hsp90 inhibitors. Chang cells were pre-treated with 17-AAG, FITC-GA for 1 hr before adding rNadA and allow internalization for one more hr at 37°C in presence of the inhibitors. Confocal microscopy analysis revealed a different cellular distribution of Rab11 in the presence of either drug (figure 9A, middle and lower panels). In the absence of inhibitors, Rab11 was mostly localized in vesicular structures (figure 9A, upper panel). In presence of 17-AAG or FITC-GA the Rab11 cellular pattern was more distributed, with a tendency to perinuclear accumulation (figure 9, middle and lower panels). Interestingly, both 17-AAG (membrane permeable Hsp90 inhibitor) and FITC-GA (membrane-impermeable inhibitor) treatment resulted in the complete loss of Rab11-rNadA colocalization (figure 9, compare right panels) whereas intracellular accumulation of rNadA was observed. Treatment with the Hsp90 inhibitors did not interfere with the co-localization of TfR with Rab11 positive endosomes (figure S3).

Taken together, these observations indicated that Hsp90 activity was not essential for the formation of vesicles containing rNadA. Rather, it may affect the recruitment of effector molecules necessary for the formation of recycling endosomes.

Due to the high level of colocalization between internalized rNadA and Rab11, we reasoned that it was surprising not to have found rNadA-TfR colocalization in our previous experiment (figure 4E–F). Indeed, TfR recycling, has been reported to rely on Rab11 endosomes [43]. To decipher this apparent contradiction, we performed a simultaneous staining of TfR and Rab11 during rNadA internalization. Chang cells were transfected with an EGFP-Rab11 expressing plasmid and 24 hours later incubated with rNadA for 2 hours. Cells were then fixed and revealed for rNadA and TfR by using specific primary antibodies. Results from confocal microscopy analysis are shown in figure 9B in which, to better discriminate colocalization between pair of proteins, two colors at the time are also shown separately. Furthermore, percentages of colocalization are reported in the graphs of figure 9B.

Confocal microscopy revealed a significant colocalization of rNadA in Rab11 vesicles (figure 9B, upper panel and left graph) although the value was lower than previous experiment. This could be due to the different method used to reveal Rab11 in the experiments of figure 9A and 9B. Whereas physiological levels of Rab11 were detected in figure 9A by specific antibody, overexpression of a fluorescent tagged protein increase the amount of endogenous GTPase and may justify the lower percentage of association. Colocalization of TfR with Rab11 was significantly higher, consistent with data from the literature (figure 9A upper panel and middle graph), but rNadA and TfR did not show contemporary presence in Rab11 vesicles (figure 9B upper panel and right graph).

These data suggest that the rNadA and TfR enter distinct population of Rab11 endosomes. Although surprising, the existence of separate sub-population of Rab11 vesicles has already been observed in a different cell type [69].

Discussion

In this report we have shown that a purified bacterial adhesin binds human epithelial cells and triggers its own internalization. Once internalized, the intracellular trafficking of rNadA avoids lysosome targeting by entering an endosome recycling pathway and is likely transported back to the extracellular space. The extracellular pool of the molecular chaperone Hsp90, a cellular chaperone previously shown to bind rNadA *in vitro* [30], has been identified as a crucial factor for the intracellular trafficking of the meningococcal adhesin.

The early steps of rNadA endocytosis remain poorly understood and, in particular, we still have not defined the cellular ligand on the membrane. Binding of rNadA to Chang cells is strongly temperature dependent and it dramatically drops below 30°C. Internalization of the adhesin is a slow event (about 30 min) compared to general mechanisms of endocytosis. In the last decade, the endocytic process has shown a surprising diversity of parallel mechanisms in a single cell [38,70]. The rNadA entry process in human epithelial cells is clathrin-independent, insensitive to general tyrosine kinases or specific Src inhibitors, and thus excludes the involvement of a caveolin dependent entry pathway [38]. Treatment with the PI3K inhibitor wortmannin, in contrast, led to rNadA accumulation at the cell periphery, consistent with the role played by this kinase in the early steps of vesicle formation [71,72]. Endocytic vesicles containing rNadA enter an ARF6-regulated trafficking pathway and over-expression of a constitutively activated form of this protein (Q67L) led to the rNadA accumulation in vacuolar structures. Activated ARF6 is described to have a dual role in cells: it promotes internalization at the level of the plasma membrane and is a targeting factor for recycling of cargo to the cell surface [43,73]. Our data suggested that rNadA could avoid the lysosomes by steering into a recycling route. Thus far, a limited number of proteins have been described to use an ARF6-regulated pathway [43,63], the list including β1-integrin and MHC-I that are continuously trafficking from endocytic vesicles back to the plasma membrane. Consistent with our results, expression of ARF6 Q67L mutant has been shown to induce vacuolar structures and blocks recycling to the cell surface of both β1-integrin and MHC-I [74]. The hypothesis that rNadA enters a recycling pathway was further reinforced by its significant colocalization with Rab11 coated vesicles. The small GTPase Rab11 is a recognized marker of recycling endosomes, and a crucial factor regulating the exocytic fusion event [43,67,68,75]. Interestingly, stimulation-dependent recycling of β1-integrin follows an Arf-6 and Rab11 dependent pathway [76]. This integrin has been recently indicated as a putative ligand for NadA by using a heterologous system of invasion [77]. In our hands, the anti-β1 antibody failed to impair both binding and internalization of rNadA in Chang cells (GB and MM unpublished results) but, due to the different experimental system used, a role for this integrin in NadA-mediated binding and/or internalization cannot be ruled out. Intriguingly, we found that Rab11 positive endosomes containing rNadA are distinct from Rab11 positive endosomes carrying TfR, although formation of both populations is PI3K dependent. Although we have not further characterized the identity of these vesicles, the existence of diversified Rab11 endosomes has been recently reported in the rat adrenal phaeochromocytoma cell line PC12 [69].

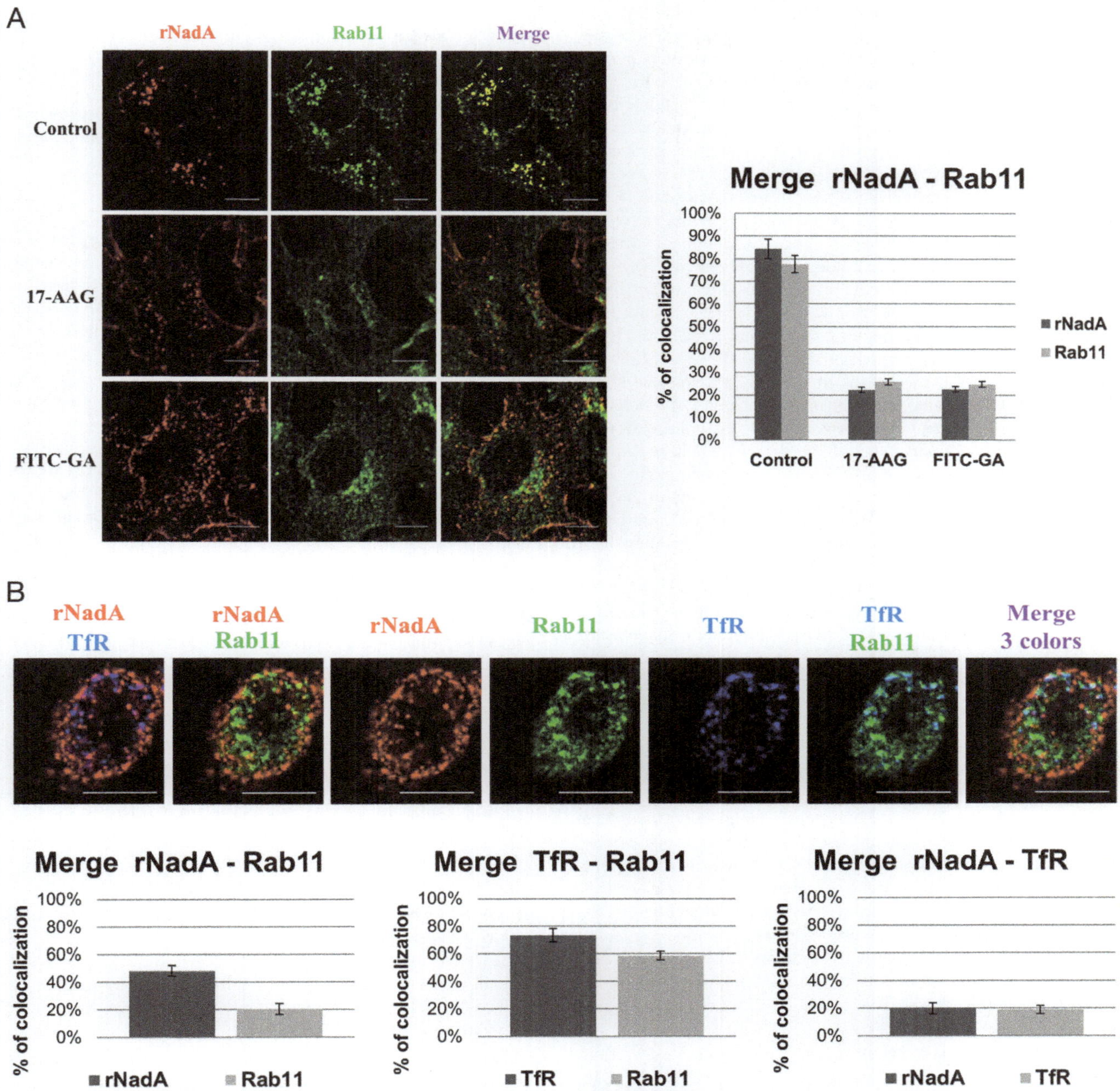

Figure 9. rNada internalization and colocalization with Rab11 and TfR. ***A: rNadA and Rab11 colocalization in presence of inhibitors.*** The control untreated Chang cells are shown in the upper panel. Pre-treatment of cells was performed for 1 hour with 10 μM 17-AAG (middle panel) or 10 μM FITC-GA (bottom panel). Cells were then incubated with 200 μg/ml rNadA at 37°C for 1 hour. Afterward, cells were fixed, permeabilized and stained for NadA. The drugs were present during the entire incubation period. *Graph*: rNadA columns indicate the percentage of rNadA immunofluorescent pixels colocalizing with Rab11 immunofluorescent pixels in untreated, 17-AAG and FITC-GA cells (from left to right) respectively. Conversely, Rab11 columns indicate the percentage of Rab11 immunofluorescent pixels colocalizing with rNadA immunofluorescent pixels in untreated, 17-AAG and FITC-GA cells (from left to right) respectively. ***B: Rab11, rNadA and TfR colocalization.*** *A:* Chang cells were transfected with EGFP-Rab11 plasmid, incubated 24 hours at 37°C. and then incubated with 200 μg/ml of rNadA at 37°C for 2 hours. Afterward, cells were fixed, permeabilized and stained for NadA (Red) and TfR (blu). Rab11 is colored in green. *Graphs*: rNadA columns indicate the percentage of rNadA immunofluorescent pixels colocalizing with Rab11 fluorescent pixels (left) or with TfR immunofluorescent pixels (right). Rab11column indicate the percentage of fluorescent pixels colocalizing with rNadA immunofluorescent pixels (left) or TfR immunofluorescent pixels (right). TfR columns indicate the percentage of TfR immunofluorescent pixels colocalizing with Rab11 fluorescent pixels (middle) or with rNadA immunofluorescent pixels (left). Data are mean ± s.e.m representative of two independent experiments, each assessing 20–25 cells.

Our attempts to identify a rNadA cell surface ligand led us to extracellular Hsp90 [30], a finding that was independently supported by another laboratory [53]. In the present paper, we have shown that external Hsp90 internalizes and colocalizes with the rNadA containing endocytic vesicles in live cells, strongly suggesting that the initial interaction between rNadA and Hsp90 takes place in the extracellular milieu. The extracellular presence of Hsp90 has been controversial for a long time, but it is now accepted that secretion of this chaperone occurs in both tumor and normal cells. Moreover, the functions of extracellular Hsp90 do not rely on the classical chaperone ATPase activity [49,51]. In our previous study we demonstrated a direct binding *in vitro* between rNadA and a complex of Hsp90 and ADP or 17-AAG. Meanwhile, in the presence of ATP, no such binding was observed [30]. Although the internalization of rNadA is not impaired by interfering with Hsp90, the ensuing intracellular trafficking is affected. Here we show that rNadA uptake in presence of the membrane-impermeable Hsp90 inhibitor FITC-GA led to intracellular accumulation of vesicles that contained the adhesin but were unable to engage Rab11. Interaction of rNadA with external Hsp90, thus, would be required to recruit a pattern of effector molecules for the targeting of the adhesin to the recycling endosomes. On the other hand, involvement of Hsp90 activity in intracellular trafficking has been poorly documented in literature. Its activity is required for activation of the Rho family of small G-protein that are known to regulate cytoskeleton rearrangement and endosomal trafficking [78,79]. Very recently, Cortese et al. reported that recycling of ErbB2 is perturbed by cell treatment with the Hsp90 inhibitor geldanamycin, leading to targeting of the receptor to mixed endosomal compartments [80]. To our knowledge, however, there are no reports on the influence of external Hsp90 on intracellular trafficking. The data presented in this work led us to hypothesize that the binding of rNadA to extracellular Hsp90 may influence a selection of the signaling involved in internalization, but further studies are required to elucidate this aspect.

The role of adhesins in meningococcal pathogenesis is thought to be confined to securing the bacteria to the host cell surface, whereas their contribution to bacterial trafficking is unaddressed [4]. We believe that the model exploited in this work, through the dissection of the pathways employed by an individual adhesin, has shed new light on the possible role that "minor" adhesion factors may have in the pathogenesis of bacterial invasion and, in the context of *N. meningitidis*, in transcellular passage through the epithelial barrier.

Supporting Information

Figure S1 Size Exclusion - HPLC coupled with a MALLS (Multi Angle Laser Light Scattering) instrument of rNadA heated at 37°C for different times. ***Panel A:*** SE-HPLC profile of rNadA at room temperature. ***Panel B, C,D, E:*** SE-HPLC profiles of rNadA after heating period at 37°C of 30 min, 1, 2 and 4 hrs respectively. ***Panel F:*** Comparison of the different SE-HPLC profiles of rNadA heated at 37°C for the indicated times. In the panel A, the diverse species eluted at different elution time can be unequivocally identified having been previously characterized [33]. Equilibrated at room temperature, rNadA preparation showed a predominant molecular weight corresponding to the native trimer (peak II) and three minor peaks corresponding to aggregates (peak I of figure 2A), monomer (peak III of figure 2A) and C-deleted monomer (peak IV of figure 2A). The NadA preparation, includes a certain number of C-deleted forms that are able to form trimers and are eluted in the peak II [33]. Incubation at 37°C did not provoke any difference in the SE-HPLC retention time and the MALLS measured MW of NadA peaks revealed at room temperature but it induced changes in their relative percentages.

Figure S2 Colocalization of Transferrin receptor with clathrin. Chang cells were fixed, permeabilized and double stained for clathrin (red) and Transferrin receptor (green). Merged image is also shown.

Figure S3 Colocalization of Transferrin receptor with Rab11 in presence of Hsp90 inhibitors. The control untreated Chang cells are shown in the upper panel. Pre-treatment of cells was performed for 1 hour with 10 µM 17-AAG (middle panel) or 10 µM FITC-GA (bottom panel). Chang cells were then fixed, permeabilized and double stained for Rab11 (green) and transferrin receptor (blu). Merged images are also shown. Graph report the percentage of colocalization obtained by 3 independent experiments.

Table S1 Relative percentages on rNadA species heated at 37°C for different times as revealed by Size Exclusion - HPLC coupled with a MALLS (Multi Angle Laser Light Scattering) in figure S1.

Movie S1 Dynamics of rNadA and MHC-I internalization in live cells. Chang cells were incubated for 1 h with Alexa488-conjugated anti-MHC-I antibody (green) and Alexa633-conjugated rNadA (red). Cells were then washed and imaged at the confocal microscope. Images were taken every 2 seconds.

Movie S2 Dynamics of rNadA and transferrin internalization in live cells. Chang cells grown overnight were incubated for 1 h with Alexa488-conjugated rNadA (green) and cy3-conjugated transferrin (red). Cells were then washed and imaged at the confocal microscope. Images were taken every 4 seconds.

Movie S3 Dynamics of rNadA in 17-AAG treated cells. Chang cells, pre-treated overnight with 0.5 µM 17-AAG (17-AAG), were incubated for 1 h with Alexa488-conjugated rNadA (green), in presence of the the drug. Cells were then washed and imaged at the confocal microscope. Images were taken every 4 seconds.

Movie S4 Dynamics of rNadA in untreated cells. Chang cells, pre-treated overnight with vehicle, were incubated for 1 h with Alexa488-conjugated rNadA (green). Cells were then washed and imaged at the confocal microscope. Images were taken every 4 seconds.

Acknowledgments

We thank Mirko Cortese (Novartis Vaccines and Diagnostics) for technical help and Elena Fontana for artwork preparation. We are particularly grateful to Robert Janulczyk for editing and critical reading of the manuscript.

Author Contributions

Conceived and designed the experiments: BA SS RR MP AL MS MM. Performed the experiments: GB MC PM BB MB VND EC RB. Analyzed the data: GB MC BB BP MS MM. Wrote the paper: GB MS MM.

References

1. Emonts M, Hazelzet JA, de Groot R, Hermans PW (2003) Host genetic determinants of Neisseria meningitidis infections. Lancet Infect Dis 3: 565–577.
2. Stephens DS (2009) Biology and pathogenesis of the evolutionarily successful, obligate human bacterium Neisseria meningitidis. Vaccine 27 Suppl 2: B71–77.
3. Hill DJ, Griffiths NJ, Borodina E, Virji M (2010) Cellular and molecular biology of Neisseria meningitidis colonization and invasive disease. Clin Sci (Lond) 118: 547–564.
4. Virji M (2009) Pathogenic neisseriae: surface modulation, pathogenesis and infection control. Nat Rev Microbiol 7: 274–286.
5. Carbonnelle E, Hill DJ, Morand P, Griffiths NJ, Bourdoulous S, et al. (2009) Meningococcal interactions with the host. Vaccine 27 Suppl 2: B78–89.
6. Stephens DS, Hoffman LH, McGee ZA (1983) Interaction of Neisseria meningitidis with human nasopharyngeal mucosa: attachment and entry into columnar epithelial cells. J Infect Dis 148: 369–376.
7. Birkness KA, Swisher BL, White EH, Long EG, Ewing EP Jr, et al. (1995) A tissue culture bilayer model to study the passage of Neisseria meningitidis. Infect Immun 63: 402–409.
8. Pujol C, Eugene E, de Saint Martin L, Nassif X (1997) Interaction of Neisseria meningitidis with a polarized monolayer of epithelial cells. Infect Immun 65: 4836–4842.
9. Sutherland TC, Quattroni P, Exley RM, Tang CM (2010) Transcellular passage of Neisseria meningitidis across a polarized respiratory epithelium. Infect Immun 78: 3832–3847.
10. Capecchi B, Adu-Bobie J, Di Marcello F, Ciucchi L, Masignani V, et al. (2005) Neisseria meningitidis NadA is a new invasin which promotes bacterial adhesion to and penetration into human epithelial cells. Mol Microbiol 55: 687–698.
11. Scarselli M, Serruto D, Montanari P, Capecchi B, Adu-Bobie J, et al. (2006) Neisseria meningitidis NhhA is a multifunctional trimeric autotransporter adhesin. Mol Microbiol 61: 631–644.
12. Pizza M, Scarlato V, Masignani V, Giuliani MM, Arico B, et al. (2000) Identification of vaccine candidates against serogroup B meningococcus by whole-genome sequencing. Science 287: 1816–1820.
13. Comanducci M, Bambini S, Brunelli B, Adu-Bobie J, Arico B, et al. (2002) NadA, a novel vaccine candidate of Neisseria meningitidis. J Exp Med 195: 1445–1454.
14. Comanducci M, Bambini S, Caugant DA, Mora M, Brunelli B, et al. (2004) NadA diversity and carriage in Neisseria meningitidis. Infect Immun 72: 4217–4223.
15. Cotter SE, Surana NK, St Geme JW 3rd (2005) Trimeric autotransporters: a distinct subfamily of autotransporter proteins. Trends Microbiol 13: 199–205.
16. Linke D, Riess T, Autenrieth IB, Lupas A, Kempf VA (2006) Trimeric autotransporter adhesins: variable structure, common function. Trends Microbiol 14: 264–270.
17. Surana NK, Cutter D, Barenkamp SJ, St Geme JW 3rd (2004) The Haemophilus influenzae Hia autotransporter contains an unusually short trimeric translocator domain. J Biol Chem 279: 14679–14685.
18. Lyskowski A, Leo JC, Goldman A (2011) Structure and biology of trimeric autotransporter adhesins. Adv Exp Med Biol 715: 143–158.
19. Dautin N, Bernstein HD (2007) Protein secretion in gram-negative bacteria via the autotransporter pathway. Annu Rev Microbiol 61: 89–112.
20. Roggenkamp A, Ackermann N, Jacobi CA, Truelzsch K, Hoffmann H, et al. (2003) Molecular analysis of transport and oligomerization of the Yersinia enterocolitica adhesin YadA. J Bacteriol 185: 3735–3744.
21. Yang Y, Isberg RR (1993) Cellular internalization in the absence of invasin expression is promoted by the Yersinia pseudotuberculosis yadA product. Infect Immun 61: 3907–3913.
22. McMichael JC, Fiske MJ, Fredenburg RA, Chakravarti DN, VanDerMeid KR, et al. (1998) Isolation and characterization of two proteins from Moraxella catarrhalis that bear a common epitope. Infect Immun 66: 4374–4381.
23. Laarmann S, Cutter D, Juehne T, Barenkamp SJ, St Geme JW (2002) The Haemophilus influenzae Hia autotransporter harbours two adhesive pockets that reside in the passenger domain and recognize the same host cell receptor. Mol Microbiol 46: 731–743.
24. Zhang P, Chomel BB, Schau MK, Goo JS, Droz S, et al. (2004) A family of variably expressed outer-membrane proteins (Vomp) mediates adhesion and autoaggregation in Bartonella quintana. Proc Natl Acad Sci U S A 101: 13630–13635.
25. O'Rourke F, Schmidgen T, Kaiser PO, Linke D, Kempf VA (2011) Adhesins of Bartonella spp. Adv Exp Med Biol 715: 51–70.
26. Serruto D, Spadafina T, Scarselli M, Bambini S, Comanducci M, et al. (2009) HadA is an atypical new multifunctional trimeric coiled-coil adhesin of Haemophilus influenzae biogroup aegyptius, which promotes entry into host cells. Cell Microbiol 11: 1044–1063.
27. El Tahir Y, Skurnik M (2001) YadA, the multifaceted Yersinia adhesin. Int J Med Microbiol 291: 209–218.
28. Hill DJ, Virji M (2003) A novel cell-binding mechanism of Moraxella catarrhalis ubiquitous surface protein UspA: specific targeting of the N-domain of carcinoembryonic antigen-related cell adhesion molecules by UspA1. Mol Microbiol 48: 117–129.
29. Riess T, Andersson SG, Lupas A, Schaller M, Schafer A, et al. (2004) Bartonella adhesin a mediates a proangiogenic host cell response. J Exp Med 200: 1267–1278.
30. Montanari P, Bozza G, Capecchi B, Caproni E, Barrile R, et al. (2012) Human heat shock protein (Hsp) 90 interferes with Neisseria meningitidis adhesin A (NadA)-mediated adhesion and invasion. Cell Microbiol 14: 368–385.
31. Bowe F, Lavelle EC, McNeela EA, Hale C, Clare S, et al. (2004) Mucosal vaccination against serogroup B meningococci: induction of bactericidal antibodies and cellular immunity following intranasal immunization with NadA of Neisseria meningitidis and mutants of Escherichia coli heat-labile enterotoxin. Infect Immun 72: 4052–4060.
32. Litt DJ, Savino S, Beddek A, Comanducci M, Sandiford C, et al. (2004) Putative vaccine antigens from Neisseria meningitidis recognized by serum antibodies of young children convalescing after meningococcal disease. J Infect Dis 190: 1488–1497.
33. Magagnoli C, Bardotti A, De Conciliis G, Galasso R, Tomei M, et al. (2009) Structural organization of NadADelta(351-405), a recombinant MenB vaccine component, by its physico-chemical characterization at drug substance level. Vaccine 27: 2156–2170.
34. Mazzon C, Baldani-Guerra B, Cecchini P, Kasic T, Viola A, et al. (2007) IFN-gamma and R-848 dependent activation of human monocyte-derived dendritic cells by Neisseria meningitidis adhesin A. J Immunol 179: 3904–3916.
35. Franzoso S, Mazzon C, Sztukowska M, Cecchini P, Kasic T, et al. (2008) Human monocytes/macrophages are a target of Neisseria meningitidis Adhesin A (NadA). J Leukoc Biol 83: 1100–1110.
36. McMahon HT, Boucrot E (2011) Molecular mechanism and physiological functions of clathrin-mediated endocytosis. Nat Rev Mol Cell Biol 12: 517–533.
37. Mayor S, Pagano RE (2007) Pathways of clathrin-independent endocytosis. Nat Rev Mol Cell Biol 8: 603–612.
38. Doherty GJ, McMahon HT (2009) Mechanisms of endocytosis. Annu Rev Biochem 78: 857–902.
39. Huotari J, Helenius A (2011) Endosome maturation. EMBO J 30: 3481–3500.
40. Saftig P, Klumperman J (2009) Lysosome biogenesis and lysosomal membrane proteins: trafficking meets function. Nat Rev Mol Cell Biol 10: 623–635.
41. Sandvig K, Pust S, Skotland T, van Deurs B (2011) Clathrin-independent endocytosis: mechanisms and function. Curr Opin Cell Biol 23: 413–420.
42. Donaldson JG, Porat-Shliom N, Cohen LA (2009) Clathrin-independent endocytosis: a unique platform for cell signaling and PM remodeling. Cell Signal 21: 1–6.
43. Grant BD, Donaldson JG (2009) Pathways and mechanisms of endocytic recycling. Nat Rev Mol Cell Biol 10: 597–608.
44. Radhakrishna H, Donaldson JG (1997) ADP-ribosylation factor 6 regulates a novel plasma membrane recycling pathway. J Cell Biol 139: 49–61.
45. D'Souza-Schorey C, Chavrier P (2006) ARF proteins: roles in membrane traffic and beyond. Nat Rev Mol Cell Biol 7: 347–358.
46. Jean S, Kiger AA (2012) Coordination between RAB GTPase and phosphoinositide regulation and functions. Nat Rev Mol Cell Biol 13: 463–470.
47. Hutagalung AH, Novick PJ (2011) Role of Rab GTPases in membrane traffic and cell physiology. Physiol Rev 91: 119–149.
48. Li J, Soroka J, Buchner J (2012) The Hsp90 chaperone machinery: Conformational dynamics and regulation by co-chaperones. Biochim Biophys Acta 1823: 624–635.
49. Li W, Sahu D, Tsen F (2012) Secreted heat shock protein-90 (Hsp90) in wound healing and cancer. Biochim Biophys Acta 1823: 730–741.
50. Tsutsumi S, Neckers L (2007) Extracellular heat shock protein 90: a role for a molecular chaperone in cell motility and cancer metastasis. Cancer Sci 98: 1536–1539.
51. Schmitt E, Gehrmann M, Brunet M, Multhoff G, Garrido C (2007) Intracellular and extracellular functions of heat shock proteins: repercussions in cancer therapy. J Leukoc Biol 81: 15–27.
52. Binder RJ, Vatner R, Srivastava P (2004) The heat-shock protein receptors: some answers and more questions. Tissue Antigens 64: 442–451.
53. Cecchini P, Tavano R, Polverino de Laureto P, Franzoso S, Mazzon C, et al. (2011) The soluble recombinant Neisseria meningitidis adhesin NadA(Delta351-405) stimulates human monocytes by binding to extracellular Hsp90. PLoS One 6: e25089.
54. Ford MG, Pearse BM, Higgins MK, Vallis Y, Owen DJ, et al. (2001) Simultaneous binding of PtdIns(4,5)P2 and clathrin by AP180 in the nucleation of clathrin lattices on membranes. Science 291: 1051–1055.
55. Rojas R, van Vlijmen T, Mardones GA, Prabhu Y, Rojas AL, et al. (2008) Regulation of retromer recruitment to endosomes by sequential action of Rab5 and Rab7. J Cell Biol 183: 513–526.
56. Delom F, Fessart D (2011) Role of Phosphorylation in the Control of Clathrin-Mediated Internalization of GPCR. Int J Cell Biol 2011: 246954.
57. Shajahan AN, Tiruppathi C, Smrcka AV, Malik AB, Minshall RD (2004) Gbetagamma activation of Src induces caveolae-mediated endocytosis in endothelial cells. J Biol Chem 279: 48055–48062.
58. Cha SK, Wu T, Huang CL (2008) Protein kinase C inhibits caveolae-mediated endocytosis of TRPV5. Am J Physiol Renal Physiol 294: F1212–1221.
59. Alvi F, Idkowiak-Baldys J, Baldys A, Raymond JR, Hannun YA (2007) Regulation of membrane trafficking and endocytosis by protein kinase C:

emerging role of the pericentrion, a novel protein kinase C-dependent subset of recycling endosomes. Cell Mol Life Sci 64: 263–270.
60. Naslavsky N, Weigert R, Donaldson JG (2003) Convergence of non-clathrin- and clathrin-derived endosomes involves Arf6 inactivation and changes in phosphoinositides. Mol Biol Cell 14: 417–431.
61. Sharma DK, Brown JC, Choudhury A, Peterson TE, Holicky E, et al. (2004) Selective stimulation of caveolar endocytosis by glycosphingolipids and cholesterol. Mol Biol Cell 15: 3114–3122.
62. Zhao X, Greener T, Al-Hasani H, Cushman SW, Eisenberg E, et al. (2001) Expression of auxilin or AP180 inhibits endocytosis by mislocalizing clathrin: evidence for formation of nascent pits containing AP1 or AP2 but not clathrin. J Cell Sci 114: 353–365.
63. Eyster CA, Higginson JD, Huebner R, Porat-Shliom N, Weigert R, et al. (2009) Discovery of new cargo proteins that enter cells through clathrin-independent endocytosis. Traffic 10: 590–599.
64. Eyster CA, Cole NB, Petersen S, Viswanathan K, Fruh K, et al. (2011) MARCH ubiquitin ligases alter the itinerary of clathrin-independent cargo from recycling to degradation. Mol Biol Cell 22: 3218–3230.
65. Zagorac GB, Mahmutefendic H, Tomas MI, Kucic N, Le Bouteiller P, et al. (2012) Early endosomal rerouting of major histocompatibility class I conformers. J Cell Physiol 227: 2953–2964.
66. Powers MV, Workman P (2006) Targeting of multiple signalling pathways by heat shock protein 90 molecular chaperone inhibitors. Endocr Relat Cancer 13 Suppl 1: S125–135.
67. Stenmark H (2009) Rab GTPases as coordinators of vesicle traffic. Nat Rev Mol Cell Biol 10: 513–525.
68. Takahashi S, Kubo K, Waguri S, Yabashi A, Shin HW, et al. (2012) Rab11 regulates exocytosis of recycling vesicles at the plasma membrane. J Cell Sci 125: 4049–4057.
69. Kobayashi H, Fukuda M (2013) Arf6, Rab11 and transferrin receptor define distinct populations of recycling endosomes. Commun Integr Biol 6: e25036.
70. Kumari S, Mg S, Mayor S (2010) Endocytosis unplugged: multiple ways to enter the cell. Cell Res 20: 256–275.
71. Amyere M, Payrastre B, Krause U, Van Der Smissen P, Veithen A, et al. (2000) Constitutive macropinocytosis in oncogene-transformed fibroblasts depends on sequential permanent activation of phosphoinositide 3-kinase and phospholipase C. Mol Biol Cell 11: 3453–3467.
72. Araki N, Egami Y, Watanabe Y, Hatae T (2007) Phosphoinositide metabolism during membrane ruffling and macropinosome formation in EGF-stimulated A431 cells. Exp Cell Res 313: 1496–1507.
73. Schweitzer JK, Sedgwick AE, D'Souza-Schorey C (2011) ARF6-mediated endocytic recycling impacts cell movement, cell division and lipid homeostasis. Semin Cell Dev Biol 22: 39–47.
74. Brown FD, Rozelle AL, Yin HL, Balla T, Donaldson JG (2001) Phosphatidylinositol 4,5-bisphosphate and Arf6-regulated membrane traffic. J Cell Biol 154: 1007–1017.
75. Ullrich O, Reinsch S, Urbe S, Zerial M, Parton RG (1996) Rab11 regulates recycling through the pericentriolar recycling endosome. J Cell Biol 135: 913–924.
76. Powelka AM, Sun J, Li J, Gao M, Shaw LM, et al. (2004) Stimulation-dependent recycling of integrin beta1 regulated by ARF6 and Rab11. Traffic 5: 20–36.
77. Nagele V, Heesemann J, Schielke S, Jimenez-Soto LF, Kurzai O, et al. (2011) Neisseria meningitidis adhesin NadA targets beta1 integrins: functional similarity to Yersinia invasin. J Biol Chem 286: 20536–20546.
78. Amiri A, Noei F, Feroz T, Lee JM (2007) Geldanamycin anisimycins activate Rho and stimulate Rho- and ROCK-dependent actin stress fiber formation. Mol Cancer Res 5: 933–942
79. Yang W, Jansen JM, Lin Q, Canova S, Cerione RA, et al. (2004) Interaction of activated Cdc42-associated tyrosine kinase ACK2 with HSP90. Biochem J 382: 199–204.
80. Cortese K, Howes MT, Lundmark R, Tagliatti E, Bagnato P, et al. (2013) The HSP90 inhibitor geldanamycin perturbs endosomal structure and drives recycling ErbB2 and transferrin to modified MVBs/lysosomal compartments. Mol Biol Cell 24: 129–144.

Inhibitory Effect of Essential Oils on *Aspergillus ochraceus* Growth and Ochratoxin A Production

Huijuan Hua[¶], Fuguo Xing[¶], Jonathan Nimal Selvaraj, Yan Wang, Yueju Zhao, Lu Zhou, Xiao Liu, Yang Liu*

Institute of Agro-Products Processing Science and Technology, Chinese Academy of Agricultural Sciences/Key Laboratory of Agro-Products Processing, Ministry of Agriculture, Beijing, P. R. China

Abstract

Ochratoxin A (OTA) is a mycotoxin which is a common contaminant in grains during storage. *Aspergillus ochraceus* is the most common producer of OTA. Essential oils play a crucial role as a biocontrol in the reduction of fungal contamination. Essential oils namely natural cinnamaldehyde, cinnamon oil, synthetic cinnamaldehyde, *Litsea citrate* oil, citral, eugenol, peppermint, eucalyptus, anise and camphor oils, were tested for their efficacy against *A. ochraceus* growth and OTA production by fumigation and contact assays. Natural cinnamaldehyde proved to be the most effective against *A. ochraceus* when compared to other oils. Complete fungal growth inhibition was obtained at 150–250 μL/L with fumigation and 250–500 μL/L with contact assays for cinnamon oil, natural and synthetic cinnamaldehyde, *L. citrate* oil and citral. Essential oils had an impact on the ergosterol biosynthesis and OTA production. Complete inhibition of ergosterol biosynthesis was observed at ≥100 μg/mL of natural cinnamaldehyde and at 200 μg/mL of citral, but total inhibition was not observed at 200 μg/mL of eugenol. But, citral and eugenol could inhibit the OTA production at ≥75 μg/mL and ≥150 μg/mL respectively, while natural cinnamaldehyde couldn't fully inhibit OTA production at ≤200 μg/mL. The inhibition of OTA by natural cinnamaldehyde is mainly due to the reduction in fungal biomass. However, citral and eugenol could significant inhibit the OTA biosynthetic pathway. Also, we observed that cinnamaldehyde was converted to cinnamic alcohol by *A. ochraceus*, suggesting that the antimicrobial activity of cinnamaldehyde was mainly attributed to its carbonyl aldehyde group. The study concludes that natural cinnamaldehyde, citral and eugenol could be potential biocontrol agents against OTA contamination in storage grains.

Editor: Richard A. Wilson, University of Nebraska-Lincoln, United States of America

Funding: This work was supported by grants from National Basic Research Program of China (973 program) (2013CB127805), from the National Natural Science Foundation of China (31000776), from the Ministry of Agriculture of China (Special Fund for Agro-Scientific Research in the Public Interest, 201203037). The funders had no role in study design, data collection and analysis, decision to publish, or preparation of the manuscript.

Competing Interests: The authors have declared that no competing interests exist.

* Email: liuyang01@caas.cn

¶ These authors contributed equally to this work.

Introduction

Aspergillus species are frequent contaminants of low-moisture foods (water activity <0.75). Increase in metabolic activity of *Aspergillus* species causes the food to spoil leading to enormous economic loss. Besides, some *Aspergillus* species can produce toxic secondary metabolites: like aflatoxins, ochratoxins and they affect the food safety. Ochratoxins are nephrotoxic, potentially carcinogenic mycotoxins in nature [1–3]. Ochratoxin is common contaminant in food products such as cereals, coffee, wine, beer, and spices. At least 20 different ochratoxin analogs have been detected in the recent years. Ochratoxin A (OTA) is the commonly produced analogue and highly toxic [4]. OTA is mainly produced by *Aspergillus ochraceus*, *Aspergillus carbonarius*, *Aspergillus westerdijkiae*, *Aspergillus niger*, *Penicillium nordicum* and *Penicillium verrucosum* [5]. *A. ochraceus* is responsible for OTA contamination in rice, wheat, oats, coffee, and beverage. *A. carbonarius* and *A. niger* are responsible for OTA in grapes, raisins, and wine [6,7].

As the most toxic and prevalent toxin of the ochratoxins group, OTA is highly neurotoxic, teratogenic, embryotoxic and genotoxic, immunosuppressive and immunotoxic in nature [8,9]. OTA was rated as potential carcinogen (group 2B) by the International Agency for Research on Cancer (IARC, 1993). European Union proposed a maximum tolerable level of 5 μg/kg OTA in cereal grains (European Commission, 2006), and it is similar to China.

Several strategies have been applied to prevent and control the growth of OTA producing fungi in grains. Chemical based control remains the common measure to reduce the incidence of post-harvest contamination in various foods. Antifungal chemicals like benzimidazoles, aromatic hydrocarbons, and sterol biosynthesis inhibitors are often used. However, the application of them increases the risk of toxic residues in foods [10–14]. Furthermore, the use of chemical fungicides often leads to fungal resistance [15]. Therefore, a huge effort is made in the recent years to limit the use of fungicides in grains and foods. Natural antimicrobial or antifungal substances are promising to replace these synthetic fungicides.

Essential oils (EOs) from plants have a broad spectrum of antifungal activity [16–20]. Bluma & Etcheverry (2008) [21] screened 41 aqueous and ethanolic extracts and 22 different EOs for their antifungal effects against *Aspergillus* section *Flavi* strains and concluded that boldo, poleo, clove, anise and thyme oils could be a potential antifungal agent. In their study, EOs screened for antifungal effect was by adding them to media followed by diffusion. Boldo, poleo and clove oils could inhibit the growth of *A. niger* and *A. carbonarius* by affecting the OTA biosynthesis pathway [22]. Smaller compounds such as monoterpenes were found to highly effective when used as headspace volatiles [23]. Vapor exposure to these oils could be an effective, non-toxic biopreservatives against OTA contamination in stored grains and foods.

The objectives of our study were to: (a) examine the efficacy of different EOs like cinnamon oil, natural cinnamaldehyde, synthetic cinnamaldehyde, *L. citrate* oil, citral, eugenol, eucalyptus, anise, peppermint and camphor oils against *A. ochraceus* and OTA accumulation on malt extract agar (MEA) and in yeast extract sucrose (YES) broth; (b) Using contact and headspace volatile assays to analyze their antifungal effect; and (c) cinnamaldehyde transformations in YES broth with *A. ochraceus*.

Materials and Methods

Essential oils

Ten EOs were used in the present study: cinnamon oil (85% cinnamaldehyde), natural cinnamaldehyde (95%), synthetic cinnamaldehyde (99%), *Litsea citrate* oil (85% citral), citral (96%), eugenol (99% eugenol), eucalyptus (80% cineole), anise (92% anethole), peppermint (50% menthol), and camphor (55% borneol) oils were purchased from Jiangxi Xue Song Natural Medicinal Oil Co., Ltd, P. R. China.

Fungal strain and culture conditions

A. ochraceus 3.4412 producing OTA was gifted by Prof. Kunlun Huang (College of Food Science and Nutritional Engineering, China Agricultural University, Beijing, P. R. China) and maintained in potato dextrose agar (PDA) for 2 weeks and stored at 4°C. The strain has previously shown to produce 1648.3 ng/g and 1536.1 ng/mL of OTA in MEA plates and YES broth, respectively.

Effect of the essential oils on *A. ochraceus* growth with fumigation

Evaluation of the antifungal activity of the EOs by fumigation was performed according to Soliman and Badeaa (2002) [24] with minor modifications. Briefly, 50 μL of 10^5 conidia/mL suspended in 0.01% Tween 20 was plated on 20 mL of MEA in a plastic Petri plate. EOs at different concentrations (25–1500 μL/L) was added to 10 mm sterile filter disks placed on cover of the Petri plate. The Petri plate was sealed by using Parafilm and incubated for 20 days at 28±1°C (five replicates for each treatment). The colonies were measured every 2 days in two directions at right angles to each other to obtain the mean diameter. Blanks were used in triplicates.

After the incubation period, the plates showing no growth were incubated for a further 7 days under the similar temperature conditions followed by replacing new filter disk without EOs. If *A. ochraceus* growth is observed, there is a fungistatic effect, whereas if no growth occurs, the effect is fungicidal.

Effect of the essential oils on *A. ochraceus* growth by contact assay

The effect of EOs on *A. ochraceus* growth on MEA was studied according to Passone et al. (2012) [23] with minor modifications. MEA plates containing 20–1500 μL/L of EO was prepared by adding EO to 20 mL autoclaved medium. 50 μL of 10^5 conidia/mL suspended in 0.01% Tween 20 was spotted on the center of each plate. Plates were sealed using Parafilm and incubated for 20 days at 28±1°C with five replicates for each treatment. Two measurements were made at right angles to each other for each colony every 2 days to obtain mean diameter. Cultures from plates showing no growth were transferred to plates containing medium without EO to determine fungistatic and fungicidal effects. Blanks were used in triplicates.

Determination of the minimum inhibitory concentration by contact assay

The minimum inhibitory concentration (MIC) of EOs was determined by the broth dilution method according to Yamamoto-Ribeiro et al. (2013) [20] with minor modifications. The EOs were diluted in 0.001% Tween-80 and tested at final concentrations of 40−40,000 μL/L. For each essential oil, 500 μL of 4×10^5 spores/mL of *A. ochraceus* conidia was added in RPMI-1640 medium (0.5 mL). Tubes were incubated at 35°C for 72 h with agitation. The lowest concentration showing no visible growth was designated as MIC. Cells from the tubes showing no growth were subcultured on PDA plates to determine if the inhibition was reversible or permanent. Positive controls were performed in medium with only the cell suspension.

Scanning electron microscopy (SEM)

SEM was performed according to a modified method from Xing et al. (2014) [25]. *A. ochraceus* was treated with natural cinnamaldehyde (100 μL/L), citral (100 μL/L) and eugenol (250 μL/L) on MEA medium with fumigation, respectively, Then the treated *A. ochraceus* mycelia were collected, and the fungi were mixed with formaldehyde, washed with PBS buffer, and dehydrated with gradient ethanol solutions (30, 60, 80, 90 and 100%, keeping the mycelia for a longer duration in 100%). The samples were subjected to critical point drying, subsequently, the fungal mycelia were prepared and observed under Hitachi S-750 SEM.

Effect of essential oils on ergosterol biosynthesis and ochratoxin A production

The effect of EOs on ergosterol biosynthesis and OTA production were determined according to a modified method from Kocić-Tanackov et al. (2012) [26]. Fifty milliliters of sterile YES broth (yeast extract 20 g, sucrose 150 g, $MgSO_4 \cdot 7H_2O$ 0.5 g, distilled water 885 mL) were poured in flask (150 mL). Natural cinnamaldehyde, citral and eugenol were aseptically added to the YES broth to obtain the following concentrations: 0, 25, 50, 75, 100, 150, or 200 μg/mL. Three 5 mm diameter discs of *A. ochraceus* mycelia from PDA medium were inoculated to the YES broth., and incubated at 28°C for 7 days under agitation in a ZWY-2102C incubator (Zhicheng, Shanghai, China). After the incubation, YES broth containing *A. ochraceus* mycelia was filtered through Whatman filter No. 1 and washed with distilled water. Mycelia were used for ergosterol determination and the filters for OTA determination. Five replicates were performed for the experimental and control group.

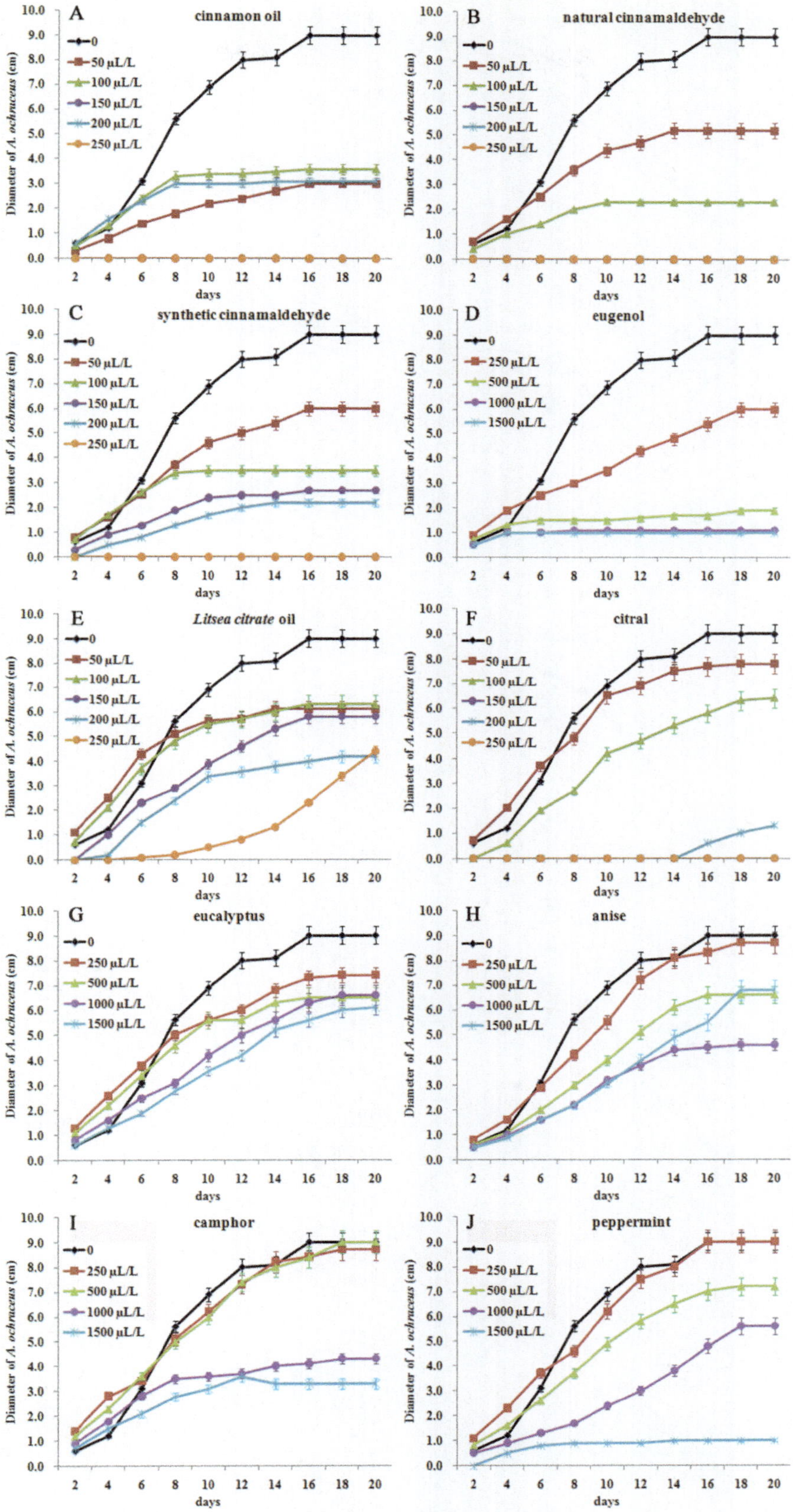

Figure 1. The effect of ten essential oils on the growth of *A. ochraceus* with fumigation.

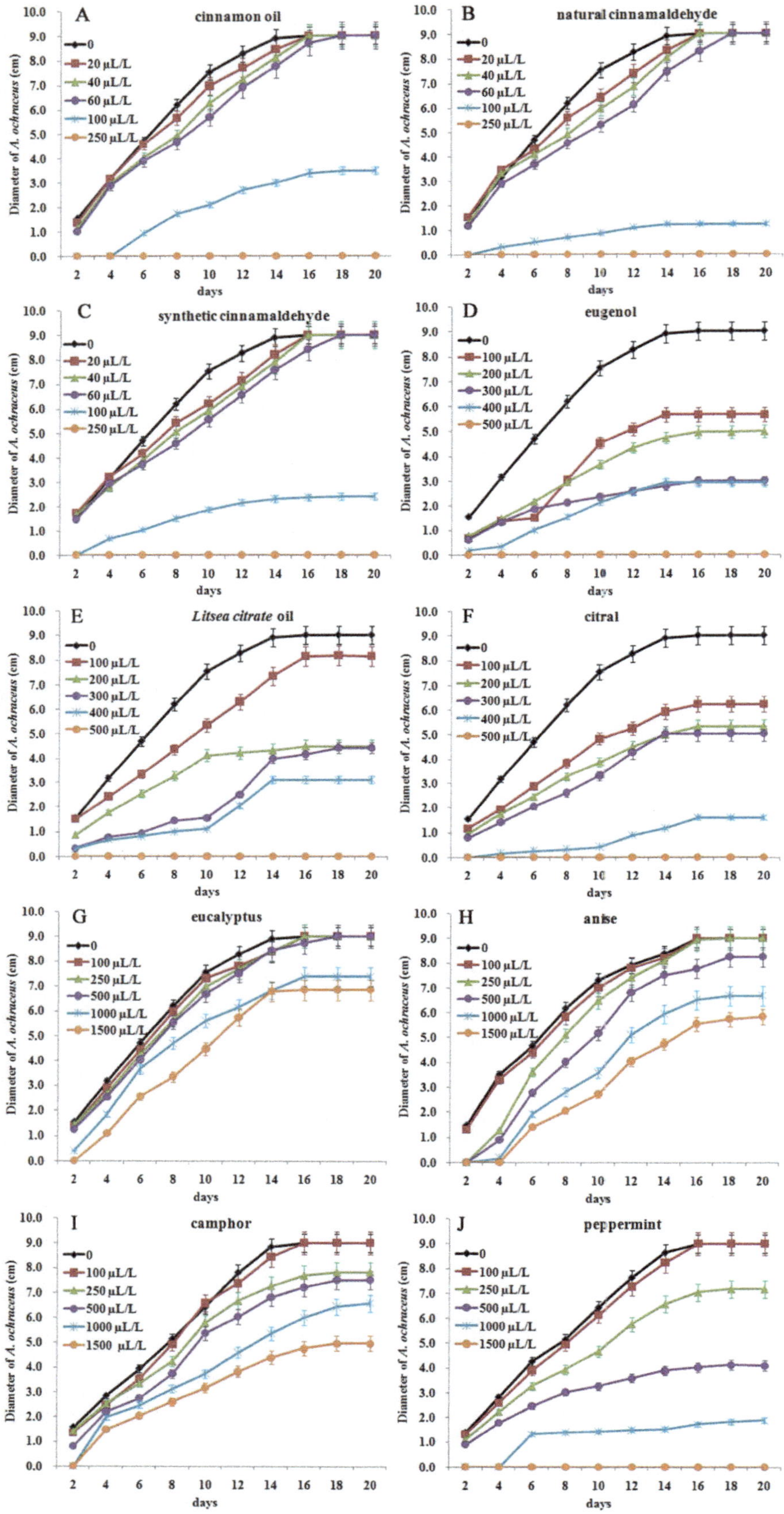

Figure 2. The effect of ten essential oils on the growth of *A. ochraceus* in contact assay.

Ergosterol and ochratoxin A standards

Ergosterol (Sigma Chemical, St. Louis, MO, USA) solutions (20 μg/mL) were prepared in absolute ethanol PA (Merck, Germany) and 100 μg/mL of OTA (Sigma Chemical, St. Louis, MO, USA) was prepared in acetonitrile:water (50:50, v/v) and stored in amber vials at −18°C.

Ergosterol extraction and determination by high-performance liquid chromatography

Ergosterol was extracted according to Silva, Corso, and Matheus (2010) [27] and transferred to a Falcon tube containing 20 mL methanol, 5 mL absolute ethanol PA, and 2 g potassium hydroxide. The solution was stirred for 5 min and placed in a water bath at 70°C for 40 min. After cooling at room temperature, 5 mL of distilled water was added and the solution was centrifuged at 1735 ×*g* for 12 min. For final extraction, *n*-hexane was added in an equal volume to the supernatant and the organic fraction was collected in an amber glass vial and evaporated with 99.99% nitrogen flow at 60°C. The residue obtained was stored at −18°C.

Ergosterol quantification was according to Yamamoto-Ribeiro et al. (2013) [20] with minor modifications. High performance liquid chromatography (HPLC) was performed with a Water 2695 (Waters Corporation, Milford, MA, USA) with an ultraviolet/visible spectrum Waters 2475 detection system at wavelength of 282 nm. Extract was resuspended in 1 mL of absolute ethanol and 100 μL was injected. For the mobile phase, methanol was used at a flow rate of 1.5 mL/min. Reverse-phase HPLC separation was on a Agilent TC-C18 (Agilent, Santa Clara, USA) with a 5 μm particle size column (250×4.6 mm). Retention time was 4.6 min and detection limit was 0.15 μg/mL.

Ochratoxin A extraction and determination by high-performance liquid chromatography

OTA was extracted according to Passone et al. (2012) [23] with minor modifications. Cell-free extracts (10 mL mixed with 10 mL methanol) were stirred in shaker tubes for 1 min, centrifuged at 14,000 rpm for 10 min and filtered (Titan 2 HPLC Filter Green; 17 mm and 0.45 μm). The eluate was evaporated to dryness with 99.99% nitrogen flow at 60°C. The residue was redissolved in mobile phase (acetonitrile/water/acetic acid, 99:99:2) and the extract was injected into a tube or stored at −18°C.

OTA production was detected and quantified by the methodology proposed by Scudamore and MacDonld (1998) [28] with minor modifications. Reversed phase HPLC with a Waters 2475 fluorescence detection (λexc 330 nm; λem 460 nm) was applied. Agilent TC-C18 column (250×4.6 mm, 5 μm particle size) was used. The mobile phase (acetonitrile/water/acetic acid, 99:99:2) was pumped at 1.0 mL/min. Injection volume was 50 μL and retention time was around 6 min. Standard curves were constructed with 5 to 100 ng/g OTA (Sigma Chemical, St. Louis, MO, USA). Toxin was quantified by correlating peak areas of sample extracts and standard curves. Mean recovery was calculated by spiking with MEA at 5 to 100 ng/g and was estimated at 89.2±9.7%. The lowest detection limit was 1 ng/g.

Cinnamaldehyde determination by high-performance liquid chromatography

The variations in concentrations of cinnamaldehyde and its possible metabolite cinnamic alcohol in YES broth with *A. ochraceus* (treatment) were determined by RP-HPLC analysis. YES broth containing 100 μg/mL of cinnamaldehyde was inoculated with *A. ochraceus* and incubated at 28°C with shaking at 180 rpm. YES broth without fungi inoculation by with cinnamaldehyde was the negative control and YES broth with fungi inoculation by without cinnamaldehyde was the positive control. Samples of 10 mL were collected at 0, 2, 4, 8, 12, 24, 48, 72, and 96 h and centrifuged at 10,000 ×*g* for 10 min at 4°C. Cell pellets were resuspended in methanol (Fisher) and sonicated for 15 min in a Bransonic bath (5510; Branson, Grass Valley, CA). The suspension was filtered through 0.2 μm nylon syringe filter units (Fisher) and analyzed for HPLC. Supernatant were mixed with 10 mL ethyl acetate (Sigma-Aldrich) to avoid interference by materials of YES broth in HPLC detection. The mixture was mixed on vortex for 3 min. After a 3 min rest, the clear ethyl acetate phase at the top was collected and the procedure was repeated twice. Ethyl acetate extracts from negative controls were not centrifuged. Pooled ethyl acetate extracts from each 10 mL sample were evaporated using a rotary evaporator (Yarong, SY2000; Shanghai, China), and the residue was dissolved in

Table 1. MICs of the essential oils against *A. ochraceus*[a].

Essential oils	MIC (μL/L)	Determine if the inhibition was reversible or permanent
Cinnamon oil (85% cinnamaldehyde)	500	Permanent (Without visible growth occurred)
Natural cinnamaldehyde (95%)	500	Permanent
Synthetic cinnamaldehyde (99%)	500	Permanent
Cinnamic alcohol (99%)	700	Permanent
Litsea citrate oil (85% citral)	800	permanent
Citral (99%)	700	Permanent
Eugenol (99%)	600	600–4000, reversible (visible growth occurred) ≥4000, permanent
Peppermint (50% methol)	2000	2000–40000, reversible ≥40000, permanent
Eucalyptus (80% cineole)	>40000	>40000, permanent
Anise (92% anethole)	>40000	Reversible
Camphor (55% bomeol)	>40000	Reversible

[a]Results are shown as minimum inhibitory concentrations (expressed as μL of essential oils added per L of liquid medium);
Reversible: visible growth occurred after subculturing on potato dextrose agar plates;
Permanent: without visible growth occurred after subculturing on potato dextrose agar plates.

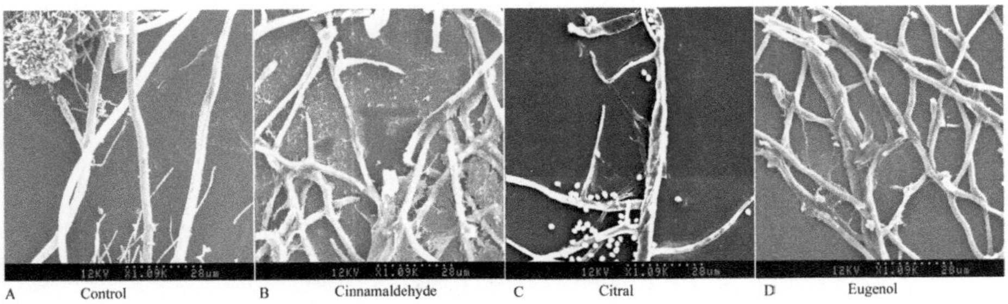

Figure 3. SEM for the nonfumigated and fumigated mycelia. A. Fumigated with 100 μL/L of natural cinnamaldehyde; B. Fumigated with 100 μL/L of citral; C. Fumigated with 250 μL/L of eugenol.

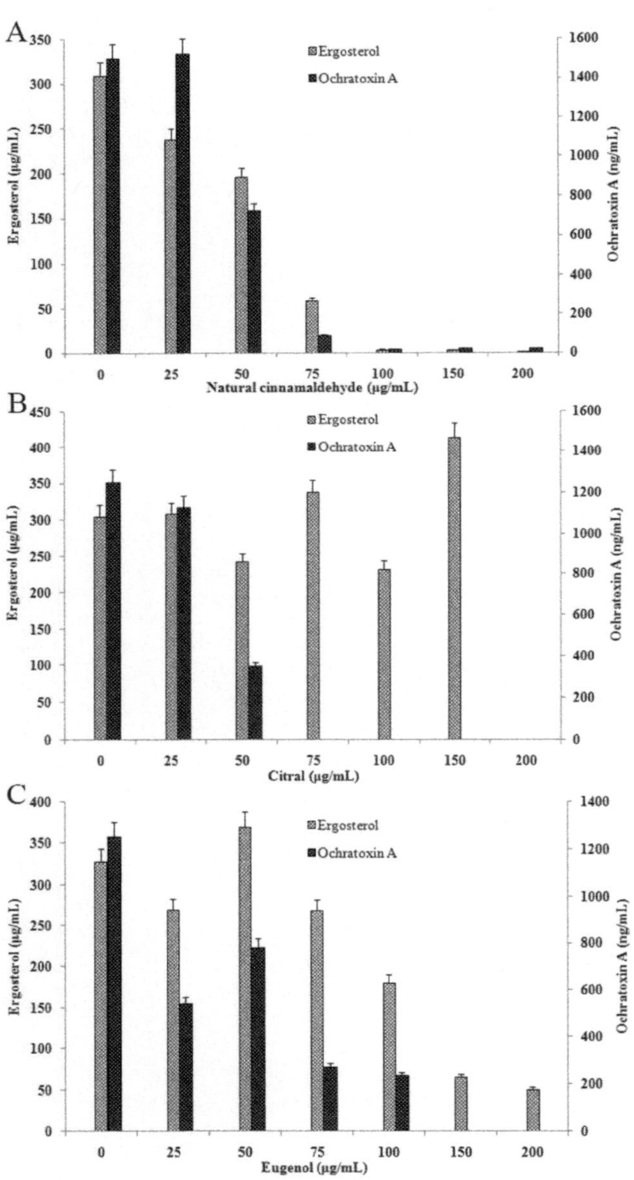

Figure 4. Effect of natural cinnamaldehyde (A), citral (B) and eugenol (C) on ergosterol biosynthesis and OTA production from *A. ochraceus* determined by HPLC. The experiment was conducted during 5 days of incubation under agitation at 28°C (n = 4).

20 mL methanol. The solution was further diluted in methanol if necessary, filtered through a 0.2 μm nylon syringe filter unit, and immediately used for HPLC analysis. Extracts from duplicate supernatants and corresponding cell pellets were analyzed separately.

Analysis used a Waters 2695 reversed-phase HPLC with a Waters 2487 UV detector at 280 nm and empower system controller. Data were processed using Waters LC-module 1 millennium software, version 32. A reverse-phase Agilent TC-C18 column was used. The mobile phases used for separation were: A, 1% acetic acid in methanol; and B, 1% acetic acid in water. A gradient flow of mobile phases (0 to 10 min at 15% A, 85% B; 10 to 13 min at 50% A, 50% B; 13 to 16 min at 60% A, 40% B; 16 to 24 min at 65% A, 35% B and 24 to 40 min at 15% A, 85% B) at 1 mL/min was used. Retention times and standard curves for cinnamaldehyde and cinnamic alcohol were obtained by analyzing 10 μL filtered standard solution (0.5 to 50 μg/mL) of each compound. Curve linearity and correlation coefficients were calculated from peak areas at each standard concentration. Ten μL extract from treatment samples or positive or negative controls was injected. Methanol extracts of cells were injected separately at the same level with supernatants and final concentrations of cinnamaldehyde or cinnamic alcohol were obtained by combining values for cell extracts and extracts from corresponding supernatants.

Statistical analyses

All the experiments results were evaluated using analysis of variance (ANOVA) for multiple comparisons followed by the Turkey test. Differences were considered significant at $p<0.05$. The growth of fungal cultures with different concentrations of the EOs was compared with growth of a control with no EO.

Results

Inhibitory effect of ten essential oils on *A. ochraceus* growth with fumigation

Cinnamon oil, natural cinnamaldehyde, synthetic cinnamaldehyde, *L. citrate* oil and citral significantly inhibited the growth of *A. ochraceus* at 50−250 μL per liter MEA medium ($P<0.01$). Natural cinnamaldehyde proved to be the most effective against *A. ochraceus*, followed by cinnamon oil, synthetic cinnamaldehyde, citral, *L. citrate* oil, eugenol, peppermint, anise, eucalyptus and camphor oils (Fig. 1). The inhibitory effect increased proportionally with concentration and was also affected by treatment duration. *A. ochraceus* growth was completely inhibited with natural cinnamaldehyde (95%) at 150 μL/L, cinnamon oil (85% cinnamaldehyde) at 250 μL/L, synthetic cinnamaldehyde (99%) at 250 μL/L, citral (96%) at 250 μL/L. After 20 days incubation, the

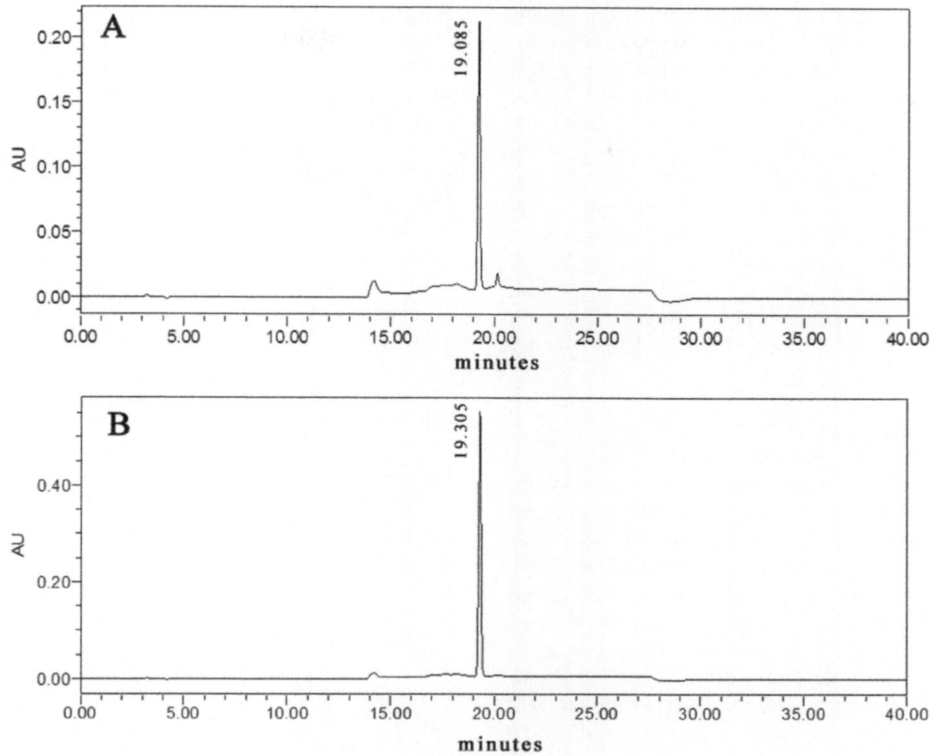

Figure 5. RP-HPLC detection of cinnamaldehyde of 50 mg/L (A) and cinnamic alcohol of 50 mg/L (B).

plates showing no growth were incubated for a further 7 days after removal of EOs, and no visible growth of *A. ochraceus* mycelia was observed. These results suggested a fungistatic effect of natural cinnamaldehyde, cinnamon oil, synthetic cinnamaldehyde and citral at lower concentrations and a fungicidal effect at higher concentrations.

Eugenol, peppermint, eucalyptus, anise and camphor oils did not completely inhibit *A. ochraceus* growth at 250–1500 µL/L. The diameters of *A. ochraceus* colonies treated with 1000 and 1500 µL/L eugenol did not increase from 1.0±0.06 cm from 4 to 20 days. Diameters of colonies treated with 1500 µL/L peppermint oil did not change from 1.0±0.07 cm from 14 to 20 days.

Inhibitory effect of ten essential oils on *A. ochraceus* growth with contact assay

Treatment with cinnamon oil, natural cinnamaldehyde, synthetic cinnamaldehyde, *L. citrate* oil, citral and eugenol significantly inhibited *A. ochraceus* growth at 100–500 µL/L on MEA

Table 2. Conversion of cinnamaldehyde to cinnamic alcohol by *A. ochraceus** in YES.

Treatment	YES plus cinnamaldehyde, uninoculated (negative control)		YES plus *A. ochracesus* plus cinnamaldehyde, inoculated (treatment)		YES plus *A. ochraceus* (positive control)	
Concentration	Cinnamaldehyde (µg/mL)	Cinnamic alcohol (µg/mL)	Cinnamaldehyde (µg/mL)	Cinnamic alcohol (µg/mL)	Cinnamaldehyde (µg/mL)	Cinnamic alcohol (µg/mL)
0 h	64.56 ± 4.26^{a}	ND	64.83 ± 5.33^{a}	ND	ND	ND
2 h	64.13 ± 3.21^{a}	ND	62.54 ± 4.26^{a}	ND	ND	ND
4 h	63.41 ± 4.65^{a}	ND	24.17 ± 1.30^{b}	36.83 ± 1.21^{a}	ND	ND
8 h	62.84 ± 5.11^{a}	ND	3.41 ± 0.34^{c}	55.23 ± 2.13^{b}	ND	ND
12 h	62.25 ± 4.32^{a}	ND	ND	55.52 ± 0.27^{b}	ND	ND
24 h	62.14 ± 4.21^{a}	ND	ND	37.33 ± 6.61^{c}	ND	ND
48 h	61.71 ± 2.26^{ab}	ND	ND	14.82 ± 2.48^{d}	ND	ND
72 h	58.85 ± 2.89^{b}	ND	ND	3.12 ± 0.39^{e}	ND	ND
96 h	55.58 ± 8.37^{c}	ND	ND	ND	ND	ND

*ND, not detected. Different superscript lowercase letters in each column indicate a significant difference ($P<0.05$).

medium ($P<0.01$). Treatment with peppermint oil significantly inhibited *A. ochraceus* growth at 500–1500 μL/L ($P<0.01$). However, eucalyptus, anise and camphor oils showed mild inhibition on *A. ochraceus* growth at 100–1500 μL/L. Natural cinnamaldehyde was highly effective against *A. ochraceus*, followed by synthetic cinnamaldehyde, cinnamon oil, citral, *L. citrate* oil, eugenol and peppermint oil (Fig. 2).

A. ochraceus growth was completely inhibited with natural cinnamaldehyde, synthetic cinnamaldehyde and cinnamon oil at 250 μL/L. For a complete antifungal effect, 500 μL/L of citral, *L. citrate* oil and eugenol were essential. At 1500 μL/L, peppermint oil completely inhibited the growth of *A. ochraceus*. However, eucalyptus, anise and camphor oils did not show complete inhibition at ≤1500 μL/L.

MIC determined by contact assay

For *A. ochraceus*, the MIC of cinnamon oil (85%), natural cinnamaldehyde (95%) and synthetic cinnamaldehyde (99%) were 500 μL/L; those of *L. citrate* oil (85%) and citral (99%) were 800 and 700 μL/L, respectively (Table 1). Cells from the tubes of the five EOs at ≥500 μL/L were subcultured on MEA plates and no visible growth was observed, suggesting a permanent growth inhibition. The MIC of eugenol (99%) was 600 μL/L which was lower than the MIC of citral (99%), however inhibition by eugenol at 600–4000 μL/L was reversible. Eucalyptus, anise and camphor oils did not completely inhibit the growth of *A. ochraceus* at ≤ 40000 μL/L.

Scanning electron microscopy (SEM)

Morphological changes which occurred in the hyphae and conidiophores of *A. ochraceus* due to growing the fungus in MEA medium fumigated with 100 μL/L of natural cinnamaldehyde, 100 μL/L of citral and 250 μL/L of eugenol are shown in Fig. 3A–D. The growing nonfumigated healthy mycelia showed a normal morphology with linear, regular, homogenous and rough cell wall hyphae (Fig. 3A). The growing apexes of the hyphae were tapered with a smooth surface. However, this morphology underwent alterations following fumigation of the culture medium with natural cinnamaldehyde, citral and eugenol. The mycelia showed evident modifications in both apical regions and throughout the length of the hyphae. These hyphae were disrupted (Fig. 3B, D) and the apical regions of the haphae were modified and abnormal with rough surfaces (Fig. 3B, D). The hyphae became slender, shrank and winding and evident craters appeared on the cell wall (Fig. 3C). The conidiophores and vesicles were disrupted, and then the conidia were dispersed in disorder (Fig. 3C).

Effect of three essential oils on ergosterol biosynthesis and OTA production

Natural cinnamaldehyde at 25–50 μg/mL caused a mild reduction in ergosterol production and significantly inhibited ergosterol production in *A. ochraceus* at 75–200 μg/mL, with inhibition ranging from 80% to 100%. Natural cinnamaldehyde significantly inhibited production of OTA by *A. ochraceus* at lower concentrations (50–75 μg/mL) and almost completely inhibited production at higher concentrations (100–200 μg/mL) (Fig. 4A). Thus, cinnamaldehyde-induced ergosterol inhibition correlated with a reduction in OTA production. The decrease in OTA production was proportional to the decrease in fungal biomass. A similar result was found by Yamamoto-Ribeiro et al. (2013) [20]. Other studies demonstrated that EOs and compounds, limonene and thymol, inhibit fumonisin production by *F. verticillioides* [29,30].

Citral at 25–150 μg/mL did not inhibit ergosterol biosynthesis and completely inhibited ergosterol production in *A. ochraceus* at 200 μg/mL (Fig. 4B). The significant oscillations were observed in ergosterol biosynthesis. Ergosterol biosynthesis was significantly induced by citral at 150 μg/mL. However, citral at 50 μg/mL significantly inhibited the OTA production and complete inhibition in OTA production was observed at 75–200 μg/mL.

Eugenol at 25–75 μg/mL caused oscillations in ergosterol production in *A. ochraceus* and effectively inhibited ergosterol production at 100–200 μg/mL, with inhibition ranging from 45% to 85% (Fig. 4C). This oscillation was reflected by an increase in ergosterol biosynthesis of *A. ochraceus* at 50 μg/mL eugenol compared with the control and significant inhibition at higher concentrations.

Conversion of cinnamaldehyde by *A. ochraceus*

The changes in concentrations of cinnamaldehyde and its possible metabolite, cinnamic alcohol, in YES with *A. ochraceus* (treatment) and without fungal inoculation but with cinnamaldehyde (negative control) were monitored by RP-HPLC analysis (Fig. 5). The concentration of cinnamaldehyde before incubation at 28°C (0 h) was 64.56±4.26 mg/L with the negative controls and 64.83±5.33 mg/L with the treatment samples (Table 2). The result showed that only about 64.83% of the cinnamaldehyde added (100 mg/L) was extracted using ethyl acetate. After 2 h the concentration of cinnamaldehyde was stable in both the negative controls and treatment cultures. After 4 h the concentration of cinnamaldehyde decreased by 40 mg/L in treatment cultures, while it remained stable in the negative control during the tested time. However, with *A. ochraceus*, the concentration of cinnamaldehyde decreased to ≤4 mg/L at 8 h and was undetected at 12 h. The concentration of cinnamic alcohol was detected at 4 h and increased to about 55 mg/L at 8–12 h, then decreased to approximate 3 mg/L at 72 h and undetected at 96 h. However, the concentration of cinnamic alcohol was undetectable in the positive control during the tested time. Since cinnamic alcohol was detected at 4–72 h in treatment samples, its MIC was determined and found to be 700 mg/L and higher than that of cinnamaldehyde.

Discussion

The 10 oils analyzed showed dose-dependent antifungal activity against *A. ochraceus* at the concentrations tested. This finding was consistent with the results obtained by Passone et al. (2012) [23] who reported that boldo (*Pëumus boldus* Mol.), poleo (*Lippia turbinate* var. *Integrifolia* [Griseb.]), clove (*Syzygium aromaticum* L.), anise (*Pimpinella anisum*) and thyme (*Thymus vulgaris*) EOs were effective non-toxic biopreservatives against *Aspergillus* section *Nigri* growth and OTA contamination *in vitro*. Boldo EO showed the greatest effect on fungal isolates with significant reductions at 1000 and 2000 μL/L and total growth inhibition at 3000 μL/L. *Matricaria chamomilla* L. oil was a potent inhibitor of *A. niger* growth *in vitro* with a growth inhibition at oil concentration of 15.62 μg/mL (~7.5%) and reached a maximum of ~92.5% at the final concentration of 1000 μg/mL [31].

Although the concentrations of oils tested in this study were not the same, the headspace volatile assays showed greater inhibition than the contact assays with the exception of eugenol and peppermint oil. Complete growth inhibition of *A. ochraceus* was verified at 150–250 μL/L by headspace volatile assays and 250–500 μL/L in the contact assays for cinnamon oil, natural cinnamaldehyde, synthetic cinnamaldehyde, *L. citrate* oil and citral. In contrast, Passone et al. (2012) [23] studies showed that boldo, poleo and clove essential oils have higher inhibition effect in contact assays than headspace volatile assays. Complete growth inhibition was observed at ≥1500 μL/L for boldo, poleo and clove oils in contact assays and ≥2000 μL/L for boldo in headspace volatile assays, while full fungal growth was not observed with 3000 μL/L of poleo and 5000 μL/L of clove oils. These results showed that both assays are necessary to evaluate the inhibitory effects of essential oils. Ergosterol has hormonal and structural effects in fungal cells and is responsible for regulating the flow and activity of membrane-bound enzymes [32]. Some antifungal agents inhibit fungal growth by interrupting ergosterol biosynthesis through binding to ergosterol in membranes. This damages the integrity and function of membrane-bound proteins and disturbs osmosis, fungal growth and proliferation [32]. In our study, natural cinnamaldehyde at 25–50 μg/mL caused a mild reduction in ergosterol production and significantly inhibited ergosterol production in *A. ochraceus* at 75–200 μg/mL, with inhibition ranging from 80% to 100%. Various research groups have reported the oscillations in ergosterol biosynthesis [20,28,33]. Yamamoto-Ribeiro et al. (2013) [20] found that *Zingiber officinale* essential oil increases ergosterol biosynthesis of *F. vertcillioides* at 100 μg/mL compared with the control group and reduces or completely inhibits ergosterol biosynthesis at higher concentrations. Lucini et al. (2006) [33] found that monoterpenes at low concentrations lead to lipid peroxidation in fungi, inducing an adaptive response and reprogramming genomic expression to protect the cell wall structure, and consequently increase ergosterol biosynthesis. Dabolena et al. (2008) [29] found that limonene, menthol, menthone, and thymol at 75 μg/mL could increase ergosterol production by *F. verticillioides*.

In general, natural cinnamaldehyde and eugenol slightly inhibited biosynthesis of ergosterol and OTA at lower concentrations and significantly or completely inhibited biosynthesis at higher concentrations. The reductions of OTA production caused by cinnamaldehyde and eugenol at 25–100 μg/mL were positively correlated with EOs-induced ergosterol inhibition. Therefore, the decrease in OTA production was proportional to the decrease in fungal biomass at some levels. This result was similar with the findings of Yamamoto-Ribeiro et al. (2013) [20] and Dambolena et al. (2010) [34]. Their studies reported that essential oils from *Zingiber officinale*, *Ocimum basilicum* L. and *Ocimum gratissimum* L. inhibited fumonisin production by *F. verticillioides* and the inhibition was proportional to the decrease in fungal biomass.

However, citral showed a different effect than natural cinnamaldehyde and eugenol. Citral at 75–150 μg/mL completely inhibited OTA production but not ergosterol biosynthesis. This result suggested that the decrease in OTA production induced by citral was not due to the decrease in fungal biomass but might be due to suppression of transcription of OTA biosynthesis genes. The polyketide synthase gene (*pks* gene) is responsible for the synthesis of the polyketide dihydroisocoumarin and involved in the first steps of the OTA biosynthetic pathway. Real-time quantify PCR (qPCR) analysis confirmed that the transcription of *pks* gene was completely inhibited by 75–150 μg/mL of citral. A similar inhibition of OTA production was observed with higher concentration of eugenol. At lower concentrations (25–100 μg/mL), the decrease in OTA production by eugenol was proportional to the decrease in fungal biomass. However, at 150–200 μg/mL OTA production was completely inhibited by eugenol but ergosterol production was not fully inhibited. This result suggested that inhibition of OTA required a lower oil concentration of EO than the inhibition of fungal growth. This finding agrees with results from Passone et al. (2012) [23], who reported complete inhibition of OTA production by *Aspergillus niger* RCP42 at ≥1000 μL/L of clove oil, whose main component is eugenol, while total fungal growth inhibition did not occur at 3000 μL/L at a_w 0.98. Complete inhibition of OTA production by *Aspergillus carbonarius* RCP203 occurred at ≥1000 μL/L of clove oil, while total fungal growth inhibition did not occur at 3000 μL/L at a_w 0.93. Murthy et al. (2009) [35] found that fungal growth and OTA production decreased progressively with increases in Ajowan ethanolic extract (AEE) concentration. At 250 μL/mL, they could see fungal growth and OTA production being inhibited. However, the inhibition towards fungal growth and OTA biosynthesis was not linear. At 50 and 150 μL/g of AEE, inhibition of OTA production was higher than inhibition of fungal growth. Jayashree and Subramanyam (1999) [36] found that eugenol up to 0.75 mmol/L inhibited aflatoxin production with no significant effect on *Aspergillus parasiticus* growth. Furthermore, they found that eugenol inhibited the lipid peroxidation in aflatoxin production with no significant inhibition of growth or primary metabolism. However, an inverse correlation was shown by Pereira et al. (2006) [37] who observed that clove oil could completely inhibit the mycelial growth of *A. ochraceus*, but no significant affect in antiochratoxigenic activity.

This is the first study to show that *A. ochraceus* can convert cinnamaldehyde to cinnamic alcohol at ≤12 h, and cinnamic alcohol was further degraded and no new metabolites were detected by RP-HPLC. The degradation might be due to *Aspergillus* genus has several enzymes such as alcohol dehydrogenase and aldehyde reductase that catalyze cinnamaldehyde to cinnamic alcohol [38,39]. Similar result was obtained by Vsivalingam et al. (2013) [40], who firstly found that *E. coli* O157:H7 could convert cinnamaldehyde to cinnamic alcohol by alcohol dehydrogenase and 2,5-diketo-D-gluconate reductase A.

Conclusions

The inhibitory effect of 10 essential oils on *A. ochraceus* growth was tested with fumigation and contact assays. Natural cinnamaldehyde proved to be the most effective against *A. ochraceus*, followed by cinnamon oil, synthetic cinnamaldehyde, citral, *L. citrate* oil, eugenol and peppermint oil. Headspace volatile assays showed higher inhibition than contact assays with exception of eugenol and peppermint oil. SEM of *A. ochraceus* cells fumigated with natural cinnamaldehyde, citral and eugenol showed alteration in the morphology of the hyphae which appeared collapsed, and abnormal branching of hyphae in apical region and loss of linearity. The inhibitory effects of natural cinnamaldehyde, citral and eugenol on ergosterol biosynthesis and OTA production indicated that natural cinnamaldehyde had a higher antifungal activity than citral and eugenol. However, natural cinnamaldehyde had a lower antiochratoxigenic effects than citral and eugenol. These results suggested that the inhibition of OTA production by natural cinnamaldehyde was mainly due to a decrease in fungal biomass. However, in addition to antifungal activity, citral and eugenol could significantly inhibit the OTA biosynthesis pathway of *A. ochraceus*.

Acknowledgments

We are sincerely grateful to the other members in the Y. L. group at Institute of Agro-Products Processing and Sciences Technology that offered helpful suggestions for the analysis.

Author Contributions

Conceived and designed the experiments: YL FGX. Performed the experiments: HJH FGX JNS XL YJZ LZ YW. Analyzed the data: HJH FGX. Contributed reagents/materials/analysis tools: HJH FGX. Contributed to the writing of the manuscript: FGX HJH YL.

References

1. Pardo E, Marín S, Ramos A, Sanchis V (2006) Ecophysiology of ochratoxigenic *Aspergillus ochraceus* and *Penicillium verrucosum* isolates. Predictive models for fungal spoilage prevention–a review. Food Addit Contam A 23: 398–410.
2. Hocking AD, Leong SL, Kazi BA, Emmett RW, Scott ES (2007) Fungi and mycotoxins in vineyards and grape products. Int J Food Microbiol 119: 84–88.
3. Bayman P, Baker JL (2006) Ochratoxins: a global perspective. Mycopathologia 162: 215–223.
4. Huffman J, Gerber R, Du L (2010) Recent advancements in the biosynthetic mechanisms for polyketide-derived mycotoxins. Biopolymers 93: 764–776.
5. Sokolić-Mihalak D, Frece J, Slavica A, Delaš F, Pavlović H, et al. (2012) The effects of wild thyme (*Thymus Serpyllum* L.) essential oil components against ochratoxin-producing *Aspergilli*. Arch Indust Hygiene Toxicol 63: 457–462.
6. Reddy K, Salleh B, Saad B, Abbas H, Abel C, et al. (2010) An overview of mycotoxin contamination in foods and its implications for human health. Toxin Rev 29: 3–26.
7. Čvek D, Markov K, Frece J, Dragičević T, Majica M, et al. (2010) Growth inhibition of *Aspergillus ochraceus* ZMPBF 318 and *Penicillium expansum* ZMPBF 565 by four essential oils. Arch Indust Hygiene Toxicol 61: 191–196.
8. O'Brien E, Dietrich DR (2005) Ochratoxin A: the continuing enigma. Crit Rev Toxicol 35: 33–60.
9. Pfohl-Leszkowicz A, Manderville RA (2007) Ochratoxin A: An overview on toxicity and carcinogenicity in animals and humans. Mol Nutr Food Res 51: 61–99.
10. Chen PJ, Moore T, Nesnow S (2008) Cytotoxic effects of propiconazole and its metabolites in mouse and human hepatoma cells and primary mouse hepatocytes. Toxicol In Vitro 22: 1476–1483.
11. Al-Omair A, Helaleh MIH (2004) Selected-Ion storage GC–MS analysis of polycyclic aromatic hydrocarbons in palm dates and tuna fish. Chromatographia 59: 715–719.
12. Baird RE, Brenneman TB, Bell DK, Murphy AP (1991) The effects of the fungicide propiconazole (Tilt) on the groundnut shell mycobiota. Mycol Res 95: 571–576.
13. Chilvers MI, Hay FS, Hills J, Dennis JJC, Wilson CR (2006) Influence of benzimidazole fungicides on incidence of *Botrytis allii* infection of onion leaves and subsequent incidence of onion neck rot in storage in Tasmania, Australia. Aust J Exp Agr 46: 1661–1664.
14. Šimko P (2005) Factors affecting elimination of polycyclic aromatic hydrocarbons from smoked meat foods and liquid smoke flavorings. Mol Nutr Food Res 49: 637–647.
15. Isaac S (1999) What is the mode of action of fungicides and how do fungi develop resistance? Mycologist 13: 38–39.
16. Thompson D (1989) Fungitoxic activity of essential oil components on food storage fungi. Mycologia 81: 151–153.
17. Tian J, Ban X, Zeng H, He J, Huang B, et al. (2011) Chemical composition and antifungal activity of essential oil from *Cicuta virosa* L. var. *latisecta* Celak. Int J Food Microbiol 145: 464–470.
18. Patil JR, Jayaprakasha GK, Chidambara Murthy KN, Tichy SE, Chetti MB, et al. (2009) Apoptosis-mediated proliferation inhibition of human colon cancer cells by volatile principles of *Citrus aurantifolia*. Food Chem 114: 1351–1358.
19. Velluti A, Sanchis V, Ramos AJ, Marín S (2004) Effect of essential oils of cinnamon, clove, lemon grass, oregano and palmarosa on growth of and fumonisin B_1 production by *Fusarium verticillioides* in maize. J Sci Food Agr 84: 1141–1146.
20. Yamamoto-Ribeiro MMG, Grespan R, Kohiyama CY, Ferreira FD, Mossini SAG, et al. (2013) Effect of *Zingiber officinale* essential oil on *Fusarium verticillioides* and fumonisin production. Food Chem 141: 3147–3152.
21. Bluma RV, Etcheverry MG (2008) Application of essential oils in maize grain: Impact on *Aspergillus* section *Flavi* growth parameters and aflatoxin accumulation. Food Microbiol 25: 324–334.
22. Passone MA, Girardi NS, Etcheverry M (2012) Evaluation of the control ability of five essential oils against *Aspergillus* section *Nigri* growth and ochratoxin A accumulation in peanut meal extract agar conditioned at different water activities levels. Int J Food Microbiol 159: 198–206.
23. Avila-Sosa R, Palou E, Jiménez Munguía MT, Nevárez-Moorillón GV, Navarro Cruz AR, et al. (2012) Antifungal activity by vapor contact of essential oils added to amaranth, chitosan, or starch edible films. Int J Food Microbiol 153: 66–72.
24. Soliman KM, Badeaa RI (2002) Effect of oil extracted from some medicinal plants on different mycotoxigenic fungi. Food Chem Toxicol 40: 1669–1675.
25. Xing F, Hua H, Selvaraj JN, Zhao Y, Zhou L, et al. (2014) Growth inhibition and morphological alterations of *Fusarium verticillioides* by cinnamon oil and cinnamaldehyde. Food Control 46: 343–350.
26. Kocić-Tanackov S, Dimić G, Tanackov I, Pejin D, Mojović L, et al. (2012) The inhibitory effect of oregano extract on the growth of *Aspergillus* spp. and on sterigmatocystin biosynthesis. LWT-Food Sci Technol 49: 14–20.
27. Silva RR, Corso CR, Matheus DR (2010) Effect of culture conditions on the biomass determination by ergosterol of *Lentinus crinitus* and *Psilocybe castanella*. World J Microbiol Biotechnol 26: 841–846.
28. Scudamore KA, MacDonald SJ (1998) A collaborative study of an HPLC method for determination of ochratoxin A in wheat using immunoaflinity column clean-up. Food Addit Contam 15: 401–410.
29. Dambolena JS, López AG, Cánepa MC, Theumer MG, Zygadlo JA, et al. (2008) Inhibitory effect of cyclic terpenes (limonene, menthol, menthone and thymol) on *Fusarium verticillioides* MRC 826 growth and fumonisin B_1 biosynthesis. Toxicon 51: 37–44.
30. Fandohan P, Gbenou JD, Gnonlonfin B, Hell K, Marasas WF, et al. (2004) Effect of essential oils on the growth of *Fusarium verticillioides* and fumonisin contamination in corn. J Agric Food Chem 52: 6824–6829.
31. Tolouee M, Alinezhad S, Saberi R, Eslamifar A, Zad SJ, et al. (2010) Effect of *Matricaria chamomilla* L. flower essential oil on the growth and ultrastructure of *Aspergillus niger* van Tieghem. Int J Food Microbiol 139: 127–133.
32. Bendaha H, Yu L, Touzani R, Souane R, Giaever G, et al. (2011) New azole antifungal agents with novel modes of action: synthesis and biological studies of new tridentate ligands based on pyrazole and triazole. Eur J Med Chem 46: 4117–4124.
33. Lucini EI, Zunino MP, López ML, Zygadlo JA (2006) Effect of monoterpenes on lipid composition and sclerotial development of *Sclerotium cepivorum* Berk. J Phytopathol 154: 441–446.
34. Dambolena JS, Zunino MP, López AG, Rubinstein HR, Zygadlo JA, et al. (2010) Essential oils composition of *Ocimum basilicum* L. and *Ocimum gratissimum* L. from Kenya and their inhibitory effects on growth and fumonisin production by *Fusarium verticillioides*. Innov Food Sci Emerg 11: 410–414.
35. Murthy PS, Borse BB, Khanum H, Srinivas P (2009) Inhibitory effects of Ajwain (*Trachyspermum ammi*) ethanolic extract on *A. ochraceus* growth and ochratoxin production. Turk J Biol 33: 211–217.
36. Jayashree T, Subramanyam C (1999) Antiaflatoxigenic activity of eugenol is due to inhibition of lipid peroxidation. Lett Appl Microbiol 28: 179–183.
37. Pereira MC, Chalfoun SM, Pimenta CJ, Angélico CL, Maciel WP (2006) Spices, fungi mycelial development and ochratoxin A production. Sci Res Essay 1: 38–42.
38. Díaz E, Ferrández A, Prieto MA, García JL (2001) Biodegradation of aromatic compounds by *Escherichia coli*. Microbiol Mol Biol Rev 65: 523–569.
39. Grahl N, Puttikamonkul S, Macdonald JM, Gamcsik MP, Ngo LY, et al. (2011) *In vivo* hypoxia and a fungal alcohol dehydrogenase influence the pathogenesis of invasive pulmonary aspergillosis. PLoS Pathog 7: e1002145.
40. Visvalingam J, Hernandez-Doria JD, Holley RA (2013) Examination of the genome-wide transcriptional response of *Escherichia coli* O157:H7 to cinnamaldehyde exposure. Appl Environ Microbiol 79: 942–950.

15

Differentiation of *Bacillus pumilus* and *Bacillus safensis* Using MALDI-TOF-MS

Raquel Branquinho[1], Clara Sousa[1,2], João Lopes[3,4], Manuela E. Pintado[5], Luísa V. Peixe[1], Hugo Osório[6,7]*

1 REQUIMTE, Laboratório de Microbiologia, Departamento de Ciências Biológicas, Faculdade de Farmácia, Universidade do Porto, Porto, Portugal, **2** CEB, Centro de Engenharia Biológica, Universidade do Minho, Braga, Portugal, **3** REQUIMTE, Laboratório de Química Aplicada, Departamento de Ciências Químicas, Faculdade de Farmácia, Universidade do Porto, Porto, Portugal, **4** CBQF, Centro de Biotecnologia e Química Fina, Escola Superior de Biotecnologia, Universidade Católica Portuguesa, Porto, Portugal, **5** Departmento de Tecnologia Farmacêutica, Faculdade Farmácia, Universidade de Lisboa, Lisboa, Portugal, **6** IPATIMUP, Instituto de Patologia e Imunologia Molecular da Universidade do Porto, Porto, Portugal, **7** Faculdade de Medicina, Universidade do Porto, Porto, Portugal

Abstract

Matrix-assisted laser desorption/ionization time-of-flight mass spectrometry (MALDI-TOF-MS) despite being increasingly used as a method for microbial identification, still present limitations in which concerns the differentiation of closely related species. *Bacillus pumillus* and *Bacillus safensis,* are species of biotechnological and pharmaceutical significance, difficult to differentiate by conventional methodologies. In this study, using a well-characterized collection of *B. pumillus* and *B. safensis* isolates, we demonstrated the suitability of MALDI-TOF-MS combined with chemometrics to accurately and rapidly identify them. Moreover, characteristic species-specific ion masses were tentatively assigned, using UniProtKB/Swiss-Prot and UniProtKB/TrEMBL databases and primary literature. Delineation of *B. pumilus* (ions at *m/z* 5271 and 6122) and *B. safensis* (ions at *m/z* 5288, 5568 and 6413) species were supported by a congruent characteristic protein pattern. Moreover, using a chemometric approach, the score plot created by partial least square discriminant analysis (PLSDA) of mass spectra demonstrated the presence of two individualized clusters, each one enclosing isolates belonging to a species-specific spectral group. The generated pool of species-specific proteins comprised mostly ribosomal and SASPs proteins. Therefore, in *B. pumilus* the specific ion at *m/z* 5271 was associated with a small acid-soluble spore protein (SASP O) or with 50S protein L35, whereas in *B. safensis* specific ions at *m/z* 5288 and 5568 were associated with SASP J and P, respectively, and an ion at *m/z* 6413 with 50S protein L32. Thus, the resulting unique protein profile combined with chemometric analysis, proved to be valuable tools for *B. pumilus* and *B. safensis* discrimination, allowing their reliable, reproducible and rapid identification.

Editor: Mickaël Desvaux, INRA Clermont-Ferrand Research Center, France

Funding: Dr. Kasthuri Venkateswaran, Dr. Irène Ouoba, Dr. Joseph W. Kloepper, Dr. Cecilie From and Dr. Maria Morea are gratefully acknowledged for providing isolates FO-36bT, SAFN-027, SAFN-037, KL-052, 51-3C and 82-2C; Bs31; SE 49 (AP3) and SE 52 (AP7); FEL 55, UNG22 and MIL46, respectively. Raquel Branquinho was supported by a PhD fellowship (Ref. SFRH/BD/61410/2009) and Clara Sousa by a post-doctoral fellowship (Ref. SFRH/BPD/70548/2010), from FCT (Fundação para a Ciência e Tecnologia, Portugal). Hugo Osório acknowledges the funding from QREN-FEDER through the Operational Program ON. 2 – O Novo Norte. IPATIMUP is an Associate Laboratory of the Portuguese Ministry of Science, Technology and Higher Education and is partially supported by the Portuguese Foundation for Science and Technology. The funders had no role in study design, data collection and analysis, decision to publish, or preparation of the manuscript.

Competing Interests: The authors have declared that no competing interests exist.

* Email: hosorio@ipatimup.pt

Introduction

Bacillus pumilus and *Bacillus safensis* represent one of the most significant and widespread terrestrial species within the *Bacillus pumilus* group [1,2,3,4,5]. A wide range of biotechnological and pharmaceutical applications has been attributed to these species, including human and animal probiotics [6] or phytosanitary-based products [7]. Moreover, they are important contaminant agents found in industrial settings, namely in food and/or pharmaceutical facilities, posing a serious problem to the quality assurance of these industrial segments [2,4]. More rarely, *B. pumilus* isolates have also been involved in foodborne poisoning [8,9] and in human infections including anthrax-like cutaneous lesions [10,11,12,13].

Studying a collection of previously identified *B. pumilus* isolates we have realized the difficulty to their differentiation from closely related species based on phenotypic and biochemical characteristics and on 16S rRNA gene sequences [4]. Sequencing housekeeping genes, such as *gyrB* (β-subunit of DNA gyrase) and *rpoB* (β-subunit of RNA polymerase) has proven to be useful for taxonomic resolution of closely related species, including *Bacillus* species. Nevertheless, are already recognized its implementation difficulties in the routine of microbiology laboratories [4,5].

Matrix-assisted laser desorption-ionization-time-of-flight mass spectrometry (MALDI-TOF-MS), is an accurate, fast, and affordable emerging technique that has been increasingly used for identification of bacteria at different taxonomical levels [14,15]. As a proteomic approach, MALDI-TOF-MS relies on the reproducible detection of microbial protein patterns, which can be used for microbial identification by comparing experimental mass spectra with a library of known reference strains or by comparing identities of species-specific biomarkers [16,17].

Using vegetative *B. pumilus* and *B. safensis* cells Farfour et al. [18] unsuccessfully attempted to discriminate these closely related species. Differentiation of these species using spore cells was proposed by Dickinson et al. [19], by MALDI-TOF-MS using a single ion at *m/z* 7620. Nevertheless, there is a need for further studies in order to assess the validity of this discriminatory tool. Moreover, Lash et al. [20,21] suggested that bacterial vegetative cells when submitted to a protein enrichment process, could provide a more informative mass spectra pattern than that obtained with bacterial spores [20,21].

In addition, the species-specific protein fingerprint constitutes a valuable approach for the assignment of species biomarkers, which usually comprise cell structure and housekeeping proteins [22]. In *Bacillus* spp., ribosomal proteins and small acid-soluble spore proteins (SASPs) have been reported as potential species-specific biomarkers [21]. Nevertheless and despite this successful characterization of *Bacillus cereus* group species [21], this was not applied to the members of the *B. pumilus* group.

In this work, through the implementation of a protocol that enhances proteins extraction [21] combined with chemometric tools, we assessed the potential of MALDI-TOF-MS fingerprinting to discriminate a comprehensive collection of *B. safensis* and *B. pumilus* isolates. Moreover, the tentative assignment of their species-specific protein biomarkers was also performed.

Materials and Methods

Isolates collection and identification

Five *B. pumilus* and twenty-two *B. safensis* isolates that were previously identified by phenotypic and genotypic (16S rRNA, *gyrB and rpoB* gene sequences) methods were studied [5,23]. They comprised diverse pulsed-field gel electrophoresis (PFGE)-types and were recovered from different geographic terrestrial locations and sources, including food samples (Norway, Italy and Africa) (n = 4), plants (USA) (n = 2), gastropods (Portugal) (n = 3), health (n = 6) and cosmetic (n = 4) products (Portugal), and clean room environments from Mars Odyssey (USA) (n = 5). Additionally, type and reference strains, *B. safensis* FO-36b^{T}, *B. pumilus* ATCC 7061^{T} and ATCC 14884 were included (Table 1).

Sample preparation

For MALDI-TOF-MS analysis, soluble proteins were extracted from a pure single colony of *Bacillus* spp. grown on LB agar (Merck, Darmstadt, Germany), which were subsequently cultured on the same medium, under aerobic conditions, for 24 h at 37°C. Bacterial cells were harvested by transferring three full loops (ca. 30 µl) from each agar plate into 20 µl of sterile water and were resuspended by vortexing. Bacterial inactivation was carried out applying the modified trifluoroacetic acid (TFA) (Sigma-Aldrich, St. Louis, MO) inactivation protocol [20], with some modifications previously established by Lash et al. (2008) [20] to improve the accuracy of the mass spectra. Briefly, 80 µl of pure TFA was added to 20 µl of each bacterial suspension. After gentle shaking (100 rpm) for 5 min at room temperature, the solution was centrifuged for 20 min at 28960 g^{-1} at 4°C. *Subsequently, the supernatant was* 10-fold *diluted with HPLC grade water (Millipore* Corp., Bedford, MA), *filtered throughout a 0.22 µm pore size filter (Millipore) and stored at −20°C until further analysis.*

Mass spectrometry methods

Snapshots of different protein composition were detected and acquired by a MALDI-TOF/TOF mass spectrometer (4800 Plus MALDI TOF/TOF Analyzer, AB SCIEX, Framingham, MA), equipped with a 200-Hz frequency Nd:YAG laser, operating at a wavelength of 355 nm. Pulse ion extraction with a 1300 ns delay time was used for collecting spectra. Measurements were carried out in linear positive mode using an acceleration voltage of 19.4 kV (Grid 1), and a lens 1 voltage of 8 kV. Each spectrum was the accumulated sum of at least 2000 laser shots within the ion range at *m/z* 2000–12000, due to the good reproducibility of the spectral profile in this interval. All the spectra were externally calibrated using a commercial mixture of angiotensin I, ACTH (adrenocorticotropic hormone) and insulin (AB SCIEX, Framingham, MA) and analyzed with the Data Explorer software (Version 4.6, AB SCIEX, Framingham, MA).

For MALDI-TOF-MS experiments, 2 µl of the filtrated microbial dilution were mixed with 2 µl of a 12-mg/ml α-cyano-4-hydroxycinnamic acid (CHCA) (Sigma-Aldrich, St. Louis, MO) solution, prepared in 100% ACN (Acetonitrile, Sigma-Aldrich) and 0.3% TFA. 1 µl of the mixture was spotted onto a 123×81 mm stainless steel MALDI sample plate (Opti-TOF 384- Well insert, AB SCIEX, Framingham, MA) and allowed to dry at room temperature. For each isolate, two biological replicates (obtained from two different agar plates) were carried out, and the mean spectra were considered for the analysis. Mass spectra were analyzed with the Data Explorer software (v3.7, build 126, AB SCIEX, Framingham, MA). Ion masses were extracted from the raw experimental mass spectra that included all the ion peaks with a relative signal to noise (S/N) ratio intensity above 2.

Chemometric methods

MALDI-TOF-MS spectra were mean-centred and analysed by partial least squares discriminant analysis (PLSDA) [24]. The PLSDA model were developed and validated based on a cross-validation strategy leave-one-out [25] where 70% of the strains were randomly selected to calibrate the model and 30% to test the model (the procedure was repeated 100 times). The PLSDA scores were the source for hierarchical cluster analysis (HCA). The purpose of HCA was the generation of dendrograms highlighting the association between isolates. Dendrograms were performed directly on unprocessed PLSDA scores using the Euclidean distance and the Median's algorithm [26]. All chemometric models were performed in Matlab version 6.5 release 13 (MathWorks, Natick, MA) and the PLS Toolbox version 3.5 for Matlab (Eigenvector Research, Manson, WA).

Biomarker identification

Intact protein masses derived from MS analysis were used to generate a pool of candidate's proteins for the identification of specific markers. The selected distinct mass information was submitted to a web-based TagIdent software tool (http://web.expasy.org/tagident/) using 1% mass error for the taxonomic selections *B. pumilus*, *B. safensis* and *B. subtilis*. No restrictions on protein isoelectric point were used. This tool allowed the identification of proteins based on the experimental masses acquired by mass spectrometry using the information available at the UniProtKB/Swiss-Prot and UniProtKB/TrEMBL protein sequence databases. Moreover, ribosomal proteins of genome-sequenced type strains, including *B. pumilus* ATCC 7061^{T} available in the database developed by Hotta et al. [27] were also included for comparison. For ion peaks matching the theoretical molecular weights we took into consideration the "N-end rule" where N-terminal methionine is cleaved when the second amino acid residue had a small side-chain, according to the previous study [28].

Table 1. Origins of *Bacillus* spp. isolates (n = 27) included in this study.

Isolate	Origin	Year/Location	References
Bacillus pumilus			
Bp ATCC14884			Reference strain
Bp ATCC 7061^{T}			Type Strain
Bp7		2005/Portugal[1]	
Bp11	Health products (n = 3)	2005/Portugal[1]	[37]
Bp15		2005/Portugal[1]	
Bacillus safensis			
Bs1		2004/Portugal[1]	
Bs2	Animals Gastropods (n = 3)	2005/Portugal[1]	
Bs3		2007/Portugal[1]	
Bs13		2005/Portugal[1]	
Bs16	Health products (n = 3)	2005/Portugal[1]	
Bs17		2005/Portugal[1]	[37]
Bs5		2002/Portugal[1]	
Bs18	Cosmetic products (n = 4)	2002/Portugal[1]	
Bs19		2002/Portugal[1]	
Bs27		2002/Portugal[1]	
Bs24		2004/Italy[2]	
Bs25	Foods/salame (n = 3)	2004/Italy[2]	[38]
Bs33		2004/Italy[2]	
Bs22	Plant Growth-Promoting Rhizobacteria (PGPR) (n = 2)	1997/USA[3]	[39]
Bs23		1997/USA[3]	
Bs31	Food/beans (n = 1)	2003/Africa[4]	[40]
Bs FO-36b^{T}	Clean-room/air particulate (n = 1)	1999/USA[5]	Type Strain, [2]
Bs35	Clean-room/floor (n = 1)	2001/USA[5]	
Bs36	Clean-room/cabinet top (n = 1)	2001/USA[5]	
Bs37	Clean-room/Mars Odyssey spacecraft surface (n = 2)	2001/USA[5]	[2]
Bs38			
Bs42	Clean-room/anteroom (n = 1)	2001/USA[5]	

^{T}Type Strain.
[1]Isolates obtained from the Quality Control Department (INFARMED), Lisbon, Portugal.
[2]Isolates FEL 55 from salame felino, UNG22 from salame ungherese and MIL46 from salame milano obtained from the Istituto di Scienze delle Produzioni Alimentari (ISPA), Bari, Italy.
[3]Isolates SE 49 (AP3) and SE 52 (AP7) from cucumber roots obtained from the Culture collection of the Department of Entomology and Plant Pathology, Auburn University, Alabama, USA.
[4]Isolates Bs31 from African locust beans for *Soumbala production* obtained from Ouagadougou, Africa.
[5]Isolates FO-36b, SAFN-027, SAFN-037, KL-052, 51-3C and 82-2C from spacecraft and assembly-facility surfaces obtained from California Institute of Technology, California, USA.

Results and Discussion

One of the current challenges in bacterial taxonomy is to integrate timely and accurate typing methods for a meaningful identification of microorganisms [14], which is particularly problematic for closely related species such as *B. pumilus* and *B. safensis* [2,5]. Moreover, because of the medical, industrial and biotechnological relevance of these species, reliable, easy and rapid methodologies for their correct differentiation are needed. Discrimination of *B. pumilus* and *B. safensis* by molecular markers sequencing (e.g. *gyrB*), a time demanding and expensive methodology, which is not readily available for routine laboratories, was recently demonstrated [4,5]. In fact, there are no comprehensive studies assessing its potential on *B. pumilus* and *B. safensis* discrimination despite the importance of the MALDI-TOF-MS application in bacterial differentiation. Moreover, combining accurate MALDI-TOF mass ion signals with the information available at a protein sequence database, such as UniProt, species-specific candidate protein biomarkers can be tentatively assigned, supporting the interest of this methodology for species identification.

Sample preparation conditions

Bacterial species identification using MALDI-TOF-MS is based on mass profiles obtained from whole bacterial cell suspensions considering proteins with low mass weight (less than 20 kDa). Additionally, for MS applications it is imperative to define a standardized protocol, including the establishment of rigorous

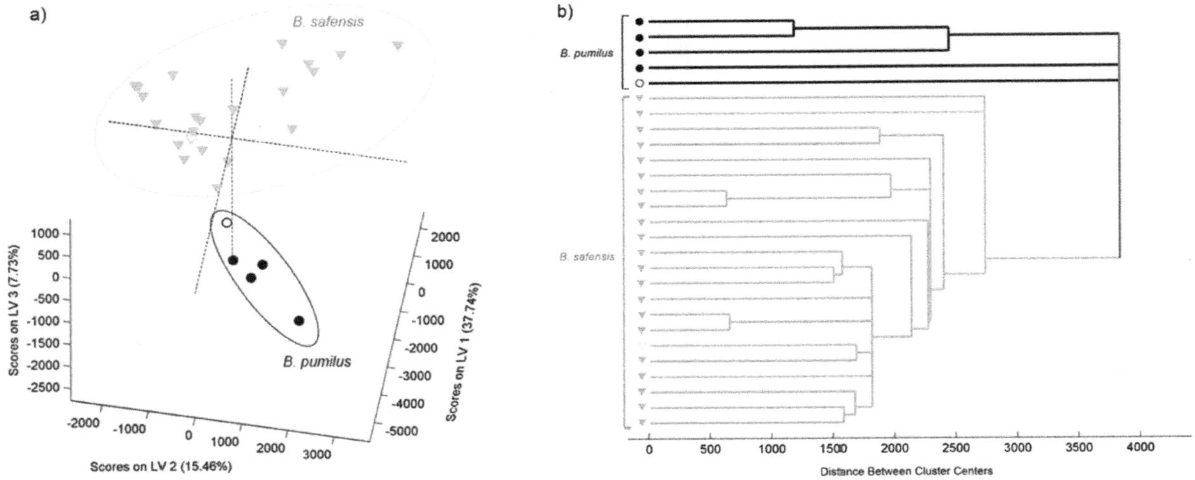

Figure 1. Average mass spectra obtained by MALDI-TOF-MS analysis of *B. pumilus* and *B. safensis* isolates. (a) *B. pumilus* (5 spectra) and (b) *B. safensis* (22 spectra) in the ion range at *m/z* 2000 to 12000. Spectra were obtained by averaging the respective experimental ion signals from all isolates.

sample preparation and cultivation conditions, including culture medium, temperature and time. Therefore, an adequate number of ion masses should be obtained to allow the identification and discrimination of closely related bacterial species. Moreover, the definition of strict parameters for spectral data acquisition is also required.

The protein enrichment protocol, reported by Lash et al. [20], for *B. cereus* group members, using a combined TFA treatment, centrifugation and filtration steps, is of relevance in the case of *Bacillus* species, since it promoted an efficient extraction of soluble microbial proteins presented in the core spore and other morphological structures [20]. The described sample preparation procedure was successfully applied to *B. pumilus* and *B. safensis* isolates, clonally diverse and collected from different terrestrial origins [4], demonstrating its suitability to generate reproducible mass spectra data with a sufficient number of ion masses and reinforcing the potential for the successful application of the MS technique for their identification in routine laboratories.

Mass spectrometry analysis

The Figure 1 displayed the average mass spectra obtained from MALDI-TOF-MS analysis for the two *Bacillus* species under study: (a) *B. pumilus* (5 spectra) and (b) *B. safensis* (22 spectra), in the ion range at *m/z* 2000 to 12000. In fact, MALDI-TOF-MS profile analysis revealed the presence of species-specific ion signals.

Table 2. Species-specific ion *m/z* values (average) of *B. pumilus* and *B. safensis* isolates.

	Experimental average ion *m/z* values[a]	
	B. pumilus	***B. safensis***
		3692.5
	3060 (BP1)[b]	3821.5
B. pumilus	3608.5 (BP2)	4305.5
	5271 (BP3)	5948.5
	6122 (BP4)	6704
		6793.5
		7415
	3692.5	
	3821.5	3396 (BS1)[b]
	4305.5	5288 (BS2)
B. safensis	5948.5	5568 (BS3)
	6704	6094 (BS4)
	6793.5	6413 (BS5)
	7415	

[a] average ion *m/z* values (±2Da).

[b] In brackets were named candidate species-specific ion masses.

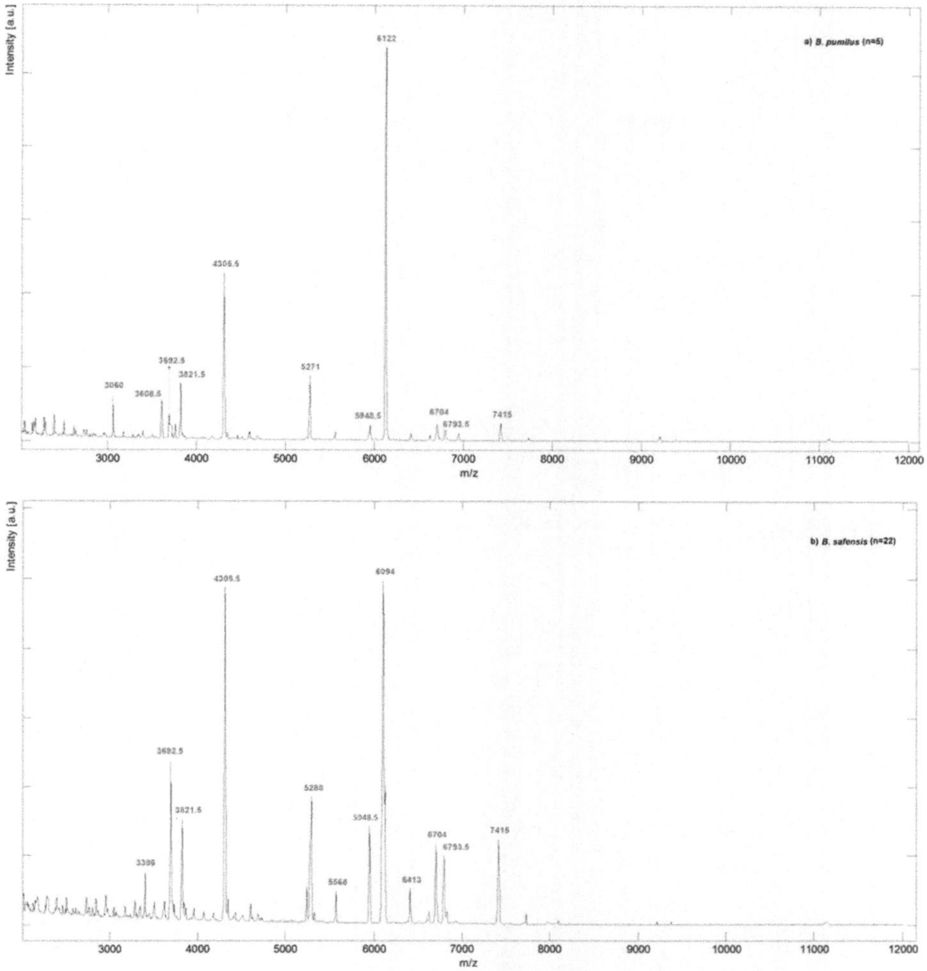

Figure 2. Score plot obtained by PLSDA regression model and the corresponding dendrogram. Score plot of the PLSDA regression model (a) and respective dendrogram (b), of *B. pumilus* and *B. safensis* isolates. Legend: ● *B. pumilus isolates* and ▼ *B. safensis isolates*. Unfilled symbols correspond to the type strains of both species.

Table 2 compiled the characteristic ion masses for each spectral group and those presented in both species.

As expected, similar fingerprints could be observed for isolates belonging to the same species, exhibiting several common ion peaks (Figure 1). For instance, the ions having *m/z* values at 3060, 3608.5, 5271 and 6122 were consistently observed in *B. pumilus* isolates, while the ions having the *m/z* values at 3396, 5288, 5568, 6094 and 6413 were related to *B. safensis*. Moreover, *B. pumilus* and *B. safensis* isolates demonstrated common ions at *m/z* 3692.5, 3821.5, 4305.5, 5948.5, 6704, 6793.5 and 7415 (Table 2), which could be considered characteristic ion peaks for both species. Therefore, the protein fingerprint similarities achieved among *B. pumilus* and *B. safensis* corroborated the closeness similarity previously verified among them [4,5].

Interestingly, some spectral variability was observed within isolates belonging to the same spectral group, including the presence or absence of some ion peaks beyond those listed in table 2 (data not shown). This observation was not surprising, since we had previously demonstrated that both species comprised a clonally diverse population [4]. Therefore, the specific peptide profile probably reflects their evolution towards an adaptation to different niches [4].

Indeed, few studies have explored the potential of MALDI-TOF-MS to properly identify *B. pumilus* or *B. safensis* isolates [18,19,29,30].

Farfour et al. attempted unsuccessfully to discriminate between *B. pumilus* and *B. safensis* vegetative cells by MALDI-TOF-MS using the Andromas database [18]. This discrepancy highlighted the need for the improvement and enlargement of this database and led eventually to the prior enrichment of the protein before MS analysis. Moreover, Böhme et al. [29], using *B. pumilus* type and reference strains (ATCC 7061^{T} and 14884) that were subjected to an extraction procedure previously to the MS analysis, suggested the presence of a series of ions at *m/z* 3620, 5297, 6617 and 7237, which were specific for this species, when compared with other *B. subtilis* group members (*B. subtilis*, *Bacillus amyloliquefaciens* and *Bacillus licheniformis*) and with members of *Bacillus cereus* group (*B. cereus*, *Bacillus megaterium* and *Bacillus thuringiensis*). In addition, the presence of this same series of ions at *m/z* 3620, 6617 and 7238 were also detected by Fernández-No et al. [30] in the same reference strains. Analysis of our MALDI-TOF-MS profiles did not reveal these ion peaks as *B. pumilus* species-specific discriminatory, when compared with *B. safensis*. Nevertheless, a closer inspection of the mass spectra of *B.*

Table 3. Overview of biomarkers tentative assignment of MALDI-TOF-MS mass signals of *B. pumilus* group species.

Specie(s)	Observed Mass (Da)	Predicted Mass (Da)	Error (%)	UniProt Accession ID	Protein Description	Peptide sequence	Organism[c]
	3060	NA	NA	Unassigned	NA		NA
	3608.5	NA	NA	Unassigned	NA		NA
		5270	0.02	A8FJG4	50S [b]RP subunit L34	MKRTFQPNNRKRSKVHGFRSRMSSKNGRLVLKRRRSKGRKKLSA	B. pumilus SAFR-032
	5271						
B. pumilus		5266.89	0.08	A8FDQ9	[a]SASP O	MTKRKANHVINGMNAAKSQGNGAGYIEDDQLVLTAEQRQNNKKRKKNQ	*B. pumilus* SAFR-032
		6114	0.2	C0H3U0	Uncharacterized membrane protein YyzG	MQTNRVILLAVMICLVSAITVFLLNGCKVDFLDIGGTIIGCFLGIFVVVRIQKKQS	*B. subtilis* subsp. *subtilis* str. 168
	6122						
		6126	0.08	P0C8M5	Transcriptional regulator SlrA	MKTHVKKDLDKGWHMLIQEARSIGLGIHDVRQFLESETASRKKNHKKTVRQD	*B. subtilis* subsp. *subtilis* str. 168
	3049	NA	NA	Unassigned	NA		NA
	3396	NA	NA	Unassigned	NA		NA
B. safensis	5288	5299.88*	0.2	A8FHD4	[a]SASP J	MSFFQKDKKAKSEKDHKQVDQLLEEASKELAGDPLQEAVQKKKNNDQ	*B. pumilus* SAFR-032
	5568	5544	0.5	A8FDQ8	[a]SASP P	MTNKNTGKDIRQNSPKEHQSGQPEPLSGSKKVKNRNHTRQKHNSHHDM	*B. pumilus* SAFR-032
	6094	NA	NA	Unassigned	NA		NA
	6413	6411.7*	0.03	A8FCW7	[b]RP subunit L32	MAVPFRRTSKMKKRLRRTHFKLQVPGMVACPECGEMKISHRVCKSCGTYKGKDVKSN	*B. pumilus* ATCC 7061[T]
	3692.5	3698	0.1	C0H3V1	UPF0752 membrane protein YczN	MSGYSNGGGYGGISSFALIVVLFILLIIVGTAFVGGF	*B. subtilis* subsp. *subtilis* str. 168
	3821.5	NA	NA	Unassigned	NA		NA
	4305.5	4305	0.01	A8F9A9	50S [b]RP subunit L36	MKVRPSVKPICEKCKVIRRKGKVMVICENPKHKQKQG	*B. pumilus* SAFR-032
B. pumilus* and *B. safensis	5948.5	5932	0.3	A8FF72	50S [b]RP subunit L33 2	MRVNITLACTECGERNYITKKNKRNNPDRVEFKKYCSRDKKQTVHRETK	*B. pumilus* SAFR-032
	6704	6691	0.3	C0H3Z1	Uncharacterized protein YjzG	MMKNGFAYKNGKLVNIFCGKEELYNELKAFLVKTFSINVKEVSRPSIYRRTKSKQLE	*B. subtilis* subsp. *subtilis* str. 168
	6793.5	6793*	0.01	B4AE40	50S [b]RP subunit L28	MARKCVITGRKTKAGNNRSHAMNSTKRTWGANLQKVRILVDGKPKR VYVSARALKSGKVERV	*B. pumilus* ATCC 7061[T]
	7415	7410.8*	0.08	B4AM90	50S [b]RP subunit L35	MPKMKTHRGSAKRFKKTGSGKLKRSHAYTSHLF ANKSTKQKRKLRKSAIVSAGDFKRIKQQLANIK	*B. pumilus* ATCC 7061[T]

Protein identity was determined by the TagIdent software and compared with ribosomal subunit proteins developed by Hotta et al. [23] described in Materials and Methods section.
[a]SASP - Small, acid-soluble spore protein.
[b]RP – Ribosomal protein.
[c]Organism – bacterial strain where the protein was described.
NA – not applicable.
*Predicted molecular weight (Mw) proposed by Hotta et al. [27], considering "N-end rule" [28] where N-terminal methionine is cleaved from specific penultimate amino acid residues such as glycine, alanine, serine, proline, valine, threonine and cysteine.

Table 4. Candidate species-specific biomarkers assignments of *B. pumilus* and *B.safensis.*

	Species	
Ion mass (*m/z*)	***B. pumilus***	***B. safensis***
5271	+	−
5288	−	+
5568	−	+
6122	+	−
6413	−	+

Ion masses are presented as *m/z* values. The presence/absence of an ion peak in each spectral group is represented by +/−, respectively.

pumilus ATCC 7061^T and 14884, revealed the presence of ions having *m/z* values at 3621, 5290 and 6624, although the ion at *m/z* 7238 was not present. Indeed, the differences found in these *B. pumilus* profiles could be justified by the distinct growth culture medium used and the sample preparation procedure employed.

Finally, in other work, Dickinson et al. [19] presented MALDI-TOF-MS as a useful taxonomic tool for differentiating spores of *B. pumilus* and *B. safensis*. Results revealed the presence of two groups of characteristic ion peaks, comprising *B. pumilus* (ions at *m/z* 6860, 7230 and 9606) and *B. safensis* (ions at *m/z* 6860, 7230, 7620 and 9606). The authors claimed the presence of the additional ion at *m/z* 7620 in the spectra profile of *B. safensis* as a species-specific biomarker, allowing the discrimination of these two species. Nonetheless, analysis of our *B. safensis* spectral data (n = 22) did not reveal the presence of this ion peak, probably because the obtained MALDI-TOF-MS profiles were from vegetative cells. Additionally, the spectral profile obtained from spores seems to be more laborious and insufficient to discriminate appropriately among these closely related species, since spectral data with few number of ion peaks were generated. For these reasons, it was not possible to compare our study with the existing ones as we have used different cultural conditions and sample preparation procedures.

On the other hand, the species identification is limited to the bacterial species spectrum presented in a specific MS database. Therefore, bacterial identification is only possible inside the frame of bacterial reference spectrum of the database used. Indeed, few well-characterized *B. pumilus* isolates are available in public databases, as the SpectraBank (http://www.spectrabank.org), namely the type strain ATCC 7061^T and the reference strain ATCC 14884, and no *B. safensis* was yet included, which constrains its identification.

Chemometric analysis

B. pumilus and *B. safensis* were clearly discriminated by a combined MALDI-TOF-MS and chemometric approach. The score plot generated by PLSDA of mass spectra of all isolates tested exhibited two individualized clusters, each one enclosing isolates belonging to a particular spectral group, which included reference and type strains (Fig. 2) of (a) *B. pumilus* and (b) *B. safensis*. The PLSDA scores were also presented as a dendrogram corroborating the species discrimination into two distinct clusters (Fig. 2). Moreover, this approach allowed the discrimination of the two *Bacillus* species with 100% of sensitivity and specificity.

Despite the recognized proficiency of chemometric tools for the analysis and identification of bacteria based on their fingerprint, in the case of *Bacillus* spp., this characterization was only previously applied in *B. cereus* group species [21]. Therefore, this is the first successful application of this approach considering members of *B. pumilus* group, stressing its relevance for discrimination among close related species.

Candidate molecular biomarkers assignment

The possibility of biomarkers identification is one of the most valuable aspects of the mass spectrometric-based identification techniques, being this approach successfully applied to different bacterial species [21,31,32].

The assignment of 17 mass signals diagnostic ions formed in the MALDI-TOF-MS of *B. pumilus* and *B. safensis* represented the first consistent evidence of the relation between these ion *m/z* signals and the specific candidate protein sequences, which were presented in Table 3. Direct bacterial discrimination by means of MALDI-TOF-MS was hampered by the absence of consistent databases supported on sufficient identified biomarkers. Indeed, only two *B. pumilus* genomes with numerous proteins defined as unknown were available in UniProtKB/Swiss-Prot and UniProtKB/TrEMBL (*B. pumilus* SAFR-032 and ATCC 7061^T) and no *B. safensis* were deposited, which hindered the candidate biomarkers identification. Therefore, this type of assignments can only be tentatively used to establish potential connections between protein sequences and ion *m/z* signals, and thus, should be prudently interpreted, although previously successfully applied in *B. cereus* group members [21].

We found evidences of *B. pumilus* and *B. safensis* specific biomarkers, associated with a series of ions at *m/z* 4305.5, 5948.5, 6793.5 and 7415, which was attributed to the 50S ribosomal subunits proteins, respectively, L36, L33, L28 and L35 of B. pumilus SAFR-032 (correspondent amino acidic sequences were also presented in Table 3). Moreover, the remaining ion detected at *m/z* of 3821.5 was not assigned, and two showed correspondences with membrane proteins of *B. subtilis* subsp. *subtilis* str 168, the YczN and YjzG at *m/z* of 3692.5 and 6704, respectively.

The tentative assignment of *B. pumilus* specific biomarkers revealed the possible correspondence of the diagnostic ion at *m/z* 5271 with the 50S ribosomal subunit protein L34 or with the SASP O. In addition, the characteristic ion at *m/z* 6122 was diagnostic for either the uncharacterized membrane protein (YyzG) and/or the transcriptional regulator – SlrA of *B. subtilis* subsp. *subtilis* str 168. Additionally, the characteristic ions at *m/z* 3060 and 3608.5 were not possible to assign.

Concerning *B. safensis* specific ions at *m/z* 5288, 5568 and 6413, potentially corresponding with two specific SASPs (SASP J and SASP P) and a 50S ribosomal subunit protein L32 were also found in *B. pumilus* SAFR-032. The remaining ion *m/z* peaks detected were not possible to designate. Therefore, the proposed characteristic biomarkers, which could be used to differentiate

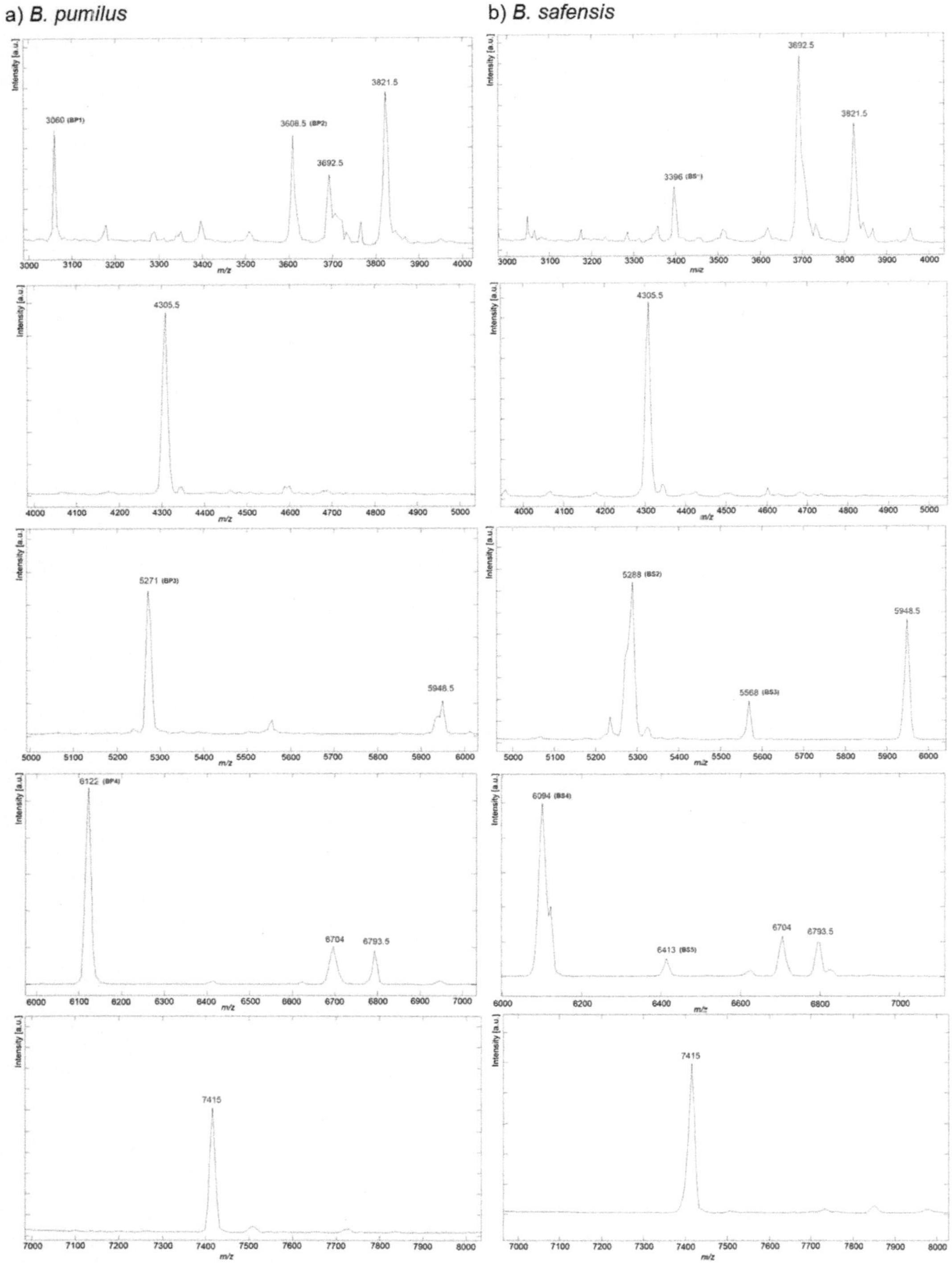

Figure 3. MS patterns of the candidate species-specific ion peaks (*m/z*) using MALDI-TOF-MS in linear mode. a) *B. pumilus* (average of 5 spectra) and b) *B. safensis* (average of 22 spectra). In brackets were named candidate species-specific ion peaks.

between *B. pumilus* and *B. safensis* are summarized in Table 4. Moreover, these differentiating series of ions were shown in Figure 3, which outlined the representative MALDI-TOF-MS biomarkers established for *B. pumilus* and *B. safensis*.

In fact, ribosomal proteins and small, acid-soluble spore proteins (SASPs) have been suggested to be responsible for many ion masses detected by MALDI-TOF-MS profiles [21]. Since up to 21% of the overall cellular protein content is ribosomal and because ribosomal proteins are part of the cellular translational machinery constitutively expressed in vegetative cells, they constitute a stable ensemble of protein biomarkers suitable for use by fingerprinting techniques [33]. Moreover SASPs, a group of species-specific proteins present in large amounts in the core region of *Bacillus* endospores, have been also proposed as biomarkers for rapid differentiation and identification of *Bacillus* spp. using mass spectrometry approaches [34,35,36].

Our results also suggested that ribosomal and spore proteins constituted most of the *B. pumilus* and *B. safensis* biomarkers. A more detailed analysis could be carried out with MS/MS peptide fragmentation of the specific proteins assigned and subsequent comparison in protein databases or even with MS/MS peptide *de novo* sequencing. Nevertheless, within the context of the present work, which aimed to establish a MALDI-TOF-MS fingerprint classification for *B. pumilus* and *B. safensis,* these results may be beneficial and improve further accuracy of MS-based detection methods in identifying these species.

Conclusion

MALDI-TOF-MS profiles combined with chemometric analysis (PLSDA) proved to be valuable tools for discrimination of *B. pumilus* and *B. safensis*, allowing its rapid identification. These high throughput approaches should be promptly considered for *Bacillus* species identification due to the inaccuracy of conventional techniques in the identification of closely related species of this genus. In this sense, it is imperative to standardize a sample preparation protocol, which should include a protein extraction and enrichment step, to provide informative and reproducible mass spectra. Moreover, tentative assignment of *B. pumilus* and *B. safensis* protein biomarkers suggested that most of them are ribosomal and spore proteins.

Author Contributions

Conceived and designed the experiments: RB CS JL LP HO. Performed the experiments: RB CS HO. Analyzed the data: RB JL CS MP LP HO. Contributed reagents/materials/analysis tools: CS JL MP LP HO. Contributed to the writing of the manuscript: RB CS LP HO.

References

1. U. S. Environmental Protection Agency (USEPA) (2004) Biopesticides Registration Action Document, *Bacillus pumilus* strain QST 2808 (PC Code 006485). Office of Pesticide Programs, Washington, DC: Government Printing Office.
2. Satomi M, La Duc MT, Venkateswaran K (2006) *Bacillus safensis* sp. nov., isolated from spacecraft and assembly-facility surfaces. Int J Syst Evol Microbiol 56: 1735–1740.
3. Pérez-Garcia A, Romero D, de Vicente A (2011) Plant protection and growth stimulation by microorganisms: biotechnological applications of *Bacilli* in agriculture. Curr Opin Biotechnol 22: 187–193.
4. Branquinho R, Meirinhos-Soares L, Carriço JA, Pintado M, Peixe LV (2014) Phylogenetic and clonality analysis of *Bacillus pumilus* isolates uncovered a highly heterogeneous population of different closely related species and clones. FEMS Microb Ecol. doi:10.1111/1574-6941.12426
5. Liu Y, Lai Q, Dong C, Sun F, Wang L, et al. (2013) Phylogenetic diversity of the *Bacillus pumilus* group and the marine ecotype revealed by Multilocus Sequence Analysis. PLoS ONE 8: 1–11.
6. Hong HA, Duc le H, Cutting SM (2005) The use of bacterial spore formers as probiotics. FEMS Microbiol Rev 29: 813–835.
7. Pérez-Garcia A, Romero D, de Vicente A (2011) Plant protection and growth stimulation by microorganisms: biotechnological applications of *Bacilli* in agriculture. Curr Opin Biotechnol 22: 187–193.
8. From C, Hormazabal V, Granum P (2007) Food poisoning associated with pumilacidin-producing *Bacillus pumilus* in rice. Int J Food Microbiol 115: 319–324.
9. From C, Pukall R, Schumann P, Hormazábal V, Granum PE (2005) Toxin-producing ability among *Bacillus* spp. outside the *Bacillus cereus* group. Appl Environ Microbiol 71: 1178–1183.
10. Haymore BR, Akers KS, Ferguson TM (2006) A case of persistent *Bacillus pumilus* bacteremia associated with cholangitis. J Infect 52: 154–155.
11. Bentur HN, Dalzell AM, Riordan FAI (2007) Central venous catheter infection with *Bacillus pumilus* in an immunocompetent child: a case report. Ann Clin Microbiol Antimicrob 6: 12.
12. Tena D, Martinez-Torres JA, Perez-Pomata MT, Sáez-Nieto JA, Rubio V, et al. (2007) Cutaneous infection due to *Bacillus pumilus*: report of 3 cases. Clin Infect Dis 44: 40–42.
13. Johnson BT, Shaw LN, Nelson DC, Mayo JA (2008) Extracellular proteolytic activities expressed by *Bacillus pumilus* isolated from endodontic and periodontal lesions. J Med Microbiol 57: 643–651.
14. Clark AE, Kaleta EJ, Arora A, Wolk DM (2013) Matrix-assisted laser desorption ionization-time of flight mass spectrometry: a fundamental shift in the routine practice of clinical microbiology. Clin Microbiol Rev 3: 547–603.
15. Keys CJ, Dare DJ, Sutton H, Wells G, Lunt M, et al. (2004) Compilation of a MALDI TOF mass spectral database for the rapid screening and characterization of bacteria implicated on human infectious diseases. Infect Genet Evol 3: 221–42.
16. Mellmann A, Bimet F, Bizet C, Borovskaya AD, Drake RR, et al. (2009) High interlaboratory reproducibility of matrix-assisted laser desorption ionization-time of flight mass spectrometry-based species identification of nonfermenting bacteria. J Clin Microbiol 47: 3732–4.
17. Emonet S, Shah HN, Cherkaoui A, Schrenzel J (2010) Application and use of various mass spectrometry methods in clinical microbiology. Clin Microbiol Infect 11: 1604–1613.
18. Farfour E, Leto J, Barritault M, Barberis C, Meyer J, et al. (2012) Evaluation of the Andromas matrix-assisted laser desorption ionization-time of flight mass spectrometry system for identification of aerobically growing Gram-positive bacilli. J Clin Microbiol 50: 2702–7.
19. Dickinson DN, La Duc MT, Satomi M, Winefordner JD, Powell DH, et al. (2004) MALDI-TOF MS compared with other polyphasic taxonomy approaches for the identification and classification of *Bacillus pumilus* spores. J Microbiol Methods 58: 1–12.
20. Lasch P, Nattermann H, Erhard M, Stämmler M, Grunow R, et al. (2008) MALDI-TOF mass spectrometry compatible inactivation method for highly pathogenic microbial cells and spores. Anal Chem 80: 2026–2034.
21. Lasch P, Beyer W, Nattermann H, Stämmler M, Siegbrecht E, et al. (2009) Identification of *Bacillus anthracis* by using matrix-assisted laser desorption ionization-time of flight mass spectrometry and artificial neural networks. Appl Environ Microbiol 75: 7229–42.
22. Welker M, Moore ER (2011) Applications of whole-cell matrix-assisted laser-desorption/ionization time-of-flight mass spectrometry in systematic microbiology. Syst Appl Microbiol 34: 2–11.
23. Wang LT, Lee FL, Tai CJ, Kasai H (2007) Comparison of *gyrB* gene sequences, 16S rRNA gene sequences and DNA–DNA hybridization in the *Bacillus subtilis* group. Int J Syst Evol Microbiol 57: 1846–1850.
24. Barker M, Rayens W (2003) Partial least squares for discrimination. J Chemom 17: 166–173.
25. Sousa C, Grosso F, Meirinhos-Soares L, Peixe L, Lopes J (2014) Identification of carbapenem-resistant *Acinetobacter baumannii* clones using infrared spectroscopy. Journal of Biophotonics 7: 287–294.
26. Næs T, Isaksson T, Fearn T, Davis T (2002) A user-friendly guide to multivariate calibration and classification. NIR Publications, Chichester UK.
27. Hotta Y, Sato J, Sato H, Hosoda A, Tamura H (2011) Classification of the genus *Bacillus* based on MALDI-TOF MS analysis of ribosomal proteins coded in S10 and *spc* operons. J Agric Food Chem 59: 5222–30.
28. Sherman F, Stewart JW, Tsunasawa S (1985) Methionine or not methionine at the beginning of a protein. Bioessays 1: 27–31.
29. Böhme K, Fernández-No IC, Barros-Velázquez J, Gallardo JM, Cañas B, et al. (2011) Rapid species identification of seafood spoilage and pathogenic Gram-positive bacteria by MALDI-TOF mass fingerprinting. Electrophoresis 32: 2951–65.
30. Fernández-No IC, Böhme K, Díaz-Bao M, Cepeda A, Barros-Velázquez J, et al. (2013) Characterization and profiling of *Bacillus subtilis*, *Bacillus cereus* and *Bacillus licheniformis* by MALDI-TOF mass fingerprinting. Food Microbiol 33: 235–242.
31. Dieckmann R, Malorny B (2011) Rapid Screening of Epidemiologically Important *Salmonella enterica* subsp. enterica Serovars by Whole-Cell Matrix-Assisted Laser Desorption Ionization–Time of Flight Mass Spectrometry. Appl Environ Microbiol 77: 4136–4146.
32. Momo RA, Povey JF, Smales CM, O'Malley CJ, Montague GA, et al. (2013) MALDI-TOF mass spectrometry coupled with multivariate pattern recognition analysis for the rapid biomarker profiling of *Escherichia coli* in different growth phases. Anal Bioanal Chem 25: 8251–8265.
33. Arnold RJ, Reilly JP (1999) Observation of *Escherichia coli* ribosomal proteins and their posttranslational modifications by mass spectrometry. Anal Biochem 269: 105–112.
34. Driks A (1999) The *Bacillus subtilis* spore coat. Microbiol Mol Biol Rev 63: 1–20.
35. Driks A (2002) Proteins of the spore core and coat. p. 527–535. In Sonenshein AL, Hoch JA, Losick R, *Bacillus subtilis* and its Closest Relatives: from Genes to Cells, ASM Press, Washington, DC.

36. Lai EM, Phadke ND, Kachman MT, Giorno R, Vazquez S, et al. (2003) Proteomic analysis of the spore coats of *Bacillus subtilis* and *Bacillus anthracis*. J Bacteriol 185: 1443–1454.
37. Branquinho R, Pintado ME, Peixe L (2012) Clonality and protein diversity of *Bacillus pumilus* isolates from different sources and geographic regions. Planta Med 78: PA5, doi:10.1055/s-0032-1320320
38. Matarante A, Baruzzi F, Cocconcelli PS, Morea M (2004) Genotyping and toxigenic potential of *Bacillus subtilis* and *Bacillus pumilus* strains occurring in industrial and artisanal cured sausages. Appl Environ Microbiol 70: 5168–5176.
39. Jetiyanon K (1997) Interaction between PGPR and cucumber during induced systemic resistance: recognition and early host defense responses. Ph.D. dissertation, Auburn University, Auburn, AL, USA.
40. Ouoba L, Diawara B, Amoa-Awua WK, Traore AS, Moller PL (2004) Genotyping of starter cultures of *Bacillus subtilis* and *Bacillus pumilus* for fermentation of African locust bean (*Parkiabiglobosa*) to produce Soumbala. Int J Food Microbiol 90: 197–205.

Bacteria Hold Their Breath upon Surface Contact as Shown in a Strain of *Escherichia coli*, Using Dispersed Surfaces and Flow Cytometry Analysis

Jing Geng[1¤], Christophe Beloin[2], Jean-Marc Ghigo[2], Nelly Henry[1*]

1 Laboratoire Jean Perrin (CNRS FRE 3231), UPMC, Paris, France, **2** Institut Pasteur, Unité de Génétique des Biofilms, Département de Microbiologie, Paris, France

Abstract

Bacteria are ubiquitously distributed throughout our planet, mainly in the form of adherent communities in which cells exhibit specific traits. The mechanisms underpinning the physiological shift in surface-attached bacteria are complex, multifactorial and still partially unclear. Here we address the question of the existence of early surface sensing through implementation of a functional response to initial surface contact. For this purpose, we developed a new experimental approach enabling simultaneous monitoring of free-floating, aggregated and adherent cells via the use of dispersed surfaces as adhesive substrates and flow cytometry analysis. With this system, we analyzed, in parallel, the constitutively expressed GFP content of the cells and production of a respiration probe—a fluorescent reduced tetrazolium ion. In an *Escherichia coli* strain constitutively expressing curli, a major *E. coli* adhesin, we found that single cell surface contact induced a decrease in the cell respiration level compared to free-floating single cells present in the same sample. Moreover, we show here that cell surface contact with an artificial surface and with another cell caused reduction in respiration. We confirm the existence of a bacterial cell "sense of touch" ensuring early signalling of surface contact formation through respiration down modulation.

Editor: Vipul Bansal, RMIT University, Australia

Funding: This work was in part supported by a grant from the 'Agence Nationale pour la Recherche,' PIRIbio program Dynabiofilm grant and the Pasteur-Curie PTR/PIC 'Nosocomial Infections.' The funders had no role in study design, data collection and analysis, decision to publish, or preparation of the manuscript.

Competing Interests: Christophe Beloin is a PLOS ONE Editorial Board member. This does not alter the authors' adherence to PLOS ONE Editorial policies and criteria.

* Email: nelly.henry@upmc.fr

¤ Current address: Danone Nutricia Research, RD 128, Palaiseau, France

Introduction

Adherent bacteria profoundly differ from planktonic bacteria in physiology and gene expression. From this collective surface-attached life mode, the bacteria gain significant adaptive advantages and exhibit increased resistance to many biocides [1,2,3]. This adhesion-induced physiological shift was suggested very early on by scientists studying bacterial populations in aqueous receptacles [4,5] and has since been confirmed on the basis of molecular biology data. Recently, abundant information on gene expression and metabolic pathway alterations in established biofilms has emerged due to the increasing spread of molecular genetics [6,7,8,9,10,11]. However, the mechanisms of such a transition are not known. The data, obtained on a several hour or day time scale, depict interfering biochemical cascades up- or downregulated in the surface-attached mode of growth compared to the free-floating mode [12]. This reinforces the idea of a surface-attached specific mode of life, but does not enable distinguishing triggering events from further developmental stages that drive biological changes on surfaces. In particular, the respective contributions of the various factors prevailing in biofilms — actual cell surface contact, cell-cell interactions, secreted soluble molecules or extracellular matrix synthesis, together with modifications in the physical and chemical environment due to confinement of cells in a 3D viscoelastic architecture — have not been identified, and their causality remains elusive. In this paper, we focused on the early stage of cell-surface contact formation. Evidence of a direct cell response upon initial adhesion is scarce. Using reporter gene technology and microscope observation in *Pseudomonas aeruginosa* individual cells, Davies and Geesey concluded that attachment of the cell to a glass surface induced *algC* upregulation as early as the first 15 min of contact [13]. In addition, Otto and Silhavy described increased expression of Cpx-regulated genes upon 1 h contact of *Escherichia coli* with artificial surfaces as compared to planktonic cells maintained in suspension; surprisingly, this regulation was observed with stationary phase cells in contact with a hydrophobic surface only [14]. Lately, Li and co-workers showed, in *Caulobacter crescentus*, that formation of physical contact between the bacterium and an artificial surface triggered "just-in-time" adhesin production [15]. These results suggest that bacterial cells possess a 'tactile' machinery which signals formation of surface contact. However, the functional responses put forward in these experiments have also been shown to be upregulated in stationary phase cell populations and in bacteria subjected to various external stresses

— e.g. nutrient deprivation, medium pH or osmolarity changes — raising the question of the direct relationship of these signals with formation of surface contact.

Here we develop an experimental approach aimed at addressing this question in a configuration which enables simultaneous detection of permanent physical contact and relevant biological activity at the single cell level. The principle of the experiments consisted in using dispersed surfaces in the form of micrometric latex particles as an adhesive substrate brought into contact with GFP-expressing bacterial cells in suspension so as to generate a microsystem in which adherent cells co-exist with single planktonic and aggregated cells. The system can then be characterized using flow cytometry, enabling multi-parametric short-time-scale analysis of the mixture [16]. To detect the impact of initial adhesion on cell metabolic activity, we used a fluorescent marker of bacterial respiration, a tetrazolium ion the fluorescence of which can be directly related to cell metabolic activity [17,18]. The experiments were performed in an *E. coli* strain constitutively expressing GFP and curli — a surface multimeric protein structure that fosters surface attachment and self-association [19]. The results indicated that bacterial metabolic activity was affected by formation of a single micrometric contact at the cell surface, either with a synthetic surface or with another cell, as early as the first ten minutes of permanent contact formation, suggesting that bacteria have developed an efficient and fast sense of touch. Interestingly, we observed that both cell-cell and cell-synthetic substrate contact triggered a similar metabolic drop. The implications of these findings on the potential existence and possible nature of a bacterial sense of touch will be discussed below. Clarification of these questions will be useful for a better understanding of the physiological shift induced by bacterial cell development on surfaces, a longstanding concern in microbiology.

Materials and Methods

Bacterial strain and growth conditions

Constitutive *E. coli* curli producers (MG1655*gfp_ompR*234) were in this study obtained by transducing, into *gfp*- tagged MG1655 (MG1655*gfp*), the *ompR*234 mutation that specifies a gain of function allele of *ompR*, a gene encoding an activator of the curli operon [20]. They were grown in lysogeny broth (LB) medium or in M63B1 medium with 0.4% glucose (M63B1Glu) at 37°C in the presence of ampicillin (Amp, 100 μg/ml).

Chemicals and particles

5-cyano-2,3-ditolyl tetrazolium chloride (CTC), Sytox-red and Syto 59 were purchased from Invitrogen; CTC was from the viability kit (B-34956). Monodispersed polystyrene microparticles of 25 μm and 3 μm diameter — (references 07313-5 and 17134-15, respectively) were purchased from Polysciences, GmbH Europe. Particles were used after extensive washing and re-suspension in M63B1 medium.

Flow cytometry

FCM analyses were performed using a Becton-Dickinson flow cytometer (Facscalibur) equipped with a 488 nm argon ion laser and 635 nm laser diode. Under 488 nm excitation, GFP emissions were recorded in fluorescence channel FL-1 (band pass 530/30 nm) and CTC in channel FL-3 (long pass >670 nm). Sytox red was excited using the 635 nm laser diode and emission was collected in channel FL-4 (band pass 661/16 nm). Data were analyzed using CellQuest (BD) multivariate analysis and FlowJo software (Tree Star). Fluorescence values were usually obtained from acquisition of 30,000 events, which optimized precision and time resolution. Series of successive acquisitions performed on the same sample exhibited standard deviation below 1% of the mean.

Confocal microscopy

Image acquisition was performed with a Nikon A1R confocal laser microscope equipped with an argon and NeHe laser using an oil objective lens (60× by 1.4 numerical aperture). Detectors and filter were set for simultaneous monitoring of GFP, CTC and Sytox red. Images were analyzed in imaging software NIS-Elements (Nikon) and Image J.

Microsystem formation and analysis

We generated a multicellular assemblage microsystem as previously described in detail [16]. Briefly, exponentially growing bacteria in suspension in M63B1Glu of an OD at 600 nm, 1-cm path length about 0.5, were brought into contact with particles at a 1:200 particle to cell ratio in a round-bottom tube and stirred at room temperature on a soft vortex (1000 $rpm.min^{-1}$). The total volume of this incubator was usually equal to 500 μl. Aliquots of 5 to 20 μl were taken at given incubation times for immediate analysis in FCM. Alternatively, the incubator was used for additional dye staining.

Fluorescent marker incubation procedures

CTC, Sytox Red and Syto 59 were incubated with cells or cell-particle mixtures at 5 mM, 5 nM and 1 μM final concentration, respectively. Fluorescent dye incubation times were usually equal to 30 min for CTC and 10 min for Sytox red and Syto 59.

Results

Combined analysis of CTC reduction and GFP expression provides a single cell respiration index to use in cell assemblages

In order to expose the early bacterial cell response to adhesion, we implemented a strategy consisting of using dispersed surfaces as the adhesive substrate and flow cytometry multiparametric analysis. This approach carried out using GFP-expressing bacteria provides large statistics data sets and time resolution on the order of seconds to determine cell-surface adhesion kinetics. To achieve quantitative monitoring of cell respiration at a single cell level, we first searched for a means of taking advantage of the multi-parametric nature of flow cytometry to design a two-color approach to bacterial respiration. For this purpose, we introduced into the experiment a fluorescent marker of bacterial respiration, a tetrazolium ion the fluorescence of which can be directly related to its level of reduction in the electron transfer chain of the bacteria, and thus to the metabolic activity of the cells under study. In its oxidized form, the CTC dye is a non-fluorescent molecule. In contrast, when the compound is reduced via the cell membrane electron transfer chain — in competition with molecular oxygen — it is converted into a water-insoluble product exhibiting characteristic red fluorescence (emission maximum around 602 nm) whose intensity inside the bacterial cell reports the amount of reduced product and thus the cell respiration level, as previously reported [17,18]. In principle, it is possible, using flow cytometry, to measure the individual cell capacity to reduce CTC and to obtain distribution of this property over a perfectly dispersed cell population. Yet, as soon as cells assemble, the single cell information is lost, except if an internal standard enables counting the number of individuals in the assemblage. For this purpose, we analyzed in parallel the flow cytometry profiles of an exponentially growing *E. coli* cell population constitutively expressing GFP and incubated with CTC. Exponentially growing bacteria cultivated at 37°C under stirring (hereafter referred to as "fresh cells") were incubated with 5 mM CTC for 30 min just before the flow cytometry test. In parallel, aliquots of the same culture were incubated with Sytox red as a dead cell marker. Incubation time was chosen on the basis of CTC reduction kinetics in cells that indicated that a pseudo-plateau had been reached at that time (see Fig. S1). The same series were prepared with cells taken out of the incubator and left at room temperature for 2 h before the test (bench cells) and with cells previously fixed in 3.7% formaldehyde solution in PBS for 1 h at room temperature (dead cells). FCM dot plots and histograms of GFP, CTC and Sytox red fluorescence intensities collected in their respective emission channels, i.e. FL1, FL3 and FL4, versus forward scattering (FSC), are shown in Fig. 1. Subcellular debris was removed from the analysis by gating the data to an SSC+ region, corresponding to SSC values higher than 10 (a.u. fixed in each channel by the instrument settings chosen at the beginning of the experiment and kept constant for all measurements). Then, we defined for each fluorescence channel a threshold value delimiting positive and negative regions with respect to the considered marker and we determined the corresponding FCM parameters, cell percentage and fluorescence intensity (Table 1). Our results showed that no significant difference in GFP expression or Sytox red labelling was observed between fresh and bench cells. In contrast, both cell distribution between CTC^+ and CTC^- regions and mean fluorescence intensity were noticeably affected (Fig. 1 and Table 1) in bench cells compared with fresh cells, indicating a drop in the reduced CTC production consistent with a cell respiring activity decline caused by a temperature and oxygenation (through sample stirring) decrease. In the extreme situation of the fixed cells, CTC reduction was abolished. Concomitantly, more than 95% of cells exhibited Sytox red labelling, and partial loss of GFP was observed. These results showed that CTC reduction sharply discriminated live from dead cells, as did Sytox red, but also measured graded respiration levels of healthy cells in which GFP content was unchanged on a 2 h time scale. Therefore, GFP intensity could be used as an internal standard to normalize reduced CTC production and define a single-cell respiration index, $f_R = fl_{CTC}/fl_{GFP}$. To check that this normalization introduced no spreading or deformation of the value distribution compared to raw fl_{CTC}, we compared the distributions of f_R and fl_{CTC} values normalized to the mean (Fig. 1D). The plot showed superimposable curves endorsing the two-color parameter f_R as a robust descriptor of respiring activity at a single cell level. In addition, the two-color assessment of bacterial respiration provided a means of deriving a single cell respiration measurement in multiple cell assemblages in which both fl_{GFP} and fl_{CTC} could be determined. In these cases, fl_{GFP} was not only an internal standard, but also served as a cell counter. An assemblage of n cells will have an FL1 fluorescence intensity, Fl^n_{GFP} enabling determination of n like $n = \frac{Fl^n_{GFP}}{\overline{fl_{GFP}}}$, with $\overline{fl_{GFP}}$ being the mean fluorescence of the co-existing single cells in the same sample. Based on the fact that only CTC but not GFP content was affected by respiring activity changes, single-cell-reduced-CTC content in the n-cell assemblage was derived using n like $\overline{fl_{CTC}} = \frac{FL_{CTC}}{n}$, where Fl_{CTC} is the FL3 fluorescence of the assemblage, enabling determination of mean single cell respiring activity, and f_R in the assemblage. This will be used hereafter to determine both surface-attached and self-associated cell respiring activity.

Flow cytometry monitoring of cell-particle microsystem-exposed multiple assemblages

To form a cell-particle microsystem in which free-floating, aggregated and adherent cells can be monitored over a short time scale using flow cytometry, we mixed, at a 1:200 ratio, 25 μm diameter polystyrene particles serving as the adhesion substrate (Fig. 2A) with curli expressing *E. coli* bacteria freshly taken from an exponentially growing culture and pipetted (Fig. 2B). At all desired time points, aliquots were removed from the stirred suspension for immediate analysis by flow cytometry. From the time of the mixture (t = 0), the cells quickly self-assembled and attached to particle surfaces. The microsystem reached its steady state in about 10 min, exhibiting multiple analyzable subgroups of single, surface-attached and self-aggregated cells at various assemblage degrees, as shown on the FL1 (GFP-fluorescence) versus FSC dot plots (Fig. 2C and D). In the example shown in Fig. 2, 98% of the particles carried between 20 to 200 cells and the self-aggregation number for 90% of the aggregates ranged from 2 to 12 cells at t = 15 min.

These results showed that a stable microsystem exhibiting multiple adhesion and self-association degrees could be reproducibly obtained in less than 10 min following mixing of bacteria with particles, with subclasses identifiable and subject to characterization on the basis of their FSC and FL1 parameters in flow cytometry. Unless otherwise stated, the experiments on bacterial respiring activity were performed using this reference microsystem in a steady state.

Differential respiring activity in single free-floating and assembled cells

In order to test the respiring activity of the bacterial cells according to their assemblage status, we introduced CTC into the microsystem previously formed by mixing cells and particles for 15 min. Incubation with CTC was completed 30 min before FCM analysis. Reduction of CTC in healthy respiring cells was visible under microscopy by formation of an intracellular red fluorescent speck, as seen in Fig. 3A. Thus GFP and reduced CTC cell content were analyzed in FCM, enabling derivation of the respiration index, using GFP as a cell counter as explained above, for the three subpopulation types present in the microsystem (Fig 3B, panel a-e). In order to compensate for all set-up fluctuations that might interfere with experiments performed some time ago, we used a respiration index normalized to 1 for free-floating cells ($\overline{f_{R,Norm}} = \frac{\overline{f_R}}{\overline{f_{R,free}}}$). Then, pooling ten experiments, we found $\overline{f_{R,Norm}}$ values equal to 0.78±0.04 and 0.75±0.06 for self-aggregated and surface-attached cells, respectively. These results indicated that both surface-attached and self-aggregated cells exhibited decreased respiring activity compared with free-floating cells. To assess the impact of cell assemblage size n on respiration down-modulation, we delineated n FL1 gates as follows: for all assemblages of n cells present in the sample, either forming an aggregate of n cells or an assemblage of n cells attached to the same particle, an R_n region was defined as including all events having an FL1 value in the interval $\langle Fl^n_1 \rangle = \left[\left(n \times \overline{fl_1} \right) \pm \frac{\overline{fl_1}}{2} \right]$ where $\overline{fl_1}$ is the mean FL1 value of single free-floating bacteria. By doing this, we obtained adjacent gates with no overlap. For the surface-attached cell population, the particle fluorescence background contribution in the FL1 channel was taken into account to define the gate. With the variation in f_R as a function of n, the assemblage size is shown in Fig. 3C for aggregates and surface-attached cells. The respiration decrease

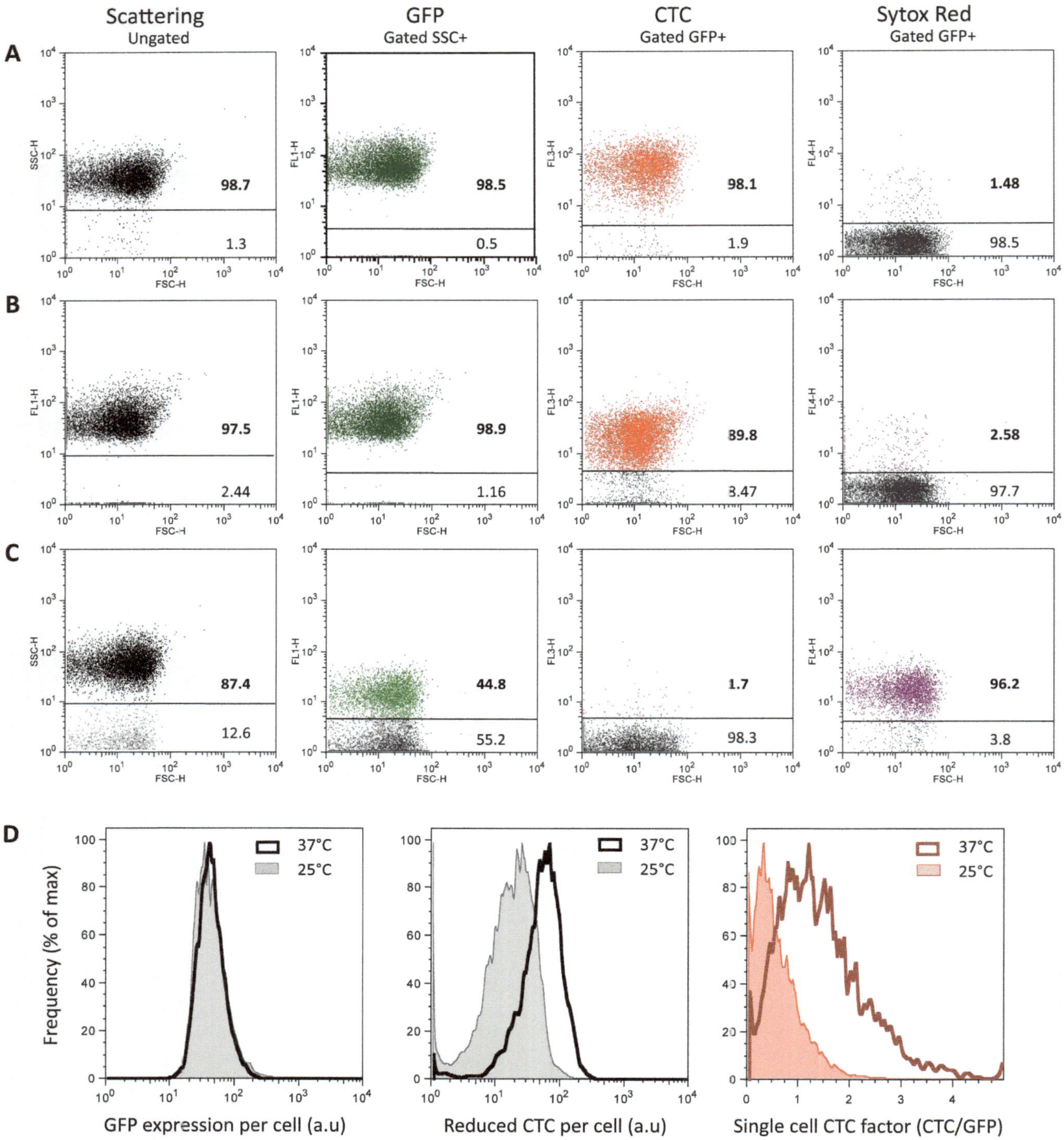

Figure 1. FCM signature of cell respiration down-modulation. Side scattering (SSC, first row) and fluorescence (FL1, second column; FL3, third column; FL4, fourth column) versus forward scattering (FSC) dot plots of bacteria constitutively expressing GFP and labelled with CTC or Sytox red are shown for cells (A) freshly taken from 37°C agitated culture, (B) left to rest at 25°C or (C) fixed in 3.7% formaldehyde. Fluorescence dot plots are from cells exhibiting a positive side scattering signal (>10 a.u). Cell frequencies below and above a fluorescence intensity threshold taken between 4 and 5 a.u. in each channel are indicated on the dot plots (see also Table 1 for fluorescence intensity data). (D) Histograms of fluorescence intensity distribution for GFP expression (left graph), CTC reduction (middle graph) and single cell respiration index (right graph) for cells freshly taken from 37°C agitated culture and left to rest at 25°C.

was detected in cell aggregates as small as n = 2 and reached its maximum for aggregates of $n = 4$ cells. This suggested that bacteria can sense even single cell-cell contact. For surface-attached cells, no assemblage size lower than 10 cells per particle with sufficient statistical significance was obtained. We observed that the respiration index decrease experienced by surface-

Table 1. Scattering and fluorescence FCM characteristics of bacterial cells in high (A) and low (B) metabolic state compared with dead cells (C).

Parameter	Gate + boundary (a.u.)	% Cells in gate+ *		Mean intensity in gate+ (a.u.)**
SSC	10.0	A	98.7	37.8±0.8
		B	97.5	36.5±0.9
		C	87.4	58.7±1.6
GFP (FL1)	4.0	A	98.5	47.8±0.7
		B	98.9	45.5±0.9
		C	44.8	15.1±0.5
CTC (FL3)	5.0	A	98.1	64.3±1.2
		B	89.8	25.1±0.9
		C	1.7	-
Sytox Red (FL4)	4.0	A	1.48	-
		B	2.58	-
		C	96.2	18.1±0.5

*A: Alive cells at 37°C; B: Alive cells at 25°C; C: Dead cells.
**Error is standard deviation over at least three measurements.

attached cells was not dependent on the number of bound cells, at least between 10 and 100 cells per particle.

To check the robustness of our approach, we used particles having large-spectrum intrinsic fluorescence, enabling a measurement signal in both FL1 and FL3 channels. We induced these particles — 2.8 μm in diameter — to form a small proportion of stable aggregates of a few individuals by centrifugation (Fig. 3D). We then determined an analogous normalized fluorescence index using the same derivations as those applied to cell assemblages. We observed that particle self-association did not affect the normalized fluorescence index (Fig. 3C, panel a). Although the particle number of aggregation did not exceed $n = 6$, this result showed that an association in itself does not induce a fluorescence shift.

We also challenged our analysis framework by examining the fluorescence of the different bacterial organizations upon labelling with 1 μM Syto 59, a dye which passively penetrates into cells and stains bacterial DNA after insertion between the base pairs without the requirement for metabolic activity. In this case, free-floating and aggregated cells exhibited the same Syto 59 (FL4 signal) to GFP (FL1 signal) fluorescence ratio (Fig. S2), also supporting the metabolic origin of the CTC fluorescence decrease in assembled cells. These results also indicated that the fluorescent dyes equally labelled free-floating and aggregated cells. The same experimental evidence could not be obtained with surface-attached cells, since Syto 59 itself was bound to the particle surface, inducing too high a fluorescence background that blurred the entire analysis. However, results obtained on cell accessibility to fluorescent dyes in aggregates could reasonably be extrapolated to surface-attached cells.

Thus, our results indicate that bacteria detected the formation of cell surface contact and responded to it by metabolism down-modulation, as confirmed by the decrease in respiring activity.

At this stage, we sought to further analyze the respiration activity decrease observed in particle-attached cells so as to determine whether the observed cell response actually resulted from cell-artificial surface anchorage or from cell-cell associations occurring on the particle surface.

Single cell response to contact

In order to discriminate between cell-cell contact on a surface and cell-surface contact, we needed to form a cell-surface assemblage of lower size. For this purpose, we designed an experimental "single-cell adhesion assay" achieved by using smaller-sized particles 3 μm in diameter and decreasing the cell-to-particle ratio to 4, which provided a majority of particles holding one single cell and a small fraction holding none or two cells. Because of the small particle radius, the cells and particles appeared on the same dot plot (Fig. 4). As previously, we defined gates on the basis of FL1 fluorescence so as to delineate single cell events, either free-floating single cells or surface-attached single cells. Again, we derived the normalized respiration index $f_{R,Norm}$ for the surface-attached cells and found it equal to 0.7 ± 0.1, which indicated that direct cell-synthetic surface contact also induced a decrease in bacterial respiring activity. The respiration index determined on the self-aggregated cells in the single-cell adhesion assay was found equal to 0.79 ± 0.06, in good agreement with previous results obtained in the presence of the 25 μm-radius particles, as expected, since self-aggregation occurred independently from surface attachment.

Short-time-scale single cell response to contact

The results shown above thus indicated that respiring activity decreased upon single cell-cell or cell-substrate contact at the surface of a bacterium after 40 min. This time scale was actually driven by the CTC response time, i.e. reduction rate, and chosen to obtain a sufficiently stable and reduced CTC signal. To evaluate whether a cell response to contact formation occurred earlier than 40 min, we performed kinetic analysis of CTC reduction over the entire period of incubation of the microsystem with CTC. The results displayed in Fig. 5 show that free-floating, aggregated and surface-attached cell respiration indexes diverged from the beginning of incubation. In contrast, the kinetic curves of the reduced CTC labelling frequencies, which report the rate of CTC inflow into the cell independently of the level of reduction, showed no difference between assembled and free-floating cells. Together, these two curve sets suggested that down-modulation of respiring activity had already started after 10 min contact with a

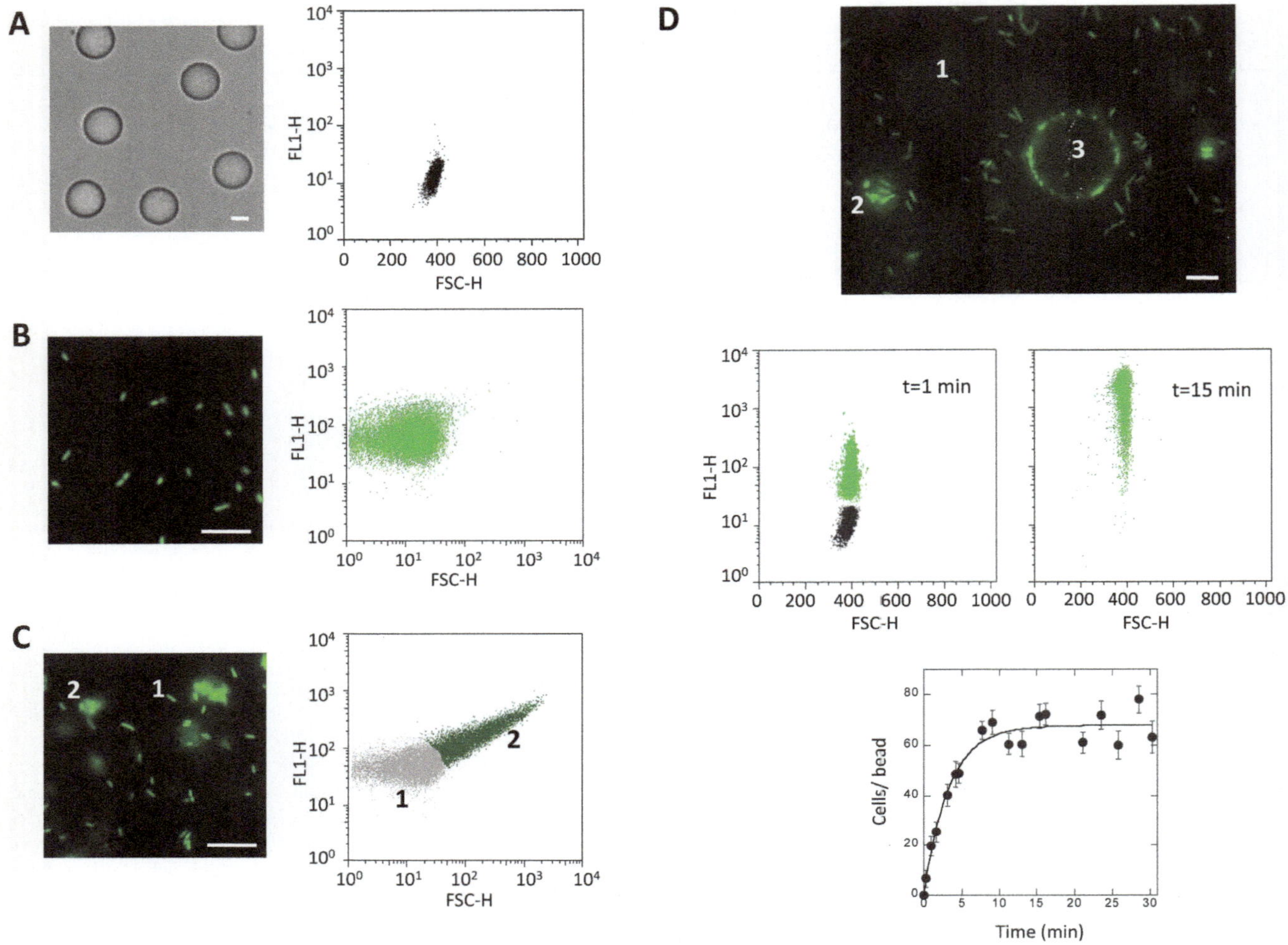

Figure 2. Cell-particle microsystem formation. Microscope image and fluorescence (FL1) versus FSC dot plots of (A) micrometric 25 μm diameter particles alone, (B) single bacteria freshly pipetted to break aggregates, (C) partially aggregated cells as recovered from a usual culture and (D) the cell-particle mixed microsystem comprising single cells (1), aggregated cells (2) and cells adhering to particle surface (3); the two FL1 versus FSC dot plots show the particle signal recorded in the microsystem 1 min and 15 min after bringing cells and particles into contact. Panel D bottom gives cell-particle association kinetics.

facing surface; elsewhere, penetration of the probe did not seem to be affected by contact of the cell either with another cell or with an artificial contact.

Discussion

Bacteria are primarily found in the form of adherent communities, where they exhibit significant remodeling of their properties compared to planktonic cells. Despite the attention that these changes have garnered, the events determining bacterial physiological shift are still not well understood [12,21].

In the work reported here, we sought to detect an early cell biological response to formation of cell-surface adhesive contact. Indeed, the phenotypic alterations observed in adherent communities were mainly examined in systems already established for hours or days during which intricate events occurred, including 3D-extracellular matrix formation [22,23,24,25], morphological changes [25] and quorum sensing communication [26], blurring the initial adhesion step and confusing clarification of the various contributions. To gain access to the initial adhesion phase, we implemented a strategy using dispersed surfaces and flow cytometry analysis, which has recently been designed to monitor bacterial short-time-scale adhesion [16]. The technique provides large statistical data sets, time resolution on the order of a few tens of seconds and simultaneous analysis of various-sized objects suspended in the same sample — the reason why flow cytometry has gained significant ground in microbial analyses lately [27,28]. In this microsystem, we found that *E. coli*, engineered to constitutively produce curli fibers on its surface, reached a steady state exhibiting stable fractions of multidimensional aggregates, planktonic and adherent cells.

To quantitatively picture cell metabolic activity within the first hour of surface colonization, we adapted this strategy to simultaneously follow not only adhesion, but also cell respiration on suspended free-floating, adherent and aggregated cells that co-exist in the microsystem formed when micrometric particles are brought into contact with a bacterial cell culture. This was achieved by introduction of a second fluorescent marker — tetrazolium ion CTC [17,18]— in addition to GFP constitutively expressed by the cells. In this way, we were able to use GFP expression as an internal standard and cell counter, which enabled

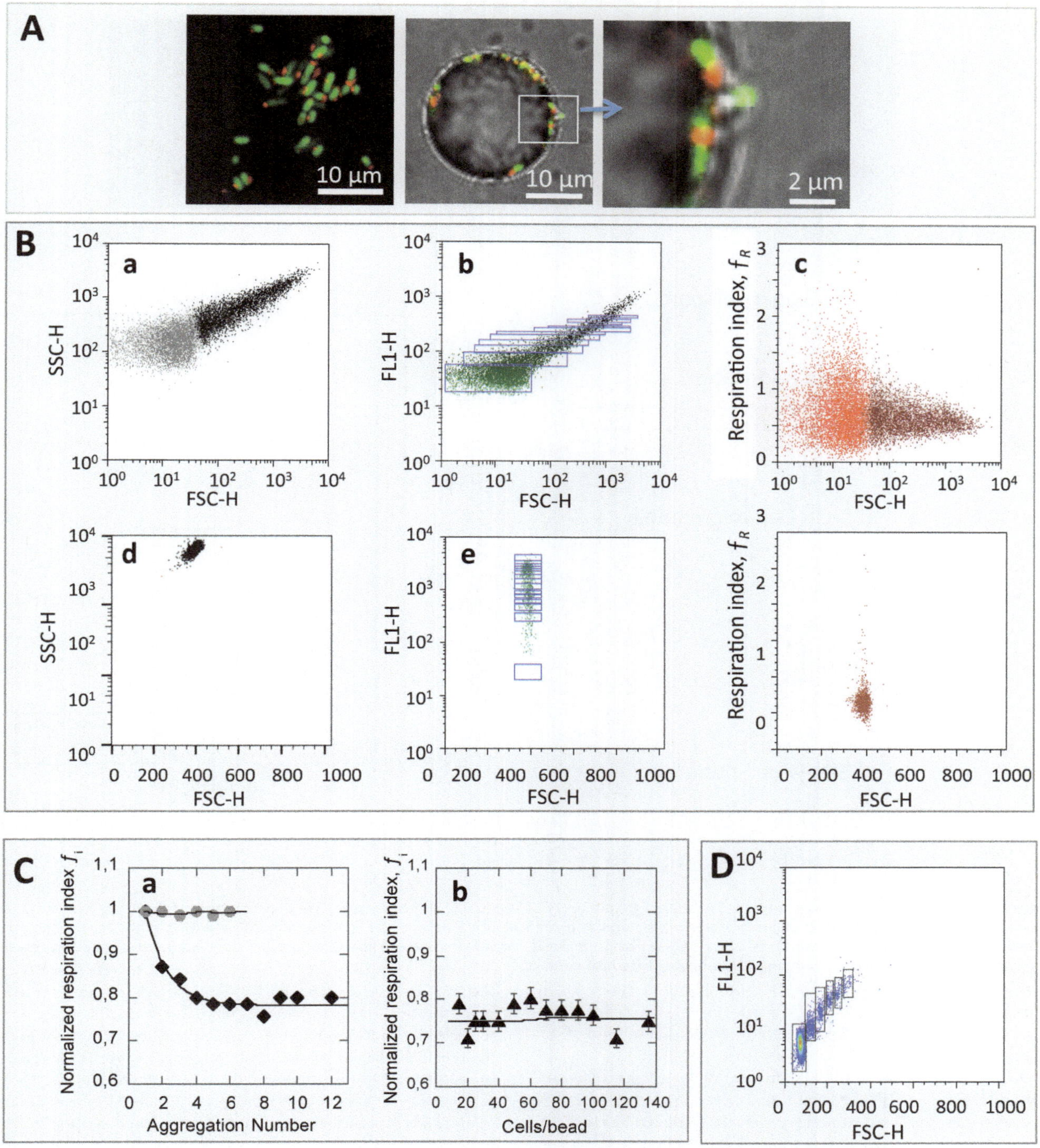

Figure 3. Adherent and aggregated cell respiration down-modulation upon contact. (A) Microscope images of CTC labelled cells in suspension and adhering to the 25 µm diameter particle surface. (B) FCM dot plots: SSC (left panels), FL1 (middle panels) and respiration index f_R (right panels) versus FSC showing cells (upper row, a to c) — single free-floating cells appear as light gray dots, aggregates as dark dots — and cells adhering to particles (lower row, d to f)); FL1 dot plots (middle panels) reporting GFP expression show cell counting gates, from R_2 to R_{12} for aggregates and from R_{10} to R_{135} for particles. Gates were defined as explained in the text based on sample single cell mean FL_1 value. (C) Graphs of normalized respiration index as a function of (a) number of cells (♦) or 2.8 µm diameter beads (●) per aggregate and (b) number of cells per particle. (D) FL1 versus FSC dot plot of 2.8 µm beads used as aggregation controls and population counting gates.

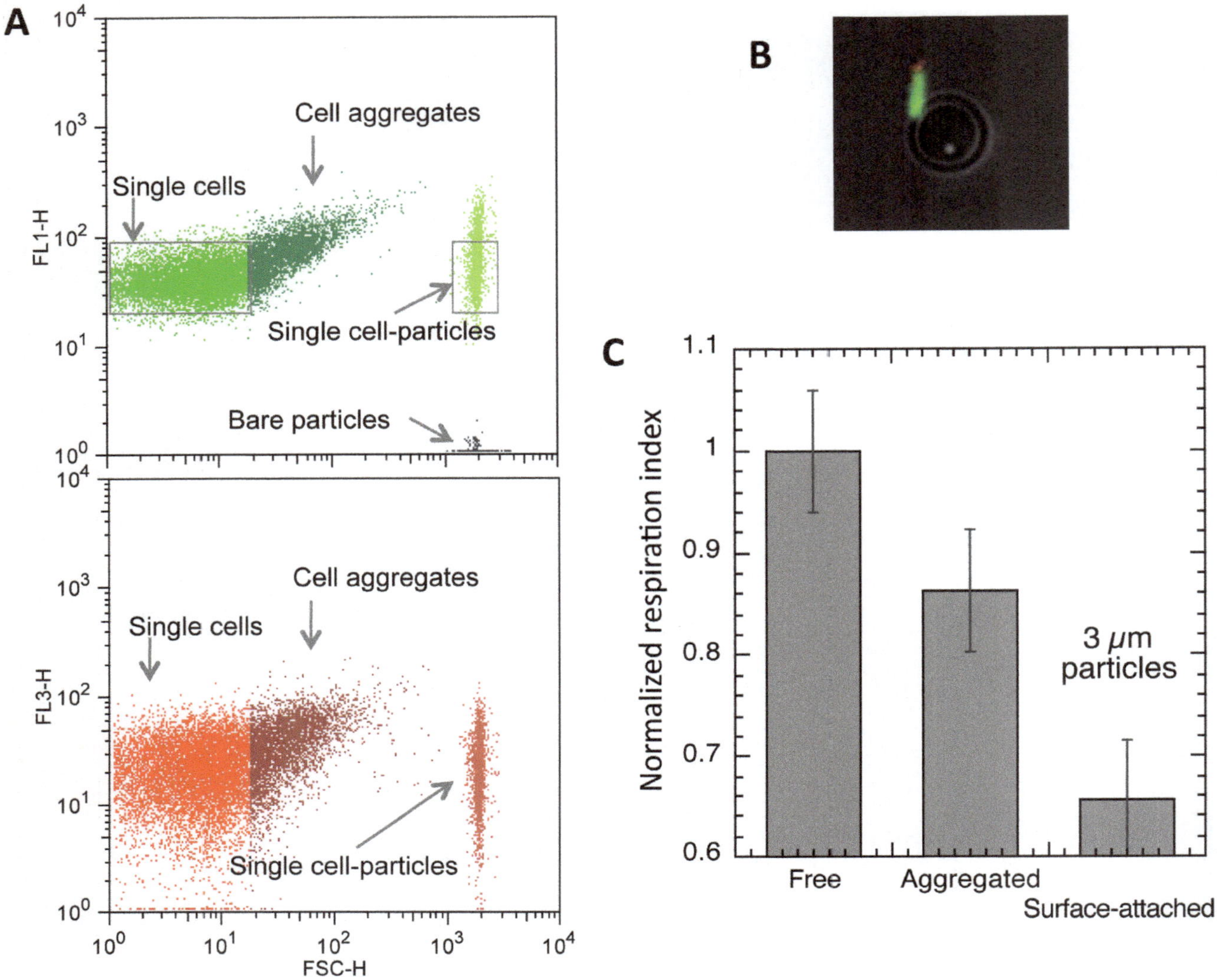

Figure 4. Single cell adhesion assay. (A) FL1 (upper graph) and FL3 versus FSC dot plots of a mixed microsystem of 3 μm particles with single free-floating and aggregated bacteria; microsystem prepared at a cell-to-particle ratio equal to 4 and incubated with 5 mM CTC. (B) Microscope image of a single cell-particle conjugate. (C) Histogram of normalized respiration index of the different cell populations comprised in the same sample, measured 40 min after bringing cells and particles into contact.

unit respiration index determination from CTC and GFP fluorescence intensity measurements.

On the basis of this two-color signature of respiring activity, we demonstrated that CTC reduction monitored at a single cell level in a population of living bacteria could report graded levels of cell respiration within a time scale of 15–20 min.

Implementing this approach in cell-particle mixtures displaying both adherent (particle-attached) and free-floating single cells, we found that the cell-surface association induced a 25 to 30% decrease in the initial cell respiration level.

Interestingly, we also found that cells living in the microsystem in the form of aggregates of various sizes exhibited a similar decrease in respiring activity. Moreover, by examining single-cell/particle associations and low aggregation number (from 2 to 10) cell clusters, we revealed that a single contact was sufficient to initiate this decrease in the respiration level.

These results point to the existence of a bacterial cell "sense of touch", i.e. the capacity to perceive an external object via formation of a physical contact, implying that bacterial cells have developed molecular pathways to elaborate a biological response to surface physical contact. Such an early single-cell level observation of a substratum-induced response — independently of interfering adhesion secondary events such as confinement, formation of chemical gradients or the onset of extracellular matrix synthesis — had been previously shown by Davies and co-workers in a small number of individuals, based on microscopy monitoring of *algC* gene activation [29]. In comparison, our results rely on data sets of thousands of cells. We show here that surface sensing involves a drop in respiration. Still, we do not know the mechanism behind this process.

Lower and co-workers [30] reported what they called a "tactile" response in *Staphylococcus aureus* after observing — on the basis of atomic force microscopy data — an accumulation of adhesins in the region of cell-substrate contact. Yet, under their experimental conditions, this behavior very likely resulted from thermodynamic equilibration of the free fraction of the binding ligands by

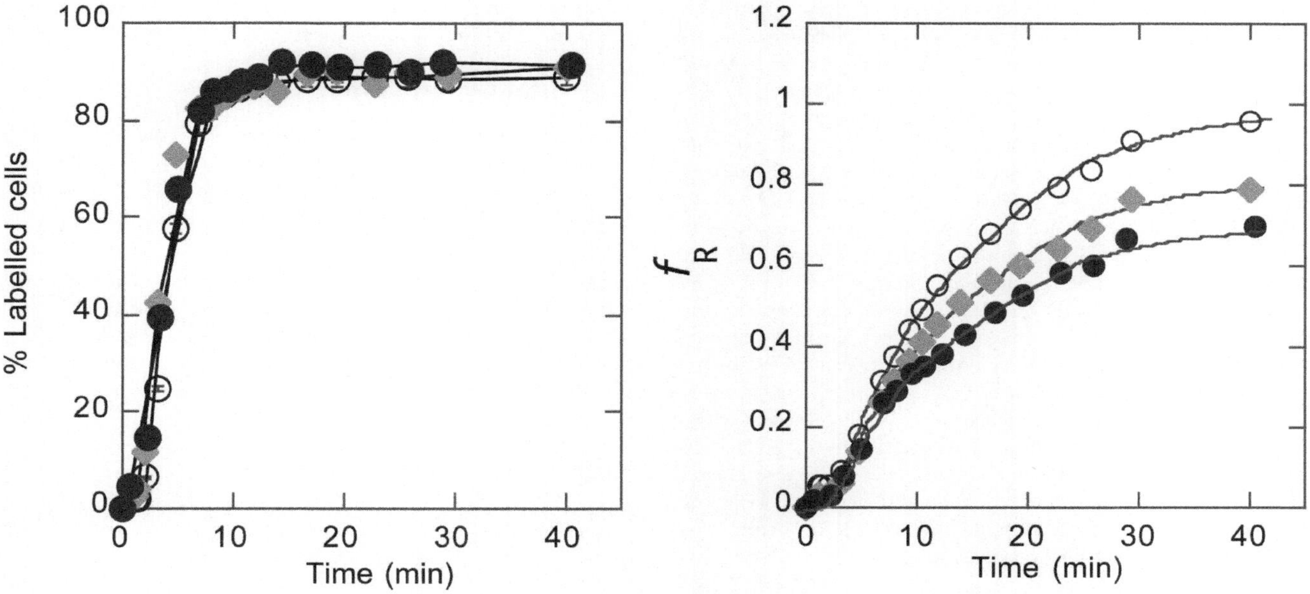

Figure 5. Kinetics of CTC cell influx and reduction. Monitoring (A) CTC positively labelled (FL3>5 a.u.) cell fraction and (B) cell respiration index, f_R as a function of CTC incubation time for free-floating (o), aggregated (◆) and adherent (●) cells. CTC reduction reported by FL3 intensity increase in CTC-labelled cells appears to be the limiting (slowest) step in the process.

diffusion, i.e. a passive process driven by the formation of specific surface receptor/ligand bonds. Our results show a striking parallel between the metabolic down-modulation induced by contact with the synthetic surface of a particle and that induced by cell-cell contact, suggesting that the molecular details of the contact play no determining role in signaling. However, there is no clear-cut conclusion concerning the fundamental difference between cell-artificial surface contact and cell-cell contact in the context of bacterial adhesion. Due to the presence in the cell culture supernatant of a variety of surface active molecules such as proteins and polyelectrolytes, the description of cell/synthetic surface interactions according to a physico-chemical model is hazardous, and the hypothesis of specific interactions between substrate-adsorbed biopolymers and cell surface receptors cannot be ruled out [31]. Again, since experimental data are often the result of biofilms established over several hour or day time scales, at a time when cell-substrate interactions are masked by the predominance of cell-cell interactions, any particular influence of the substrate upon development of the adhesive community is difficult to detect. Here, since we individually and simultaneously monitored each type of contact in the same sample, we were able to demonstrate that formation of cell-cell contact induced a cell response similar to the formation of cell-synthetic surface contact. Recently, based on growth rate, tolerance to antibiotics and extracellular matrix property criteria, Alhede and co-workers [32] pointed out that *Pseudomonas aeruginosa* floating aggregates and surface-attached biofilm shared most phenotypes, suggesting that the substrate per se does not play a determining role in establishment of biofilm properties. These behaviors might, of course, depend on the nature of the adhesive surface, although this point remains difficult to clarify in the context of bacterial adhesion. In our experiments, the polystyrene surface of the particles is expected to be neutral; in fact, it exhibited a slight negative charge (zeta-potential ≈ -10 mV) when suspended in cell culture minimal medium (data not shown). In previous work [16], we showed that the particle surface was quickly converted to being negatively charged in the presence of cell culture medium independently of the initial physico-chemical properties of the particles, explaining the similar colonization kinetics observed using different particles. This indicates that the substratum initial properties are often masked by a conditioning film in the presence of bacterial cell culture products.

The similarity of the cell response to a single contact with another cell and with a synthetic surface also enables hypothesizing surface sensing mechanisms driven by physical forces. Indeed, the formation of a contact on the cell surface is likely to induce local mechanical stress of cell surface appendages, which could serve to signal contact formation and trigger the biochemical cascade — as shown for flagella [33]. Other issues such as breaking of cell axial symmetry by formation of surface contact or a reduction in the amount of cell surface exposed to external medium and fluid flows could also be taken into consideration when formulating new working hypotheses for elucidating surface sensing mechanisms.

Our results, which associate respiration reduction with surface sensing, also raise the question of the role of that event in surface contact signaling. Recently, a contact-dependent growth inhibition process involving metabolic down-modulation upon cell-cell contact was discovered in *E. coli* [34,35]. This phenomenon — mediated by surface-specific protein recognition in an asymmetric inhibitor/target contact involving the immunity protein and toxin translocation across target cell envelopes [36] — is assuredly different from the one observed here. However, it provides an example of a contact-driven bacterial response pathway in which metabolic parameters are reduced in the absence of nutrient deprivation. Moreover, Aoki and co-workers suggested that a respiration decrease could be a general response to formation of physical contact at the cell surface; they proposed that respiration reduction could enable survival under conditions of oxidative stress. Thus far, to the best of our knowledge, no experimental evidence supporting a molecular mechanism linking respiration reduction and contact formation has been published.

Conversely, other authors detected metabolism stimulation events upon surface adhesion [37]. However, their experiments were performed under 12 h starvation conditions, thus significantly differing from our own experimental conditions.

Here we provide new tools for investigating in more detail this metabolic shift induced by surface contact at the single-cell level. Our approach enables discriminating between sensing mediated by soluble secreted factors and direct physical cell surface contact by monitoring planktonic and attached cells exposed in parallel to the same medium. This is a key issue in the better understanding of the mechanisms driving adherent cell transition towards an altered physiological state. Our results indicate that the reduction in the cell respiration level participates in early signaling triggered upon cell aggregation or settlement on an artificial substrate.

These results raise many questions for further investigation. What is the impact of this reduction on overall cell functioning, and, for instance, on the cell division rate? Is this a way for cells growing in the form of attached communities to achieve greater tolerance to oxidative stress or to various antimicrobials? What are the molecular bases for this contact-dependent behavior? The design of specific short-lifetime genetic reporters, to be monitored in our multiparametric dispersed surface approach, will undoubtedly help to answer these questions.

Supporting Information

Figure S1 CTC reduction kinetics. (A) Fraction of CTC-positively-labelled (FL3>5 a.u.) cells and (B) cell respiration index, f_R as a function of CTC incubation time. Incubation with 5 mM CTC at 37°C under stirring.

Figure S2 A neutral labelling index in free-floating and aggregated cells. Syto 59 (FL4 signal) to GFP (FL1 signal) fluorescence ratio calculated from fluorescence dot plots for free-floating (light grey triangle) and aggregated (black diamond) cells. Incubation with 1 μM Syto59 at 37°C under stirring.

Author Contributions

Conceived and designed the experiments: JG CB JMG NH. Performed the experiments: JG. Analyzed the data: JG NH. Contributed reagents/materials/analysis tools: JG NH. Wrote the paper: NH.

References

1. Stewart PS, Costerton JW (2001) Antibiotic resistance of bacteria in biofilms. Lancet 358: 135–138.
2. Hoiby N, Bjarnsholt T, Givskov M, Molin S, Ciofu O (2010) Antibiotic resistance of bacterial biofilms. Int J Antimicrob Agents 35: 322–332.
3. Anderson GG, O'Toole GA (2008) Innate and induced resistance mechanisms of bacterial biofilms. Curr Top Microbiol Immunol 322: 85–105.
4. van Loosdrecht MC, Lyklema J, Norde W, Zehnder AJ (1990) Influence of interfaces on microbial activity. Microbiol Rev 54: 75–87.
5. Zobell CE (1943) The Effect of Solid Surfaces upon Bacterial Activity. J Bacteriol 46: 39–56.
6. Prigent-Combaret C, Vidal O, Dorel C, Lejeune P (1999) Abiotic surface sensing and biofilm-dependent regulation of gene expression in *Escherichia coli*. J Bacteriol 181: 5993–6002.
7. Kuchma SL, O'Toole GA (2000) Surface-induced and biofilm-induced changes in gene expression. Curr Opin Biotechnol 11: 429–433.
8. Schembri MA, Kjaergaard K, Klemm P (2003) Global gene expression in *Escherichia coli* biofilms. Mol Microbiol 48: 253–267.
9. Beloin C, Valle J, Latour-Lambert P, Faure P, Kzreminski M, et al. (2004) Global impact of mature biofilm lifestyle on *Escherichia coli* K-12 gene expression. Mol Microbiol 51: 659–674.
10. Domka J, Lee J, Bansal T, Wood TK (2007) Temporal gene-expression in *Escherichia coli* K-12 biofilms. Environ Microbiol 9: 332–346.
11. Wood TK (2009) Insights on *Escherichia coli* biofilm formation and inhibition from whole-transcriptome profiling. Environ Microbiol 11: 1–15.
12. Petrova OE, Sauer K (2012) Sticky situations: key components that control bacterial surface attachment. J Bacteriol 194: 2413–2425.
13. Davies DG, Geesey GG (1995) Regulation of the alginate biosynthesis gene *algC* in Pseudomonas aeruginosa during biofilm development in continuous culture. Appl Environ Microbiol 61: 860–867.
14. Otto K, Silhavy TJ (2002) Surface sensing and adhesion of Escherichia coli controlled by the *Cpx*-signaling pathway. Proc Natl Acad Sci U S A 99: 2287–2292.
15. Li G, Brown PJ, Tang JX, Xu J, Quardokus EM, et al. (2012) Surface contact stimulates the just-in-time deployment of bacterial adhesins. Mol Microbiol 83: 41–51.
16. Beloin C, Houry A, Froment M, Ghigo JM, Henry N (2008) A short-time scale colloidal system reveals early bacterial adhesion dynamics. PLoS Biol 6: e167.
17. Kim DH, Yim SK, Kim KH, Ahn T, Yun CH (2009) Continuous spectrofluorometric and spectrophotometric assays for NADPH-cytochrome P450 reductase activity using 5-cyano-2,3-ditolyl tetrazolium chloride. Biotechnol Lett 31: 271–275.
18. Rodriguez GG, Phipps D, Ishiguro K, Ridgway HF (1992) Use of a fluorescent redox probe for direct visualization of actively respiring bacteria. Appl Environ Microbiol 58: 1801–1808.
19. Barnhart MM, Chapman MR (2006) Curli biogenesis and function. Annu Rev Microbiol 60: 131–147.
20. Vidal O, Longin R, Prigent-Combaret C, Dorel C, Hooreman M, et al. (1998) Isolation of an *Escherichia coli* K-12 mutant strain able to form biofilms on inert surfaces: involvement of a new *ompR* allele that increases curli expression. J Bacteriol 180: 2442–2449.
21. Monds RD, O'Toole GA (2009) The developmental model of microbial biofilms: ten years of a paradigm up for review. Trends Microbiol 17: 73–87.
22. Sutherland IW (2001) The biofilm matrix—an immobilized but dynamic microbial environment. Trends Microbiol 9: 222–227.
23. Branda SS, Vik S, Friedman L, Kolter R (2005) Biofilms: the matrix revisited. Trends Microbiol 13: 20–26.
24. Flemming HC, Wingender J (2010) The biofilm matrix. Nat Rev Microbiol 8: 623–633.
25. Serra DO, Richter AM, Klauck G, Mika F, Hengge R (2013) Microanatomy at cellular resolution and spatial order of physiological differentiation in a bacterial biofilm. MBio 4: e00103–00113.
26. Irie Y, Parsek MR (2008) Quorum sensing and microbial biofilms. Curr Top Microbiol Immunol 322: 67–84.
27. Muller S, Nebe-von-Caron G (2010) Functional single-cell analyses: flow cytometry and cell sorting of microbial populations and communities. FEMS Microbiol Rev 34: 554–587.
28. Davey HM, Davey CL (2011) Multivariate data analysis methods for the interpretation of microbial flow cytometric data. Adv Biochem Eng Biotechnol 124: 183–209.
29. Davies DG, Chakrabarty AM, Geesey GG (1993) Exopolysaccharide production in biofilms: substratum activation of alginate gene expression by *Pseudomonas aeruginosa*. Appl Environ Microbiol 59: 1181–1186.
30. Lower SK, Yongsunthon R, Casillas-Ituarte NN, Taylor ES, DiBartola AC, et al. (2010) A tactile response in *Staphylococcus aureus*. Biophys J 99: 2803–2811.
31. Geng J, Henry N (2011) Short Time-Scale Bacterial Adhesion Dynamics Advances in Experimental Medicine and Biology pp. 315–331.
32. Alhede M, Kragh KN, Qvortrup K, Allesen-Holm M, van Gennip M, et al. (2011) Phenotypes of non-attached *Pseudomonas aeruginosa* aggregates resemble surface attached biofilm. PLoS One 6: e27943.
33. Anderson JK, Smith TG, Hoover TR (2010) Sense and sensibility: flagellum-mediated gene regulation. Trends Microbiol 18: 30–37.
34. Aoki SK, Pamma R, Hernday AD, Bickham JE, Braaten BA, et al. (2005) Contact-dependent inhibition of growth in *Escherichia coli*. Science 309: 1245–1248.
35. Aoki SK, Webb JS, Braaten BA, Low DA (2009) Contact-dependent growth inhibition causes reversible metabolic downregulation in *Escherichia coli*. J Bacteriol 191: 1777–1786.
36. Webb JS, Nikolakakis KC, Willett JL, Aoki SK, Hayes CS, et al. (2013) Delivery of CdiA nuclease toxins into target cells during contact-dependent growth inhibition. PLoS One 8: e57609.
37. Hong Y, Brown DG (2009) Variation in bacterial ATP level and proton motive force due to adhesion to a solid surface. Appl Environ Microbiol 75: 2346–2353.

17

NAD(P)H-Hydrate Dehydratase- A Metabolic Repair Enzyme and Its Role in *Bacillus subtilis* Stress Adaptation

Miroslava Petrovova[1☯], **Jan Tkadlec**[1,2☯], **Lukas Dvoracek**[1], **Eliska Streitova**[1,2], **Irena Licha**[1*]

1 Department of Genetics and Microbiology, Faculty of Science, Charles University, Prague, Czech Republic, 2 Department of Medical Microbiology, 2nd Faculty of Medicine, Charles University, Prague, Czech Republic

Abstract

Background: One of the strategies for survival stress conditions in bacteria is a regulatory adaptive system called general stress response (GSR), which is dependent on the SigB transcription factor in *Bacillus* sp. The GSR is one of the largest regulon in *Bacillus* sp., including about 100 genes; however, most of the genes that show changes in expression during various stresses have not yet been characterized or assigned a biochemical function for the encoded proteins. Previously, we characterized the *Bacillus subtilis*168 osmosensitive mutant, defective in the *yxkO* gene (encoding a putative ribokinase), which was recently assigned *in vitro* as an ADP/ATP-dependent NAD(P)H-hydrate dehydratase and was demonstrated to belong to the SigB operon.

Methods and Results: We show the impact of YxkO on the activity of SigB-dependent P*ctc* promoter and adaptation to osmotic and ethanol stress and potassium limitation respectively. Using a 2DE approach, we compare the proteomes of WT and mutant strains grown under conditions of osmotic and ethanol stress. Both stresses led to changes in the protein level of enzymes that are involved in motility (flagellin), citrate cycle (isocitrate dehydrogenase, malate dehydrogenase), glycolysis (phosphoglycerate kinase), and decomposition of Amadori products (fructosamine-6-phosphate deglycase). Glutamine synthetase revealed a different pattern after osmotic stress. The patterns of enzymes for branched amino acid metabolism and cell wall synthesis (L-alanine dehydrogenase, aspartate-semialdehyde dehydrogenase, ketol-acid reductoisomerase) were altered after ethanol stress.

Conclusion: We performed the first characterization of a *Bacillus subtilis*168 knock-out mutant in the *yxkO* gene that encodes a metabolite repair enzyme. We show that such enzymes could play a significant role in the survival of stressed cells.

Editor: Vasu D. Appanna, Laurentian University, Canada

Funding: The study was supported by projects of Ministry of Education, Youth and Sports of the Czech Republic (LC06066) and by the Charles University by project of Specific University Research (SVV-2014-260081). The funders had no role in study design, data collection and analysis, decision to publish, or preparation of the manuscript.

Competing Interests: The authors have declared that no competing interests exist.

* Email: licha@natur.cuni.cz

☯ These authors contributed equally to this work.

Introduction

In an effort to understand the global adaptation network that evolved in *Bacillus* sp., several recent studies were carried out, focused on the genome-wide transcriptional profiling of the stress response of *Bacillus subtilis* 168 [1–4]. Several physiological analyses of the *Bacillus subtilis* 168 proteome during the adaptation to various environmental stresses have been published as well [5–7]. These studies identified stress specific regulons that are involved in stress function and confirm that the synthesis of most vegetative proteins is repressed, with the exception of enzymes that take part in adaptive responses.

One of the important strategies for survival in the genus *Bacillus* is a regulatory adaptive system called general stress response (GSR). It occurs as the large expression of stress proteins and is induced by a wide range of stresses, including high and low temperature; osmotic, ethanol, oxidative, and acidic stress; the addition of some antibiotics; starvation for glucose, phosphate, and oxygen; and blue or red light [2], [8–12] It is also induced on the transition into the stationary phase [13] and provides cells unspecified, multiple, and preventive resistance and gives the cells sufficient time for the induction of specific stress responses.

The general stress regulon, dependent on the SigB factor, is one of the largest operons in *Bacillus* sp., including about 100 genes [4]. However, most of the genes that show changes in expression during various stresses have not yet been characterized or assigned a biochemical function for the encoded proteins, and the evidence of the contribution of individual proteins from the general stress regulon to stress resistance of *Bacillus subilis* 168 cells is not complete.

Many genes of this regulon are putative regulatory factors, and all are under complex regulation by the control of other sigma factors and other regulatory proteins or RNAs, which allows their complex networking. It is assumed that their role is to protect DNA, proteins, metabolites, and lipids against the harmful effects of stress and to repair them.

Most recently, it was shown by Young [14] that the extent of stress determines response specificity and that the general stress response pathway activates different genes to a variety of stress conditions.

With the aim of elucidating the mechanism of adaptation of *Bacillus subtilis* to limited concentrations of potassium, we previously isolated a mutant with reduced salt tolerance only at a limited potassium concentration [15] in which the *yxkO* gene was interrupted. The product of this gene was formerly predicted to have a ribokinase activity based on sequence and structural homologies and the presence of ATP- and Mg^{2+}-binding sites [16]. Most recently, while experiments of this work were completed, the biochemical activity of the YxkO protein was assigned *in vitro* as an ADP/ATP-dependent NAD(P)H-hydrate dehydratase (EC 4.2.1.93). This enzyme convert abnormal metabolite NAD(P)H hydrate (NAD(P)HX) to NAD(P)H and is conserved over the kingdoms [17]. NAD(P)HX is slowly catalyzed from NAD(P)H by glyceraldehyde 3-phosphate dehydrogenase [18] or is produced non enzymatically in the course of the non-physiological conditions respectively [19], [20]. NAD(P)HX is unable to react as cofactor and it inhibits several dehydrogenases with detrimental effect on a cell [20], [21]. Enzymes with such activity are called metabolite repair or metabolite-proofreading enzymes and play a role similar to the proofreading activities of DNA polymerases and aminoacyl-tRNA synthetases [22].

The increased transcriptional activity of this gene after osmotic, heat, and ethanol stress was observed in the transcriptomic study of Petersohn [3], as well as in a recent extensive systematic and quantitative exploration of transcriptome changes in *Bacillus subtilis* [4]. The mutant in this gene was included in a phenotype screening study determining the contribution of individual SigB-dependent genes of unknown function to stress resistance, showing a lower survival rate following severe ethanol, heat, and osmotic stresses [23]. Most recently, it was shown to be under exclusive SigB regulation [24].

In accordance with our prior study and in addition to the reduced tolerance to the environmental stresses mentioned above, the mutant in the *yxkO* gene exhibits reduced growth under potassium limitation and altered motility under hyperosmotic conditions. This multiple effect of the gene disruption on phenotype led us originally to the hypothesis that the product of the *yxkO* gene has a regulatory function.

The present study aimed to determine the contribution of the *yxkO* gene product to stress adaptation by estimating the transcriptional activity of the P*ctc* promoter, as the Ctc protein is considered a marker of general stress response [11], and by discovering the changes of the cytoplasmic protein level pattern in the mutant and a wild-type strain when exposed to salt and ethanol stress.

Materials and Methods

Bacterial strains

The *Bacillus subtilis* strains and plasmids used in this study are listed in Table 1. *Escherichia coli* DH5α strain (*deoR endA1 gyrA96 hsd R17* (r_k^-, m_k^-) *recA1 relA1 supE44 thi-1* (*lacZYA-arg F*) *U169 φ80lacZ M15* F$^-$λ$^-$) (Clontech) was used for propagation of plasmid constructs.

Growth conditions

For genetic manipulations, *Escherichia coli* and *Bacillus subtilis* strains were cultivated routinely in LB medium.

For growth rate measurements, the *Bacillus subtilis* strains (WT, LD1 and MP2) were cultivated under vigorous agitation at 37°C and synchronized in exponential growth by reinoculation from overnight cultures (grown in LB medium) to LB medium for ethanol stress experiments or to MM medium with 0.5 mM K^+ (as described previously [15]) for both ethanol and osmotic stress experiments, respectively. The salt stress conditions were performed by the exposure of exponentially growing cells (OD_{600} of ~0.3) to 0.6 M NaCl. For the ethanol stress setup, ethanol to a final concentration of 4% (v/v) was added at the same growth condition as for the salt stress. For the MP2 strain, erythromycin (0.3 μg/ml) was added.

For transcriptional activity, *Bacillus subtilis* strains (WT and MP2) were cultivated in LB or MM medium with the defined concentration of potassium as described above for growth rate measurements; the cell samples were collected in intervals before and after stress exposure, as indicated in the particular experiment (see *Results and Discussion*).

For 2DE analysis, *Bacillus subtilis* strains (WT and MP2) were cultivated in MM medium with 0.5 mM K^+ as described above for growth rate measurements and harvested 60 min after stress exposure.

B. subtilis yxkO knock-out mutant strain construction

For transfer of the insertional mutation of the mini-Tn*10* transposon to the *yxkO* gene from an asporogenic genetic background from a previously prepared mutant [15], the PBS1 lysate was prepared from the L-42 mutant, and *Bacillus subtilis* 168 was transduced. The transductants were selected on LB chloramphenicol plates (5 μg/ml). The classical transduction protocol was used [25]. Insertion of mini-Tn*10* into the *yxkO* gene in a particular clone was confirmed by PCR using primers designed for the *yxkO* gene and the transposon region, as well. The respective PCR product was confirmed by sequencing, and the mutant was named LD1.

For inactivation of the *yxkO* gene, an integrative vector, pMUTIN4 from *Bacillus* Genetic Stock Center (BGSC - ECE 139), was used. A fragment of the *yxkO* allele with a ribosome binding site was generated by PCR with the primers 5′-*GGAGGATCCATAACAGGACAATCAGCC*-3′ and 5′-*CCCGAATTCAGGAAAAGAAAGCAGAGGAG*-3′ (the *Eco*RI and *Bam*HI restriction sites for direct cloning into pMUTIN4 are underlined) and ligated into *Eco*RI-*Bam*HI-digested pMUTIN4. The ligation mixture was transformed into *E. coli* DH5α, and clones with pMP2 plasmid were selected on LB ampicillin plates (100 μg/ml). Presence of the *yxkO* allele fragment in the plasmid was confirmed by sequencing.

Circular pMP2 plasmid was then used for transformation into *Bacillus subtilis* SG64, and single-crossover recombinants were selected on LB plates supplemented with erythromycin (0.3 μg/ml). Correct insertion into the chromosome was confirmed in a selected isolate by PCR and sequencing and named MP2.

Bacillus subtilis P*ctc* promoter probe mutant strain construction

The promoter region of the *ctc* gene [26] was generated by PCR with the primers 5′- *CATAGAATTCCCATTTTTCGAGGTTTAAATCCTT*-3′ and 5′- *TTTAGGATCCCGAGTAAAGTCCGTTCTTTCTT*-3′ (the *Eco*RI and *Bam*HI restriction sites for direct cloning into plasmid are underlined) and

Table 1. *Bacillus subtilis* strains, phage, and plasmids used in this study.

Name or code	Relevant genotype or description	Reference, Source, or construction
Strains		
B.s. 168 (WT)	*trpC2*	BGSC (1A1)
B.s. SG4 (WT)	*xglA1, xglR1*	BGSC (1A680)
L-42	*pheA1, spo0F221, trpC2*, mini-Tn*10::yxkO*	[15]
LD1	*trpC2*, mini-Tn*10::yxkO*	this study
MP2	Mutin4::*yxkO, trpC, xglA1, xglR1*	this study
WT/Pctc	(PctcΦ *spoVG-lacZ, cat*)::*amy, trpC, xglA1, xglR1*	this study
MP2/Pctc	Mutin4::*yxkO*, (PctcΦ spoVG-*lacZ, cat*)::*amy, trpC, xglA1, xglR1*	this study
Phage PBS1	Bacillusphage PBS1	BGSC (1P1)
Plasmids		
pDG1661	*spoVG-lacZ, spc, cat, bla, amy* 5′ and 3′ segment	BGSC (ECE112)
pMUTIN	*spoVG-lacZ, Pspac, lacI, erm, bla*	BGSC (ECE139)
pJT2	pDG1661 with *Pctc* insert transcriptionally fused to *spoVG-lacZ*	this study
pMP2	pMUTIN with 5′ end segment of *yxkO* with SD site fused to *spoVG-lacZ*	this study

ligated to the same sites of the suicide pDG1661 plasmid, which possesses an insertion site to the *amy* locus (BGSC – ECE112).

The ligation mixture was transformed into *E. coli* DH5 α strain, yielding pJT2 plasmid by selecting on LB plates supplemented by ampicillin (100 μg/ml).

For *Bacillus subtilis* mutant strain (WT/*Pctc* and MP2/*Pctc*) construction, the plasmid was linearized by *Xho*I restriction enzyme (RE) and transformed into the *Bacillus subtilis* SG4 strain and the MP2 mutant strain prepared as above. Correct insertion into the *amy* locus and right orientation towards the reporter *lacZ* gene were confirmed by PCR and sequencing in double-crossover recombinants selected on LB plates supplemented with chloramphenicol (5 μg/ml).

β - Galactosidase assay of *lacZ* transcriptional fusions

Samples were cultivated and collected at specified intervals. Cell samples were permeabilized with lysozyme, and galactosidase activities were measured at OD_{420} and expressed in Miller units (M.U.) according to the protocol from the *Bacillus* Genetic Stock Center (BGSC) Catalog of Integration Vectors (http://www.bgsc.org/_catalogs/Catpart4.pdf), page 15.

2DE sample preparation

For one experiment, WT and MP2 mutant were grown and stressed in parallel, and cells were cultivated as specified in the *Growth condition* section. Cells were harvested by centrifugation (4000×g, 4°C, 5 min), washed with 0.01 M Tris/KCl buffer pH 8.0, resuspended in 1 ml of the same buffer with protease inhibitors (Sigma), and disrupted by sonication. The cell debris was removed by centrifugation (15,000×g, 4°C, and 10 min), and protein concentration in the supernatant was estimated with the BCA Protein Assay Kit – Reducing Agent Compatible (Pierce). The cell lysates were then treated with the Bio-Rad Ready Prep 2-D Cleanup Kit and diluted in an appropriate volume of rehydration buffer up to 100 μg of total protein per sample. From each growth condition (WT, WT stressed, MP2, MP2 stressed), samples in technical triplicates were prepared.

2D polyacrylamide gel electrophoresis and protein visualization

Each sample prepared above was loaded onto an IPG strip pH 4–7 (Bio-Rad). Isoelectric focusing (IEF) was performed after passive overnight rehydration at room temperature using voltage that linearly increased to steady state [100 V for 2 h (slow), 300 V for 2 h (slow), 8000 V for 2 h slow, 8000 V for 7 h (rapid), and finished at 500 V] using the Bio-Rad Protean IEF system. After IEF, the strips were washed in equilibration solution (50 mM, Tris-HCl (pH 6.8), 6 M urea, 30% glycerol and 2% SDS) containing 0.02 M DTT for 10 min, followed by a second 10-min wash in equilibrium buffer containing 0.025 M iodoacetamide and bromophenol blue for gel staining.

The separation in the second dimension was carried out using precast gradient gels (Criterion Precast Gel (10.5%–14%) Bio-Rad). Gels were run on the Bio-Rad Criterion Dodeca Cell device at 5 V for 30 min and at 100 V for 2 hours at room temperature. Gels from one independent experiment in technical triplicates for each stress condition were run together (12 gels).

After staining with Colloidal Coomassie G-250 (Simply Blue Safestain, Invitrogen Life technologies, Paisley, UK), the gels were scanned with a GS-800 calibrated densitometer.

Digitalization of gel images and data analysis

The 2DE image analysis was performed using PDQuest 8.0 software (Bio-Rad, Hercules, CA). For matching and quantification, raw images were smoothed to remove noise, background was subtracted, and a spot-by-spot visual validation of automated analysis was done from then on to increase the reliability of the matching. Only spots that exhibited similar intensity in each gel of the technical triplicates in particular growth condition and in both biological replicates were taken for further analysis. Data from PDQuest analysis are published in Table S1. Identified protein spots were manually cut out of the gels and analyzed by MS.

MS analysis

The bands of interest were cut out of the gels and chopped into 1×1×1-mm pieces. The pieces were destained; to reduce and block cysteines, DTT and iodoacetamide were applied. The samples were trypsinized as described previously [27]. The dried-droplet method of sample preparation was employed, and spectra were acquired on a 4800 Plus MALDI TOF/TOF analyzer (AB Sciex). The data were analyzed using in-house running Mascot server 2.2.07 and matched against the current release of the NCBI protein sequence database: taxonomy: Bacillus (*Taxonomy ID:* 1386), 833049 sequences, and protein scores greater than 72 were significant ($p<0.05$). Cysteine carbamidomethylation, methionine oxidation, and N, Q deamination were set as fixed or variable modifications, respectively. One missed cleavage site was allowed. Precursor accuracy was set to 50 ppm, and the accuracy for MS/MS spectra was set to 0.25 Da. Detailed results are collected in Table 2.

Results and Discussion

Confirmation of *yxkO* knock-out mutant phenotype

Firstly, to verify the *yxkO* gene phenotype previously described in a sporulation-proficient strain (*spo0F221*) [15], we transduced the cassette of mini-Tn*10* from the L-42 mutant using PBS1 bacteriophage to *Bacillus subtilis* 168, and in parallel, we constructed an insertional mutant using the pMUTIN4 plasmid, as well (see *Materials and Methods*). Both mutants revealed the previously described phenotype, an increase in generation time from 55 min to 115 min and osmosensitivity in low K^+ conditions (15) but also lower viability, even under stress conditions, as we demonstrated in the long-term growth measurements (Figure 1A).

By doing so, we confirmed that the previously described phenotypic changes in mutant L-42 corresponded to the inactivation of the *yxkO* gene and not to the *spo0F* mutation. Both types of the mutant revealed prolonged lag phase (Figure 1B), and lower viability when exposed to ethanol and salt

Table 2. MALDI-TOF peptide mapping identification.

Protein	Protein Accession	Mascote Score	Best Protein Description
Osmotic stress			
Pgk	gi|16080446	104	phosphoglycerate kinase [Bacillus subtilis subsp. subtilis str. 168]
CitC	gi|16079965	885	isocitrate dehydrogenase [Bacillus subtilis subsp. subtilis str. 168]
Mdh	gi|16079964	105	malate dehydrogenase [Bacillus subtilis subsp. subtilis str. 168]
Mdh deg	gi|16079964	150	malate dehydrogenase [Bacillus subtilis subsp. subtilis str. 168]
AtpD iso1	gi|16080734	93	FOF1 ATP synthase subunit beta [Bacillus subtilis subsp. subtilis str. 168]
AtpD iso 2	gi|16080734	215	FOF1 ATP synthase subunit beta [Bacillus subtilis subsp. subtilis str. 168]
AtpD iso 3	gi|16080734	335	FOF1 ATP synthase subunit beta [Bacillus subtilis subsp. subtilis str. 168]
FrlB	gi|16080314	97	hypothetical protein BSU32610 [Bacillus subtilis subsp. subtilis str. 168]
Hag iso1	gi|16080589	83	flagellin [Bacillus subtilis subsp. subtilis str. 168]
Hag iso2	gi|16080589	78	flagellin [Bacillus subtilis subsp. subtilis str. 168]
Hag iso 3	gi|16080589	83	flagellin [Bacillus subtilis subsp. subtilis str. 168]
GroEL	gi|16077670	74	chaperonin GroEL [Bacillus subtilis subsp. subtilis str. 168]
GlnA	gi|16078809	123	glutamine synthetase [Bacillus subtilis subsp. subtilis str. 168]
Ethanol stress			
Pgk	gi|16080446	329	phosphoglycerate kinase [Bacillus subtilis subsp. subtilis str. 168]
CitC	gi|16079965	118	isocitrate dehydrogenase [Bacillus subtilis subsp. subtilis str. 168]
Mdh	gi|16079964	129	malate dehydrogenase [Bacillus subtilis subsp. subtilis str. 168]
Mdh deg	gi|16079964	209	malate dehydrogenase [Bacillus subtilis subsp. subtilis str. 168]
Atp iso1	gi|16080734	125	FOF1 ATP synthase subunit beta [Bacillus subtilis subsp. subtilis str. 168]
Atp iso 2	gi|16080734	155	FOF1 ATP synthase subunit beta [Bacillus subtilis subsp. subtilis str. 168
AtpD iso 3	gi|16080734	341	FOF1 ATP synthase subunit beta [Bacillus subtilis subsp. subtilis str. 168
FrlB	gi|16080314	119	fructoselysine-6-P-deglycase [Bacillus subtilis subsp. subtilis str. 168]
Hag iso1	gi|16080589	157	flagellin [Bacillus subtilis subsp. subtilis str. 168]
Hag iso2	gi|16080589	175	flagellin [Bacillus subtilis subsp. subtilis str. 168]
Hag iso 3	gi|16080589	89	flagellin [Bacillus subtilis subsp. subtilis str. 168]
GroEL	gi|16077670	140	chaperonin GroEL [Bacillus subtilis subsp. subtilis str. 168]
Ilvc iso1	gi|16079881	194	ketol-acid reductoisomerase [Bacillus subtilis subsp. subtilis str. 168]
Ilvc iso2	gi|16079881	183	ketol-acid reductoisomerase [Bacillus subtilis subsp. subtilis str. 168]
Asd	gi|16078738	81	aspartate-semialdehyde dehydrogenase [Bacillus subtilis subsp. subtilis str. 168]
Ald	gi|16080244	79	L-alanine dehydrogenase [Bacillus subtilis subsp. subtilis str. 168]

Detailed data from MALDI-TOF peptide mapping identification of proteins, that differed in levels when exposec to osmotic and ethanol stress.

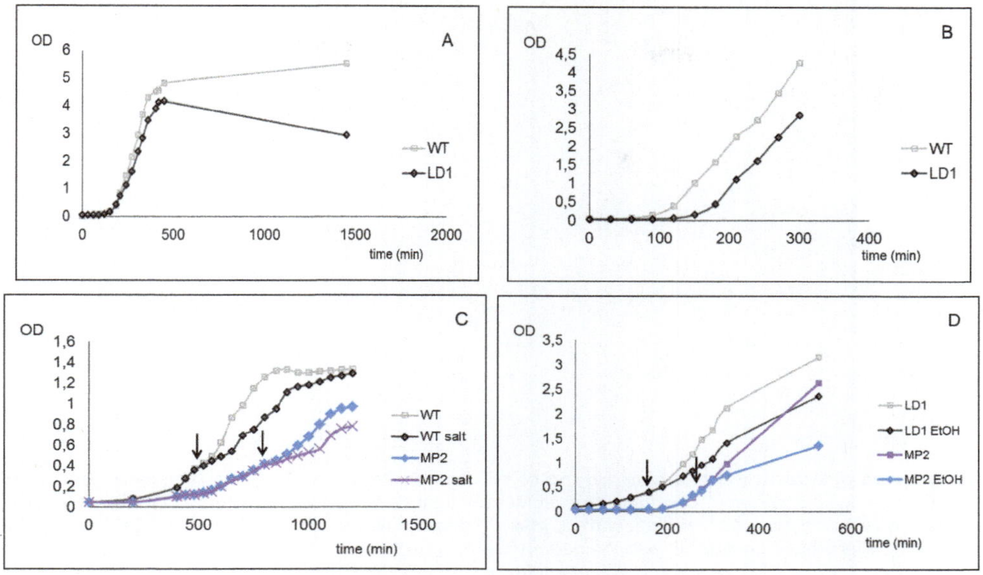

Figure 1. Growth characterization of WT, LD1, and MP2 mutants of *Bacillus subtilis* in long-term cultivation and in response to stress. For long-term growth measurements, the cells were grown in LB medium (A). Effect of MM medium and K^+ limitation (0.5 mM K^+) to growth rate of mutant (LD1) (B). Osmotic stress performed only for MP2 mutant in MM medium and K^+ limitation is demonstrated (C). Effect of ethanol stress was measured in MM medium and K+ limitation for both mutants (LD1, MP2) (D). (WT - *Bacillus subtilis* 168 or SG4, LD1 and MP2– *yxkO* knock-out mutants, for details see Table 1). Exposure of stress is marked by arrows. Measurements were done with cells synchronized in exponential growth, and stress conditions were set up as is described in *Material and Methods*. The typical growth rate curve is shown from measurements made in triplicate for each condition.

stress when cultivated in media with a potassium concentration lower than 1 mM (0.5 mM) (Figure 1C, D). Both types of mutants revealed a similar course of the growth curves.

Characterization of *yxkO* gene promoter transcriptional activity under stress conditions

As the next routine step in characterization of the *yxkO* mutant, we tested activity of the P*yxkO* promoter when exposed to several stress conditions. For this experiment, we used the previously prepared MP2 mutant, as the plasmid pMUTIN4 was constructed to be used for gene knock-out and as a promoter probe with the *lacZ* reporter, as well, with the benefit of complementation of the WT allele. Surprisingly, no significant increase of transcription was observed under any cultivation condition and induction of the WT allele with IPTG, respectively (low, high concentration of K^+, salt stress), except for ethanol stress, when a slight increase of transcription was detected after the stress exposure (data not shown). The same result was observed previously, when transcriptional activity was tested from solely promoter cloned to pDG1661promoter probe vector (unpublished data). These data indicate that the protein is expressed only at low levels or undergoes more complex regulation of transcription.

Testing of YxkO's impact on activation of general stress response

When direct estimation of the promoter activity of the P*yxkO* promoter had failed, we tested the impact of the YxkO on the induction of SigB-dependent genes. For this purpose, we generally used the well-characterized SigB-dependent P*ctc* promoter, which is transcribed under the most studied stress conditions, and the P*ctc* promoter fused to *lac*Z (P*ctc-lac*Z) beneficially monitors SigB activity in the WT promoter context, as was shown previously [11]. The promoter-probe strains in both genetic backgrounds were prepared as described in the *Materials and Methods* section, and the promoter activity of P*ctc* was monitored. The tested conditions were ethanol stress in complex media and salt stress when cultivated in high (10 mM) and low (0.5 mM) concentrations of K^+, respectively. Results are shown in Figure 2.

When exposed to ethanol stress, the promoter activity of P*ctc* in the MP2 mutant was markedly lower (Figure 2A). In the high K^+ concentration condition assay, the activity of P*ctc* increased in WT and the MP2 mutant after the salt stress, as well, albeit at a lower level in the mutant (Figure 2B). When cells were cultivated in medium with low K^+ concentration, the transcription from P*ctc* revealed a completely different course. The transcription level was high in WT, even before salt stress, and remained low in the mutant, even after the stress exposure (Figure 2C).

Therefore, we monitored the promoter activity of P*ctc* when cells were cultivated first in the medium with high K^+ concentration and then shifted to medium with low K^+ concentration. In both WT and the mutant strain, an increase in the P*ctc* activity occurred after the K^+ concentration shift but in the mutant the promoter activity was systematically lower at all-time points (Figure 2D). As mentioned above, a 1 mM concentration of K^+ limits the potassium KtrAB transport system of *Bacillus subtilis* 168, which is predicted to be involved in osmoadaptation [28].

These results indicated that in *Bacillus subtilis* 168, a K^+ concentration in the medium below 1 mM is a stress condition and causes the general stress response.

The different level of P*ctc* transcription in the presented experiments is in agreement with the recent observation of Young et al. that the extent of the stress causes a differential effect on the adaptation response in subpopulations of *Bacillus subtilis* [14]. It is obvious that potassium deficiency coupled with osmotic stress causes a distinct adaptive response than the osmotic stress itself. As the product of the *yxkO* gene is assumed to be NADH hydrate

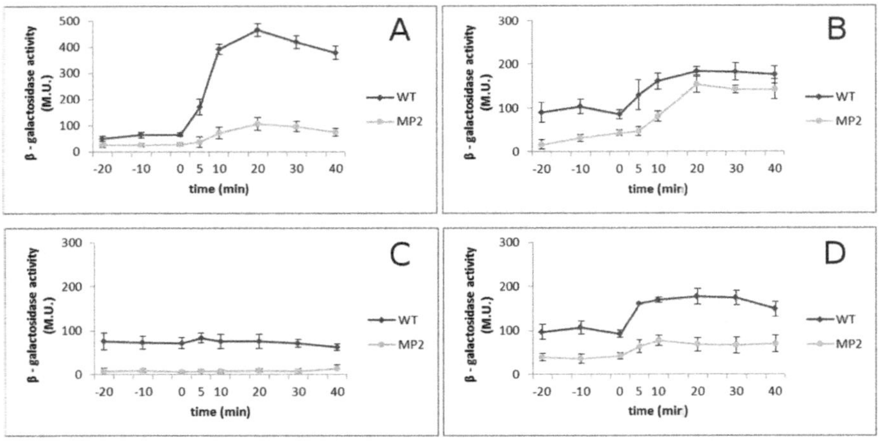

Figure 2. Transcription level of P*ctc* on genetic background of WT and MP2 mutant. P*ctc* activation measurements of WT and MP2 were performed under ethanol stress in LB medium (A), osmotic stress in MM medium with 10 mM K^+ concentration (B), osmotic stress in MM medium with 0.5 mM K^+ concentration (C), shift from MM medium with 10 mM K^+ concentration to MM medium with 0.5 mM K^+ concentration (D). Time 0 indicates application of stress. Details of transcription activity measurements, growth, and stress conditions are described in *Material and Methods.*

dehydratase, it is tempting to speculate about the role of energy imbalance in the activation of the SigB regulon and SigB itself.

Comparative 2DE analysis of YxkO impact on an adaptive response to stress

We compared protein levels of cytoplasmic proteins in the pI range of 4–7 in exponentially growing cells of WT and the yxkO knock-out mutant (MP2) of *Bacillus subtilis* in medium with limited potassium and exposed to osmotic and ethanol stress, respectively. Cells were grown and analyzed by comparative 2DE as described in *Materials and Methods*.

The results, summarized in Figures 3–6, show that disruption of the *yxkO* gene caused changes in the level of seven proteins belonging to various metabolic pathways upon both stresses (Figure 3 and Figure 4), and we detected only one protein with a different intensity solely under osmotic stress (Figure 5), three enzymes revealed an altered pattern merely under ethanol stress (Figure 6). Figure S1 illustrates the pattern of the 2DE spot distribution.

Proteins with levels that differ under both stresses

Metabolic enzymes. It is described elsewhere that after osmotic stress, a prompt decrease in expression of most of the glycolytic and citrate cycle enzymes occurs and, in intervals to 60 min, increases to initial levels after resumption of growth arises [5]. Our results are in agreement with this, as we determined the different levels of phosphoglycerate kinase, an enzyme representing the glycolytic pathway, and as we detected changes in the protein level of isocitrate dehydrogenase and malate dehydrogenase from citrate cycle enzymes, as well. In the mutant, 60 minutes after stress exposure, the protein level remained low in the case of all three mentioned proteins, and we detected degradation products of malate dehydrogenase (Mdh-deg, MP2 stress sample) in the case of EtOH stress (Figure 3). There is no study in the literature about the mechanism that is involved in the cessation and recovery of expression of these three vegetative proteins after osmotic stress, but the role of the *yxkO* gene is evident from our experiment.

This effect highlights the recently recognized biochemical activity of YxkO protein as an enzyme that regenerates the hydrated form of NAD(P)H emerging in cells during stress [17]. The absence of dehydration of NADHX and the regeneration of activities of both isocitrate dehydrogenase and malate dehydrogenase may result in energetic imbalance and, subsequently, growth retardation and reduced stress adaptation in the MP2 mutant, as we documented by extension of the lag phase of stressed cultures of mutant cells.

We also detected changes in the protein level of the ATPase subunit, similar to Höper's results [5], and we determined three isoforms of the beta subunit, AtpD. The protein levels of two isoforms (iso1, iso2) manifested the same pattern as in the case of metabolic enzymes, and recovery in the mutant did not occur, either. After stress, there was a significant decrease in the protein level of both isoforms in the mutant (Figure 3).

Stress adaptation. Another protein that was determined to be differentially accumulated under both stresses was fructosamine-6-phosphate deglycase (FrlB, formerly YurP); the protein level increased after the stress in the WT and this increase was absent in the mutant, regardless of stress exposure (Figure 4). This gene was also detected to be upregulated under most stress conditions at the transcription level [4] and was shown to be under direct regulation of the CodY general regulator [29]. This enzyme catalyzes the cleavage of fructosamine-6-phosphate to glucose-6-phosphate and the corresponding amines in *Escherichia coli* and *Bacillus subtilis* and is involved in enzymatic deglycation of Amadori products [30]. Intracellular glycation has been described to occur in *Escherichia coli* [31], and it is very likely to occur in *Bacillus subtilis*, as well. DNA-binding activity of CodY is regulated by the intracellular level of GTP [32], and it was evidenced that during nutrient starvation, the ratio of ATP/GTP is changed [33]. In environmental stress adaptation, this phenomenon has not yet been studied, but it could be supposed that inactivation of NADH-synthesizing enzymes gives rise to an energy imbalance, changing the ATP/GTP ratio, as well.

Changes in the protein value caused by *yxkO* disruption were also recorded for GroEL (Figure 4). In the mutant, the increase of protein level occurred in non-stressed conditions, as well when compared to the WT. This can be explained by extension of the lag phase and the decline of the renewing of isocitrate dehydrogenase levels in the mutant according to WT after stress exposure, which denotes to failure of stress adaptation and triggers

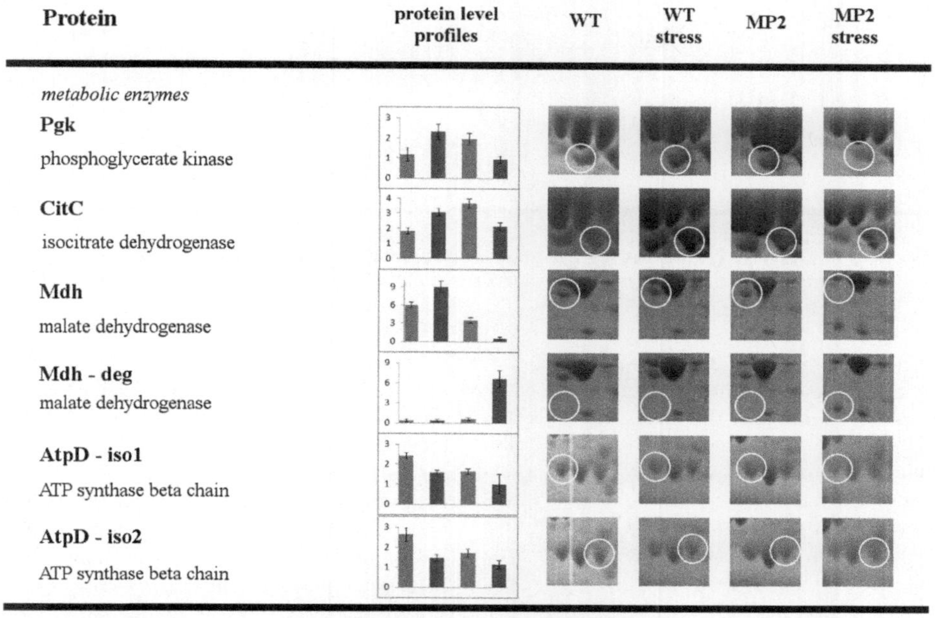

Figure 3. Comparative 2DE analysis of WT versus MP2 mutant exposed to osmotic and ethanol stress – metabolic enzymes. Proteins with protein level profiles that aresimilar for both stresses. For experimental conditions and data evaluation, see *Material and Methods.* Separate columns of the bar charts show the protein level of respective proteins, as calculated from the quantification of the spot volume by PDQuest 8.0 software; y-axes are scaled in intensity for each particular protein. Bars represent each strains and conditions, and there are in the same order as the protein level profiles are presented.

increased levels of GroEL as a result of the devastating effects of both stresses on cellular proteins.

Motility. Another protein that differed in protein level in WT versus MP2 and the changes of which corresponded to the mutant phenotype that the *yxkO* gene disruption affected is flagellin (Figure 4). When the cells were cultivated under a limited concentration of potassium, there was massive expression of flagellin protein. This corresponds to the rapid movement and substantial flagellation of cells observed in light and electron microscopy, respectively. In native microscopic preparates, we also observed different motility of WT and mutant cells before and after being subjected to salt stress. It has been described that after

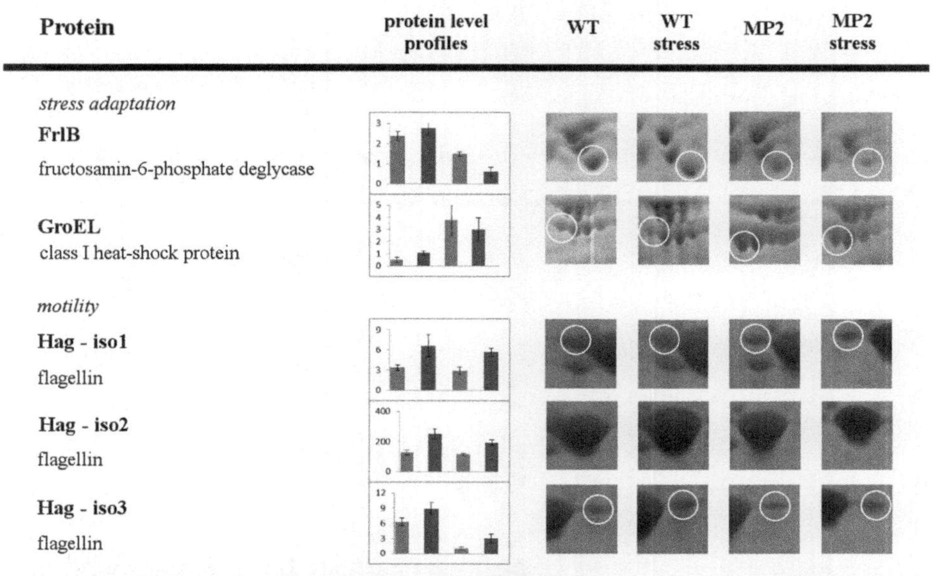

Figure 4. Comparative 2DE analysis of WT versus MP2 mutant exposed to osmotic and ethanol stress – stress adaptation and motility. Proteins with protein level profiles that are similar for both stresses. For experimental conditions and data evaluation, see *Material and Methods.* The picture description is same as for Figure 3.

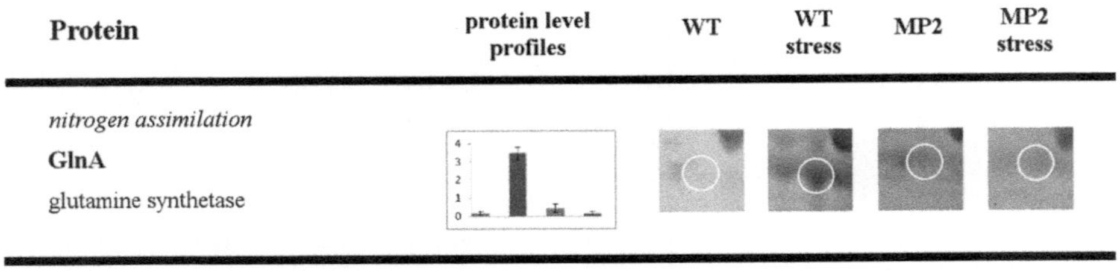

Figure 5. Comparative 2DE analysis of WT versus MP2 mutant exposed to osmotic stress. Proteins with protein level profiles that are unique for osmotic stress. For experimental conditions and data evaluation, see *Material and Methods*. The picture description is same as for Figure 3.

osmotic shock, a decrease in the transcription of genes coding for motility apparatus components occurs [1]. In agreement with Steils results, we have observed that after the shock, the WT cells ceased to move in the adaptation phase, whereas the mutant cells did not show any significant changes in their motility, regardless of the stress (data not shown).

On our gels, we detected three isoforms of flagellin, with the evidence that the formation of quantitatively less represented isoforms (iso1 and iso3) was affected by the disruption of the *yxkO* gene. A drop in other proteins of the flagellar apparatus, such as basal body components, was not detected. This could indicate that the product of the *yxkO* gene does not influence the intracellular amount of flagellin protein but causes its modification, which allows or prevents the transport of monomers out of the cell and its degradation in the cytoplasm, respectively. The origins and nature of these isoforms of flagellin have not been described in the literature to date and require further work to be elucidated. We can speculate that non-enzymatic glycation (see FrlB above) is involved.

Protein with levels that differ solely under osmotic stress

Nitrogen assimilation. Glutamine synthetase, the key enzyme of nitrogen assimilation, is a protein, the level of which was affected by mutation in the *yxkO* gene only after osmotic stress (Figure 5). We determined a massive increase in the protein amount of this enzyme in WT after the salt stress, while in the MP2 mutant, the protein stayed at a low level, even after the stress. Our results differ from those of Höper [5], which can be explained by the fact that we used a mineral medium with ammonium sulphate as the sole source of nitrogen and a limited concentration of potassium, conditions under which nitrogen metabolism are also significantly influenced by CodY and ATP assistance, as was recently reported by Gunka [34].

Proteins with a different level under ethanol stress

A comprehensive proteomic study of *Bacillus subtilis* under ethanol stress with which we could compare our results has not yet been published. After the ethanol stress, we observed changes in a similar set of proteins as in the salt stress experiment, with a comparable intensity profile (see above). However, we identified proteins, the accumulation of which was influenced only after the ethanol stress (Figure 6). Changes were detected in the protein amount of three enzymes: ketol-acid reductoisomerase (IlvC), aspartate-semialdehyde dehydrogenase (Asd) and amino acid L-alanine dehydrogenase (Ald).

Valin, leucine, and isoleucine metabolism. The first enzyme, ketol-acid reductoisomerase (IlvC), is known to be involved in the biosynthesis of valine, leucine, and isoleucine, and its encoding gene is a part of the *ilv-leu* operon [35]. We identified that the cellular level of IlvC decreased in the mutant after the stress, while in WT, it remained unchanged. Valine, leucine, and isoleucine are the precursors for the biosynthesis of iso- and anteiso-branched fatty acids, which represent the major fatty acid species of membrane lipids in *Bacillus* sp. The increase in their expression was proposed as a long-term adaptation mechanism to cold and ethanol stresses [36]. It is worth mentioning that as already reported by Molle [37], the *ilv-leu*

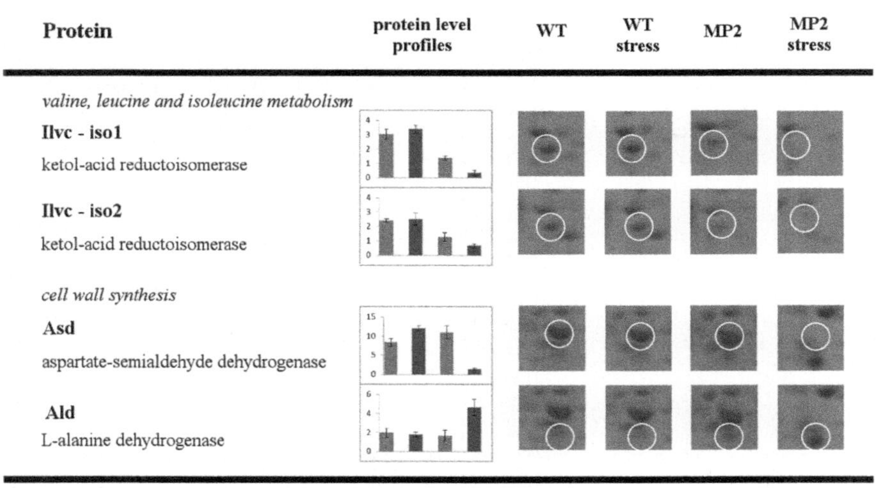

Figure 6. Comparative 2DE analysis of WT versus MP2 mutant exposed to ethanol stress. Proteins with protein level profiles that are unique for ethanol stress. For experimental conditions and data evaluation, see *Material and Methods*. The picture description is same as for Figure 3.

operon was shown to be controlled by the global transcriptional regulator CodY.

Cell wall synthesis. The second enzyme, aspartate-semialdehyde dehydrogenase (Asd), the expression of which was lower after the stress in the mutant, is involved in the metabolism of alanine, aspartate, and glutamate. The third enzyme, amino acid L-alanine dehydrogenase (Ald), catalyzes deamination of alanine to pyruvate. Its level in the mutant strain was markedly higher after the stress, while in WT, it was not affected. Both are the precursors for the synthesis of diaminopimelic acid, which is a constituent of peptidoglycan. It has been reported that ethanol stress increases the expression of proteins involved in cell wall synthesis [38], and phosphorylation of aspartate-semialdehyde dehydrogenase by PtkA kinase was described [39]. An imbalance in expression of this protein may cause a defect in cell wall synthesis in the mutant strain and consequently decrease its adaptability to ethanol stress.

Conclusions

In this study, we performed the first characterization of a knock-out mutant in the *yxkO* gene, which biochemical activity was classified *in vitro* as an NAD(P)H hydrate dehydratase, and provided evidence that it is involved in stress adaptation (to osmotic and ethanol stress and under potassium limitation).

Our results correspond to the observed phenotype of the mutant and show that stress adaptation reactions in the mutant are merely suppressed or not activated, as reflected by the observed decline in the growth and extension of the lag phase, inhibition of recovery of NADH-dependent proteins during the lag phase after stress exposure (Idh, Mdh, etc.), energy imbalance and repression of CodY-regulated proteins (FrlB, IlvC), and inhibition of activation of a general stress protein (Ctc) in the mutant. We have also shown that potassium limitation is a stress condition for *Bacillus subtilis* that activates the general stress response. This reflects the complexity and interconnection of stress adaptation processes and is evidence that either adaptation mechanism is important in stress resistance. Thus metabolite repair enzymes play a significant role in the survival of stressed cells.

Supporting Information

Figure S1 The 2-DE image of cytoplasmic proteins from WT Bacillus subtilis 168 cultivated in the ethanol stress condition, illustrating the pattern of the 2DE spot distribution. The patterns of the 2DE spot distributions in the studied culture conditions of WT and MP2 mutant were similar. Proteins that exhibited changes in protein abundances and were identified by MS MALDI-TOF are pointed out. Details of identified proteins are described in *Results and Discussion*, Figures 3–6, and Table 2.

Acknowledgments

Mass spec measurements were done in the Service Laboratories of the Biology Section, Laboratory of Mass Spectrometry, Charles University in Prague, Faculty of Science, Vinicna 7, 128 44 Praha 2. https://www.natur.cuni.cz/biologie/servisni-laboratore/laborator-hmotnostni-spektrometrie.

Author Contributions

Conceived and designed the experiments: MP JT LD IL. Performed the experiments: MP JT LD ES. Analyzed the data: MP JT LD IL. Contributed reagents/materials/analysis tools: MP JT LD ES IL. Contributed to the writing of the manuscript: MP JT LD IL.

References

1. Steil L, Hoffmann T, Budde I, Volker U, Bremer E (2003) Genome-wide transcriptional profiling analysis of adaptation of Bacillus subtilis to high salinity. J Bacteriol 185: 6358–6370.
2. Price CW, Fawcett P, Ceremonie H, Su N, Murphy CK, et al. (2001) Genome-wide analysis of the general stress response in Bacillus subtilis. Mol Microbiol 41: 757–774.
3. Petersohn A, Brigulla M, Haas S, Hoheisel JD, Volker U, et al. (2001) Global analysis of the general stress response of Bacillus subtilis. J Bacteriol 183: 5617–5631.
4. Nicolas P, Mäder U, Dervyn E, Rochat T, Leduc A, et al. (2012) Condition-Dependent Transcriptome Reveals High-Level Regulatory Architecture in Bacillus subtilis. Science 335: 1103–1106.
5. Hoper D, Bernhardt J, Hecker M (2006) Salt stress adaptation of Bacillus subtilis: a physiological proteomics approach. Proteomics 6: 1550–1562.
6. Hecker M, Pane-Farre J, Volker U (2007) SigB-dependent general stress response in Bacillus subtilis and related gram-positive bacteria. Annu Rev Microbiol 61: 215–236.
7. Hahne H, Mader U, Otto A, Bonn F, Steil L, et al. (2010) A comprehensive proteomics and transcriptomics analysis of Bacillus subtilis salt stress adaptation. J Bacteriol 192: 870–882.
8. Mascher T, Margulis NG, Wang T, Ye RW, Helmann JD (2003) Cell wall stress responses in Bacillus subtilis: the regulatory network of the bacitracin stimulon. Mol Microbiol 50: 1591–1604.
9. Gaidenko TA, Kim TJ, Weigel AL, Brody MS, Price CW (2006) The blue-light receptor YtvA acts in the environmental stress signaling pathway of Bacillus subtilis. J Bacteriol 188: 6387–6395.
10. Boylan SA, Redfield AR, Brody MS, Price CW (1993) Stress-induced activation of the sigma B transcription factor of Bacillus subtilis. J Bacteriol 175: 7931–7937.
11. Benson AK, Haldenwang WG (1992) Characterization of a regulatory network that controls sigma B expression in Bacillus subtilis. J Bacteriol 174: 749–757.
12. Avila-Perez M, van der Steen JB, Kort R, Hellingwerf KJ (2010) Red light activates the sigmaB-mediated general stress response of Bacillus subtilis via the energy branch of the upstream signaling cascade. J Bacteriol 192: 755–762.
13. Boylan SA, Redfield AR, Price CW (1993) Transcription factor sigma B of Bacillus subtilis controls a large stationary-phase regulon. J Bacteriol 175: 3957–3963.
14. Young JW, Locke JC, Elowitz MB (2013) Rate of environmental change determines stress response specificity. Proc Natl Acad Sci U S A 110: 4140–4145.
15. Ulanova D, Holanova V, Prenosilova L, Naprstek J, Licha I (2007) Mutation of a gene encoding a putative ribokinase leads to reduced salt tolerance under potassium limitation in Bacillus subtilis. Folia Microbiol (Praha) 52: 203–208.
16. Zhang RG, Grembecka J, Vinokour E, Collart F, Dementieva I, et al. (2002) Structure of Bacillus subtilis YXKO-a member of the UPF0031 family and a putative kinase. J Struct Biol 139: 161–170.
17. Shumilin I, Cymborowski M, Chertihin O, Jha KN, Herr JC, et al. (2012) Identification of unknown protein function using metabolite cocktail screening. Structure 20: 1715–1725.
18. Oppenheimer NJ, Kaplan NO (1974) Glyceraldehyde-3-phosphate dehydrogenase catalyzed hydration of the 5–6 double bond of reduced beta-nicotinamide adenine dinucleotide (beta NADH). Formation of beta-6-hydroxy-1,4,5,6-tetrahydronicotinamide adenine dinucleotide. Biochemistry 13: 4685–4694.
19. Marbaix AY, Noel G, Detroux AM, Vertommen D, Van Schaftingen E, et al. (2011) Extremely conserved ATP- or ADP-dependent enzymatic system for nicotinamide nucleotide repair. J Biol Chem 286: 41246–41252.
20. Yoshida A, Dave V (1975) Inhibition of NADP-dependent dehydrogenases by modified products of NADPH. Arch Biochem Biophys 169: 298–303.
21. Prabhakar P, Laboy JI, Wang J, Budker T, Din ZZ, et al. (1998) Effect of NADH-X on cytosolic glycerol-3-phosphate dehydrogenase. Arch Biochem Biophys 360: 195–205.
22. Van Schaftingen E, Rzem R, Marbaix A, Collard F, Veiga-da-Cunha M, et al. (2013) Metabolite proofreading, a neglected aspect of intermediary metabolism. J Inherit Metab Dis 36: 427–434.
23. Hoper D, Volker U, Hecker M (2005) Comprehensive characterization of the contribution of individual SigB-dependent general stress genes to stress resistance of Bacillus subtilis. J Bacteriol 187: 2810–2826.
24. Young JW, Locke JC, Elowitz MB (2013) Rate of environmental change determines stress response specificity. Proc Natl Acad Sci U S A 110: 4140–4145.

25. Cutting SM, Vander Horn PB (1990) Genetic analysis, PBS1 Geneneralized transduction. In: Harwood CR, Cutting SM editors. Molecular Biological Methods for Bacillus subtilis: John Wiley and sons.
26. Ray C, Hay RE, Carter HL Jr., Moran CP Jr. (1985) Mutations that affect utilization of a promoter in stationary-phase Bacillus subtilis. J Bacteriol 163: 610–614.
27. Shevchenko A, Tomas H, Havlis J, Olsen JV, Mann M (2006) In-gel digestion for mass spectrometric characterization of proteins and proteomes. Nat Protoc 1: 2856–2860.
28. Holtmann G, Bakker EP, Uozumi N, Bremer E (2003) KtrAB and KtrCD: two K^+ uptake systems in Bacillus subtilis and their role in adaptation to hypertonicity. J Bacteriol 185: 1289–1298.
29. Belitsky BR, Sonenshein AL (2008) Genetic and biochemical analysis of CodY-binding sites in Bacillus subtilis. J Bacteriol 190: 1224–1236.
30. Deppe V, Bongaerts J, O'Connell T, Maurer KH, Meinhardt F (2011) Enzymatic deglycation of Amadori products in bacteria: mechanisms, occurrence and physiological functions. Appl Microbiol Biotechnol 90: 399–406.
31. Mironova R, Niwa T, Hayashi H, Dimitrova R, Ivanov I (2001) Evidence for non-enzymatic glycosylation in Escherichia coli. Mol Microbiol 39: 1061–1068.
32. Ratnayake-Lecamwasam M, Serror P, Wong KW, Sonenshein AL (2001) Bacillus subtilis CodY represses early-stationary-phase genes by sensing GTP levels. Genes Dev 15: 1093–1103.
33. Zhang S, Haldenwang WG (2005) Contributions of ATP, GTP, and Redox State to Nutritional Stress Activation of the Bacillus subtilis σB Transcription Factor J Bacteriol 7554–7560.
34. Gunka K, Commichau F (2012) Control of glutamate homeostasis in Bacillus subtilis: a complex interplay between ammonium assimilation, glutamate biosynthesis and degradation. Mol Microbiol 85: 213–224.
35. Mader U, Hennig S, Hecker M, Homuth G (2004) Transcriptional organization and posttranscriptional regulation of the Bacillus subtilis branched-chain amino acid biosynthesis genes. J Bacteriol 186: 2240–2252.
36. De Mendoza D, Schujman GE, Aguilar PS (2002) Biosynthesis and function of membrane lipids. In: A. L. Sonenshein JAH, and R Losick editor. Bacillus and its closest relatives. Washington, D.C.: ASM Press.
37. Molle V, Nakaura Y, Shivers RP, Yamaguchi H, Losick R, et al. (2003) Additional targets of the Bacillus subtilis global regulator CodY identified by chromatin immunoprecipitation and genome-wide transcript analysis. J Bacteriol 185: 1911–1922.
38. Eiamphungporn W, Helmann JD (2008) The Bacillus subtilis sigma(M) regulon and its contribution to cell envelope stress responses. Mol Microbiol 67: 830–848.
39. Jers C, Pedersen MM, Paspaliari DK, Schutz W, Johnsson C, et al. (2010) Bacillus subtilis BY-kinase PtkA controls enzyme activity and localization of its protein substrates. Mol Microbiol 77: 287–299.

18

The Hybrid Type Polyketide Synthase SteelyA Is Required for cAMP Signalling in Early *Dictyostelium* Development

Takaaki B. Narita[1,2], Zhi-hui Chen[3], Pauline Schaap[3], Tamao Saito[4]*

1 Graduate School of Science and Technology, Sophia University, Tokyo, Japan, **2** Research Fellow of Japan Society for the Promotion of Science (DC2), Tokyo, Japan, **3** College of Life Sciences, University of Dundee, Dundee, United Kingdom, **4** Department of Materials and Life Sciences, Sophia University, Tokyo, Japan

Abstract

Background: In our previous study we found that the expression of *stlA* showed peaks both in the early and last stages of development and that a product of SteelyA, 4-methyl-5-pentylbenzene-1,3-diol (MPBD), controlled *Dictyostelium* spore maturation during the latter. In this study we focused on the role of SteelyA in early stage development.

Principal Findings: Our *stlA* null mutant showed aggregation delay and abnormally small aggregation territories. Chemotaxis analysis revealed defective cAMP chemotaxis in the *stlA* null mutant. cAMP chemotaxis was restored by MPBD addition during early stage development. Assay for cAMP relay response revealed that the *stlA* null mutant had lower cAMP accumulation during aggregation, suggesting lower ACA activity than the wild type strain. Exogenous cAMP pulses rescued the aggregation defect of the *stlA* null strain in the absence of MPBD. Expression analysis of cAMP signalling genes revealed lower expression levels in the *stlA* null mutant during aggregation.

Conclusion: Our data indicate a regulatory function by SteelyA on cAMP signalling during aggregation and show that SteelyA is indispensable for full activation of ACA.

Editor: Gareth Bloomfield, MRC Laboratory of Molecular Biology, United Kingdom

Funding: This work was supported by JSPS KAKENHI Grant Number 24510306 to TS. TBN is a research fellow of Japan Society for the Promotion of Science (DC2). This work was supported by Grant-in-Aid for JSPS Fellows to TBN. The funders had no role in study design, data collection and analysis, decision to publish, or preparation of the manuscript.

Competing Interests: The authors have declared that no competing interests exist.

* Email: tasaito@soph .ac.jp

Introduction

Dictyostelium cells live in the soil as unicellular amoebae and consume bacteria and yeast as a food source. When their food source is depleted, *Dictyostelium* cells aggregate to form a multicellular organism. Cells in the aggregate start to differentiate into prestalk and prespore cells in a heterogeneous mixture which resembles salt and pepper [1]. Aggregates then transform into migrating slugs in which the differentiated cells sort to produce a prestalk-prespore pattern. After the slug stage, cells complete differentiation into their terminal types to form the stalk and spore cells of the mature fruiting body [1,2]. Several small molecules are reported to control *Dictyostelium* development, amongst which the second messenger cAMP.

In *Dictyostelium*, cAMP acts not only as an intracellular second messenger but also as a secreted extracellular signal. During aggregation, secreted cAMP acts as a chemoattractant, which causes *Dictyostelium* cells to move together by chemotaxis [3]. In addition to inducing chemotaxis, the activation of cAMP receptor cAR1 also initiates a signal cascade that results in cAMP synthesis. Synthesized cAMP is then secreted to attract neighboring cells [4,5]. cAMP also has an intracellular role in development, controlling many aspects of the developmental programme through the activation of PKA [6–10].

cAMP is produced by three distinct adenylyl cyclases during *Dictyostelium* development. These are ACA, ACB, and ACG. ACA produces the cAMP required for cell aggregation in early development [5,11], ACB is involved in the maturation of stalk and spore cells [12,13], and ACG controls prespore differentiation in the slug stage and also acts as an osmosensor that controls spore germination in the fruiting body [10,13,14].

The genome sequence of *Dictyostelium* was published in 2005 [15]. *Dictyostelium*, it turned out, has a massive potential for the production of polyketides, with 45 polyketide synthase (PKS) genes found in its genome [15,16]. Two are novel types of hybrid PKSs called "Steely" [17], which to our best knowledge, are only found in the cellular slime molds. These novel enzymes are a fusion of iterative type I and type III PKSs. Iterative type I PKSs are large single polypeptides including multifunctional domains encoded by only a single gene and iteratively utilize some of their functional domains for a polyketide biosynthesis, whereas type III PKSs are single enzymes comprising of only *keto*-synthase domain [18,19]. One of the Steely enzymes, SteelyB, was reported to produce

Table 1. Summary of the primer sets used in qRT-PCR.

Primer name	Sequence
ig7 F	5′-TTACATTTATTAGACCCGAAACCAAGCG-3′
ig7 R	5′-AACAGCTATCACCAAGCTTGATTAGCC-3′
carA F	5′-CCAGCACTCAATATTCTCC-3′
carA R	5′-ATGATGATAAAGAAGATGAAGATGAACC-3′
acaA F	5′-ATGCAATCCAATGCTCAAGATAATG-3′
acaA R	5′-AATGAGCCAATTTCACCCAAGAG-3′
pdsA F	5′-ATCCTCTGTCGCTTGTGATTGG-3′
pdsA R	5′-CCATTATTATTTGCTTCTTTTAATTGTTG-3′
regA F	5′-CCAAAACCACAAGAGCAAGAATCG-3′
regA R	5′-GAAGTTGAAGTTAAAGGAGCGGTCG-3′
piaA F	5′-GGTAATTTATCTTCACATATCACTG-3′
piaA R	5′-AATGAATCTTCTAGCACCTAAACG-3′
dagA F	5′-ACTACTAAAAGATCTCATCCAACTAC-3′
dagA R	5′-AAGCATCTTGAATGTATTGAGTCATTG-3′
gpaB F	5′-CAGCTTCAAAAACGACACG-3′
gpaB R	5′-CGAAAGAGCCTCAAACTATATCAAAG-3′
csA F	5′-CTGTTTCAGTTGGAGTTGGAGATG-3′
csA R	5′-CAGGTATCTTTTTAAACAATGCCC-3′

Differentiation Inducing Factor-1 (DIF-1), an inducer of prestalk cell differentiation [17]. We showed earlier that a mutant that lacked the *stlB* gene made morphologically normal but structurally weak stalk and did not produce a basal disc or lower cup. These defects were recovered by the addition of DIF-1 in the medium, indicating that DIF-1 induces basal disc and lower cup formation among prestalk cells [17,20]. DIF-1 has an additional role in developmental regulation [21], and both DIF-1 and the related molecule DIF-2 function as modulators of cAMP-induced chemotaxis [22]. This suggests that DIFs govern multiple aspects of the developmental programme.

Another Steely enzyme, SteelyA, was reported to make 4-methyl-5-pentylbenzene-1,3-diol (MPBD) *in vivo* and we showed that a *stlA* knockout mutant has defective spore maturation, which was recovered by the addition of MPBD to the mutant cells [23]. Previously, two distinct patterns of *stlA* gene expression were reported: one of them was that the gene was expressed during the early stage development [17], whereas another one was that the gene was expressed during only late stage [24]. We resolved these conflicting reports concerning the gene expression pattern, by re-examining the expression pattern with by RT-PCR using two different primer sets. It appeared that the expression of *stlA* peaked in the pre-aggregation stage, followed by smaller peak during the late culmination stage [23]. We also confirmed that the function of SteelyA during this second peak was the control of spore maturation [23,25]. In subsequent research we specified the timing of MPBD-regulated spore maturation and found that MPBD was critical in the later stage of development [25].

In this study, we performed a detailed analysis of the function of SteelyA during early stage development and found that MPBD is

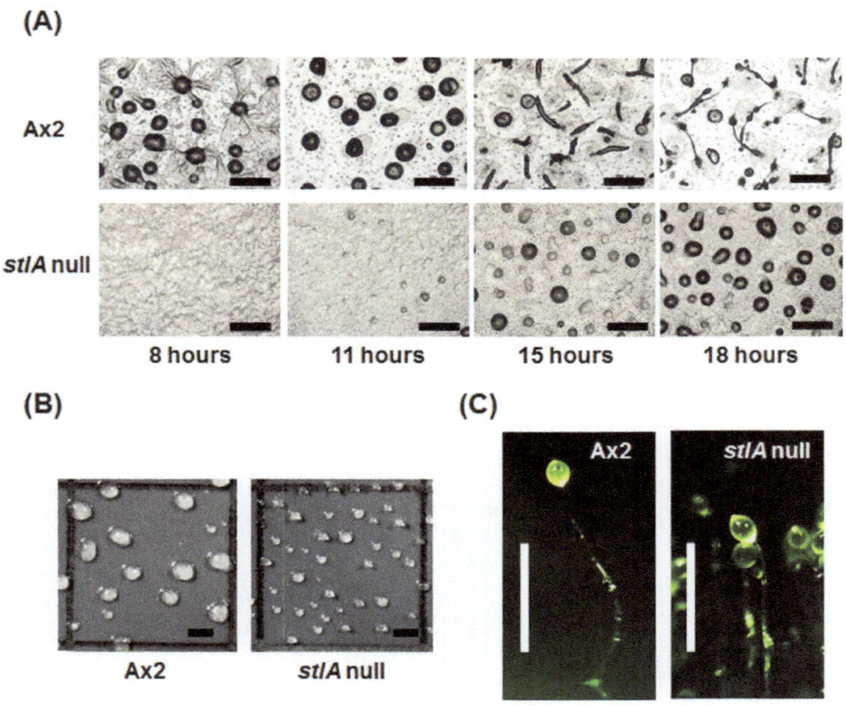

Figure 1. Developmental phenotype of *stlA* null mutant in the early stage. (A) Cells were developed on the PB agar plate at a density of 1×10^6 cells/cm^2. *StlA* null cells (bottom panels) showed aggregation delay compared to wild type strain, Ax2 cells (top panels). (B) Cells were developed on nitrocellulose filters at 1×10^6 cells/cm^2. The cells were observed when they finished forming multicellular aggregates. (C) Fruiting body of Ax2 and *stlA* null mutant developed on the PB agar plate at a density of 1×10^6 cells/cm^2. Bars indicate 500 µm (A and B) and 1 mm (C), respectively.

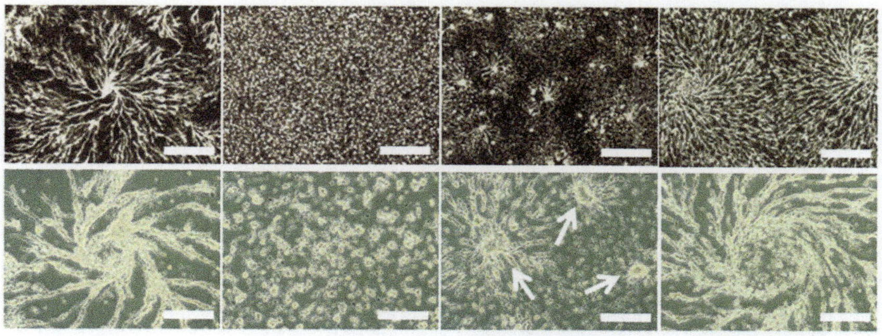

Figure 2. Aggregation stream formation under the submerged condition. Ax2 or *stlA* null cells were plated at 5×10^5 cells/cm^2 under the submersion. Photographs were taken at the indicated time. *StlA* null cells showed shorter aggregation streams and smaller aggregation territories than these of Ax2 cells (see arrows). In the presence of 200 nM MPBD, *stlA* null cells showed normal aggregation streams. These observations were confirmed in experiments repeated more than 3 independent times. Bars indicate 500 µm (top panels) and 200 µm (bottom panels).

required for normal expressions of cAMP signalling genes for aggregation.

Materials and Methods

Cell culture and development

The *Dictyostelium discoideum* Ax2 strain was grown in an axenic medium (HL-5) at 22°C [26]. The *stlA* null strain (Strain ID: DBS0236953) was grown in HL-5 medium supplemented with 10 µg/ml blasticidin S [17]. To initiate multicellular development, the axenically grown cells were harvested at a density of approximately 5 × 10^6 cells/ml, washed twice in the phosphate buffer (PB) (2.7 mM Na_2HPO_4/10.7 mM K_2HPO_4, pH 6.2), and plated on PB agar plates (1.5% agar in PB) or nitrocellulose filters (0.45 µm pore size, Black gridded, 47 mm diameter, Millipore) at a density of 1×10^6 cells/cm^2. For gene expression analysis, developing cells were harvested every 3 hours until 12 h and used for RNA purification. For submerged streaming, the axenically grown cells were harvested and washed twice, and then submerged in 1 ml of PB at 5×10^5 cells/cm^2 in a culture dish and the depth of PB was about 3 mm.

Chemotaxis assay and cAMP pulsing

Chemotaxis assay towards cAMP was performed as described [22] with a few modifications. Axenically grown cells in shaking culture were harvested in the log phase, washed in PB, and were starved at a density of 2×10^7 cells/ml in PB in the presence or absence of MPBD for 6.5 h. Then, starved cells were washed twice in PB and resuspended in PB to a final density of 3×10^5 cells/ml. Small portion of agar (~10 g) was washed with water (500 ml each time) repeatedly and extensively. Then the agar was dried completely. The agar powder was dissolved in PB and boiled to prepare the hydrophobic agar plate. Three droplets of starved cells, 2 µl (3 mm diameter) each, were placed on 1% hydrophobic agar plate in the petri dish. Chemotaxis towards cAMP was tested by placing a 2 µl droplet of various concentrations of cAMP at 3 mm distance from the cell-filled droplet. The distribution of cells within their droplet was observed after 1 h, and droplets were scored 'positive' if at least twice as many cells had migrated to the side closer to the cAMP droplet than to the other side. The chemotaxis rate was calculated as the percentage of the number of 'positive' droplets to that of total droplets.

To find out the chemotactic response to folate, axenically grown cells in shaking culture were harvested in the log phase, washed twice in PB, and were resuspended in PB to a final density of 2×10^7 cells/ml. Droplets of cells, 1 µl of the cell suspension, were placed on a hydrophobic agar plate at 3 mm distance from a 1 µl droplet of different concentrations of folate and were observed after 4 h from placing on the plate.

To examine the effect of cAMP pulsing on the *stlA* null strain, Ax2 or *stlA* null cells were harvested from the exponential growth phase and washed twice in PB. Washed cells were resuspended at a density of 2×10^7 cells/ml in PB and then stimulated with 20 nM cAMP pulses every 6 min for 6 h. The pulsed cells were washed and then used for the chemotaxis assay or developed on nitrocellulose filters.

Assay for cAMP relay response in intact cells

Cells were harvested in the log phase, washed twice, and distributed on non-nutrient agar plates (1.5% agar in 5 mM $Na_2HPO_4{\cdot}2H_2O$/5 mM KH_2PO_4 buffer, pH 6.5) at 1×10^8 cells/plate. Cells were incubated for 16 h at 4°C and some hours at 22°C until aggregation territories started to form (2 h for the case of wild type cells and 3–4 h for the mutant cells) and then were harvested in the phosphate buffer (5 mM $Na_2HPO_4{\cdot}2H_2O$/5 mM KH_2PO_4, pH 6.5) and resuspended at a density of 1×10^8 cells/ml. Cells were stimulated with 5 µM 2′-deoxycAMP in the presence of 5 mM DTT. Stimulation was terminated by adding an equal volume of 3.5% perchloric acid, and lysates were neutralized with 50% saturated $KHCO_3$. cAMP levels were analyzed by isotope dilution assay, with [^{3}H]cAMP as the competitor for binding to the bovine PKA regulatory subunit [27,28].

Gene expression analysis by qRT-PCR

Specific primer sets for qRT-PCR were designed to amplify a unique sequence located near the end of 3′ region of each gene (Table.1). BLAST similarity searches were used to confirm that each primer sequence amplified the unique sequence. Before using each primer set for qRT-PCR, genomic DNA was used as a template to evaluate the quality of primer set.

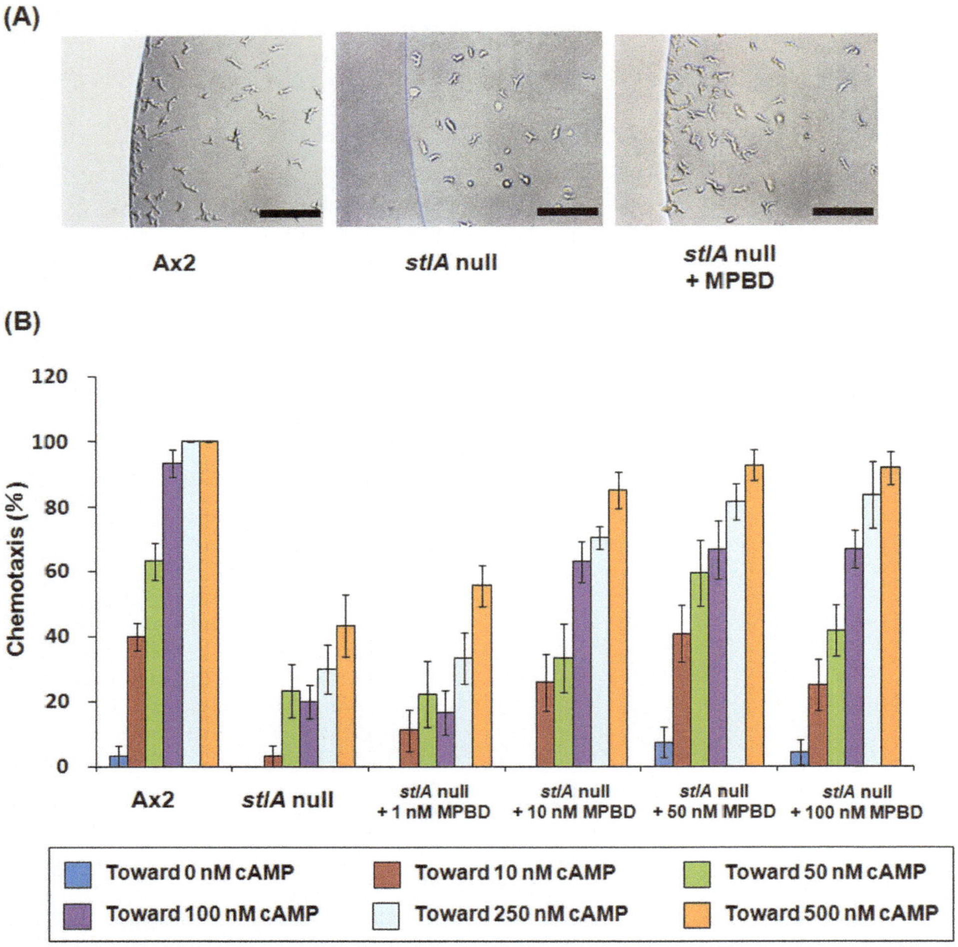

Figure 3. Chemotaxis defects in *stlA* null cells and the restoration by MPBD addition. (A) Cells were starved for 6.5 h in PB in the presence or absence of 200 nM MPBD and then the cell droplets were spotted next to 250 nM cAMP droplets on a hydrophobic agar plate. Cell droplets did not contain MPBD as starved cells were washed twice to eliminate MPBD before using this assay. The distribution of cells within droplets was observed and photographs were taken after 1 h. Bars: 100 µm. (B) Chemotaxis rate of the *stlA* null mutant with or without various doses of MPBD during starvation. The rate was calculated as the percentage of the number of 'positive' droplets to that of total droplets by the two droplet chemotaxis assay (see Materials and Methods). Cells were assayed for chemotaxis towards indicated concentrations of cAMP. Values are the means and SEM (bars) of 10 independent experiments (n = 10). The statistical analysis was performed by one-way ANOVA with Bonferroni's post-test. Significant differences ($p<0.05$) were shown in pairs as follows: Ax2–*stlA* null (towards 10, 50, 100, 250, and 500 nM cAMP), Ax2–*stlA* null with 1 nM MPBD (towards 100, 250, and 500 nM cAMP), Ax2–*stlA* null with 10 nM MPBD (towards 100 and 250 nM cAMP), *stlA* null–*stlA* null with 10 nM MPBD (towards 100, 250 and 500 nM cAMP), *stlA* null–*stlA* null with 50 nM MPBD (towards 100, 250 and 500 nM cAMP), *stlA* null–*stlA* null with 100 nM MPBD (towards 100, 250 and 500 nM cAMP), *stlA* null with 1 nM MPBD–*stlA* null with 10 nM MPBD (towards 100 nM cAMP), *stlA* null with 1 nM MPBD–*stlA* null with 50 nM MPBD (towards 100, 250 and 500 nM cAMP), and *stlA* null with 1 nM MPBD–*stlA* null with 100 nM MPBD (towards 100 and 500 nM cAMP).

Total RNA was extracted from cells at the indicating time point using the RNeasy mini kit (Qiagen) according to the manufacturer's instructions and then the contaminating genomic DNA was eliminated using Recombinant DNase I (TaKaRa). DNase treatment was performed as follows: 50 µg total RNA was mixed with 10 units Recombinant DNase I and DNase I buffer (2 mM Tris-HCl; pH 7.5, 5 mM NaCl, 0.01 mM $CaCl_2$, 5% glycerol) and incubated for 30 min at 37°C. The reaction was stopped by incubating with 25 mM EDTA at 80°C for 2 min. cDNA was synthesized by reverse transcription of 1 µg total RNA with oligo dT_{15} using ImProm-II Reverse Transcription System (Promega). qPCR was carried out with ABI PRISM 7000 Sequence Detection System (Applied Biosystems) and the amplifications were performed using iTaq Universal SYBR Green Supermix (Bio-Rad). The qPCR program was performed as follows: One cycle of 30 sec at 95°C followed by 40 cycles of 15 sec at 95°C, 1 min at 60°C for annealing, extension, and plate reading. Results of qRT-PCR were analyzed using the comparative C_T method [29] with the amplification of *ig7* (mitochondrial large rRNA) as a control.

Results and Discussion

In previous studies we found that the expression profile of *stlA* peaks in the early stage of development and shows a weak peak in the last stage of development [23]. In this study we chose to understand SteelyA's function in the early stage of development by analyzing the *stlA* null mutant.

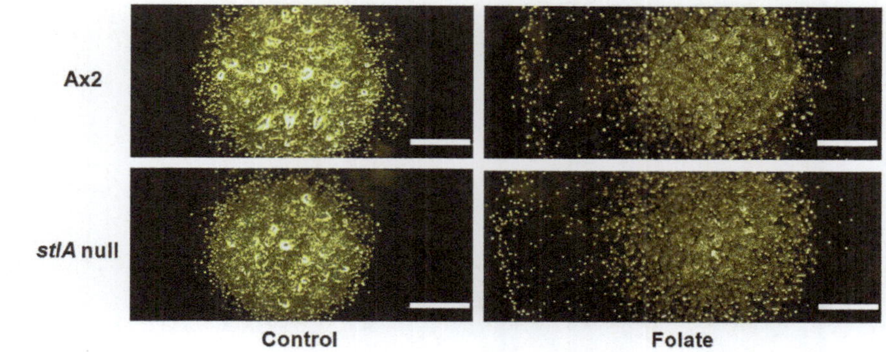

Figure 4. Representative result of chemotactic response to folate. Cells were harvested, washed twice, and resuspended in PB (2×10^7 cells/ml). The cell droplet was spotted next to a droplet of 500 µM folate on a hydrophobic agar plate. Photographs were taken after 4 h from placing the cell droplet by phase- contrast microscopy. We took photographs of the cell droplet dividing into two pieces and then connected those photographs because we were not able to take it with one piece. Pictures presented are the connected photographs. Bars: 1 mm.

StlA null strain shows small aggregation territories

We first studied cell aggregation in the *stlA* null mutant and found that this process was delayed by about 3 h when compared with its parental strain Ax2. In some cases, the *stlA* null strain showed developmental delay by more than 7 h (Fig. 1A).

In addition to the aggregation delay, *stlA* null cells formed smaller aggregates than those of Ax2 cells. When cells were developed on a nitrocellulose filter, the *stlA* null cells made smaller and more numerous aggregates compared to Ax2 cells (Fig. 1B). As the result of forming small aggregates, fruiting bodies of the *stlA* null strain were smaller in size than those of the Ax2 strain (Fig. 1C).

To analyze the streaming behavior of *stlA* null cells in more detail, we observed aggregation under submerged conditions (Fig. 2). Under these conditions, *stlA* null cells made shorter aggregation streams and showed smaller aggregation territories than Ax2 cells. Since these aggregates were homogeneously distributed on the substratum the smaller aggregate sizes were not due to the uneven distribution of cells. To investigate whether these defects were due to the lack of MPBD, the major product of SteelyA enzyme, we added 200 nM MPBD directly to the mutant cells because 200 nM MPBD was needed to rescue the spore maturation defect in the mutant [23]. We found that the aggregation of *stlA* null cells was rescued in the presence of MPBD during starvation (Fig. 2). This observation suggests that MPBD is involved in cAMP-mediated aggregation.

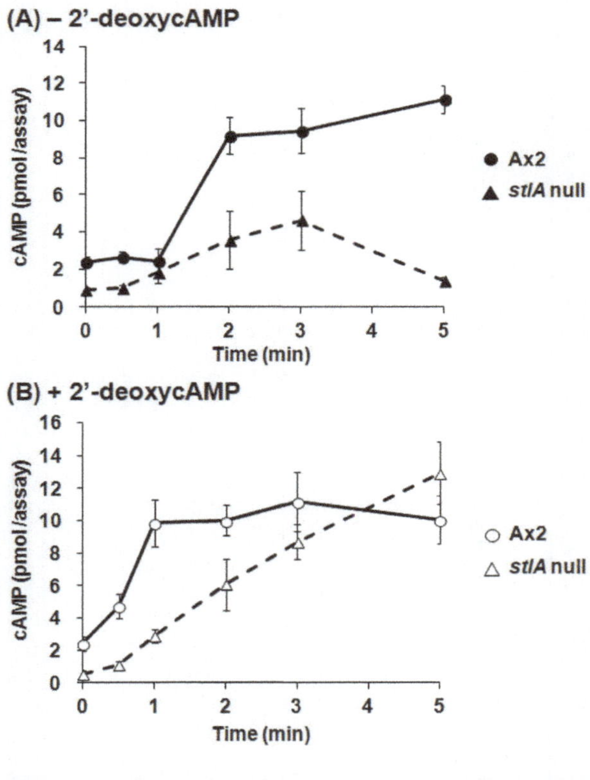

Figure 5. cAMP relay response in *stlA* null mutant. Cells were developed on non-nutrient agar plates until aggregation territories started to form. Cells were resuspended in PB and stimulated with 5 µM 2′-deoxycAMP and 5 mM DTT or with 5 mM DTT only. (A) cAMP accumulation in Ax2 or *stlA* null cells without 2′-deoxycAMP stimulation. The cAMP accumulation in *stlA* null cells was less than half of that in Ax2 cells. (B) cAMP accumulation in each cell that ACA was activated by 2′-deoxycAMP stimulation. *StlA* null cells showed a slower production rate than Ax2 cells. However, the accumulation in *stlA* null cells was almost the same as that in Ax2 cells eventually. Values are the means and SD (bars) of 3 independent experiments (n = 3).

StlA null cells are impaired in chemotaxis towards cAMP

Small aggregation territories and delayed aggregation could result from defects in chemotaxis, cell adhesion or cAMP synthesis [30–32]. The observation that the *stlA* null mutant showed aggregation defect suggested that the mutant had a defect in chemotactic response to cAMP or cAMP production. To find out whether the aggregation defect was due to either chemotactic response to cAMP or cAMP production, we first investigated chemotaxis towards cAMP using the two droplet chemotaxis assay. As shown in Figure 3A, *stlA* null cells showed a chemotaxis defect towards 250 nM cAMP. In contrast, Ax2 cells showed clear cAMP chemotaxis under the same condition. This cAMP chemotaxis defect of the *stlA* null strain was recovered by addition of 200 nM MPBD in the buffer during 6.5 h starvation. This result indicates that the presence of MPBD during starvation induces competence for chemotaxis towards cAMP. We next examined the effect of various MPBD concentrations on chemotaxis. As shown in Figure 3B, MPBD improved cAMP chemotaxis in a dose dependent manner in the *stlA* null mutant. The minimal concentration for MPBD recovery of cAMP chemotaxis is about

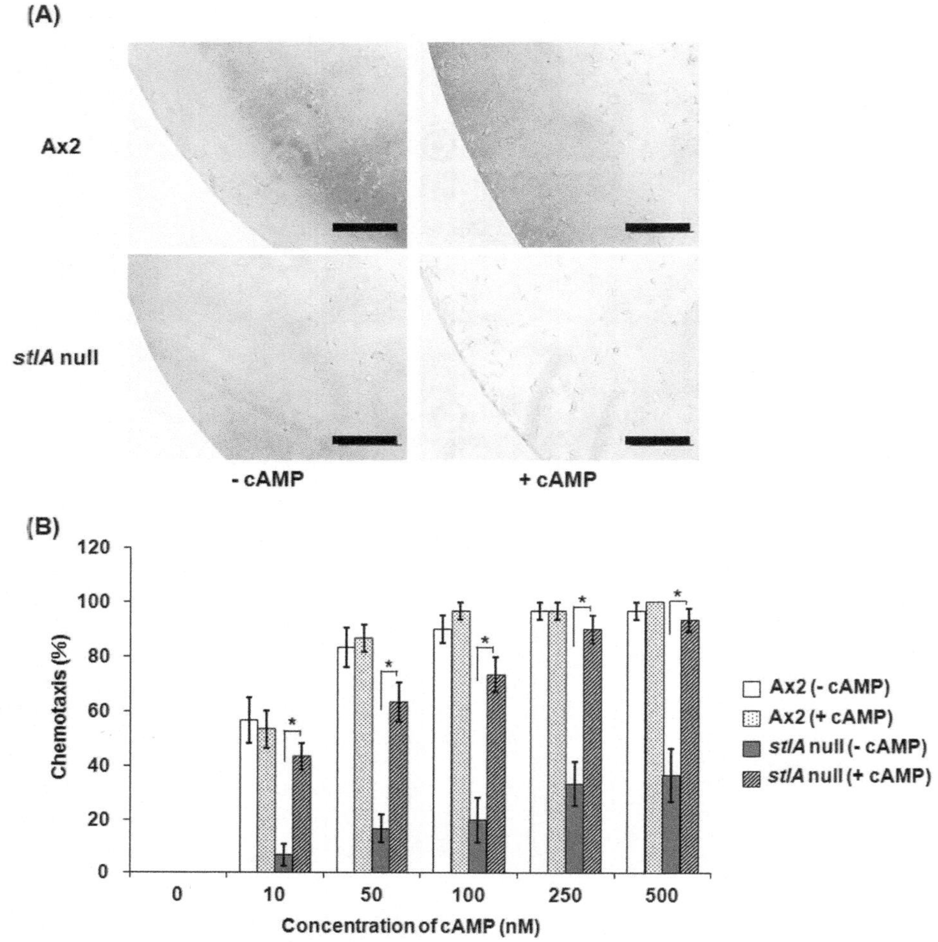

Figure 6. Effect of exogenous cAMP pulses on chemotaxis in *stlA* null cells. (A) Cells were starved at 2×10^7 cells/ml in PB with or without 20 nM cAMP pulses every 6 min for 6 h. Droplets of pulsed cells were spotted next to 500 nM cAMP droplets on hydrophobic agar plates. Photographs were taken after 1 h. Bars: 200 μm. (B) Chemotaxis rate of the *stlA* null mutant with or without cAMP pulsing. Cells were assayed for chemotaxis towards indicated concentrations of cAMP using the two droplet chemotaxis assay and their chemotaxis rates were calculated (see Materials and Methods). Values are the means and SEM (bars) of 10 independent experiments (n = 10). $^*p<0.001$ (two paired t-test).

100 nM and this concentration is similar to the recovery concentration of spore encapsulation in the *stlA* null mutant [23]. The defect in cAMP-induced chemotaxis was not due to the defect in cell motility because the *stlA* null strain showed the same chemotactic response to folate as the parent Ax2 (Fig. 4). Figure 4 shows a representative result of chemotactic response to 500 μM folate of each cell. We also examined the response to 250 μM or 100 μM folate and the same results were obtained (data not shown).

cAMP-induced cAMP production is impaired in stlA null mutant

Next, we focused on cAMP-induced cAMP production in the *stlA* null mutant. When cells were stimulated by extracellular cAMP during the aggregation stage, they soon started to produce extracellular cAMP to transmit the aggregation signal to adjacent cells. Figure 5 shows the results of cAMP relay response assay. When cells were stimulated with cAMP analogue, 2′-deoxy cAMP, cAMP receptor was stimulated and then ACA was activated. We measured the cAMP accumulation in the 2′-deoxy cAMP stimulated cells that ACA was activated. Figure 5 shows the result with 2′-deoxy cAMP stimulation. Without 2′-deoxy cAMP stimulation, *stlA* null cells accumulated less than half of cAMP in Ax2 cells (Fig. 5A). On the other hand, with 2′-deoxy cAMP stimulation, *stlA* null cells eventually showed almost the same cAMP accumulation, but had slower production kinetics than Ax2 cells (Fig. 5B).

These results strongly suggested that the *stlA* null mutant had a defect in the cAMP signal relay, especially related to the production of cAMP in the early aggregation stage. To confirm this, the cells were starved and stimulated with periodical addition of cAMP. Figure 6A shows the rescue of cAMP chemotaxis in *stlA* null cells by cAMP pulsing. When *stlA* null cells were stimulated with periodical cAMP addition during starvation, they showed the normal cAMP chemotaxis. This recovery of cAMP chemotaxis by cAMP pulsing indicates that *stlA* null cells have a defect in periodical cAMP production. Figure 6B indicates the dose dependency of cAMP chemotaxis in cAMP pulsed cells examined by the two droplet chemotaxis assay. Also, cAMP pulsed *stlA* null cells showed normal development without aggregation delay, that is, the aggregation defect of the mutant was recovered by cAMP pulsing in the absence of MPBD (Fig. 7). These results suggested

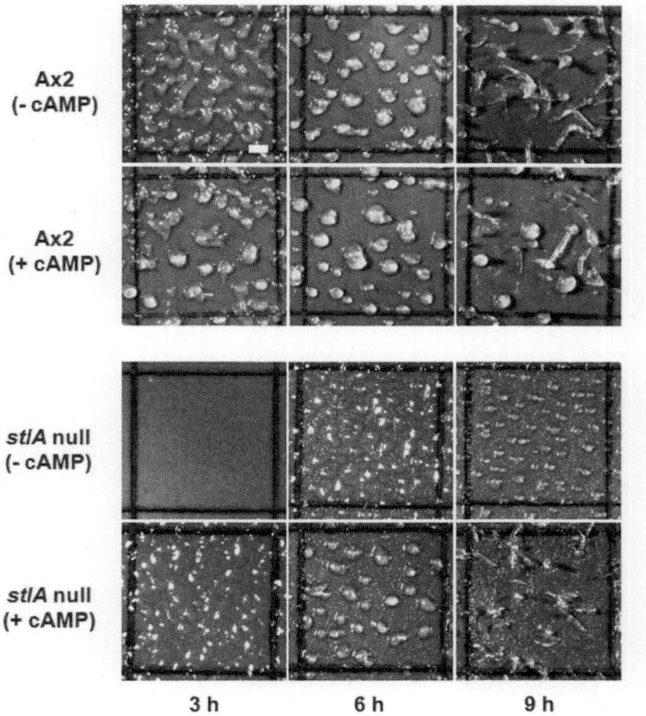

Figure 7. cAMP pulsed *stlA* null cells show normal aggregation in the absence of MPBD. Cells were stimulated by 20 nM cAMP pulses every 6 min for 6 h and then were developed on nitrocellulose filters at 1×10^6 cells/cm^2. Time points on panels indicate time elapsed since cells were washed following the completion of cAMP pulsing. cAMP pulsed *stlA* null cells show normal development without developmental delay. We observed the same phenomena in experiments repeated 3 independent times. Bar: 400 μm.

that MPBD was involved in cAMP signal relay and regulates the synthesis and/or release of cAMP in the early stage of development.

cAMP signalling genes are down regulated in stlA null mutant

The observation that the *stlA* null mutant responds chemotactically to folate but not to cAMP suggests a defect in cAMP signal transduction. cAMP-induced chemotaxis is mediated by complex parallel signal transduction cascade. The observation that pretreatment with MPBD restores cAMP-induced chemotaxis in the *stlA* null mutant suggests that MPBD may induce the expression of genes required for this response.

To determine which step(s) caused the defective chemotaxis in the *stlA* null mutant we examined the expression pattern of cAMP signalling genes in *stlA* null and Ax2 cells. Figure 8 shows the gene expression patterns of key genes in cAMP chemotaxis by quantitative RT-PCR. We analyzed the genes coding for the cAMP receptor (*carA*), adenylyl cyclase (*acaA*), secreted cAMP phosphodiesterase (*pdsA*) [33], intracellular cAMP phosphodiesterase (*regA*) [34], positive regulators of adenylyl cyclase (*piaA* and *dagA*) [35,36] and an α subunit of G-protein coupled with a cAMP receptor (*gpaB*) [37]. The mRNA levels of these genes were lower in the mutant cells than in Ax2 cells, with the sole exception of the *piaA* gene. The reason for this difference in *piaA* gene expression from other cAMP signalling genes will be discussed later.

This decreased expression pattern was especially clear at 3 h when aggregation genes started to express. Although the maximal mRNA levels of these genes were initially lower in the *stlA* null mutant than in the Ax2, their levels at 9 h were almost the same as those of the Ax2 cells (9 h in Figure 8). At this point (9 h), the Ax2 cells had reached the late aggregation stage, while *stlA* null cells were early aggregation. We also examined the expression levels of a gene coding for cell-cell adhesion protein (*csA*) [38]. This gene is strictly developmentally regulated and cAMP inducible [39]. Although the expression of *csA* was lowered and delayed in the *stlA* null mutant compared to Ax2, its mRNA level in the *stlA* null mutant was the same as Ax2 by 9 h. These results indicate that at least at 9 h after starvation (t9), the expression levels of cAMP signalling genes in the *stlA* null mutant are enough to induce the expression of cAMP inducible genes, and as a result, *csA* is expressed normally. We also confirmed that MPBD rescued *carA* and *acaA* expression defects in the *stlA* null mutant (data not shown).

As mentioned above, the expression of *piaA*, a positive regulator of adenylyl cyclase, was not down regulated as other genes in the *stlA* null mutant but rather up regulated. This regulation of *piaA* may be due to the difference of pathways activating ACA between CRAC (Cytosolic Regulator of Adenylyl Cyclase) and Pia (coded by *dagA* and *piaA*, respectively). Pia is a component of the target of rapamycin complex 2 (TORC2) [40]. TORC2 plays roles in activation of ACA and PKB [40–43]. By reconstitution experiments using cells lacking both CRAC and Pia, it has shown that both CRAC and Pia are absolutely required for ACA activity [35]. However, it has also reported that CRAC is not required for TORC2 formation [40]. In our *stlA* null mutant, *dagA* was down regulated but expressed partially. This means that the CRAC pathway in the *stlA* null mutant is partially active but has only a small effect on ACA. We propose that the *stlA* null mutant should make use of the TORC2 pathway to compensate for reduced ACA activity caused by down regulated *dagA*. As a result, the expression of *piaA* may be up regulated in the *stlA* null mutant.

From above data, we concluded that the function of SteelyA during early stage is to make starving cells aggregation competent by regulating the cAMP signalling gene expression. This means both the chemotactic response to cAMP and adenylyl cyclase activity required for aggregation are controlled by MPBD, a major product of SteelyA.

In this study, we focused on the role of SteelyA during the early stage of development and showed that the polyketide MPBD produced by SteelyA is required for the expression of cAMP signalling genes during early development. A *stlA* null mutant had aggregation defects because their expression in the mutant was down regulated.

The *stlA* null mutant showed similar phenotype to the *padA*$^-$ mutant [30,44]. In both mutant, chemotactic cAMP response was impaired and cAMP signalling genes were down regulated. However, the aggregation defect in the *padA*$^-$ mutant is not rescued by cAMP pulses because cAMP pulses are not able to fully induce cAMP relay genes in the mutant. In addition, accumulated extracellular cAMP levels in the *padA*$^-$ mutant are almost the same as that in wild type unlike the *stlA* null mutant [30]. In contrast, the aggregation defect in the *stlA* null mutant was restored by cAMP pulses and the mutant showed lower accumulated cAMP levels. MPBD signalling cascade for induction of cAMP signalling genes, therefore, is probably different pathway from PadA signalling cascade. MPBD may act on the same pathway as cAMP pulses for stimulation of cAMP signalling gene expression because cAMP pulsed *stlA* null cells show normal aggregation in the absence of MPBD.

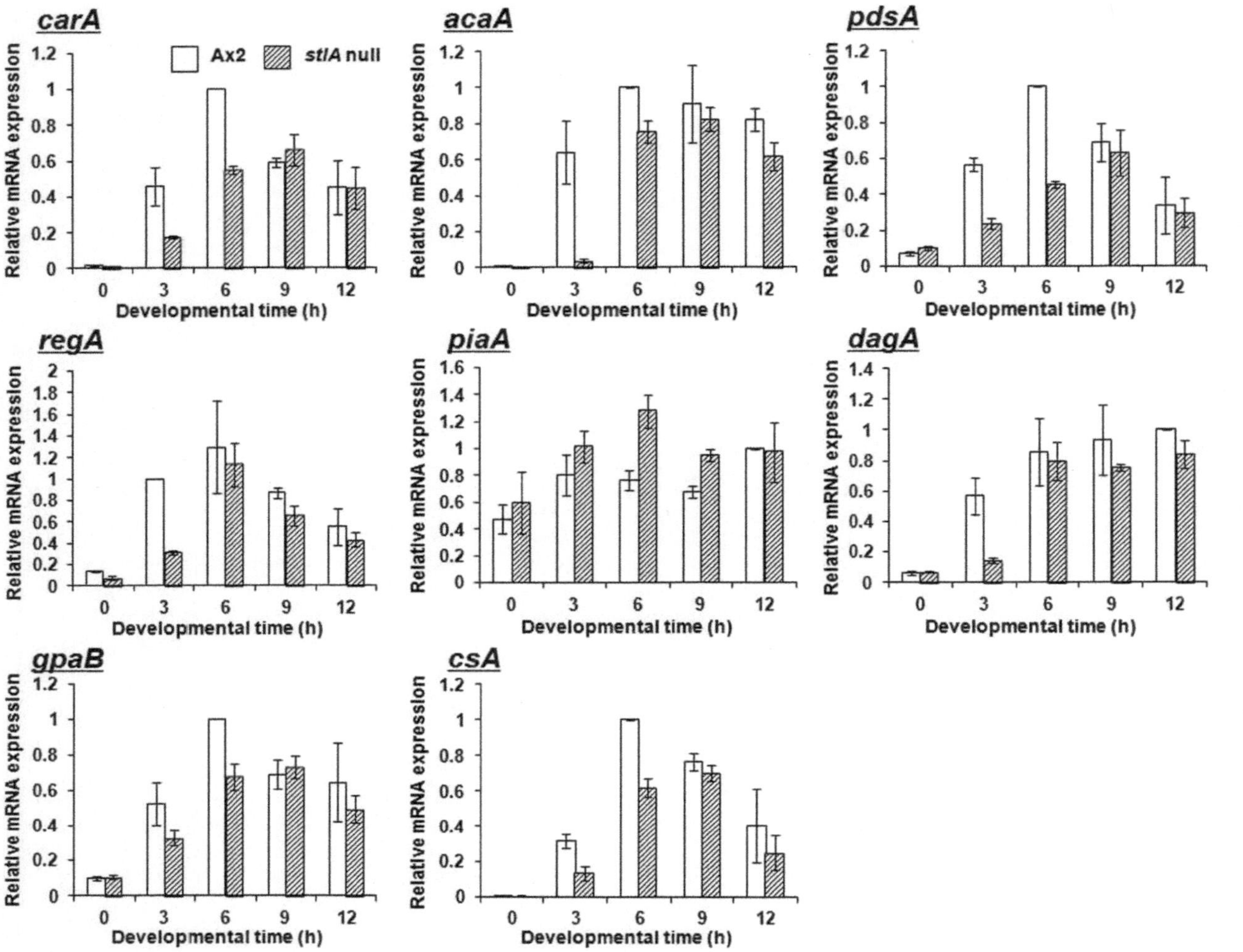

Figure 8. Expression patterns of cAMP signalling genes in *stlA* null mutant. Expression levels of each gene were analyzed by qRT-PCR. The levels were normalized to that of the endogenous control *ig7* (mitochondrial large rRNA). In addition, the expression value of Ax2 at some time point (*carA, acaA, pdsA, gpaB* and *csA*: 6 h, *regA*: 3 h, *piaA* and *dagA*: 12 h) in each gene was used as an internal standard and values were normalized it. Values indicated in graphs are the means and SD (bar) of 3 independent experiments, which were performed by using independent total RNA.

MPBD acts as a spore maturation factor during the late stage of development [23,25]. Anjard et al. reported that MPBD might bind to CrlA, the GPCR coupled to Gα1, and inhibit GskA to induce spore differentiation by SDF-1 [45]. CrlA, suggested as an MPBD receptor, is also expressed during early stage [46,47]. However, the *crlA* knockout mutant was reported as forming larger aggregates than wild type cells and exhibited normal chemotaxis to cAMP [46]. Although there is the difference of parental strains between *crlA* and *stlA* mutant (KAx3 and Ax2, respectively), the phenotype of *crlA* mutant differs from that of *stlA* null mutant. Therefore, MPBD–CrlA pathway might be not used to modulate the expression of cAMP signalling genes during the early stage of development. MPBD might regulate those expressions by acting through some receptor other than CrlA.

Author Contributions

Conceived and designed the experiments: TBN PS TS. Performed the experiments: TBN ZHC. Analyzed the data: TBN ZHC PS TS. Contributed reagents/materials/analysis tools: TBN ZHC PS TS. Contributed to the writing of the manuscript: TBN PS TS.

References

1. Kay RR, Thompson CRL (2009) Forming patterns in development without morphogen gradients: Scattered differentiation and sorting out. *Cold Spring Harbor Perspectives in Biology* 1: a001503.
2. Kessin RH (2001) *Dictyostelium* - Evolution, cell biology, and the development of multicellularity. Cambridge, UK: Cambridge Univ. Press.
3. Konijn TM, Van De Meene JGC, Bonner JT, Barkley DS (1967) The acrasin activity of adenosine-3′,5′-cyclic phosphate. *Proc Natl Acad Sci U S A* 58: 1152–1154.
4. Dinauer MC, MacKay SA, Devreotes PN (1980) Cyclic 3′,5′-AMP relay in *Dictyostelium discoideum* III. The relationship of cAMP synthesis and secretion during the cAMP signaling response. *J. Cell Biol* 86: 537–544.
5. Mahadeo DC, Parent CA (2006) Signal relay during the life cycle of *Dictyostelium*. *Curr Top Dev Biol* 73: 115–140.
6. Harwood AJ, Hopper NA, Simon MN, Driscoll DM, Veron M, et al. (1992) Culmination in *Dictyostelium* is regulated by the cAMP-dependent protein kinase. *Cell* 69: 615–624.

7. Hopper NA, Anjard C, Reymond CD, Williams JG (1993) Induction of terminal differentiation of *Dictyostelium* by cAMP-dependent protein kinase and opposing effects of intracellular and extracellular cAMP on stalk cell differentiation. *Development* 119: 147–154.
8. Hopper NA, Harwood AJ, Bouzid S, Véron M, Williams JG (1993) Activation of the prespore and spore cell pathway of *Dictyostelium* differentiation by cAMP-dependent protein kinase and evidence for its upstream regulation by ammonia. *EMBO J.* 12: 2459–2466.
9. Stepanovic V, Wessels D, Daniels K, Loomis WF, Soll DR (2005) Intracellular role of adenylyl cyclase in regulation of lateral pseudopod formation during *Dictyostelium* chemotaxis. *Eukaryot Cell* 4: 775–786.
10. van Es S, Virdy KJ, Pitt GS, Meima M, Sands TW, et al. (1996) Adenylyl cyclase G, an osmosensor controlling germination of *Dictyostelium* spores. *J Biol Chem* 271: 23623–23625.
11. Pitt GS, Milona N, Borleis J, Lin KC, Reed RR, et al. (1992) Structurally distinct and stage-specific adenylyl cyclase genes play different roles in *Dictyostelium* development. *Cell* 69: 305–315.
12. Chen ZH, Schilde C, Schaap P (2010) Functional dissection of adenylate cyclase R, an inducer of spore encapsulation. *J Biol Chem* 285: 41724–41731.
13. Söderbom F, Anjard C, Iranfar N, Fuller D, Loomis WF (1999) An adenylyl cyclase that functions during late development of *Dictyostelium*. *Development* 126: 5463–5471.
14. Alvarez-Curto E, Saran S, Meima M, Zobel J, Scott C, et al. (2007) cAMP production by adenylyl cyclase G induces prespore differentiation in *Dictyostelium* slugs. *Development* 134: 959–966.
15. Eichinger L, Pachebat JA, Glöckner G, Rajandream MA, Sucgang R, et al. (2005) The genome of the social amoeba *Dictyostelium discoideum*. *Nature* 435: 43–57.
16. Zucko J, Skunca N, Curk T, Zupan B, Long PF, et al. (2007) Polyketide synthase genes and the natural products potential of *Dictyostelium discoideum*. *Bioinformatics* 23: 2543–2549.
17. Austin MB, Saito T, Bowman ME, Haydock S, Kato A, et al. (2006) Biosynthesis of *Dictyostelium discoideum* differentiation-inducing factor by a hybrid type I fatty acid-type III polyketide synthase. *Nat Chem Biol* 2: 494–502.
18. Austin MB, Noel JP (2003) The chalcone synthase superfamily of type III polyketide synthases. *Nat Prod Rep* 20: 79–110.
19. Shen B (2006) Polyketide biosynthesis beyond the type I, II and III polyketide synthase paradigms. *Curr Opin Chem Biol* 7: 285–295.
20. Saito T, Kato A, Kay RR (2008) DIF-1 induces the basal disc of the *Dictyostelium* fruiting body. *Dev Biol* 317: 444–453.
21. Kuwayama H, Oyama M, Kubohara Y, Maeda M (2000) A novel role of differentiation-inducing factor-1 in *Dictyostelium* development, assessed by the restoration of a developmental defect in a mutant lacking mitogen-activated protein kinase ERK2. *Dev Growth Differ* 42: 531–538.
22. Kuwayama H, Kubohara Y (2009) Differentiation-inducing factor-1 and -2 function also as modulators for *Dictyostelium* chemotaxis. *PLoS One* 4: e6658.
23. Narita TB, Koide K, Morita N, Saito T (2011) *Dictyostelium* hybrid polyketide synthase, SteelyA, produces 4-methyl-5-pentylbenzene-1,3-diol and induces spore maturation. *FEMS Microbiol Lett* 319: 82–87.
24. Ghosh R, Chhabra A, Phatale PA, Samrat SK, Sharma J, et al. (2008) Dissecting the functional role of polyketide synthases in *Dictyostelium discoideum*: biosynthesis of the differentiation regulating factor 4-methyl-5-pentylbenzene-1,3-diol. *J Biol Chem* 283: 11348–11354.
25. Sato YG, Kagami HN, Narita TB, Fukuzawa M, Saito T (2013) Steely enzymes are involved in prestalk and prespore pattern formation. *Biosci Biotechnol Biochem* 77: 2008–2012.
26. Watts DJ, Ashworth JM (1970) Growth of myxameobae of the cellular slime mould *Dictyostelium discoideum* in axenic culture. *Biochem J* 119: 171–174.
27. Gilman AG (1972) Protein binding assays for cyclic nucleotides. In "Advances in Cyclic Nucleotides Research, New Assay Methods for Cyclic Nucleotides" (P. . Greengard and G. A. . Robinson, Eds.), Vol. 2: pp. 9–24. Raven Press, New York.
28. Van Haastert PJM (1984) A method for studying cAMP relay in *Dictyostelium discoideum*: The effect of temperature on cAMP relay. *J Gen Microbiol* 130: 2559.
29. Livak KJ, Schmittgen TD (2001) Analysis of relative gene expression data using real-time quantitative PCR and the 2(-Delta Delta C(T)) Method. *Methods* 25: 402–408.
30. Garciandia A, Suarez T (2013) The NMRA/NMRAL1 homologue PadA modulates the expression of extracellular cAMP relay genes during aggregation in *Dictyostelium discoideum*. *Dev Biol* 381: 411–422.
31. Roisin-Bouffay C, Jang W, Caprette DR, Gomer RH (2000) A precise group size in *Dictyostelium* is generated by a cell-counting factor modulating cell-cell adhesion. *Mol Cell* 6: 953–959.
32. Tang Y, Gomer RH (2008) A protein with similarity to PTEN regulates aggregation territory size by decreasing cyclic AMP pulse size during *Dictyostelium discoideum* development. *Eukaryot Cell* 7: 1758–1770.
33. Faure M, Franke J, Hall AL, Podgorski GJ, Kessin RH (1990) The cyclic nucleotide phosphodiesterase gene of *Dictyostelium discoideum* contains three promoters specific for growth, aggregation, and late development. *Mol Cell Biol* 10: 1921–1930.
34. Shaulsky G, Escalante R, Loomis WF (1996) Developmental signal transduction pathways uncovered by genetic suppressors. *Proc Natl Acad Sci U S A* 93: 15260–15265.
35. Chen MY, Long Y, Devreotes PN (1997) A novel cytosolic regulator, Pianissimo, is required for chemoattractant receptor and G protein-mediated activation of the 12 transmembrane domain adenylyl cyclase in *Dictyostelium*. *Genes Dev* 11: 3218–3231.
36. Lilly PJ, Devreotes PN (1994) Identification of CRAC, a cytosolic regulator required for guanine nucleotide stimulation of adenylyl cyclase in *Dictyostelium*. *J Biol Chem* 269: 14123–14129.
37. Kumagai A, Hadwiger JA, Pupillo M, Firtel RA (1991) Molecular genetic analysis of two G alpha protein subunits in *Dictyostelium*. *J Biol Chem* 266: 1220–1228.
38. Noegel A, Gerisch G, Stadler J, Westphal M (1986) Complete sequence and transcript regulation of a cell adhesion protein from aggregating *Dictyostelium* cells. *EMBO J.* 5: 1473–1476.
39. Ma PCC, Siu CH (1990) A pharmacologically distinct cyclic AMP receptor is responsible for the regulation of gp80 expression in *Dictyostelium discoideum*. *Mol Cell Biol* 10: 3297–3306.
40. Lee S, Comer FI, Sasaki A, McLeod IX, Duong Y, et al. (2005) TOR complex 2 integrates cell movement during chemotaxis and signal relay in *Dictyostelium*. *Mol Biol Cell* 16: 4572–4583.
41. Cai H, Das S, Kamimura Y, Long Y, Parent CA, et al. (2010) Ras-mediated activation of the TORC2-PKB pathway is critical for chemotaxis. *J Cell Biol* 190: 233–245.
42. Charest PG, Shen Z, Lakoduk A, Sasaki AT, Briggs SP, et al. (2010) A Ras signaling complex controls the RasC-TORC2 pathway and directed cell migration. *Dev Cell* 18: 737–749.
43. Kamimura Y, Xiong Y, Iglesias PA, Hoeller O, Bolourani P, et al. (2008) PIP3-independent activation of TorC2 and PKB at the cell's leading edge mediates chemotaxis. *Curr Biol* 18: 1034–1043.
44. Núñez-Corcuera B, Serafimidis I, Arias-Palomo E, Rivera-Calzada A, Suarez T (2008) A new protein carrying an NmrA-like domain is required for cell differentiation and development in *Dictyostelium discoideum*. *Dev Biol* 321: 331–342.
45. Anjard C, Su Y, Loomis WF (2011) The polyketide MPBD initiates the SDF-1 signaling cascade that coordinates terminal differentiation in *Dictyostelium*. *Eukaryot Cell* 10: 956–963.
46. Raisley B, Zhang M, Hereld D, Hadwiger JA (2004) A cAMP receptor-like G protein-coupled receptor with roles in growth regulation and development. *Dev Biol* 265: 433–445.
47. Rot G, Parikh A, Curk T, Kuspa A, Shaulsky G, et al. (2009) dictyExpress: a *Dictyostelium discoideum* gene expression database with an explorative data analysis web-based interface. *BMC Bioinformatics* 10: 265.

A Conserved Cysteine Residue of *Bacillus subtilis* SpoIIIJ Is Important for Endospore Development

Luísa Côrte[1]ⓒ, Filipa Valente[1]ⓒ¤, Mónica Serrano[1], Cláudio M. Gomes[1], Charles P. Moran Jr[2], Adriano O. Henriques[1]*

1 Instituto de Tecnologia Química e Biológica, Universidade Nova de Lisboa, Oeiras, Portugal, **2** Department of Microbiology and Immunology, Emory University School of Medicine, Atlanta, Georgia, United States of America

Abstract

During sporulation in *Bacillus subtilis*, the onset of activity of the late forespore-specific sigma factor σ^G coincides with completion of forespore engulfment by the mother cell. At this stage, the forespore becomes a free protoplast, surrounded by the mother cell cytoplasm and separated from it by two membranes that derive from the asymmetric division septum. Continued gene expression in the forespore, isolated from the surrounding medium, relies on the SpoIIIA-SpoIIQ secretion system assembled from proteins synthesised both in the mother cell and in the forespore. The membrane protein insertase SpoIIIJ, of the YidC/Oxa1/Alb3 family, is involved in the assembly of the SpoIIIA-SpoIIQ complex. Here we show that SpoIIIJ exists as a mixture of monomers and dimers stabilised by a disulphide bond. We show that residue Cys134 within transmembrane segment 2 (TM2) of SpoIIIJ is important to stabilise the protein in the dimeric form. Labelling of Cys134 with a Cys-reactive reagent could only be achieved under stringent conditions, suggesting a tight association at least in part through TM2, between monomers in the membrane. Substitution of Cys134 by an Ala results in accumulation of the monomer, and reduces SpoIIIJ function *in vivo*. Therefore, SpoIIIJ activity *in vivo* appears to require dimer formation.

Editor: Dirk-Jan Scheffers, University of Groningen, Groningen Institute for Biomolecular Sciences and Biotechnology, Netherlands

Funding: This work was supported by grants POCTI/BIA-BCM/60855/2004 and PEst-OE/EQB/LA0004/2011 from "Fundação para a Ciência e a Tecnologia" (FCT) to A.O.H., and by National Institute of General and Medical Sciences Public Health Services R01 Grant GM54395-29 to C.P.M. LC (SFRH/BD/6489/2001) and M.S (SFRH/BPD/36328/2007) and F.V. (SFRH/BPD/26470/2006) were the recipients of a PhD (LC) and post doctoral fellowships (MS and FV) from the FCT. The funders had no role in study design, data collection and analysis, decision to publish, or preparation of the manuscript.

Competing Interests: Dr. Cláudio Gomes, a co-author of the manuscript, is a PLOS ONE Editorial Board member. This does not alter the authors' adherence to PLOS ONE Editorial policies and criteria.

* Email: aoh@itqb.unl.pt

ⓒ These authors contributed equally to this work.

¤ Current address: IPATIMUP, Rua Dr. Roberto Frias, s/n, Porto, Portugal

Introduction

Protein secretion and membrane protein insertion are fundamental processes in all living organisms, and several pathways are known to serve those purposes, namely the Sec, Tat, and YidC/Oxa1/Alb3 pathways [1,2,3]. These pathways have been extensively studied in bacteria, specifically in the model organisms *Escherichia coli* and *Bacillus subtilis*. The most prominent one is the Sec pathway. Its core is the SecYEG channel, which is utilised both for secretion and membrane protein insertion, using distinct piloting factors that guide each class of proteins to SecYEG. The SecB chaperone in *E. coli* (possibly CsaA in *B. subtilis*) guides proteins to be secreted. In contrast, nascent membrane proteins are delivered to SecYEG by the signal recognition particle (SRP) [reviewed in 1,4]. The twin-arginine translocation (Tat) pathway has the peculiarity of permitting the transport of previously folded proteins [reviewed in 2]. YidC/Oxa1/Alb3 family members, found in archaea, bacteria and in eukaryotic organelles, participate in the biogenesis of membrane proteins. The first protein from this family to be identified was Oxa1 in mitochondria of *Saccharomyces cerevisiae*. Members of this family are polytopic membrane proteins that share a core architecture of five transmembrane (TM) segments. They perform critical roles in membrane insertion and folding and also in the assembly of energy-transducing multimeric membrane complexes [reviewed in 3]. The best-studied proteins from this family in bacteria are YidC (*E. coli*) and SpoIIIJ (*B. subtilis*), respectively, which function with the Sec channel or independently of it. The Sec-independent function of these proteins is conserved and similar to that of Oxa1 in mitochondria, which lack the Sec system [3,5,6]. The large periplasmic region of YidC (absent in SpoIIIJ and from the conserved core that characterises the family) was found to be monomeric [7,8]. However, full-length YidC was observed as a dimer through cryo-electron microscopy [9], suggesting that the determinants for dimerisation lie within the conserved, membrane-associated, regions of the protein. SpoIIIJ is required for spore formation in *B. subtilis* [10]. The conversion of vegetative cells into highly resistant spores is one of the responses that this organism uses to cope with highly stressful situations [11,12,13]. Besides SpoIIIJ, another member of this family exists in *B. subtilis*, YqjG. Although they can substitute for each other during growth, only SpoIIIJ supports efficient sporulation [10,14,15,16,17].

Spore differentiation takes place in a cell divided into two unequally-sized compartments, between which successive waves of gene expression in one compartment or the other are activated in a coordinated manner and in register with the course of morphogenesis [11,12,13]. Transcription is controlled by the successive activities of cell type-specific RNA polymerase sigma factors. From the forespore, σ^F signals the activation of σ^E in the larger mother cell, which in turn is required for activity of σ^G in the forespore following engulfment completion. σ^G is in turn responsible for the activation of σ^K in the mother cell [18,19]. A key process of endosporulation is the engulfment of the forespore by the mother cell. The onset of σ^G activity is normally coupled to engulfment completion, when the forespore becomes isolated from the extracellular milieu and surrounded by two membranes [13,19]. At this stage of development, assembly of a specialised secretion system that spans the two forespore membranes is thought to allow the mother cell to feed the forming spore, allowing the continuation of macromolecular synthesis [10,20,21,22,23,24]. Eight mother cell-specific proteins encoded by the *spoIIIA* operon, and the forespore-specific SpoIIQ protein form the complex. SpoIIIJ is needed specifically during this step of sporulation possibly because it participates in the biogenesis of the secretion complex [10,20,23,24,25,26]. Evidence suggests that SpoIIIJ interacts with one of the *spoIIIA* proteins, SpoIIIAE, in the context of the Sec translocon to promote its correct insertion into the forespore membranes [20,24].

Here we have purified SpoIIIJ from *E. coli* cells and shown that it forms a dimer. We also provide evidence that a cysteine residue predicted to be in the second transmembrane segment of SpoIIIJ facilitates dimer formation by forming a disulphide bond at the dimer interface. Replacement of the cysteine by an alanine residue results in accumulation of the monomer and under certain conditions impairs the activity of σ^G and reduces sporulation. Our results suggest a model in which a disulphide-bond contributes, with additional non-covalent interactions, to the formation of a SpoIIIJ dimer, and that this dimer is important for SpoIIIJ activity during sporulation.

Materials and Methods

Media, bacterial strains and general techniques

The *B. subtilis* strains used in this work are congenic derivatives of the Spo$^+$ strain MB24 (*trpC2 metC3*) and are listed in Table S1. Luria-Bertani (LB) medium was used for growth or maintenance of *E. coli* and *B. subtilis*, and sporulation was induced in Difco sporulation medium (DSM) [27,28].

Strains and plasmids

A *spoIIIJ*^{C134A} allele for expression in *E. coli* was constructed in two steps: an initial amplification from chromosomal DNA of the wild-type *B. subtilis* strain MB24 using primers pairs J112D with JC134A_D, and JC134A_R with Jhis (Table S2); next, the PCR fragments were joined using the external primers through splicing by overlap extension (SOE) [29]. The final PCR product was cleaved with *Nco*I and *Eco*RI, ligated to pETDuet-1 (Novagen, Darmstadt, Germany) similarly digested to obtain pFiV1 (plasmids are listed in Table S3), resulting in a fusion to a His$_6$ tag at the 3′ of *spoIIIJ*^{C134A}.

For controlled expression of His$_6$ fusions to *spoIIIJ* and *spoIIIJ*^{C134A} in *B. subtilis* from a promoter that can be induced with isopropyl-β-D-thiogalactopyranoside (IPTG), chromosomal DNA of MB24 was amplified with primers J174D and JhisR (for wild-type *spoIIIJ*) and primer pairs J174D with JC134A_R, and JC134A_D with JhisR, being the last two products joined by PCR (SOE) with the external primers (for the *spoIIIJ*^{C134A} mutant). The final PCR products were digested with *Bgl*II and *Sph*I and ligated to pDH88 [30] similarly digested, generating pFiV2 and pFiV3, respectively. Both pFiV2 and pFiV3 were digested with *Eco*RI and *Bam*HI, the inserts were recovered and ligated to pDG1664 [31] similarly digested to produce pFiV4 and pFiV5, respectively; these plasmids were inserted at the non-essential locus *thrC* of JOB44 [16] through a double recombination event, producing AH5425 (Δ*spoIIIJ* Δ*thrC*::P$_{spac}$-*spoIIIJ*- *his*$_6$) and AH5426 (Δ*spoIIIJ* Δ*thrC*::P$_{spac}$-*spoIIIJ*^{C134A}-*his*$_6$, respectively. The two strains were transformed with Δ*yycR*::P$_{sspE}$-*cfp* from AH6566 [24] (constructed with DNA from BTD2633 (kindly provided by D. Rudner) whilst selecting for Cm^r, resulting in AH5431 and AH5432, respectively (Table S1). JOB44 was transformed with DNA of BTD2633 resulting in AH5433 (Δ*spoIIIJ* Δ*yycR*::P$_{sspE}$-*cfp*).

To construct pLC138, two PCR products were obtained with primer pairs PYqjG-460D with YA50C_R, and YA50C_D with YqjG-His-R, which were joined by PCR (SOE) using the external primers. Both the final PCR product and pLC111 (Côrte and Henriques, unpublished data) were digested with *Bam*HI and ligated. To construct the *yqjG*$^{A50C/C142A}$-*his*$_6$ derivative, primers PYqjG-460D with YC142A_R were used to amplify the first half of the gene from pLC138, and primers YC142A_D with YqjG-His-R for the second half, using pLC115 (Côrte and Henriques, unpublished data) as a template. The products were joined by PCR (SOE) with the external primers, digested with *Bam*HI and cloned into pLC111 similarly digested, resulting in pLC155. Transformation of JOB44 with pLC155 resulted in AH5382 (Δ*spoIIIJ* Δ*amyE*::P$_{yqjG}$-*yqjG*$^{A50C/C142A}$-*his*$_6$).

Production of SpoIIIJ-His$_6$ and of SpoIIIJ^{C134A}-His$_6$

E. coli strain C43(DE3) bearing pMS266 or pFiV1, were grown in LB and induced with IPTG for 3 hours at 37°C [32]. Cells were broken in a French pressure cell (19,000 lb/in^2) in lysis buffer [20 mM Tris-HCl pH 7.6, 50 mM NaCl, 20% glycerol, 1 mM phenylmethylsulphonyl fluoride (PMSF), DNAse I].

The whole-cell extract was separated in a soluble and a membrane fraction by ultracentrifugation (100,000 × *g*, 45 min, at 4°C). Membrane proteins were extracted from the pellet with 2% dodecyl maltoside (DDM), 20 mM Tris-HCl pH 7.6, 20% glycerol, 0.5 M NaCl and 10 mM imidazole, for 1 h on ice, and again ultracentrifuged (as above). Purification proceeded using a Ni^{2+}-NTA affinity column (Qiagen, Hilden, Germany) previously equilibrated with a buffer composed of 0.5 M NaCl, 0.1% DDM 20 mM Tris-HCl pH 8 with 20 mM imidazole. The column was washed five times with the same buffer. Bound proteins were was eluted with a similar buffer containing 25 mM, 50 mM or 100 mM imidazole.

Circular dichroism

Purified SpoIIIJ-His$_6$ or SpoIIIJ^{C134A}-His$_6$ were first dialysed in 20 mM Tris-HCl pH 7.6, 10% glycerol, 0.5 M NaCl, 0.1% DDM. Far-UV CD spectra were measured at 20°C on a Jasco J-815D CD spectrometer using a quartz polarised 1 mm path length cuvette, from 200 to 260 nm.

Size exclusion chromatography

SpoIIIJ-His$_6$ or SpoIIIJ^{C134A}-His$_6$ purified from *E. coli* were loaded onto a Superose 12 HR 10/30 column (GE Healthcare) previously equilibrated with a buffer composed of 20 mM Tris-HCl pH 7.6, 10% glycerol, 0.5 M NaCl, 0.1% DDM, and 100 mM imidazole. The column was calibrated with the gel filtration molecular markers Dextran Blue, aldolase (158 kDa), ovalbumin (43 kDa), chymotryspinogen A (25 kDa), and lysozyme (14 kDa), in the same buffer used for equilibration. Fractions were subjected to immunoblot analysis.

BN-PAGE

Blue-native polyacrylamide gel electrophoresis (BN-PAGE) was performed using gradient gels (5–15%) as described previously [33]. An elution fraction of 9 μg of His-tagged purified SpoIIIJ containing 100 mM imidazole and 500 mM NaCl was used. HMW-Native (Amersham Biosciences) was used as a molecular weight marker.

Fluorescence microscopy

Samples (0.6 ml) of DSM cultures were collected at 4 and 6 h after the onset of sporulation, from cultures with (0.5 mM) or without IPTG, and resuspended in 0.2 ml of phosphate-buffered saline (8 mM sodium phosphate [pH 7.5], 150 mM NaCl) supplemented with 10 μg ml^{-1} of the lipophilic membrane dye FM4-64 (Molecular Probes) and with 0.2 μg ml^{-1} of the DNA dye 4-,6-diaminodino-2-phenylindole (DAPI). The cells were then observed by fluorescence microscopy [34,35]. For the quantitative analysis of P_{sspE}-*cfp* expression at least 200 cells were scored for the fluorescence patterns designated by low (class a) or high (class b). ImageJ (http://imagej.nih.gov/ij/) was used for quantification of the fluorescence signal. Low and high fluorescence refers to the intensity of the fluorescent signal as determined by a threshold defined with ImageJ, using σ^G-inactive cells and σ^G-active cells as references.

MalPEG labelling of *B. subtilis* SpoIIIJ-His_6

Cells grown in LB were collected by centrifugation, resuspended in 10 mM Tris-HCl pH 7.4, 1 mM EDTA, 1 mM PMSF, 1 mM Tris(2-carboxyethyl)phosphine (TCEP) and lysed with a French pressure cell (19 000 lb/in^2). Membranes were isolated by a 60-min centrifugation at 100,000× *g*, resuspended in 50 mM Tris-HCl pH 6.8, 1 mM TCEP and further incubated either at 37°C or 80°C, with or without methoxypolyethylene glycol 5000 maleimide (malPEG), in the presence of 6 M urea and SDS (1 or 2%) or with no SDS nor urea. Other conditions for labelling of SpoIIIJ-His_6 were as follows: 10% SDS; 2% SDS and 8 M Urea; 1% Triton and 4 M GdnHCl. Note that GdnHCl and SDS form a precipitate and could not be used in conjunction for this analysis.

Whole-cell lysates and immunoblot analysis

B. subtilis strains were grown in DSM and samples collected one hour before the end of the growth phase (defined as the onset of sporulation, or T_0), and two and four hours thereafter. Cells were resuspended in 10 mM Tris-HCl pH 7.4, 1 mM EDTA, 1 mM PMSF, 0.1 mM dithiothreitol (DTT) and lysed with a French pressure cell as above. Proteins (30 μg) were electrophoretically resolved on 12.5% polyacrylamide gels (SDS-PAGE) [36], transferred to nitrocellulose membranes and subjected to immunoblot analysis with a mouse anti-His_6 antibody (Novagen, Darmstadt, Germany) for the detection of SpoIIIJ-His_6 and YqjG-His_6. The proteins were visualised with the ECL detection system (Amersham Biosciences) as described by the manufacturer.

Topology prediction: SpoIIIJ's Cys134

TMPRED (http://www.ch.embnet.org/software/TMPRED_form.html) and TOP PRED (http://mobyle.pasteur.fr/cgibin/portal.py?form = toppred) were used to predict the localisation of SpoIIIJ Cys134 relative to the membrane plane (cytoplasmic/exterior/in the membrane).

Results

A role for Cys134 in the oligomerisation of SpoIIIJ

We purified SpoIIIJ with a C-terminal His_6 tag, SpoIIIJ-His_6, from cells of *E. coli* C43(DE3). SpoIIIJ-His_6 was solubilised with DDM from the membrane fraction obtained by ultracentrifugation and further purified over a Ni^{2+}-column (Fig. S1A). About 0.8 mg/L of essentially pure SpoIIIJ-His_6, as judged by SDS-PAGE analysis, was obtained (Fig. S1). SpoIIIJ is a polytopic membrane protein with 5 predicted transmembrane (TM) α-helices and a calculated molecular mass of 27 kDa (Fig. 1A). In agreement, analysis by far-UV circular dichroism indicates that the purified and detergent-solubilised SpoIIIJ-His_6 corresponds to a folded α-helical rich protein (Fig. S2). Under reducing SDS-PAGE conditions, the most abundant form of the protein migrates with an apparent mass of 23 kDa (Fig. 1B) (see also [16]). However, higher molecular weight forms of SpoIIIJ-His_6 were detected following Coomassie staining of SDS-PAGE gels, in the absence of the reducing agent DTT (Fig. 1B). Multimeric forms of SpoIIIJ-His_6 have been observed before in the membrane fraction of *E. coli* cells over-producing this protein [24]. The high molecular weight forms of SpoIIIJ-His_6 in Fig. 1B, have an apparent size of 30 and 45 kDa (Fig. 1B, bands "a" and "b", respectively; these are indicated by red arrows). Immunoblotting with an anti-His_6 antibody did not reveal additional forms (Fig. 1B; shorter forms of the protein are discussed below). This suggested that disulphide bonds between two cysteine residues of SpoIIIJ could be responsible for the formation of the 30 and 45 kDa forms.

In addition to the higher molecular weight forms of SpoIIIJ-His_6 visualised in Fig. 1B (bands "a" and "b"), species migrating closely together at around 15 kDa were also seen, particularly by immunoblot analysis (Fig. 1B). The apparent molecular weight of band "a" is compatible with a dimer of full-length mature SpoIIIJ (that is, cleaved by signal peptidase II). Band "b" could, in turn, represent a heterodimer formed by full-length mature SpoIIIJ-His_6 and one of the species found around the 15 kDa region of the gel. To test this hypothesis, bands "a" and "b" were excised from an SDS-PAGE gel and re-run separately. Starting with band "a", both "a" and monomeric SpoIIIJ-His_6 were detected by immunoblot analysis (Fig. 1C, red and black arrows, respectively). Thus, band "a" seems to be composed of two full-length mature SpoIIIJ monomers. Starting with band "b", both "b" and monomeric SpoIIIJ-His_6 were detected, along with this band, "c", that seems to correspond to the slower migrating species of the doublet presumably resulting from processing of SpoIIIJ-His_6 migrating just above the 15 kDa region of the gel (Fig. 1B and C). N-terminal sequencing of these two species of about 15 kDa, indicates that both start at residue L122 (Fig. 1B). These species are possibly generated by proteolytic cleavage of SpoIIIJ-His_6 at L122, located in the cytoplasmic loop connecting TM1 and TM2, and removing TM1 (Fig. 1A). Cleavage at this position would generate fragments of 11 kDa (from Cys23 to part of the first cytoplasmic loop (including TM1), fragment "d" in Fig. 1A) and 16 kDa (the remainder of the first cytoplasmic loop to the end of the protein (including TM2), fragment "c" in Fig. 1A). The faster migrating species is not generated by a second cleavage event close to C-terminal of the larger species, as both are detected with an antibody against the C-terminally located His_6 tag (Fig. 1B). Possibly, the difference in mobility of the doublet is caused by some other modification of the proteolytic product of SpoIIIJ-His_6. In any event, these observations support the idea that SpoIIIJ-His_6 purified from *E. coli* forms dimeric species. Moreover, they suggest

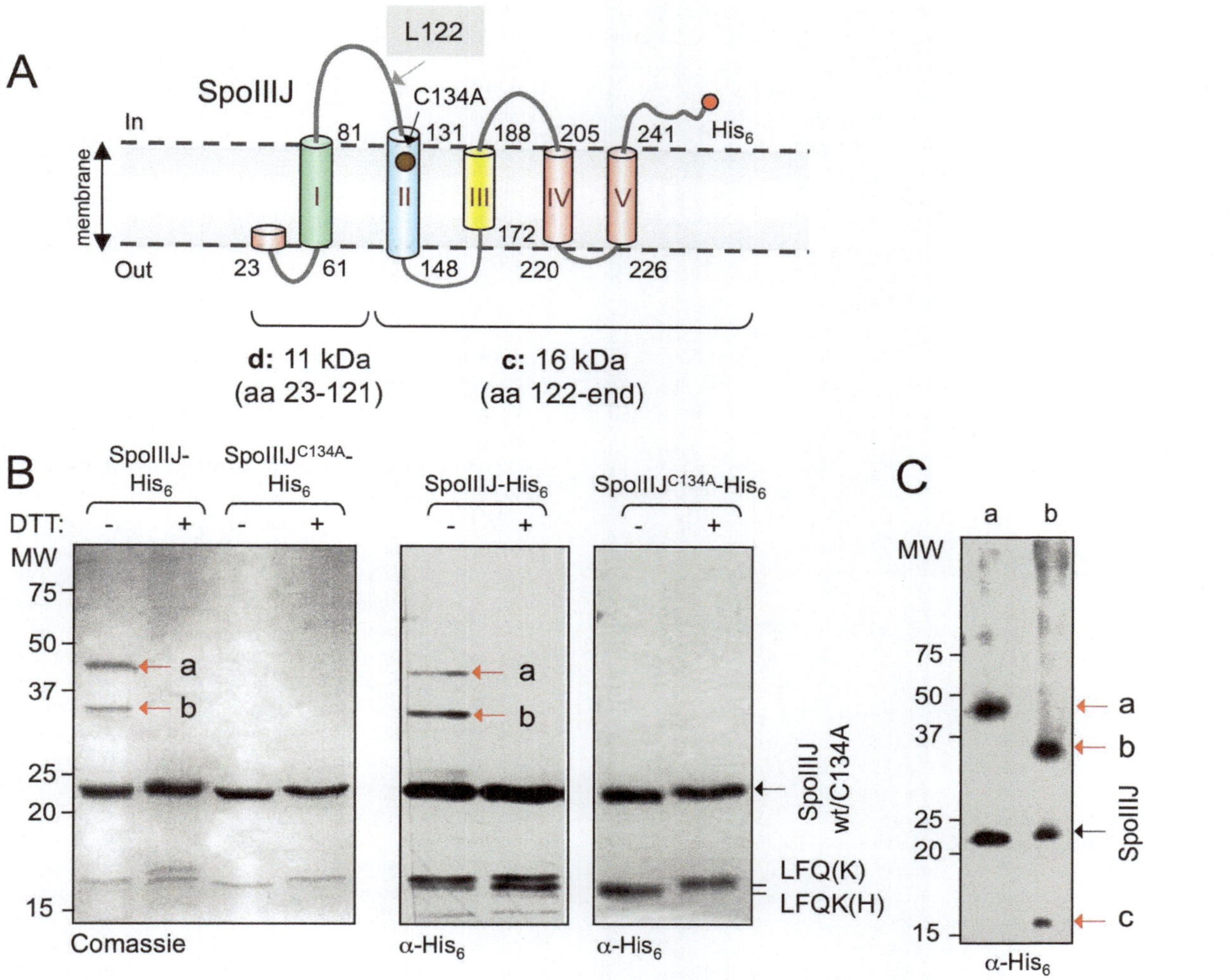

Figure 1. Dimerisation of SpoIIIJ-His$_6$ purified from *E. coli*. (A) Topological model of SpoIIIJ in the membrane. The numbers refer to the amino acid residues that delimit the transmembrane (TM) segments, depicted as cylinders. Cys23, in the signal peptide is indicated, as well as Cys134 in TM2 (red circle). (B) SDS-PAGE and Coomassie-staining (left panel) or immunoblotting (right) with an anti-His$_6$ antibody of purified SpoIIIJ-His$_6$ and SpoIIIJ^{C134A}-His$_6$ in loading buffer with (100 mM) or without DTT. The N-terminal sequence of two processing products of SpoIIIJ-His$_6$ is indicated. (C) Immunoblot analysis of SpoIIIJ-His$_6$ species "a" and "b", isolated from the gel in panel B (left), using an anti-His$_6$ antibody. In panels B–C, the position of molecular weight markers (in kDa) is shown; black arrows show the position of full-length SpoIIIJ-His$_6$ and red arrows the position of other SpoIIIJ forms.

that SpoIIIJ-His$_6$ can form a heterodimer with a form of the protein lacking TM1.

SpoIIIJ-His$_6$ has two Cys residues. One is located in the lipoprotein-type signal peptide and is modified during insertion of the protein into the membrane (Côrte and Henriques, unpublished data). A second, unmodified in mature SpoIIIJ-His$_6$, is Cys134, located in TM2 (Fig. 1A). Cys134 in SpoIIIJ is conserved among orthologues of the protein, as shown in the alignment of Fig. S3A. A cysteine residue is found at the homologous position of YqjG (Cys142), the other member of the YidC/Oxa1/Alb3 family found in *B. subtilis*, and also in the *E. coli* YidC protein. As shown with several membrane topology prediction programs (see Materials and Methods), this residue seems to be always located within a TM segment (Fig. S3A). A helical wheel projection of the second TM segment of SpoIIIJ (Fig. S3B) showed it is a typical hydrophobic helix [37,38]. This residue was the likely candidate for formation of disulphide bonds between SpoIIIJ monomers. To test this, we overproduced and purified a C-terminal His-tagged variant of SpoIIIJ in which Cys134 was substituted by an alanine (Fig. S1B). In contrast to the wild-type, no higher molecular weight bands of SpoIIIJ^{C134A}-His$_6$ were detected either by Coomassie-staining or immunoblotting of SDS-PAGE gels in the presence or absence of DTT (Fig. 1B). Far-UV CD analysis showed that the C134A substitution is not deleterious for the SpoIIIJ fold, as the spectrum obtained for this purified and detergent solubilised variant is essentially identical to that of the wild-type protein (Fig. S2A and B). Nevertheless, whereas reduction of SpoIIIJ-His$_6$ with DTT yields minor secondary structure interconversions compatible with structural rearrangements upon reduction of the presumed C134-C134 disulphide, the same is not observed in the case of SpoIIIJ^{C134A}-His$_6$ (Fig. S2C).

As an independent test of oligomer formation by SpoIIIJ-His$_6$, we used blue-native (BN) PAGE and size exclusion chromatography (SEC). By BN-PAGE, three species were observed that exhibited molecular weights compatible with a monomer, a dimer (the most represented species) and traces of a possible hexamer (Fig. 2A). SEC of purified SpoIIIJ-His_6 revealed two main peaks (Fig. 2B) both of which contained the protein (as show by the immunoblot of the peak fractions; insert in Fig. 2B). The larger peak corresponds to a species with a calculated size (75 kDa) closer to that of a SpoIIIJ-His_6 dimer (Fig. 2B). The smaller peak is caused by a species with a calculated size (47 kDa) closer to that of monomeric SpoIIIJ-His_6 (Fig. 2B). The differences in these sizes from that estimated from BN-PAGE analysis probably results from the presence of dodecyl maltoside (DDM), the amount of which we were unable to estimate by SEC-MALLS (data not shown). Peaks with the same estimated sizes were also observed for $SpoIIIJ^{C134A}$-His_6 but in this case, the size of peak B, corresponding to the smaller species, increased relative to peak A (Fig. 2B). Quantification of the peak areas shows ratios of B to A of 1:9 for SpoIIIJ-His_6 and 1:2.5 for $SpoIIIJ^{C134A}$-His_6.

Formation of the dimer size species (peak A) does not require Cys134 but taken all together our results suggest that Cys134 stabilises a multimeric (possibly a dimeric) form of SpoIIIJ-His_6, in line with the suggestion that this residue is involved in formation of a disulphide bond between two monomers of the protein.

Cys134 of SpoIIIJ is part of a protein-protein interface in *B. subtilis*

Cys134 is thought to be located in the plane of the membrane (Fig. 1A and S3), and involved in formation of a disulphide bond between SpoIIIJ-His_6 monomers (Fig. 1). We wanted to examine whether Cys134 is located within the membrane and if it forms part of a protein-protein interaction interface. The membrane fraction of whole cell lysates prepared from *B. subtilis* cultures was isolated by ultracentrifugation and the available Cys residues in the proteins present in the sample were labelled with the Cys-modifying reagent methoxypolyethylene glycol 5000 maleimide (malPEG for simplicity). This reagent binds free sulfhydryl groups in cysteine residues forming thioether bonds, whilst adding an extra 5 kDa per available cysteine residue. As malPEG induces distortion in gel migration, the minimum concentration required to obtain appreciable labelling was determined and found to be 1 mM (data not shown). As malPEG is membrane-impermeable, SDS was also used to solubilise the membranes. If the cysteine is in the membrane plane, labelling was expected only in the presence of SDS. To increase the availability of free sulfhydryl groups by reducing eventual disulphide bonds, Tris(2-carboxyethyl)phosphine (TCEP) was used. Unlike DTT, this reagent does not compete with the cysteine residues for malPEG. As a positive control for labelling we used a substituted form of YqjG in which a Cys residue was introduced at position 50. As for the highly similar SpoIIIJ (Fig. 1A), this position is predicted to be located outside the cytoplasmic membrane, and facing the extracellular space (Fig. 3A). This residue should be more readily labelled with malPEG than a residue located in the plane of the membrane. Cys142 of YqjG (Fig. 3A) is homologous to Cys134 of SpoIIIJ and was substituted by an alanine to facilitate interpretation of the labelling pattern. Labelling of SpoIIJ with malPEG was not detected at 37°C even in the presence of 2% SDS (Fig. 3B) or at higher concentrations of the detergent (not shown). The level of SpoIIIJ in mixtures incubated at 80°C in the absence of SDS is strongly reduced (Fig. 3C, lanes 6 and 8), because under these conditions the protein failed to enter the gel (not shown). The combined action of detergents and denaturants has been shown to disclose normally hidden residues [39,40]. However, none of the combinations herein tested (GdnHCl, urea, Triton and SDS), result in labelling of Cys134 of SpoIIIJ-His_6 (see Materials and Methods). However, SpoIIIJ-His_6 was labelled at 80°C in the presence of 2% SDS (Fig. 3C). Thus, high temperature and the presence of a detergent are required for labelling Cys134 of SpoIIIJ-His_6 with malPEG. In contrast, labelling of the positive control $YqjG^{A50C/C142A}$-His_6 was obtained in the presence of 1% SDS and also in its absence (Fig. 3B and C). We observed a shift in the apparent molecular weight for malPEG-derivatised $YqjG^{A50C/C142A}$-His_6 larger than the expected 5 kDa, which has also been reported previously [41].

Together, these results are compatible with the view that Cys134 is located within the membrane plane, and is part of a tight protein-protein interaction surface.

Accumulation of wild-type and substituted SpoIIIJ during sporulation in *B. subtilis*

spoIIIJ is required for sporulation [10]. To evaluate the ability of $SpoIIIJ^{C134A}$ to support efficient sporulation, both the wild-type and C134A form of SpoIIIJ were expressed from the IPTG-inducible P_{spac} promoter, placed at the non-essential *thrC* locus in a *spoIIIJ* deletion background (Fig. 4A). When measured 24 hours after the onset of sporulation in DSM, the non-induced C134A strain produced very low levels of heat-resistant spores when compared to its wild-type counterpart (Table 1). Importantly, the level of heat-resistant spores for the non-induced wild-type allele was close to that found for a wild-type strain (*i.e.*, expressing *spoIIIJ* from its normal locus) (Table 1). In contrast, upon addition of IPTG both the wild-type and the C134A mutant exhibited sporulation levels similar to a wild-type strain (Table 1). The fact that a low spore titre was observed for the non-induced mutant but not for the non-induced wild-type protein, suggested that under these conditions $SpoIIIJ^{C134A}$-His_6 accumulates to lower levels or is less functional than the wild-type form. To distinguish between these possibilities, we analysed the levels of either protein under inducing and non-inducing conditions. When induced, both the wild-type and C134A forms of SpoIIIJ accumulate to levels close to those of the protein produced from its native locus (Fig. S4). When non-induced, the levels of SpoIIIJ or $SpoIIIJ^{C134A}$ are both barely detectable, but similar to one another (Fig. 4B). Hence, it seems unlikely that different expression levels account for the difference in spore titre observed for the two forms of the protein (Table 1). Note that a species with the expected size for a dimer is seen for both $SpoIIIJ^{wt}$-His_6 and $SpoIIIJ^{C134A}$-His_6 (Fig. 4B, red arrows). The representation of this species did not increase when the experiment was repeated in the absence of DTT, suggesting that Cys134 is not sensitive to the reducing agent (data not shown), in agreement with the idea that this residue is part of a tight protein-protein interaction surface (Fig. 3; above). Because the level of heat-resistant spores for the non-induced wild-type allele was close to that found for a wild-type strain (Table 1, see also above), the data shows that low levels of wild-type SpoIIIJ are sufficient for efficient sporulation. Together, these results suggest that $SpoIIIJ^{C134A}$ is less functional than the wild-type protein, and that this reduced functionality emerges at low expression levels of the corresponding alleles.

Requirement of Cys134 of SpoIIIJ for σ^G activity

Since SpoIIIJ is required for late forespore-specific gene expression during sporulation [10], we wanted to assess the impact of the C134A substitution on the activity of σ^G. We used the σ^G-controlled transcriptional fusion P_{sspE}-*cfp* [23] to monitor the activity of σ^G in single cells by fluorescence microscopy (Fig. 4C). DSM cultures of strains expressing either *spoIIIJ* or $spoIIIJ^{C134A}$ from P_{spac} in a *spoIIIJ* mutant background were sampled at hours 4 and 6 after the onset of sporulation. The cells

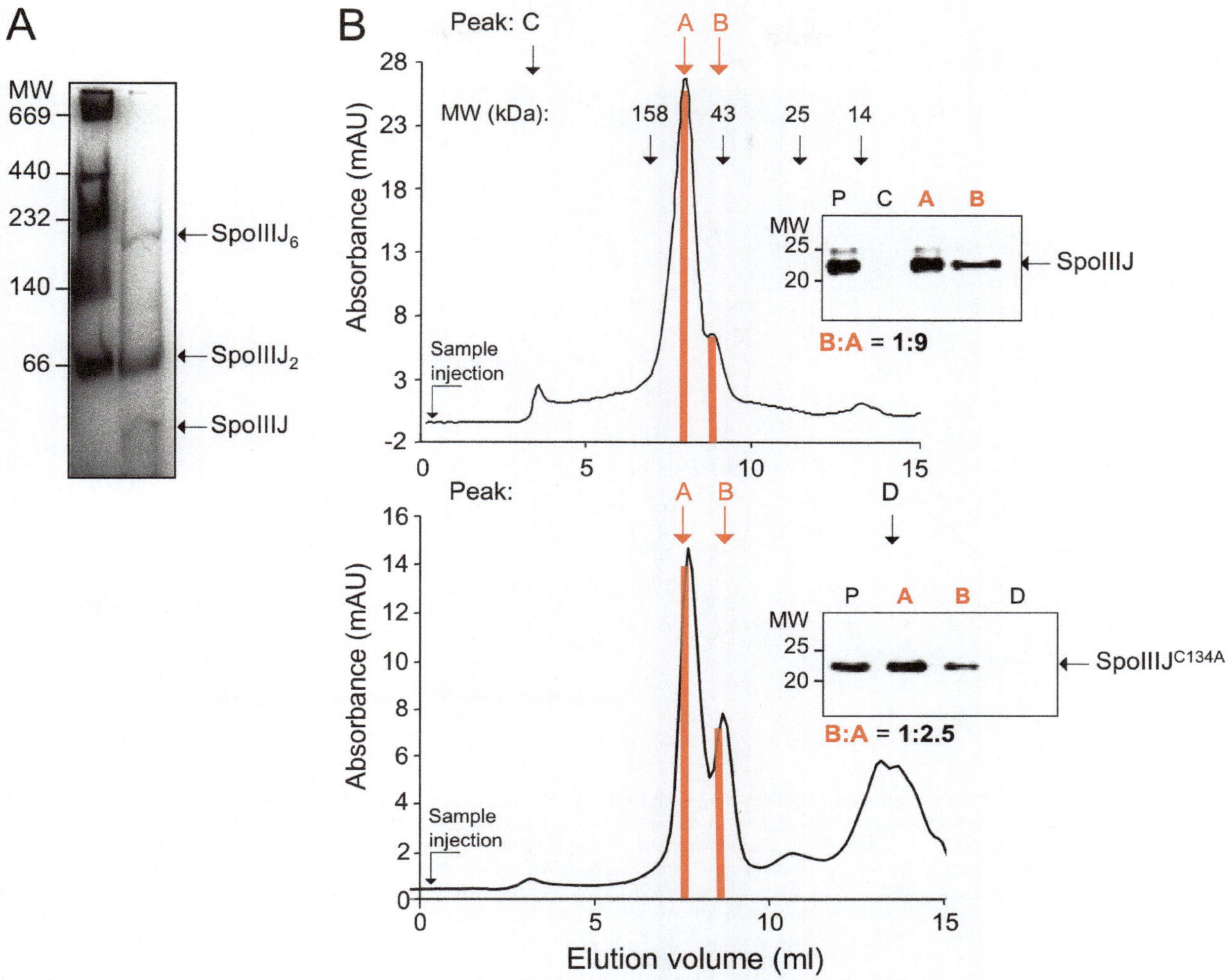

Figure 2. Oligomeric state of SpoIIIJ and SpoIIIJ^{C134A}. (A) Blue-Native PAGE of purified SpoIIIJ-His$_6$. Presumed hexameric (SpoIIIJ$_6$), dimeric (SpoIIIJ$_2$) and monomeric (SpoIIIJ) species are indicated. MW markers are shown in kDa. (B) Size exclusion chromatography of SpoIIIJ-His$_6$ (top) or SpoIIIJ^{C134A}-His$_6$ (bottom) in the presence of 0.1% DDM and 500 mM NaCl at pH 7.6 (see the Material and Methods section for details). The vertical black arrows indicate the elution volumes of the size standards and additional peaks. The inserts show the immunoblot analysis of the peaks indicated in the two panels. Purified SpoIIIJ-His$_6$ (top) or SpoIIIJ^{C134A}-His$_6$ (bottom) were included as a migration control (lane P). The areas of the A and B peaks in the two panels were estimated using the ImageJ software (http://imagej.nih.gov/ij/) and their ratio indicated.

were observed by fluorescence microscopy after staining with the membrane dye FM4-64 and the nucleoid stain DAPI (see Materials and Methods), to identify and score the morphological stage of sporulation. Sporulating cells were assigned to two classes according to the intensity of the CFP signal: with a low (*a*) and with a high (*b*) fluorescence signal (Fig. 4C). High-fluorescent cells completed engulfment and activated σ^G, thus leading to high expression of the reporter fusion P_{sspE}-*cfp*; on the other hand, low fluorescence mainly results from the weaker utilization of the *sspE* promoter by the preceding sigma factor, σ^F, in the forespore line of gene expression [24].

For the non-induced strain bearing the P_{spac}-*spoIIIJ* allele 23.4% of the cells at hour 4 of sporulation, and 23.5% of the cells at hour 6 were scored as highly fluorescent (class *b*), an indication of active σ^G; in contrast, the P_{spac}-*spoIIIJ*^{C134A}-expressing strain under non-inducing conditions showed a large proportion of low-fluorescence cells (class *a*; 38.1% at hour 4 and 58.8% at hour 6), whilst the high-fluorescence cells were scarce (class *b*; 6.8% at hour 4 and 11.6% at hour 6) (Fig. 4C). A distribution similar to that of the non-induced P_{spac}-*spoIIIJ*^{C134A} strain was obtained for a *spoIIIJ* null mutant (not shown), as expected, as in these cells only σ^F is active [10]. Upon addition of IPTG (to 0.5 mM), similar percentages of highly fluorescent, class *b* cells were obtained for the strains expressing either *spoIIIJ*^{C134A} (26.8% at hour 4 and 34% at hour 6) or its wild-type counterpart (23.6% and 41.3% at hour 6, respectively) (Fig. 4C). Together, the results show that expression of *spoIIIJ*^{C134A} does not cause a delay in sporulation. Rather, the results suggest that the C134A substituted form of SpoIIIJ is less active than the wild-type protein, but the lower activity of *spoIIIJ*^{C134A} is evident *in vivo* only when lower than normal levels of the proteins is produced.

Discussion

B. subtilis SpoIIIJ, a member of the YidC/Oxa1/Alb3 family of insertases, forms a dimer. The Cys134 residue, predicted to reside within the TM2 segment of SpoIIIJ, has a role in dimer formation, presumably by establishing a disulphide bond. Several lines of

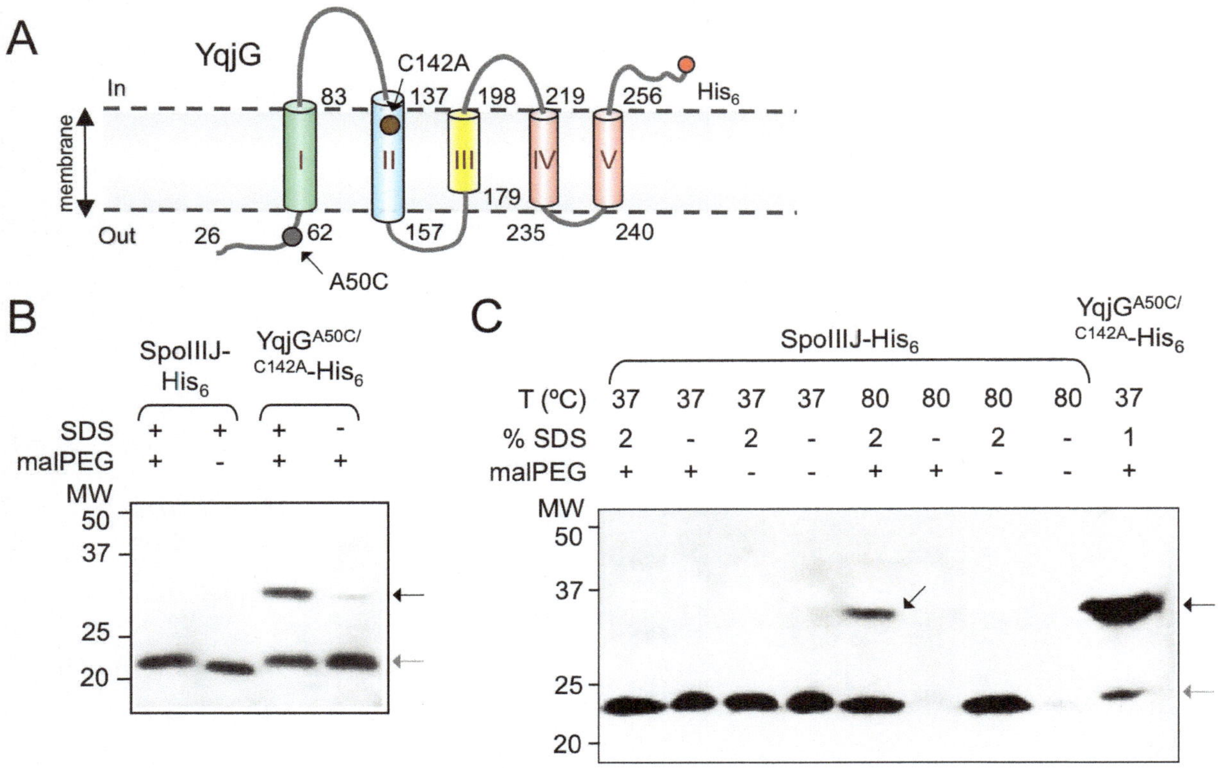

Figure 3. Labelling of YqjG$^{A50C/C142A}$-His$_6$ and SpoIIIJ-His$_6$ with malPEG. (A) Topological model of YqjG in the membrane. The numbers refer to the amino acid residues that delimit the transmembrane (TM) segments I to V. The positions of the A50C and C142A mutations are indicated. (B) Strains producing SpoIIIJ-His$_6$ or YqjG$^{A50C/C142A}$-His$_6$ in a *spoIIIJ* mutant background were grown in liquid LB, and samples withdrawn. Cells were resuspended in a buffer containing 1 mM TCEP, lysed and membranes isolated. The membranes were resuspended in the presence of 1 mM TCEP and further incubated with or without malPEG in the presence of 2% SDS (lanes 1–2), 1 or 0% SDS (lanes 3 and 4, respectively). Proteins (30 µg) were electrophoretically resolved by SDS-PAGE and immunoblotted with an anti-His$_6$ antibody for the detection of SpoIIIJ-His$_6$ and YqjG$^{A50C/C142A}$-His$_6$. (C) Samples from cultures of strains producing SpoIIIJ-His$_6$ or YqjG$^{A50C/C142A}$-His$_6$ were withdrawn at the onset of the stationary growth phase in LB medium. Samples were treated as in (B), except that no TCEP was added to the French press buffer. Incubation proceeded at the indicated temperatures in the presence of 1 mM TCEP and with or without 1 mM malPEG and SDS, as indicated. In both B and C, grey and black arrows indicate SpoIIIJ-His$_6$ or YqjG$^{A50C/C142A}$-His$_6$ or labeled (shifted) reaction products, respectively. The position of molecular weight markers (in kDa) is shown.

evidence suggest that formation of an intramembrane disulphide bond, involving Cys134 promotes formation of, or stabilises a SpoIIIJ dimer. SpoIIIJ was detected as a monomer and a dimer by SDS-PAGE, but only a monomeric species was seen in the presence of the reducing agent DTT (Fig. 5A). In contrast, SpoIIIJ^{C134A}, in which the only Cys residue in mature SpoIIIJ was changed to an Ala, migrated as a monomeric species, even in the absence of DTT (Fig. 5A). Also, the monomer:dimer ratio, as estimated by SEC, was higher for SpoIIIJ^{C134A} in comparison to the wild-type protein. Circular dichroism spectroscopy analysis suggested that DDM-solubilised SpoIIIJ is a folded, mainly α-helical protein, and we found that the addition of DTT caused an alteration in the content of α-helices and anti-parallel β-sheets for SpoIIIJ but not for SpoIIIJ^{C134A}. This change in secondary structure may be explained, at least in part, by a change in the monomer:dimer ratio. Clearly, despite its importance, Cys134 is not absolutely required for dimer formation. Rather, the dimer appears to be sustained by a disulphide bond along with other non-covalent interactions. Oligomerisation is a widespread mechanism for regulating protein function [42]. For SpoIIIJ, dimer formation is important, at least for sporulation (see also below), and an inter-chain disulphide bond may be advantageous as it might stabilise the functional form of the protein. More generally, inter-chain disulphide bonds may help coping with naturally occurring mutations that reduce the potential for dimerisation or oligomerisation [43]. The Cys134 residue is predicted to occur within the membrane plane, in TM2. Conservation of a Cys residue at a homologous position in the YqjG paralogue, and among SpoIIIJ orthologues, underscores the importance of this residue, and presumably of disulphide bond formation for optimal functionality.

It is often assumed that the membrane environment does not favour formation of disulphide bonds, but an increasing number of reports argue in favour of the existence of such intramembrane disulphide bonds [*e.g.*, 44,45,46,47]. The experiments using malPEG were designed to test whether a free cysteine residue was accessible for modification in SpoIIIJ. We have found that labelling with malPEG only took place by exposing the protein to high temperature in the presence of high concentrations of SDS in (Fig. 3). The harsh conditions required for labelling of Cys134 suggest that this residue is present at a tight interface, as the bulky malPEG molecule may have only limited access to the cysteine residue [39]. This may contribute to the incomplete labelling observed (Fig. 3). Since the labelling experiments were implemented with extracts prepared from *B. subtilis* cultures, another factor contributing for incomplete labelling could be the existence of a mixed population of SpoIIIJ in several oligomeric states and/

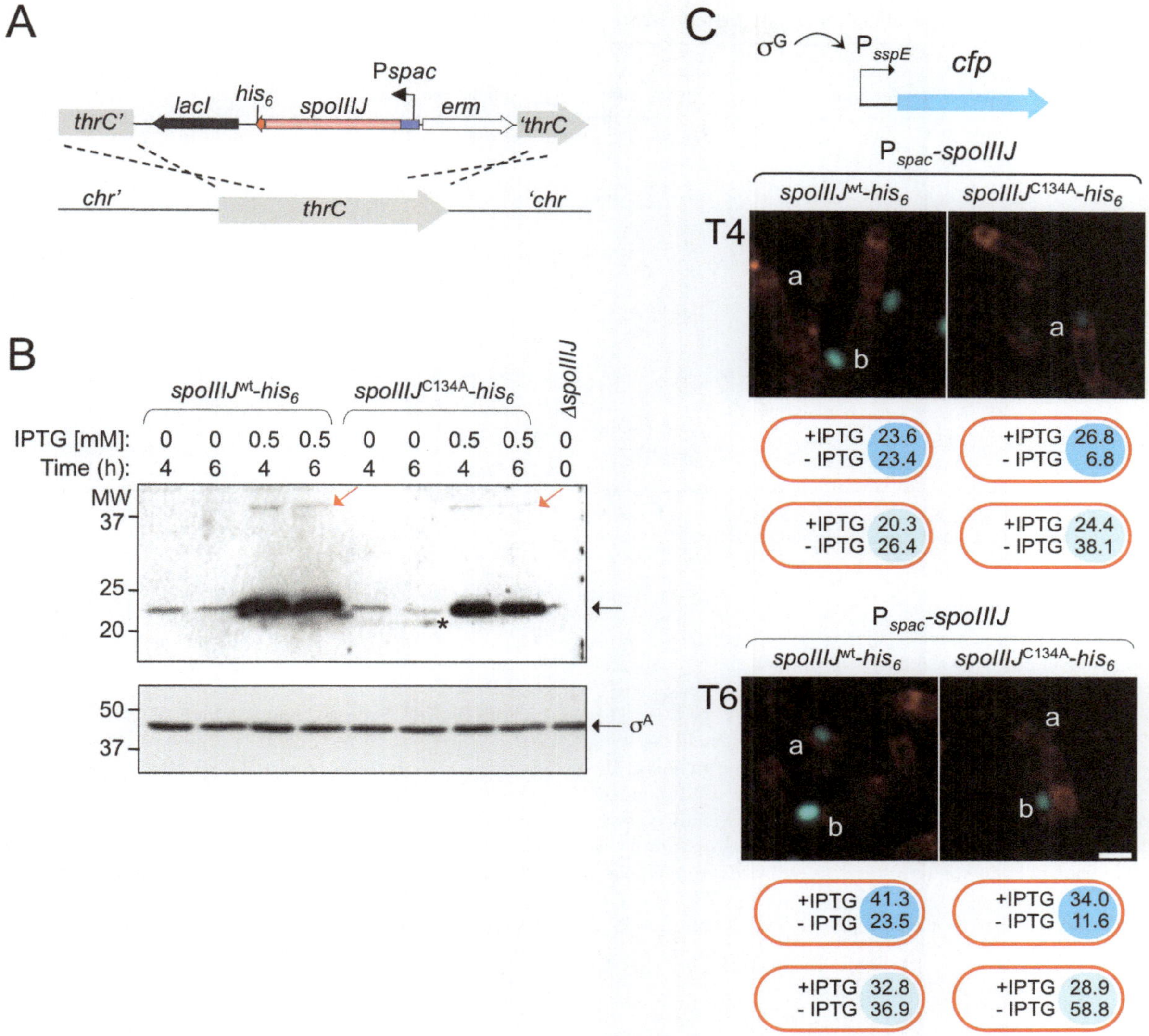

Figure 4. Effect of *spoIIIJ*^{C134A} on sporulation. (A) Schematic representation of the double crossover integration of a P_{spac}-*spoIIIJ-his*$_6$ fusion at the non-essential *thrC* locus. (B) Strains expressing SpoIIIJ-His$_6$ or SpoIIIJ^{C134A}-His$_6$ from P_{spac} at *thrC* (as shown in A) in a Δ*spoIIIJ* background, were grown in parallel with a Δ*spoIIIJ* mutant (last lane) in liquid DSM with or without IPTG as indicated and samples withdrawn at hours 4 and 6 of sporulation. Proteins in whole-cell extracts were subject to immunoblot analysis with an anti-His$_6$ (upper panel) or an anti-σ^A antibody (lower panel). The samples were resolved by SDS-PAGE under reducing conditions. The position of SpoIIIJ-His$_6$, SpoIIIJ^{C134A}-His$_6$ or σ^A is indicated by black arrows. The red arrows show the position of presumed SpoIIIJ-His$_6$ or SpoIIIJ^{C134A}-His$_6$ dimers. The asterisk indicates a non-specific cross-reacting species. The position of molecular weight markers (in kDa) is shown. (C) Expression of a P_{sspE}-*cfp* fusion, used as a reporter for σ^G activity, in DSM cultures of strains expressing wild-type *spoIIIJ* or the *spoIIIJ*^{C134A} allele from the P_{spac} promoter at *thrC* in a Δ*spoIIIJ* background. The strains were grown in the absence or in the presence of IPTG (0.5 mM). Only images for the non-induced strains are shown. Samples were collected at the indicated times (in hours) after the onset of sporulation, stained with the membrane dye FM4-64 and observed by fluorescence microscopy. CFP and FM4-64 fluorescence are shown in blue and red, respectively. The fluorescence patterns were designated by low (class *a*) or high (class *b*) fluorescence. Scale bar, 1 µm. The cartoons below each panel show the percentage of cells in class *a* (light blue) or class *b* (darker blue) for each of the indicated strains grown in the absence ("−") or in the presence ("+") of 0.5 mM IPTG, as indicated. At least 200 cells were scored for each of the *a* or *b* classes (see the Material and Methods section).

or belonging to distinct protein complexes allowing different degrees of malPEG accessibility.

Distinct oligomeric states have been reported for members of the YidC/Oxa1/Alb3 family of membrane protein insertases. Oxa1 from *Neurospora crassa* was suggested to form tetramers [48] and Oxa1 from *S. cerevisiae* was observed to form only dimers [9], only tetramers [49], or dimers and tetramers [50]. Alb3 from *Arabidopsis thaliana* is able to form dimers [51]. Finally, YidC from *E. coli* was reported to form dimers [9,52,53,54] and was also purified as a mixture of monomers and dimers [55,56]. We cannot exclude that forms of SpoIIIJ of order higher than the dimer are formed and have physiological relevance. The BN-PAGE, for examples, suggests that a SpoIIIJ hexamer can be formed, and at least part of the dimer population could serve as an intermediate for the formation of this species. However, we cannot presently exclude that the hexamer is only formed due to overproduction of the protein in *E. coli*.

Table 1. Efficiency of sporulation[a] of strains bearing various *spoIIIJ* alleles.

Strain	Relevant Genotype	[IPTG] mM	Viable cell count[b]	Heat[R] cell count	Spo (%)
MB24	wild-type	0	3.4×10^{8}	2.3×10^{8}	68.3
			$\pm2.3\times10^{8}$	$\pm3.0\times10^{8}$	
AH5425	*spoIIIJ* Δ*spoIIIJ::km*	0	5.6×10^{7}	1.3×10^{7}	23.7
			$\pm7.1\times10^{6}$	$\pm3.8\times10^{6}$	
AH5426	*spoIIIJ*^{C134A} Δ*spoIIIJ::km*	0	4.6×10^{7}	1.2×10^{6}	2.5
			$\pm8.7\times10^{6}$	$\pm1.8\times10^{5}$	
AH5425	*spoIIIJ* Δ*spoIIIJ::km*	0.5	$\pm1.8\times10^{8}$	1.2×10^{8}	68
			$\pm1.5\times10^{7}$	$\pm1.7\times10^{7}$	
AH5426	*spoIIIJ*^{C134A} Δ*spoIIIJ::km*	0.5	7.9×10^{7}	3.5×10^{7}	43.5
			$\pm1.4\times10^{7}$	$\pm8.1\times10^{6}$	
JOB44	Δ*spoIIIJ::km*	0	4.8×10^{7}	0	0
			$\pm7.6\times10^{6}$		

[a]The titre of viable cells and heat-resistant spores was measured 24 h after the onset of sporulation in DSM (see Materials and Methods).
[b]The values are averages ± SD, for three independent experiments.

A role for TM1 of SpoIIIJ (corresponding to TM2 of YidC) in SpoIIIJ-SpoIIIJ interactions is hinted from the fact that a heterodimer composed of a full-length, mature, and a truncated SpoIIIJ monomer (lacking TM1) was observed but not a homodimer of two truncated SpoIIIJ monomers (Fig. 1). This role for TM1 may be direct or indirect. TM1 could itself be involved in protein-protein interactions, or it could maintain the regions directly interacting (possibly TM2, with C134) in the correct position (Fig. 5B). Interestingly, TM2 of YidC (corresponding to SpoIIIJ's TM1) was suggested to have a structural role as well, in helix-helix interactions [57]. In addition, an interaction between TM2 and 3 of YidC (corresponding to TM1 and 2 of SpoIIIJ, respectively) was also suggested [58]. Importantly, a recent study by Geng and co-authors in which a series of SpoIIIJ/YqjG chimeras were analysed [59], also highlights the critical role of TM2 of SpoIIIJ (containing Cys134) for its role during sporulation.

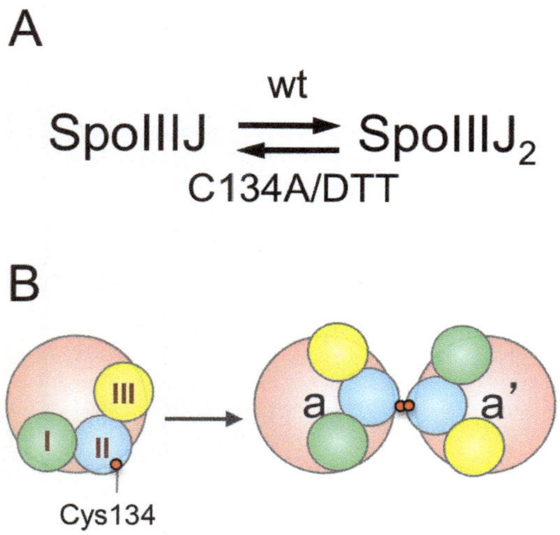

Figure 5. Dimerisation of SpoIIIJ. (A) SpoIIIJ occurs as a monomer and a dimer. The dimer is thought to be stabilised by a disulphide bond involving residue Cys134. However, this residue is not essential for dimer formation. (B) A disulphide bond between cysteine residues (red circles), along with other interactions promotes the formation and/or maintenance of the dimeric form of SpoIIIJ. TM segments 1, 2 and 3 are depicted as green, blue, and yellow circles, respectively.

Both SpoIIIJ and YidC function together with or independently of the Sec system [reviewed in 3,60]. For instance, SpoIIIJ participates in the biogenesis of the SpoIIIAE polytopic membrane protein, one of the components of the secretion system required for normal σ^G activity after engulfment completion, possibly in conjunction with the Sec pathway [20,24]. Depending on either mode of action, in conjunction with or independently of the Sec pathway, SpoIIIJ may function mostly as a monomer or mostly as a dimer. The precise role fulfilled by SpoIIIJ during membrane protein biogenesis could also vary regarding different substrates, either in guiding membrane protein insertion or later, in the folding or the assembly of membrane protein complexes [6]. Our observation that at low expression levels, SpoIIIJ^{C134A} but not the wild-type protein, shows a reduced ability to support σ^G activity and sporulation (Fig. 4 and Table 1), may suggest that a higher concentration of the mutant protein is needed to promote dimer formation, as the disulphide bond cannot be formed. Dimeric (or oligomeric) SpoIIIJ may be the form of the protein that best functions in the assembly of the SpoIIIA-SpoIIQ-based secretion system.

The basis for the observation that SpoIIIJ but not its paralogue YqjG supports efficient sporulation is not fully understood. Since a positional homologue of Cys134 is found in YqjG, it seems possible that this residue is also involved in disulphide bond formation in this protein. If so, formation of a disulphide bond between proximal TM2 segments does not distinguish SpoIIIJ and YqjG with respect to sporulation. Other regions of SpoIIIJ may be important, as a class of suppressor mutations that permit σ^G activation and sporulation in the absence of *spoIIIJ* map to TM5 and to the loop between TM4 and TM5 of *yqjG* [20].

Supporting Information

Figure S1 Overproduction and purification of SpoIIIJ-His$_6$ and SpoIIIJ^{C134A}-His$_6$. Overproduction and purification of SpoIIIJ-His$_6$ (A) or SpoIIIJ^{C134A}-His$_6$ (B) from *E. coli* strain C43(DE3) carrying pMS266 or pFV1, respectively. The cells were grown in LB to mid log phase, split into two cultures, and one was induced with IPTG. The cells were lysed and fractionated into a soluble and a membrane fraction. Proteins in the membrane fraction were solubilised with 2% DDM and the extract applied onto a Ni^{2+}-NTA column. (A) The fractions analysed by SDS-PAGE are as follows: lanes 1 and 2, crude extract of non-induced and induced cells, respectively; lanes 3 and 4, membrane fraction extracted with 2% DDM from non-induced and induced cells, respectively; lane 5, column flow through; lane 6, column wash; lane 7, 50 mM imidazole elution fraction; lane 8, 100 mM imidazole fraction. (B) lanes 1 and 2, crude extract of non-induced and induced cells, respectively; lane 3, membrane extract solubilised with 2% DDM; lane 4, flow through; lane 5, column wash; lane 6, 25 mM imidazole elution fraction; lanes 7 and 8, 50 mM imidazole elution. The position of molecular weight markers (in kDa) is shown; arrows show the position of full-length SpoIIIJ-His_6 or $SpoIIIJ^{C134A}$-His_6.

Figure S2 Circular dichroism spectroscopy of SpoIIIJ. Far UV-CD spectra of purified SpoIIIJ-His_6 (A) or $SpoIIIJ^{C134A}$-His_6 (B) (0.2 mg/ml of purified protein in 20 mM Tris-HCl pH 8, 0.1 M NaCl, 10% glycerol) in the presence or the absence of 1 mM DTT, dotted and solid lines, respectively. The spectra are typical of folded α-helical rich proteins with minima at 208 and 222 nm, consistent with the predicted structure of SpoIIIJ. Addition of DTT affects the spectrum of SpoIIIJ-His_6, but has no effect on the spectrum of $SpoIIIJ^{C134A}$-His_6. (C) Relative estimates of the secondary structure of SpoIIIJ-His_6 (with or without 1 mM DTT, dark grey and black bars, respectively) and $SpoIIIJ^{C134A}$-His_6 (with or without 1 mM DTT, white and light grey bars, respectively). Addition of DTT decreases the α-helix content by 6% and increases the content of antiparallel β-sheets (more 8%) of SpoIIIJ-His_6, but does not significantly alter $SpoIIIJ^{C134A}$-His_6.

Figure S3 Conservation of Cys134 among SpoIIIJ orthologues. (A) Alignment of the transmembrane (TM) segment 2 of SpoIIIJ (red) and YqjG (green) proteins from several *Bacillus* species and of TM3 of YidC (blue) from *E. coli*. Conserved residues are shaded in grey except for the cysteine (yellow). The sequences were aligned with ClustalW [61]. (B) Helical wheel projection of TM2 of SpoIIIJ from *B. subtilis* (http://rzlab.ucr.edu/scripts/wheel/wheel.cgi). Circles denote hydrophilic residues and diamonds hydrophobic ones. The color code is as follows: green, hydrophobic residues, with the amount of green decreasing proportionally to the hydrophobicity; yellow, zero hydrophobicity; red, the most hydrophilic (uncharged) residue, the amount of red decreasing proportionally to the hydrophilicity.

Figure S4 Expression of SpoIIIJ-His_6 and $SpoIIIJ^{C134A}$-His_6 in *B. subtilis*. The figure compares the levels of SpoIIIJ-His_6 and $SpoIIIJ^{C134A}$-His_6 expressed from the *thrC* locus under the control of P_{spac}, with the level of SpoIIIJ-His_6 expressed from the *spoIIIJ* locus under the control of its normal promoter (identified by the "*" symbol). The cultures were grown in liquid DSM in the presence (for the fusions at *thrC*) or in the absence of IPTG (for the fusion at the *spoIIIJ* locus), and samples withdrawn at hours 4 and 6 of sporulation. Proteins in whole-cell extracts were subject to immunoblot analysis with an anti-His_6 (upper panel) or an anti-σ^A antibody (lower panel). The gels were run under reducing conditions. The positions of SpoIIIJ-His_6 (wt or C134A) and σ^A are indicated by black arrows; red arrows show the position of presumed SpoIIIJ dimers. The position of molecular weight markers (in kDa) is shown. Note that the SpoIIIJ-His_6* fusion, expressed from the *spoIIIJ* locus under the control of its normal promoter, has a linker longer than the fusions expressed under the control of P_{spac} at *thrC* locus, and shows a slightly higher apparent mass [24].

Acknowledgments

We thank D. Rudner (Harvard Medical School) for the gift of strains.

Author Contributions

Conceived and designed the experiments: AOH CPM. Performed the experiments: LC FV MS CMG. Analyzed the data: LC FV MS CMG CPM AOH. Wrote the paper: LC FV AOH.

References

1. Du Plessis DJF, Nouwen N, Driessen AJM (2011) The Sec translocase. Biochim Biophys Acta 1808: 851–865.
2. Natale P, Brüser T, Driessen AJM (2008) Sec- and Tat-mediated protein secretion across the bacterial cytoplasmic membrane-distinct translocases and mechanisms. Biochim Biophys Acta 1778: 1735–1756.
3. Saller MJ, Wu ZC, de Keyzer J, Driessen AJM (2012) The YidC/Oxa1/Alb3 protein family: common principles and distinct features. Biol Chem 393: 1279–1290.
4. Tjalsma H, Bolhuis A, Jongbloed JD, Bron S, van Dijl JM (2000) Signal peptide-dependent protein transport in *Bacillus subtilis*: a genome-based survey of the secretome. Microbiol Mol Biol Rev 64: 515–547.
5. Glick BS, Von Heijne G (1996) *Saccharomyces cerevisiae* mitochondria lack a bacterial-type sec machinery. Protein Sci 5: 2651–2652.
6. Saller MJ, Fusetti F, Driessen AJ (2009) *Bacillus subtilis* SpoIIIJ and YqjG function in membrane protein biogenesis. J Bacteriol 191: 6749–6757.
7. Oliver DC, Paetzel M (2008) Crystal structure of the major periplasmic domain of the bacterial membrane protein assembly facilitator YidC. J Biol Chem 283: 5208–5216.
8. Ravaud S, Stjepanovic G, Wild K, Sinning I (2008) The crystal structure of the periplasmic domain of the *Escherichia coli* membrane protein insertase YidC contains a substrate binding cleft. J Biol Chem 283: 9350–9358.
9. Kohler R, Boehringer D, Greber B, Bingel-Erlenmeyer R, Collinson I, et al. (2009) YidC and Oxa1 form dimeric insertion pores on the translating ribosome. Mol Cell 34: 344–353.
10. Errington J, Appleby L, Daniel RA, Goodfellow H, Partridge SR, et al. (1992) Structure and function of the *spoIIIJ* gene of *Bacillus subtilis*: a vegetatively expressed gene that is essential for sigma G activity at an intermediate stage of sporulation. J Gen Microbiol 138: 2609–2618.
11. Piggot PJ, Hilbert DW (2004) Sporulation of *Bacillus subtilis*. Curr Opin Microbiol 7: 579–586.
12. Stragier P, Losick R (1996) Molecular genetics of sporulation in *Bacillus subtilis*. Annu Rev Genet 30: 297–341.
13. Hilbert DW, Piggot PJ (2004) Compartmentalization of gene expression during *Bacillus subtilis* spore formation. Microbiol Mol Biol Rev 68: 234–262.
14. Murakami T, Haga K, Takeuchi M, Sato T (2002) Analysis of the *Bacillus subtilis spoIIIJ* gene and its Paralogue gene, *yqjG*. J Bacteriol 184: 1998–2004.
15. Rubio A, Jiang X, Pogliano K (2005) Localization of translocation complex components in *Bacillus subtilis*: enrichment of the signal recognition particle receptor at early sporulation septa. J Bacteriol 187: 5000–5002.
16. Serrano M, Côrte L, Opdyke J, Moran CP Jr, Henriques AO (2003) Expression of *spoIIIJ* in the prespore is sufficient for activation of sigma G and for sporulation in *Bacillus subtilis*. J Bacteriol 185: 3905–3917.

17. Tjalsma H, Bron S, van Dijl JM (2003) Complementary impact of paralogous Oxa1-like proteins of *Bacillus subtilis* on post-translocational stages in protein secretion. J Biol Chem 278: 15622–15632.
18. Rudner DZ, Losick R (2001) Morphological coupling in development: lessons from prokaryotes. Dev Cell 1: 733–742.
19. Kroos L, Zhang B, Ichikawa H, Yu YT (1999) Control of sigma factor activity during *Bacillus subtilis* sporulation. Mol Microbiol 31: 1285–1294.
20. Camp AH, Losick R (2008) A novel pathway of intercellular signalling in *Bacillus subtilis* involves a protein with similarity to a component of type III secretion channels. Mol Microbiol 69: 402–417.
21. Meisner J, Wang X, Serrano M, Henriques AO, Moran CP Jr (2008) A channel connecting the mother cell and forespore during bacterial endospore formation. Proc Natl Acad Sci U S A 105: 15100–15105.
22. Camp AH, Losick R (2009) A feeding tube model for activation of a cell-specific transcription factor during sporulation in *Bacillus subtilis*. Genes Dev 23: 1014–1024.
23. Doan T, Morlot C, Meisner J, Serrano M, Henriques AO, et al. (2009) Novel secretion apparatus maintains spore integrity and developmental gene expression in *Bacillus subtilis*. PLoS Genetics 5: e1000566–e1000566.
24. Serrano M, Vieira F, Moran CP Jr, Henriques AO (2008) Processing of a membrane protein required for cell-to-cell signaling during endospore formation in *Bacillus subtilis*. J Bacteriol 190: 7786–7796.
25. Illing N, Errington J (1991) The *spoIIIA* operon of *Bacillus subtilis* defines a new temporal class of mother-cell-specific sporulation genes under the control of the σE form of RNA polymerase. Mol Microbiol 5: 1927–1940.
26. Kellner EM, Decatur A, Moran CP Jr (1996) Two-stage regulation of an anti-sigma factor determines developmental fate during bacterial endospore formation. Mol Microbiol 21: 913–924.
27. Henriques AO, Beall BW, Roland K, Moran CP Jr (1995) Characterization of *cotJ*, a sigma E-controlled operon affecting the polypeptide composition of the coat of Bacillus subtilis spores. J Bacteriol 177: 3394–3406.
28. Henriques AO, Melsen LR, Moran CP Jr (1998) Involvement of superoxide dismutase in spore coat assembly in *Bacillus subtilis*. J Bacteriol 180: 2285–2291.
29. Horton RM, Hunt HD, Ho SN, Pullen JK, Pease LR (1989) Engineering hybrid genes without the use of restriction enzymes: gene splicing by overlap extension. Gene 77: 61–68.
30. Henner DJ (1990) Inducible expression of regulatory genes in *Bacillus subtilis*. Methods Enzymol 185: 223–228.
31. Guerout-Fleury AM, Frandsen N, Stragier P (1996) Plasmids for ectopic integration in *Bacillus subtilis*. Gene 180: 57–61.
32. Miroux B, Walker JE (1996) Over-production of proteins in *Escherichia coli*: mutant hosts that allow synthesis of some membrane proteins and globular proteins at high levels. J Mol Biol 260: 289–298.
33. Schägger H, von Jagow G (1991) Blue native electrophoresis for isolation of membrane protein complexes in enzymatically active form. Anal Biochem 199: 223–231.
34. Pogliano J, Osborne N, Sharp MD, Abanes-De Mello A, Perez A, et al. (1999) A vital stain for studying membrane dynamics in bacteria: a novel mechanism controlling septation during *Bacillus subtilis* sporulation. Mol Microbiol 31: 1149–1159.
35. Setlow B, Magill N, Febbroriello P, Nakhimovsky L, Koppel DE, et al. (1991) Condensation of the forespore nucleoid early in sporulation of *Bacillus* species. J Bacteriol 173: 6270–6278.
36. Laemmli UK (1970) Cleavage of structural proteins during the assembly of the head of bacteriophage T4. Nature 227: 680–685.
37. Bowie JU (1997) Helix packing in membrane proteins. J Mol Biol 272: 780–789.
38. von Heijne G (2011) Membrane proteins: from bench to bits. Biochem Soc Trans 39: 747–750.
39. Koide K, Maegawa S, Ito K, Akiyama Y (2007) Environment of the active site region of RseP, an *Escherichia coli* regulated intramembrane proteolysis protease, assessed by site-directed cysteine alkylation. J Biol Chem 282: 4553–4560.
40. Neale EJ, Rong H, Cockcroft CJ, Sivaprasadarao A (2007) Mapping the membrane-aqueous border for the voltage-sensing domain of a potassium channel. J Biol Chem 282: 37597–37604.
41. Jefferies KC, Forgac M (2008) Subunit H of the vacuolar (H+) ATPase inhibits ATP hydrolysis by the free V1 domain by interaction with the rotary subunit F. J Biol Chem 283: 4512–4519.
42. Ali MH, Imperiali B (2005) Protein oligomerization: how and why. Bioorg Med Chem 13: 5013–5020.
43. Bolliger L, Johansson B (1999) Identification and functional characterization of the zeta-chain dimerization motif for TCR surface expression. J Immunol 163: 3867–3876.
44. Arien H, Wiser O, Arkin IT, Leonov H, Atlas D (2003) Syntaxin 1A modulates the voltage-gated L-type calcium channel (Ca(v)1.2) in a cooperative manner. J Biol Chem 278: 29231–29239.
45. Chiang W-C, Knowles AF (2008) Inhibition of human NTPDase 2 by modification of an intramembrane cysteine by p-chloromercuriphenylsulfonate and oxidative cross-linking of the transmembrane domains. Biochemistry 47: 8775–8785.
46. Rutledge T, Cosson P, Manolios N, Bonifacino JS, Klausner RD (1992) Transmembrane helical interactions: zeta chain dimerization and functional association with the T cell antigen receptor. EMBO J 11: 3245–3254.
47. Sousa VL, Brito C, Costa T, Lanoix J, Nilsson T, et al. (2003) Importance of Cys, Gln, and Tyr from the transmembrane domain of human alpha 3/4 fucosyltransferase III for its localization and sorting in the Golgi of baby hamster kidney cells. J Biol Chem 278: 7624–7629.
48. Nargang FE, Preuss M, Neupert W, Herrmann JM (2002) The Oxa1 protein forms a homooligomeric complex and is an essential part of the mitochondrial export translocase in *Neurospora crassa*. J Biol Chem 277: 12846–12853.
49. Reif S, Randelj O, Domanska G, Dian EA, Krimmer T, et al. (2005) Conserved mechanism of Oxa1 insertion into the mitochondrial inner membrane. J Mol Biol 354: 520–528.
50. Krüger V, Deckers M, Hildenbeutel M, van der Laan M, Hellmers M, et al. (2012) The mitochondrial oxidase assembly protein1 (Oxa1) insertase forms a membrane pore in lipid bilayers. J Biol Chem 287: 33314–33326.
51. Dünschede B, Bals T, Funke S, Schünemann D (2011) Interaction studies between the chloroplast signal recognition particle subunit cpSRP43 and the full-length translocase Alb3 reveal a membrane-embedded binding region in Alb3 protein. J Biol Chem 286: 35187–35195.
52. Boy D, Koch HG (2009) Visualization of distinct entities of the SecYEG translocon during translocation and integration of bacterial proteins. Mol Biol Cell 20: 1804–1815.
53. Lotz M, Haase W, Kuhlbrandt W, Collinson I (2008) Projection structure of YidC: a conserved mediator of membrane protein assembly. J Mol Biol 375: 901–907.
54. Nouwen N, Driessen AJ (2002) SecDFyajC forms a heterotetrameric complex with YidC. Mol Microbiol 44: 1397–1405.
55. van der Laan M, Houben EN, Nouwen N, Luirink J, Driessen AJ (2001) Reconstitution of Sec-dependent membrane protein insertion: nascent FtsQ interacts with YidC in a SecYEG-dependent manner. EMBO Rep 2: 519–523.
56. van der Does C, de Keyzer J, van der Laan M, Driessen AJ (2003) Reconstitution of purified bacterial preprotein translocase in liposomes. Methods Enzymol 372: 86–98.
57. Klenner C, Kuhn A (2012) Dynamic disulfide scanning of the membrane-inserting Pf3 coat protein reveals multiple YidC substrate contacts. J Biol Chem 287: 3769–3776.
58. Yuan J, Phillips GJ, Dalbey RE (2007) Isolation of cold-sensitive *yidC* mutants provides insights into the substrate profile of the YidC insertase and the importance of transmembrane 3 in YidC function. J Bacteriol 189: 8961–8972.
59. Geng Y, de Keyzer J, Scheffers D-J, Driessen AJM (2014) Defining the region of *Bacillus subtilis* SpoIIIJ that is essential for its sporulation specific function. J Bacteriol 196: 1318–1324.
60. Wang P, Dalbey RE (2011) Inserting membrane proteins: the YidC/Oxa1/Alb3 machinery in bacteria, mitochondria, and chloroplasts. Biochim Biophys Acta 1808: 866–875.
61. Thompson JD, Higgins DG, Gibson TJ (1994) CLUSTAL W: improving the sensitivity of progressive multiple sequence alignment through sequence weighting, position-specific gap penalties and weight matrix choice. Nucleic Acids Res 22: 4673–4680.

20

Atomic Force Microscopy Reveals a Morphological Differentiation of *Chromobacterium violaceum* Cells Associated with Biofilm Development and Directed by N-Hexanoyl-L-Homoserine Lactone

Anara A. Kamaeva[1,3], Alexey S. Vasilchenko[1,2], Dmitry G. Deryabin[1]*

1 Department of Microbiology, Orenburg State University, Orenburg, Russia, **2** Laboratory of Disbiosis, Institute of Cellular and Intracellular Symbiosis, RAS, Orenburg, Russia, **3** RSE «Republican Collection of Microorganisms», Astana, Republic of Kazakhstan

Abstract

Chromobacterium violaceum abounds in soil and water ecosystems in tropical and subtropical regions and occasionally causes severe and often fatal human and animal infections. The quorum sensing (QS) system and biofilm formation are essential for *C. violaceum's* adaptability and pathogenicity, however, their interrelation is still unknown. *C. violaceum's* cell and biofilm morphology were examined by atomic force microscopy (AFM) in comparison with growth rates, QS-dependent violacein biosynthesis and biofilm biomass quantification. To evaluate QS regulation of these processes, the wild-type strain *C. violaceum* ATCC 31532 and its mini-Tn5 mutant *C. violaceum* NCTC 13274, cultivated with and without the QS autoinducer N-hexanoyl-L-homoserine lactone (C_6-HSL), were used. We report for the first time the unusual morphological differentiation of *C. violaceum* cells, associated with biofilm development and directed by the QS autoinducer. AFM revealed numerous invaginations of the external cytoplasmic membrane of wild-type cells, which were repressed in the mutant strain and restored by exogenous C_6-HSL. With increasing bacterial growth, polymer matrix extrusions formed in place of invaginations, whereas mutant cells were covered with a diffusely distributed extracellular substance. Thus, quorum sensing in *C. violaceum* involves a morphological differentiation that organises biofilm formation and leads to a highly differentiated matrix structure.

Editor: Etienne Dague, LAAS-CNRS, France

Funding: This work was supported by The Ministry of Education and Science of the Russian Federation as the government assignment to perform research at the Orenburg State University. The funders had no role in study design, data collection and analysis, decision to publish, or preparation of the manuscript.

Competing Interests: The authors have declared that no competing interests exist.

* Email: dgderyabin@yandex.ru

Introduction

Chromobacterium violaceum is a Gram-negative, facultative anaerobic, non-sporing beta-proteobacterium that is abundant in soil and water ecosystems in tropical and subtropical regions [1–3]. As a typical saprophyte, this bacterium occasionally becomes an aggressive opportunistic pathogen causing severe and often fatal human and animal infections [4–8].

The remarkable versatility and adaptability of *C. violaceum* have been explained by complete genome sequencing of this bacterium [9]; this revealed 4,431 open reading frames (ORFs) associated with energy generation, transport, signal transduction, cell motility, secretion, and secondary metabolism [10,11], such as is important for mammalian pathogenicity proteins [12]. Finally, a large number of ORFs related to the regulation of gene expression were also found, and widespread utilisation of inducible quorum sensing systems to detect and respond to changes in cell population density have been demonstrated.

C. violaceum uses a LuxIR-type quorum sensing system [13–15] in which CviI is an N-hexanoyl-L-homoserine lactone (C_6-HSL) synthase, and CviR is a cytoplasmic receptor protein (DNA-binding transcription factor) that activates gene expression following binding to the diffusible autoinducer C_6-HSL [16]. Among a number of quorum sensing-controlled genes, the most well known is the *vioABCDE*-operon [17]. Expressed at high cell density, VioA – E proteins carry out the enzymatic oxidation and coupling of two molecules of tryptophan to give a rearranged pyrrolidone-containing scaffold in the final purple pigment violacein [18].

Another important activity of *C. violaceum* is self-produced polymer matrix (biofilm) formation [19] that structures the bacterial consortia and is linked to virulence through resistance to antibiotics, disinfectives and phagocytosis. Like several pathogenic bacteria, *C. violaceum* has the hms*HFR*-CV2940 operon, whose products synthesise the *N*-acetyl-D-glucosamine monomer and polymerise it in the biofilm exopolysaccharide [20]. Furthermore, bacteria communicate with each other in biofilms by diffusible autoinducers, though the role of cell-to-cell signalling in *C. violaceum* biofilm development is not characterised.

Using *in vivo* data generated from a library of point mutations in a CviR-regulated promoter, Stauff and Bassler [16] scanned the

C. violaceum genome to predict quorum sensing-regulated CviR binding sites. It has been confirmed that CviR controls *cvi*I (see below) and chitinase [16,21], and it also participates in the expression of transcriptional regulators, guanine deaminase, and the type VI secretion-related gene. However, the DNA motif bound by CviR is absent in the *hmsHFR*-CV2940 operon. Thus, a canonical quorum-sensing control of *C. violaceum* biofilm formation is still unknown.

These inconsistent data have led us to investigate biofilm formation by a wild-type *C. violaceum* strain and its mini-Tn5 mutant, the latter of which is deficient in autoinducer production but retains the QS mechanisms in response to exogenous C_6-HSL. Using atomic force microscopy (AFM), we have revealed for the first time a morphological differentiation of *C. violaceum* cells associated with biofilm formation and directed by C_6-HSL.

Materials and Methods

Bacterial strains and culture conditions

The wild-type strain *C. violaceum* ATCC 31532 was acquired from the American Type Culture Collection (LGC Standarts, UK), and *C. violaceum* NCTC 13274 from the National Collection of Type Cultures (Health Protection Agency, UK). The mini-Tn5 mutant of the wild-type strain has an insertion in the *cvi*I gene, and it is deficient in the production of its own autoinducer but retains the ability to respond on N-hexanoyl-L-homoserine lactone and a variety of other (C_4–C_8) short-chain HSLs [22,23]. Therefore, *C. violaceum* NCTC 13274 is colourless, but produces a characteristic purple pigment violacein by incubation with C_6-HSL, which is widely used as an indicator in the study of quorum sensing mechanisms.

C. violaceum ATCC 31532 was grown in 3 ml of Luria–Bertani (LB) broth (AppliChem GmbH, Germany) with 10 mg/ml glucose at 30°C. *C. violaceum* NCTC 13274 was cultivated in the same conditions with and without 0.1 μM of C_6-HSL obtained from the Cayman Chemical Company (Ann Arbor, USA). At 24, 48 and 72 h intervals, the biomass from single vials was collected, and bacterial growth was observed by measuring the absorbance at 450 nm (OD_{450}) using a Stat Fax 303 6VIS microstrip reader (Awareness Technology Inc., (Palm, USA).

Violacein extraction

Bacterial cells were harvested by centrifugation at 10,000× g for 5 min (MiniSpin, Eppendorf, 1 ml), and the supernatant was removed. To extract violacein, 200 μl of 95% ethanol was added and samples were vortexed at room temperature for 5 min. After centrifugation (10,000× g for 5 min), the upper phase containing violacein was collected, and the culture was re-extracted with a second volume of ethanol. The extracts were combined, and the relative concentrations of violacein were measured using a microstrip reader at 575 nm (OD_{575}) [24].

Crystal violet binding assay

The biofilm biomass quantification of *C. violaceum* was determined by crystal violet staining [25]. Because the dye and violacein have very similar wavelength spectra (maximal absorbance 540 and 575 nm for crystal violet and violacein, respectively), the protocol for biofilm quantification was slightly modified, and ethanol extracted colourless air-dryed biomass (see above) was used for the binding assay. A total of 200 μl of an aqueous solution of crystal violet (0.1%) was added to each well and incubated at room temperature. The staining solution was removed, and probes were washed twice (10,000× g for 5 min) with 200 μl of deionised water to remove excess dye. The dye that was strongly bounded to the biofilm was solubilised with 200 μl of 95% ethanol for 5 min, washed and re-extracted with a second volume of ethanol. Relative quantitative determinations of crystal violet were measured in combined samples by reading the optical density at 540 nm (OD_{540}) [26].

Atomic force microscopy

For AFM, the planktonic and biofilm-forming *C. violaceum* were individually imaged. Planktonic bacteria were pipetted out by touching the top of the tips to the corner of each well, washing from the cultivation media (3,000× g for 5 min), re-suspending in deionised water; then droplets of 15-μl bacterial suspensions were applied to a freshly cleaved mica surface (5×5 mm). Remaining biofilms were carefully transferred under sterile conditions to a mica surface immediately. Samples were dried with controlled 93% humidity according to [27].

The observations was obtained with an atomic force microscope SMM-2000 (JSC "Proton-MIET Plant", Russia) in contact mode in air. The instrument was equipped with silicon nitride cantilever MSCT-AUNM (Veeco Instruments Inc., USA) with a pyramidal (V-shaped) tip with a typical radius of ~10 nm and a spring constant of 0.01 N/m.

During AFM imaging, the root-mean-square roughness (R_{rms} – the standard deviation of the Z values) for the height images were conducted over 5- μm by 5- μm areas on different biofilm surfaces and was calculated using SMM-2000 software. The images were flattened and plane fitted prior to analysis.

Statistical analysis

All the experiments were performed at least three times in triplicate. The data are reported as means ± standard deviation. Two-tailed *t*-test statistics were used at significance levels of 0.05 and 0.01.

Results

Growth, violacein production and biofilm formation by wild-type strain *C. violaceum* ATCC 31532

In order to evaluate if violacein production and biofilm formation were dependent on cell density, *C. violaceum* ATCC 31532 was grown in LB broth for 24, 48 and 72 h (Fig. 1). This experiment showed a typical growth (OD_{450}) curve with maximal progression in 24 h followed by a gradual increase to the late stationary phase. Production of violacein was detected at 24 h, but the most growth of characteristic OD_{575} was established at high cell density between 48 and 72 h where this value increased 3.4-fold. On the other hand, the biofilm production was quite similar to the growth rate, and the OD_{540} after crystal violet staining increased proportionally to an OD_{450}. Thus, violacein and biofilm in the wild-type strain *C. violaceum* ATCC 31532 differentially corresponded to cell density and only the first developed in a fashion similar to a quorum-dependent parameter.

AFM imaging of wild-type strain *C. violaceum* ATCC 31532 cells and biofilm

The planktonic and biofilm-forming bacteria growing for 24, 48 and 72 h were immobilised on a mica surface to observe the cells' general morphology and detailed surface topography (Fig. 2).

The 24-h individual *C. violaceum* cells had a characteristic elongated shape, and based on AFM, the cells were 2.45±0.31 μm length, 0.85±0.09 μm wide and 0.34±0.07 μm high and without distinctions between planktonic and biofilm-forming bacteria. However, the surface topography of these cells was significantly

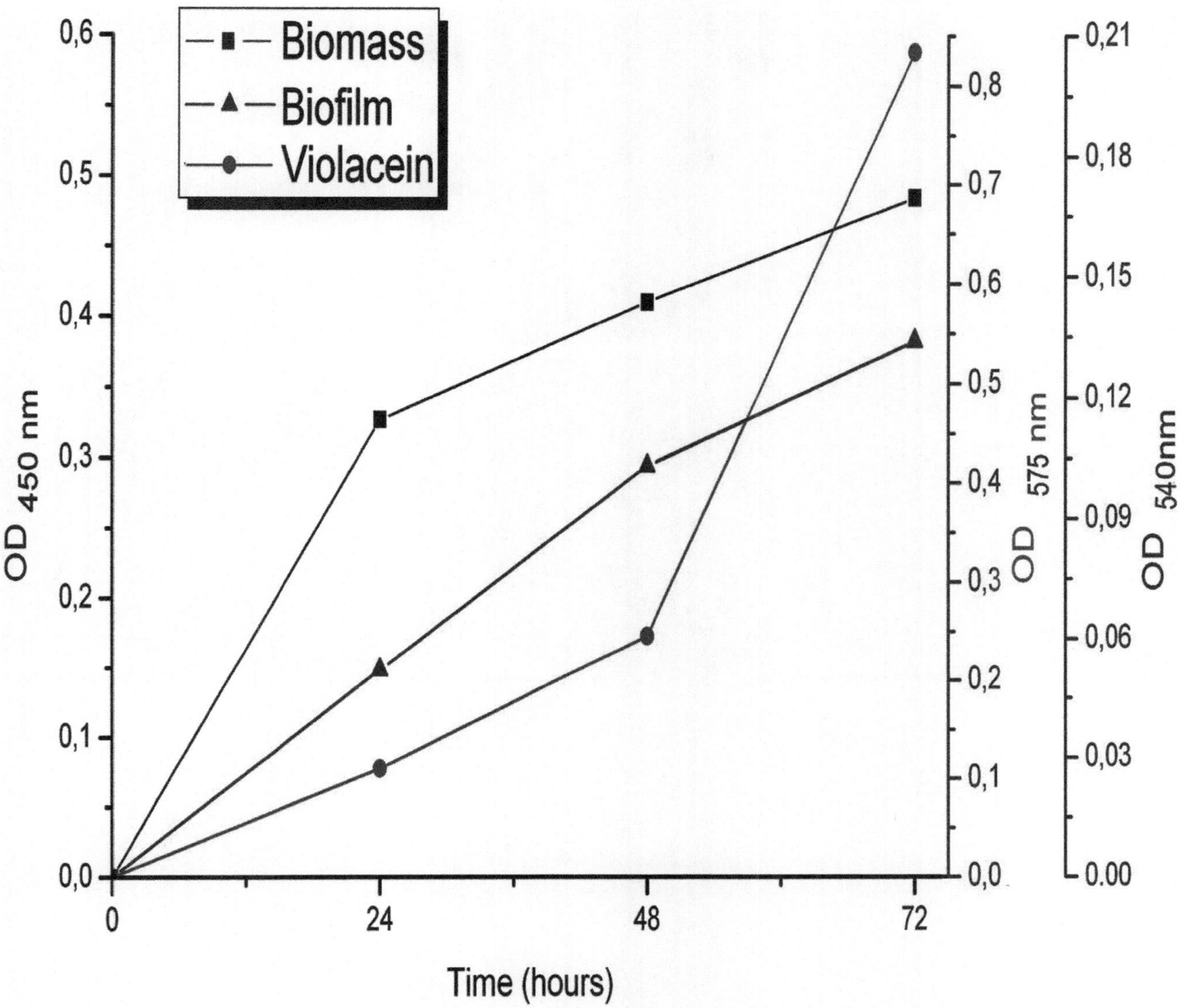

Figure 1. Time course (a) of the cell density (OD $_{450}$) for violacein (OD $_{575}$) and biofilm (OD $_{540}$) production by wild-type strain *C. violaceum* ATCC 31532, growing in LB broth for 24, 48 and 72 h.

different, and unusual invaginations (about 30–40 per cell taking into consideration 50% cell surface visibility) of the external membrane in biofilm-forming bacteria were observed (fig. 2 a). Based on AFM profiles, the invaginations had a 171±43 nm diameter (Fig. 3 c).and a 31.1±7.5 nm depth, whereas the distance from their centres were nearly 288±80 nm (Fig. 3 e).

At 48 h invaginations similar to these observed with biofilm-forming cells at 24 h (fig. 2 a) are observed on planktonic cells (Fig. 2 e), while biofilm-forming bacteria became essentially different at the same time (Fig. 2 b). High-resolution AFM images (Fig. 2 b) of biofilm-forming bacterial cells surfaces showed that the bacteria were covered with granular structures, and hill-like extrusions had appeared.

The extrusions' average diameter measured by AFM profiles was 250±61 nm (Fig. 3 d), and the average distance from their tops was 289±97 nm (Fig. 3 f). The arrangement and distance centre-to-centre of the 24-h invaginations and 48-h extrusions ($P = 0.4673$) have assured us of their spatial interrelation, that the extrusions formed in place of the previous invaginations. On the other hand, comparison of *C. violaceum* growth and AFM images has allowed us to explain processes occurring by biofilm's matrix production.

At 72 h, AFM revealed the granulated biofilm surface and contours of biofilm-forming bacterial cells (Fig. 2 c). At the same time, the planktonic bacteria changed and also developed extrusions on their surface (Fig. 2 f). In addition, flagella (approx. 20 nm in diameter) and some other thinner fimbrial structures were detected in planktonic samples.

Thus, using AFM, we report for the first time the morphological differentiation of C. violaceum ATCC 31532 cells increasing with cultivation time and preliminarily developed in biofilm-forming cells, whereas similar changes in the planktonic cells have developed late. However, the quorum-sensing control of *C. violaceum* cell differentiation and biofilm formation remains unclear.

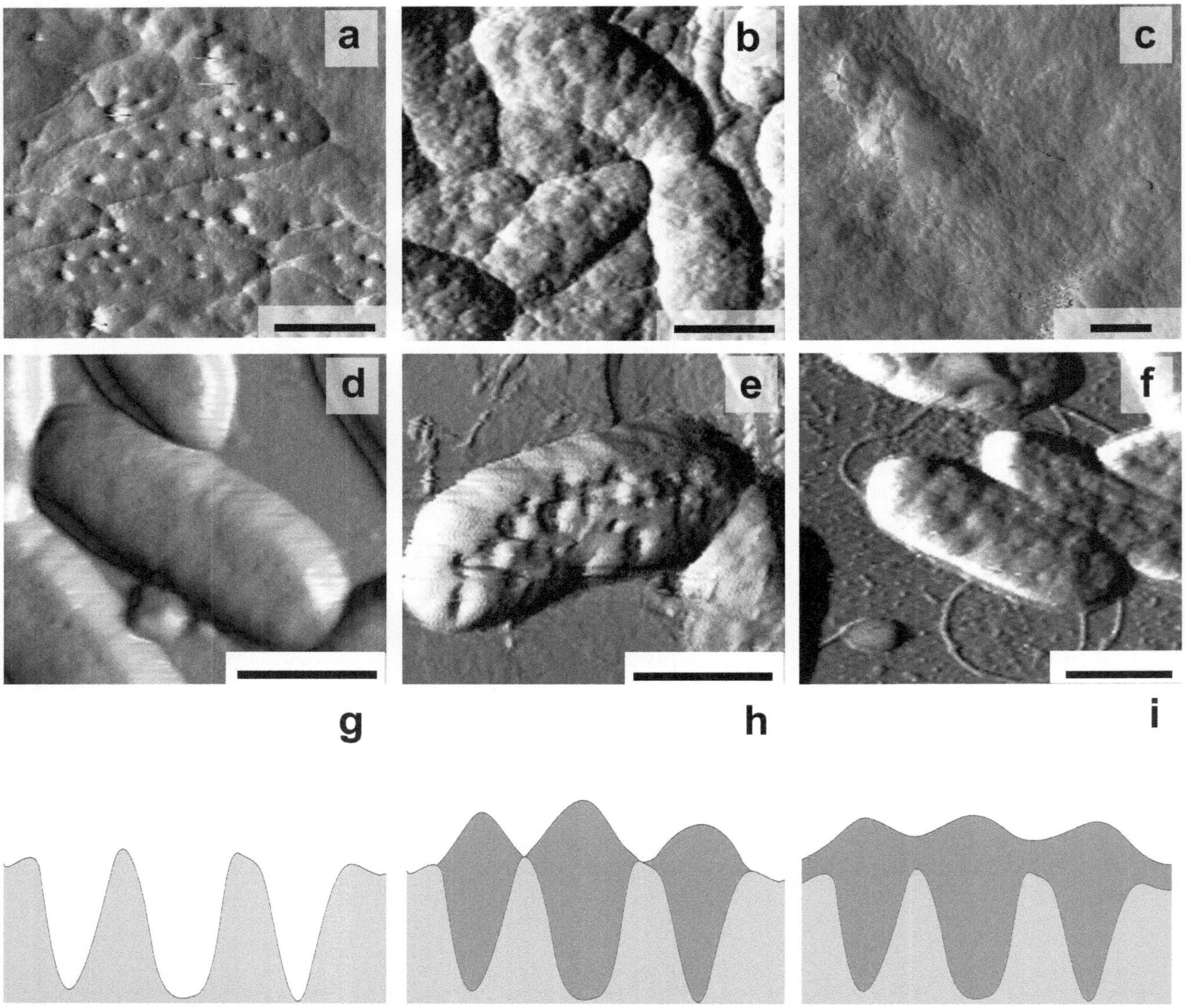

Figure 2. AFM images (a–f) and schematic model of extracellular matrix secretion (g–i) by wild-type strain *C. violaceum* ATCC 31532, growing for 24 (a, d, g), 48 (b, e, h) and 72 h (c, f, i) in biofilmic (a–c; g–i) and planktonic (d–f) conditions. Scale bar – 1 μm.

Growth, violacein production and biofilm formation by the mutant strain *C. violaceum* NCTC 13274

In order to evaluate the effects of quorum sensing on biofilm formation, we investigated the mini-Tn5 mutant strain *C. violaceum* NCTC 13274 cultivated without and with the natural autoinducer N-hexanoyl-L-homoserine lactone (Fig. 4).

When *C. violaceum* NCTC 13274 were grown for 24, 48 and 72 h, both in LB broth and in LB broth with 0.1 μM C_6-HSL, violacein production differed cardinally ("no" or "yes"), whereas growth rates and growth curves were comparable. After 24 h of incubation, the OD_{450} value of a colourless culture, grown in the absence of autoinducer, was approximately 11.0% higher than that in the presence of C_6-HSL, and approximately 22.9% higher than that of the wild-type strain *C. violaceum* ATCC 31532. In our opinion, this could be due to energy savings on violacein biosynthesis in comparison with HSL-deficient strain, but 48 and 72 h growth rates of the colourless cultures became less than those of pigmented ones. On the other hand, violacein production by *C. violaceum* NCTC 13274 incubated with C_6-HSL was significantly higher than that of the wild-type strain *C. violaceum* ATCC 31532, and the characteristic OD_{575} in comparison with a wild strain increased 4.2-fold in 24 h, 5.3-fold in 48 h and 1.9-fold in 72 h, respectively. We suspect it was because exogenous C_6-HSL was available at the growth start, and *C. violaceum* NCTC 13274 developed violacein biosynthesis irrespective of the cell density: growth curves for wild-type and mutant strains were very similar.

Contrary to the qualitative distinctions of violacein production, biofilm formation by *C. violaceum* NCTC 13274 with and without C_6-HSL only differed quantitatively. These processes were quite similar in *C. violaceum* NCTC 13274 growing with C_6-HSL and in the wild-type strain *C. violaceum* ATCC 31532, while staining of colourless culture using the crystal violet revealed only 64.0–70.3% biofilm biomass in comparison with the pigmented culture.

In brief, if the mini-Tn5 insertion in the *cvi*I gene led to the block in violacein biosynthesis, and the exogenous autoinducer

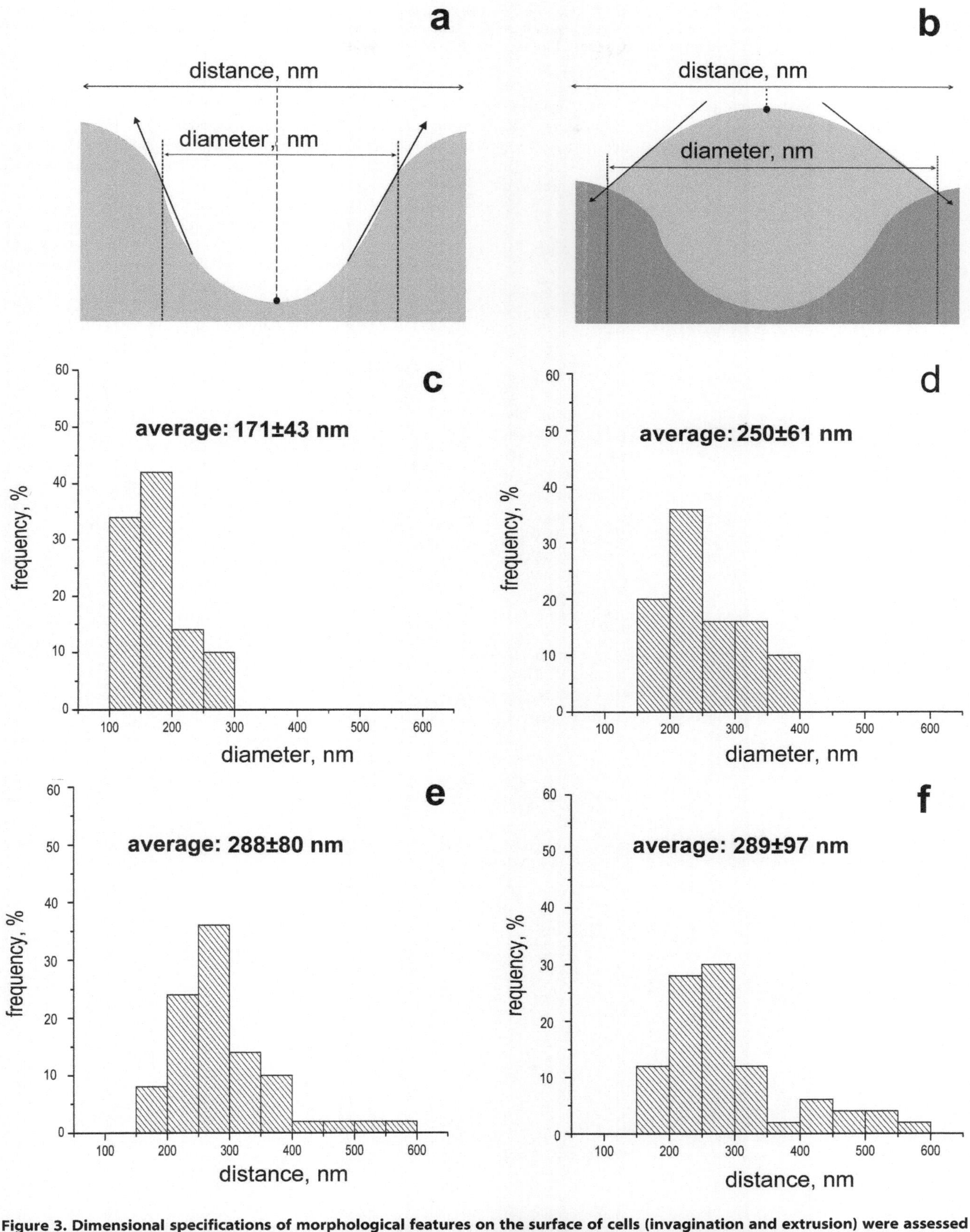

Figure 3. Dimensional specifications of morphological features on the surface of cells (invagination and extrusion) were assessed on the basis of AFM profile data carried out through two adjoined invaginations or extrusion. Distance was defined as a spacing between points located: in center of two adjoined invaginations (a) or on the top of two adjoined extrusions (b). Diameter of invaginations and

extrusions is defined as a distance between two points that are located on two crossing lines (solid line and dotted line). Precise morphometry of *C. violaceum* ATCC 31532 cells surfaces (in biofilm conditions), revealed: diameters of invaginations (c) and extrusions (d); distances between centres of two adjoined invaginations (e) and two adjoined extrusions (f) (histograms and average value).

restored and strengthened it considerably, then biofilm formation decreased quantitatively, and the autoinducer only restored it. Thus, violacein production developed like a quorum-dependent parameter, whereas biofilm formation occurred in a non-canonical fashion, i.e. biofilm formation occurred without direct control by a QS-inducer, but was associated with luxR/luxI genes.

AFM imaging of *C. violaceum* NCTC 13274 cells and comparative study of biofilms' surfaces

To confirm or reject the quorum-sensing control of biofilm formation, *C. violaceum* NCTC 13274 cells cultivated with and without C_6-HSL were imaged by the AFM technique (Fig. 5 and Fig. S1).

Surprisingly, 24-h biofilm-forming cells growing without autoinducer displayed a smooth surface (Fig. 5 a), whereas cells in the presence of C_6-HSL had numerous invaginations (Fig. 5 d). The topography of invaginations of *C. violaceum* NCTC 13274 cells growing with autoinducer was very similar in depth and diameter to that in the the wild-type strain *C. violaceum* ATCC 31532. At the same time, there were more invaginations (about 50–80 per cell taking into consideration 50% cell surface visibility), the distance from their centres was less (155±21 nm), and the arrangement was more ordered. These data strongly suggest that *C. violaceum* cells differentiate in a quorum-dependent manner that is naturally developed in culture growth, and which is cancelled without autoinducer and resumes with exogenous C_6-HSL.

At 48 h, AFM revealed the numerous extrusions on the *C. violaceum* NCTC 13274 cell's surfaces growing with C_6-HSL. This extrusions were like wild-type biofilm-forming cells (see above), and the number per cell, such as distance from extrusion tops, was very similar to the 24-h invaginations.

On the other hand, *C. violaceum* NCTC 13274 cultivated without C_6-HSL had no invaginations or extrusions, while high-resolution AFM revealed that the bacterial cell surfaces had been covered with a diffusely distributed extracellular substance.

The study of the *C. violaceum* NCTC 13274 planktonic cells without C_6-HSL also shown their smooth surface (Fig. S1 a–c) but addition of autoinducer restored their morphological differentiation (Fig. S1 d–f), however it was less pronounced compared with biofilm-forming cells.

Because AFM is capable of surface morphology imaging with nanometre resolution, we used this technique for comparative study of mature 72-h biofilms formed by the wild-type strain *C. violaceum* ATCC 31532 and mini-Tn5 mutant strain *C. violaceum* NCTC 13274 cultivated with and without C_6-HSL.

Figure 6 shows that *C. violaceum* ATCC 31532 produced relatively smooth biofilm consisting of tightly associated polymer blocks, and the average surface roughness was only 13.0±2.65 nm.

Noticeably, deficiency in autoinducer production *C. violaceum* NCTC 13274 exhibited a more irregular biofilm surface structure with a average roughness of 70.2±12.5 nm ($P = 0.0007$) that was probably caused by breakage of the quorum-dependent secretion machinery. At last, topography AFM images of biofilms formed by *C. violaceum* NCTC 13274 cultivated with exogenous C_6-HSL shows that there is no significant difference between roughness in the wild-type and mutant strains (average value 5.08±0.73 nm; $P = 0.0018$).

Thus, quorum sensing in C. violaceum does not control intracellular exopolysaccharide biosynthesis but probably involves the secretion of these molecules as extracell biofilm structure development, and AFM shows unusual bacterial cell morphological differentiation, which is associated with biofilm development and is directed by N-hexanoyl-L-homoserine lactone.

Discussion

Biofilm formation is the result of a complex process that involves mono- and polysaccharide biosynthesis, its transport through a microbial cell wall and a subsequent arrangement of structure and architecture of the biofilm matrix [28–30]. Among diverse regulatory mechanisms postulated for biofilm formation, the significant system is quorum sensing that involves production,

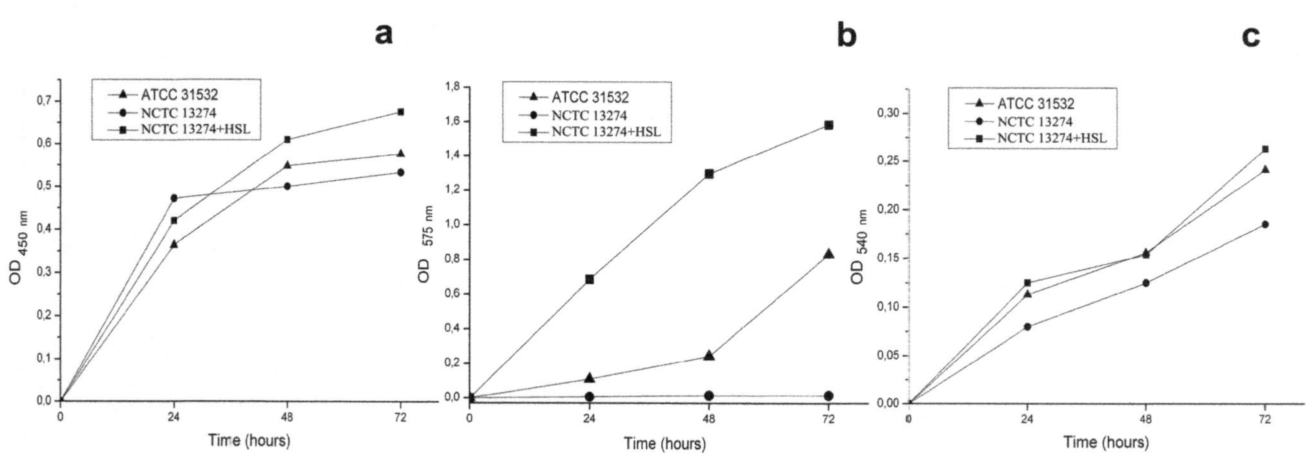

Figure 4. Time course of cell density (a), for violacein production (b) and biofilm formation (c) by wild-type strain *C. violaceum* ATCC 31532 in comparison with mutant strain *C. violaceum* NCTC 13274, growing in LB broth without and with 0.1 μM C_6-HSL.

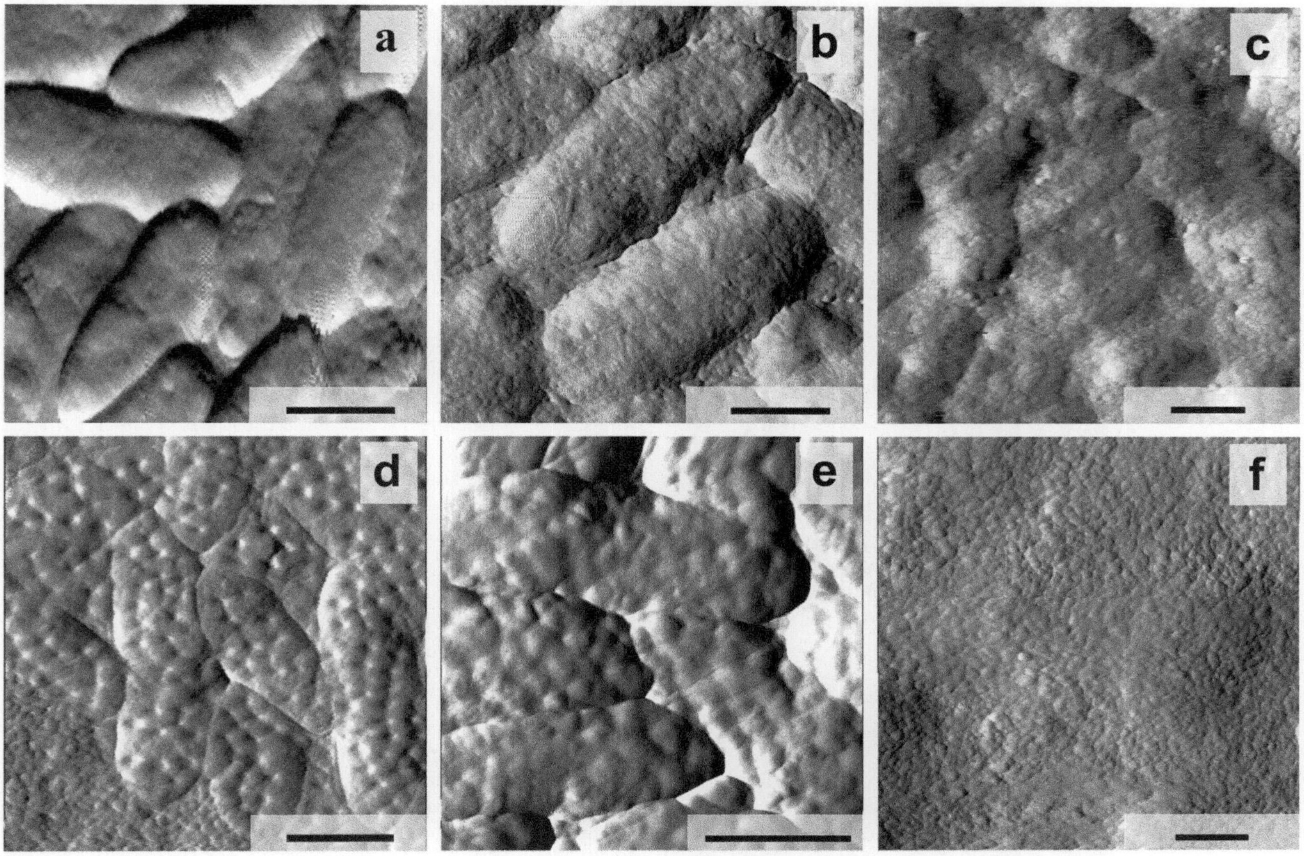

Figure 5. AFM images of biofilm-forming *C. violaceum* NCTC 13274, growing in LB broth without (a–c) and with 0.1 μM C_6-HSL (d–f) for 24 (a, d), 48 (b, e) and 72 h (c, f). Scale bar – 1 μm.

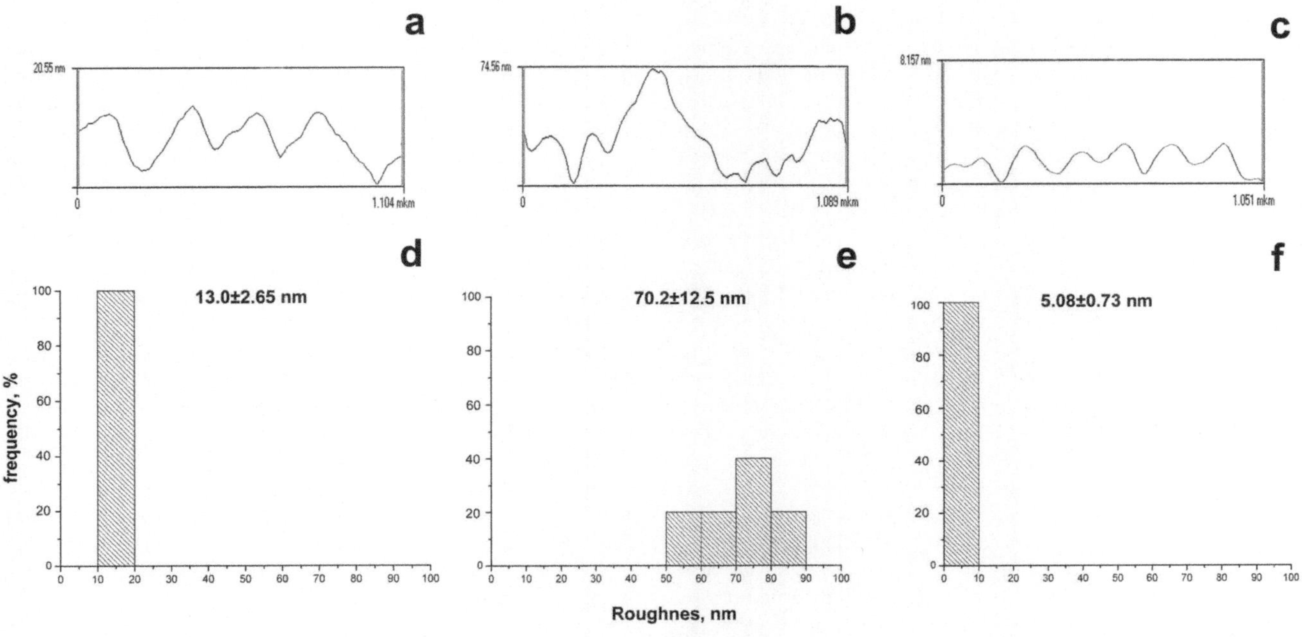

Figure 6. Cross sections (a–c), where x-axis is a heights of relief points and y-axis is a cross sections distance; and roughness values (d–f) of biofilm surfaces, produced by wild-type strain *C. violaceum* ATCC 31532 (a, d) and mutant strain *C. violaceum* NCTC 13274, growing without (b, e) and with 0.1 μM C_6-HSL (c, f).

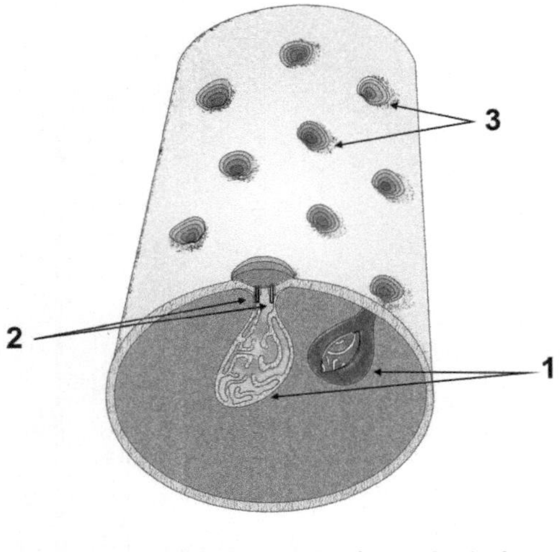

Figure 7. Predicted model of topological spatial relations between mesosomal-like structures (1), type VI secretion proteins (2) and membrane invaginations (3) in C_6-HSL induced *Chromobacterium violaceum* cells. This art was generated by Corel Draw 13 (Corel Corp.) and Adobe Photoshop CS 5 (Adobe Sys. Inc.).

release and detection of small signalling molecules (autoinducers). This allows microbial cells to activate target gene expression in a cell density-dependent manner, including genes involved in biofilm development and maturation [31–34]. However, because these mechanisms vary between different bacterial species, the role of quorum sensing in biofilm formation cannot be described in general terms [35].

Several independent research groups have demonstrated that quorum sensing is essential for biofilm formation by *Pseudomonas aeruginosa* to express virulence factors. This microorganism has hierarchically arranged quorum sensing, with the LasIR on top, controlling the RhlIR, and the additional PQS/HHQ/MvfR systems [36,37]. In particular, *lasI* mutants that did not produce the specific autoinducer 3-oxo-C12-HSL formed thinner and undifferentiated biofilms than in wild-type, and mutant biofilms appeared normal when *P. aeruginosa* grew in the presence of a synthetic signalling molecule [38]. However, in some recent studies, there was no apparent difference in the biofilms formed by the quorum-sensing mutants and the wild-type strains [39,40]. Finally, the detailed *P. aeruginosa* virulence regulatory network did not show the direct quorum-sensing control of biofilm formation, delocalised on cell regulatory pathways under the key BfiSR, MifR and BfmSR activators [41].

C. violaceum has a simple unicomponent quorum-sensing system [13–15] where the *cviI* gene encodes the autoinducer (C_6-HSL) synthase, and *cviR* encodes a cytoplasmic receptor protein that binds C_6-HSL and activates expression of target genes [16]. However quorum-sensing control of biofilm formation by *C. violaceum* is not clear because specific CviR binding sites are absent in the exopolysaccharide biosynthesis operon [16].

Here, we describe the use of atomic force microscopy (AFM) combined with classical bacteriological techniques to reveal and quantify *C. violaceum* quorum sensing and biofilm formation. In these experiments, wild-type and *cviI* mutant bacteria allow to estimate both phenomena in violacein production and crystal violet staining tests, respectively. In turn, the use of AFM allows visualisation of cell morphology, matrix distribution and biofilm formation that can be carried out in their native state and under physiological conditions. Thus, AFM requires minimal sample preparation and creates 3D images with nanometer or sub-nanometer resolutions.

Our basic conclusion was that dependence between quorum sensing and biofilm formation in *C. violaceum* is not typical for luxI/luxR manner. The evidence for this comparison was as follows: (i) different dynamics of strictly quorum-dependent violacein and biofilms production during wild-type strain *C. violaceum* ATCC 31532 growth; (ii) the absolute block of violacein biosynthesis in the *C. violaceum* NCTC 13274 insertion mutant, whereas biofilm production is few lower than in cases of wild type and mutant strain with inducer; (iii) the considerable strengthening in *C. violaceum* NCTC 13274 violacein biosynthesis by exogenous C_6-HSL in comparision with wild-type strain *C. violaceum* ATCC 31532, whereas biofilm formation was slightly increased to reach its wild type value only. Thus, like *P. aeruginosa* [39,40], the *C. violaceum* wild-type strain and quorum-sensing mutants did not show apparent quantitative differences in the biofilm formation.

Surprisingly, AFM revealed an unusual *C. violaceum* cell morphological differentiation, associated with biofilm development and directed by a quorum-sensing autoinducer. AFM revealed numerous invaginations of the external cytoplasmic membrane of wild-type cells, which was repressed in the mutant strain and restored by exogenous C_6-HSL. With increasing bacterial growth, the extrusions composed apparently from biofilm matrix were formed in place of invaginations, whereas mutant cells were covered with a diffusely distributed extracellular substance. Finally, quorum sensing-controlled C. violaceum formed a fine biofilm structure that differed from non-QS-controlled *C. violaceum*. In summary, our results suggest that quorum sensing controls biofilm quality but not quantity, and without this control bacteria produce undifferentiated biofilms, unlike wild-type biofilm. Therefore, authors assume that biofilm under QS-control become strong and steady against mechanical and chemical stress as shown peviously in [38].

As part of our experimental result, we were interested in determining the mechanisms of morphological differentiation and biofilm matrix development controlled by quorum sensing. The present study by AFM confirmed the existence of *C. violaceum's* cytoplasmic membrane invaginations as revealed by transmission electron microscopy previously [42]. On this basis, we propose the association of the invaginations with bacterial mesosomal-like structures that are rich in biosynthetic enzymes (Fig. 7), although we do not yet know which quorum-sensing controlled genes are involved in this process. For example, quorum-sensing controlled type VI secretion [16], whose multi-component machine is implicated in biofilm formation [43] and related biofilm-specific antibiotic resistance [44].

Thus, it is possible that the intracellular mesosomes [42], the surface's invaginations (present work), and the outer membrane vesicles [45–47] are simultaneous manifestations of one phenomenon of bacterial cell differentiation associated with QS-controlled biofilm development

Future investigation of this mechanism could reveal insights into both the function of quorum sensing in biofilm formation and the features of bacterial cell morphological differentiation associated with biofilm development and directed by quorum-sensing autoinducers.

Supporting Information

Figure S1 AFM images (height signal) of *C. violaceum* NCTC 13274 growing for 24 (a, d), 48 (b, e) and 72 h (c, f) in planktonic conditions without (a–c) and with 0.1 μM C_6-HSL (d–f). Scale bar – 1 μm.

Author Contributions

Conceived and designed the experiments: DGD. Performed the experiments: AAK ASV. Analyzed the data: AAK ASV DGD. Contributed reagents/materials/analysis tools: DGD. Contributed to the writing of the manuscript: DGD ASV.

References

1. Gallis M, Logan NF (2005) *Chromobacterium bergonzini* 1881, 153 AL. In DJ Brenner, NR Krieg, JT Staley (ed) Bergey's manual of systematic bacteriology, vol 2, 2nd edn. Springer, New York, pp 824–827.
2. Antunes LCM, Ferreira RBR, Buckner MMC, Finlay BB (2010) Quorum sensing in bacterial virulence. Microbiology 156:2271–2282.
3. Rajalakshmi G, Sankaravadivoo A, Prabhakaran S (2011) Characterization of *Chromobacterium violaceum* isolated from spoiled vegetables and antibiogram of violacein. J Adv Lab Res Bio II:22–27.
4. Richard C (1993) *Chromobacterium violaceum*, opportunist pathogenic bacteria in tropical and subtropical regions. Bull Soc Pathol Exot 86:169–173.
5. Lee J, Kim JS, Nahm CH, Choi JW, Kim J, et al. (1999) Two cases of *Chromobacterium violaceum* infection after injury in a subtropical region. Journal of Clinical Microbiology 37:2068–2070.
6. Dyer NW, Krogh DF, DeVold R, Wilson SL, White DG (2000) Chromobacteriosis in a Chinese red panda (*Ailurus fulgens styani*). J Vet Diagn Invest 12:177–179.
7. Durán N, Menck CF (2001) *Chromobacterium violaceum*: a review of pharmacological and industiral perspectives. Critical Reviews in Microbiology 27:201–222.
8. Yang C-H, Li Y-H (2011) *Chromobacterium violaceum* infection: A clinical review of an important but neglected infection. Journal of the Chinese Medical Association 74:435–441.
9. Brazilian National Genome Project Consortium (2003) The complete genome sequence of *Chromobacterium violaceum* reveals remarkable and exploitable bacterial adaptability. Proc Natl Acad Sci U S A 100:11660–11665.
10. Silva R, Araripe JR, Rondinelli E, Ürményi TP (2004) Gene expression in *Chromobacterium violaceum*. Genet Mol Res 3:64–75.
11. Hungria M, Nicolás MF, Guimarães CT, Jardim SN, Gomes EA, et al. (2004) Tolerance to stress and environmental adaptability of *Chromobacterium violaceum*. Genet Mol Res 3:102–116.
12. Alves de Brito CF, Carvalho CMB, Santos FR, Gazzinelli RT, Oliveira SC, et al. (2004) *Chromobacterium violaceum* genome: molecular mechanisms associated with pathogenicity. Genet Mol Res 3:148–161.
13. Fuqua C, Winans SC, Greenberg EP (1994) Quorum sensing in bacteria: the LuxR-LuxI family of cell-density responsive transcriptional regulators. J Bacteriol 176:269–275.
14. Fuqua C, Winans SC, Greenberg EP (1996) Census and consensus in bacterial ecosystems: the LuxR-LuxI family of quorum-sensing transcriptional regulators. Annu Rev Microbiol 50:727–751.
15. Fuqua C, Parsek MR, Greenberg EP (2001) Regulation of gene expression by cell-to-cell communication: acyl-homoserine lactone quorum sensing. Annu Rev Genet 35:439–468.
16. Stauff DL, Bassler BL (2011) Quorum sensing in *Chromobacterium violaceum*: DNA recognition and gene regulation by the CviR receptor. J Bacteriol 193:3871–3878.
17. Hoshino T (2011) Violacein and related tryptophan metabolites produced by *Chromobacterium violaceum*: biosynthetic mechanism and pathway for construction of violacein core. Appl Microbiol Biotechnol 91:1463–1475.
18. Balibar CJ, Walsh CT (2006) In vitro biosynthesis of violacein from L-tryptophan by the enzymes VioA-E from *Chromobacterium violaceum*. Biochemistry 45:15444–15457.
19. Hoiby N, Johansen HK, Ciofu O, Jensen RØ, Bjarnsholt T, et al. (2007) Foreign body infections - biofilms and quorum sensing. Ugeskrift for laeger 169:4163–4166.
20. Becker S, Soares C, Porto LM (2009) Computational analysis suggests that virulence of *Chromobacterium violaceum* might be linked to biofilm formation and poly-NAG biosynthesis. Genet Mol Res 32:640–644.
21. Chernin LS, Winson MK, Thompson JM, Haran S, Bycroft BW, et al. (1998) Chitinolytic activity in *Chromobacterium violaceum*: substrate analysis and regulation by quorum sensing. J Bacteriol 180:4435–4441.
22. Fuqua C, Greenberg P (2002) Listening in on bacteria: acyl-homoserine lactone signalling. Molecular cell biology 3:685–695.
23. Steindler L & Venturi V (2007) Detection of quorum-sensing N-acyl homoserine lactone signal molecules by bacterial biosensors. FEMS Microbiol Lett 266:1–9.
24. Wang Y, Ikawa A, Okaue S, Taniguchi S, Osaka I, et al. (2008) Quorum sensing signaling molecules involved in the production of violacein by *Pseudoalteromonas*. Biosci Biotechnol Biochem 72:1958–1961.
25. O'Toole GA (2011) Microtiter dish biofilm formation assay. Journal of Visualized Experiments 47:1–2.
26. O'Toole GA, Kolter R (1998) The initiation of biofilm formation in *Pseudomonas fluorescens* WCS365 proceeds via multiple, convergent signalling pathways: a genetic analysis. Mol Microbiol 30:295–304.
27. Nikiyan HN, Vasilchenko AS, Deryabin DG (2010) Humidity-dependent bacterial cells functional morphometry investigations using atomic force microscope. Int J Microbiol. doi:10.1155/2010/704170.
28. Davey ME, O'Toole GA (2000) Microbial biofilms: from ecology to molecular genetics. Microbiol Mol Biol Rev 64:847–867.
29. Donlan RM (2002) Biofilms: Microbial life on surfaces. Emerging Infect Dis 8:881–890.
30. Vu B, Chen M, Crawford RJ, Ivanova EP (2009) Bacterial extracellular polysaccharides involved in biofilm formation. Molecules 14:2535–2554.
31. de Kievit TR, Iglewski H (2000) Bacterial Quorum sensing in pathogenic relationships. Infection and Immunity 68:4839–4849.
32. Waters CM, Bassler BL (2005) Quorum sensing: cell-to-cell communication in bacteria. Annu Rev Cell Dev Biol 21:319–346.
33. Nadell CD, Xavier JB, Levin SA, Foster KR (2008) The evolution of Quorum sensing in bacterial biofilms. PLOS Biology 6:e14.
34. Antunes AA, Silva MLRB, da Silva CAA, de Campos-Takaki GM (2006) Characterization of *Chromobacterium violaceum* isolated from Paca River, Pernambuco, Brazil. Revista de Biologia e Ciências da Terra 1:48–55.
35. Hooshangi S, Bentley WE (2008) From unicellular properties to multicellular behavior: bacteria quorum sensing circuitry and applications. Curr Opin Biotechnol 19:550–555.
36. Schuster M, Losrtoh CP, Ogi T, Greenberg EP (2003) Identification, timing, and signal specificity of *Pseudomonas aeruginosa* quorum-controlled genes: a transcriptome analysis. Journal of Bacteriology 185:2066–2079.
37. Wagner VE, Li LL, Isabella VM, Iglewski BH (2007) Analysis of the hierarchy of quorum-sensing regulation in *Pseudomonas aeruginosa*. Anal Bioanal Chem 387:469–479.
38. Davies DG, Parsek MR, Pearson JP, Iglewski BH, Costerton JW, et al. (1998) The involvement of cell-to-cell signals in the development of a bacterial biofilm. Science 280:295–298.
39. Purevdorj B, Costerton JW, Stoodley P (2002) Influence of hydrodynamics and cell signaling on the structure and behavior of *Pseudomonas aeruginosa* biofilms. Appl Environ Microbiol 68:4457–4464.
40. Heydorn A, Ersboll B, Kato J, Hentzer M, Parsek MR, et al. (2002) Statistical analysis of *Pseudomonas aeruginosa* biofilm development: impact of mutations in genes involved in twitching motility, cell-to-cell signaling, and stationary-phase sigma factor expression. Appl Environ Microbiol 68:2008–2017.
41. Balasubramanian D, Schneper L, Kumari H, Mathee K (2013) A dynamic and intricate regulatory network determines *Pseudomonas aeruginosa* virulence. Nucleic Acids Research 41:1–20.
42. Rucinsky TE, Cota-Robles EH (1974) Mesosome structure in *Chromobacterium violaceum*. Journal of Bacteriology 118:717–724.
43. Aschtgen MS, Gavioli M, Dessen A, Lloubès R, Cascales E. (2010) The SciZ protein anchors the enteroaggregative *Escherichia coli* Type VI secretion system to the cell wall. Mol Microbiol. 75(4):886–99.
44. Zhang ML, Hinz AJ, Nadeau J-P, Mah T-F (2010) *Pseudomonas aeruginosa tssC1* links type VI secretion and biofilm-specific antibiotic resistance. Journal of Bacteriology 193:5510–5513.
45. Zhurina MV, Kostrikina NA, Parshina EY, Strelkova EA, Yusipovich AI, et al. (2013) Visualization of the extracellular polymeric matrix of *Chromobacterium violaceum* biofilms by microscopic methods. J Microbiology 82:517–524.
46. Yonezawa H, Osaki T, Kurata S, Fukuda M, Kawakami H, et al. (2009) Outer Membrane vesicles of *Helicobacter pylori* TK1402 are involved in biofilm formation. BMC Microbiology 9:197 doi:10.1186/1471-2180-9-197.
47. Baumgarten T, Sperling S, Seifert J, von Bergen M, Steiniger F, et al. (2012) Membrane vesicle formation as a multiple-stress response mechanism enhances *Pseudomonas putida* DOT-T1E cell surface hydrophobicity and biofilm formation. Appl. Environ. Microbiol 78: 6217–6224.

PERMISSIONS

All chapters in this book were first published in PLOS ONE, by The Public Library of Science; hereby published with permission under the Creative Commons Attribution License or equivalent. Every chapter published in this book has been scrutinized by our experts. Their significance has been extensively debated. The topics covered herein carry significant findings which will fuel the growth of the discipline. They may even be implemented as practical applications or may be referred to as a beginning point for another development.

The contributors of this book come from diverse backgrounds, making this book a truly international effort. This book will bring forth new frontiers with its revolutionizing research information and detailed analysis of the nascent developments around the world.

We would like to thank all the contributing authors for lending their expertise to make the book truly unique. They have played a crucial role in the development of this book. Without their invaluable contributions this book wouldn't have been possible. They have made vital efforts to compile up to date information on the varied aspects of this subject to make this book a valuable addition to the collection of many professionals and students.

This book was conceptualized with the vision of imparting up-to-date information and advanced data in this field. To ensure the same, a matchless editorial board was set up. Every individual on the board went through rigorous rounds of assessment to prove their worth. After which they invested a large part of their time researching and compiling the most relevant data for our readers.

The editorial board has been involved in producing this book since its inception. They have spent rigorous hours researching and exploring the diverse topics which have resulted in the successful publishing of this book. They have passed on their knowledge of decades through this book. To expedite this challenging task, the publisher supported the team at every step. A small team of assistant editors was also appointed to further simplify the editing procedure and attain best results for the readers.

Apart from the editorial board, the designing team has also invested a significant amount of their time in understanding the subject and creating the most relevant covers. They scrutinized every image to scout for the most suitable representation of the subject and create an appropriate cover for the book.

The publishing team has been an ardent support to the editorial, designing and production team. Their endless efforts to recruit the best for this project, has resulted in the accomplishment of this book. They are a veteran in the field of academics and their pool of knowledge is as vast as their experience in printing. Their expertise and guidance has proved useful at every step. Their uncompromising quality standards have made this book an exceptional effort. Their encouragement from time to time has been an inspiration for everyone.

The publisher and the editorial board hope that this book will prove to be a valuable piece of knowledge for researchers, students, practitioners and scholars across the globe.

LIST OF CONTRIBUTORS

Ruchi Jain
Department of Internal Medicine (Infectious Diseases), Yale University School of Medicine, New Haven, Connecticut, United States of America

Barbara I. Kazmierczak
Department of Internal Medicine (Infectious Diseases), Yale University School of Medicine, New Haven, Connecticut, United States of America
Department of Microbial Pathogenesis, Yale University School of Medicine, New Haven, Connecticut, United States of America

Dagbjort H. Petursdottir, Olga D. Chuquimia., Raphaela Freidl and Carmen Fernández
Department of Molecular Biosciences, The Wenner-Gren Institute, Stockholm University, Stockholm, Sweden

Renato Pariz Maluta, Wanderley Dias da Silveira and Thaís Cabrera Galvão Rojas
Bacterial Molecular Biology Laboratory, Department of Genetics, Evolution and Bioagents, Institute of Biology, State University of Campinas (UNICAMP), Campinas, SP, Brazil

Catherine Mary Logue, Ting Meng and Lisa Kay Nolan
Department of Veterinary Microbiology and Preventive Medicine, College of Veterinary Medicine, Iowa State University, Ames, Iowa, United States of America

Monique Ribeiro Tiba Casas
Instituto Adolfo Lutz, São Paulo, SP, Brazil

Elisabete Aparecida Lopes Guastalli
Instituto Biológico, CAPTAA, Unidade de Pesquisa e Desenvolvimento de Bastos, Bastos, SP, Brazil

Augusto Cezar Montelli
Departamento de Clínica Médica, Faculdade de Medicina, Universidade Estadual Paulista, Botucatu, SP, Brazil

Teruê Sadatsune
Departamento de Microbiologia e Imunologia, Instituto de Biociências, Universidade Estadual Paulista, Botucatu, SP, Brazil

Marcelo de Carvalho Ramos
Department of Internal Medicine, UNICAMP, Campinas, Brazil

Sílvio B. Santos, Carla Carvalho, Joana Azeredo and Eugénio C. Ferreira
Centre of Biological Engineering, Universidade do Minho, Braga, Portugal

Runu Ghosh, Bipransh Kumar Tiwary, Anoop Kumar and Ranadhir Chakraborty
OMICS Laboratory, Department of Biotechnology, University of North Bengal, Siliguri, West Bengal, India

Rebekah F. Hare
Department of Biology and Wildlife, Institute of Arctic Biology, University of Alaska Fairbanks, Fairbanks, Alaska, United States of America

Karsten Hueffer
Department of Veterinary Medicine, University of Alaska Fairbanks, Fairbanks, Alaska, United States of America

Albert Mayola, Oihane Irazoki, Jordi Barbé and Susana Campoy
Departament de Genética i de Microbiologia, Universitat Autónoma de Barcelona, Bellaterra (Cerdanyola del Vallès), Spain

Ignacio A. Martínez and Dmitri Petrov
ICFO-Institut de Ciències Fotóniques, Castelldefels, Spain

Filippo Menolascina and Roman Stocker
Department of Civil and Environmental Engineering, Massachusetts Institute of Technology, Cambridge, Massachusetts, United States of America

José A. Reyes-Darias and Tino Krell
Department of Environmental Protection, Estación Experimental del Zaidín-Consejo Superior de Investigaciones Científicas, Granada, Spain

Pingping Zhang, Ruifu Yang, Lei Zhou and Yong Zhao
Laboratory of Analytical Microbiology, State Key Laboratory of Pathogen and Biosecurity, Beijing Institute of Microbiology and Epidemiology, Beijing, People's Republic of China

Xiao Liu
Laboratory of Analytical Microbiology, State Key Laboratory of Pathogen and Biosecurity, Beijing Institute of Microbiology and Epidemiology, Beijing, People's Republic of China
School of Public Health and Management, Chongqing Medical University, Chongqing, People's Republic of China

Chengbin Wang and Chunfeng Li
Department of Clinical Laboratory, Chinese People's Liberation Army General Hospital, Beijing, People's Republic of China

Fei Hua
Department of Etiology, Taishan Medical University, Tai'an, People's Republic of China

Loredana Scalschi, Gemma Camañes, Eugenio Llorens, Emma Fernández-Crespo, Pilar García-Agustín and Begonya Vicedo
Grup de Bioquímica i Biotecnologia, Area de Fisiologia Vegetal, Departament de Cie`ncies Agra`ries i del Medi Natural, ESTCE, Universitat Jaume I, Castellón, Spain

María M. López
Departamento de Protección Vegetal y Biotecnología, Instituto Valenciano de Investigaciones Agrarias (IVIA), Apartado Oficial, Moncada, Valencia, Spain

Ramón Alberto Batista-García, Edgar Balcázar-Ló pez, Ayixón Sánchez-Reyes and Laura Cuervo-Soto
Facultad de Ciencias, Universidad Autónoma del Estado de Morelos, Cuernavaca, Morelos, Mexico
Centro de Investigación en Biotecnología, Universidad Autónoma del Estado de Morelos, Cuernavaca, Morelos, México

Estefan Miranda-Miranda
Centro Nacional de Investigación Disciplinaria en Parasitología Veterinaria, Instituto Nacional de Investigaciones Forestales Agrícolas y Pecuarias, Cuernavaca, Morelos, Mexico

Denise Aceves-Zamudio, Catalina Morales-Herrera, Rocío Rodríguez-Hernández and Jorge Folch-Mallol
Centro de Investigación en Biotecnología, Universidad Autónoma del Estado de Morelos, Cuernavaca, Morelos, México

Karina Atriztán-Hernández
Facultad de Ciencias Biológicas, Universidad Autónoma del Estado de Morelos, Cuernavaca, Morelos,
Mexico

Liu-Hui Fu, Kang-Di Hu, Lan-Ying Hu, Yan-Hong Li, Yong-Sheng Liu and Hua Zhang
School of Biotechnology and Food Engineering, Hefei University of Technology, Hefei, China

Liang-Bin Hu
School of Food Science, Henan Institute of Science and Technology, Xinxiang, China

Hong Yan
College of Chemical and Environmental Engineering, Harbin University of Science and Technology, Key Laboratory of Green Chemical Technology of College of Heilongjiang Province, Harbin, China

Alberto Luini
Institute of Protein Biochemistry, CNR, Naples, Italy

Lydia M. Bogomolnaya
Department of Microbial Pathogenesis and Immunology, College of Medicine, Texas A&M University, Bryan, Texas, United States of America
Institute of Fundamental Medicine and Biology, Kazan Federal University, Kazan, Russia

Lindsay Aldrich, Yuri Ragoza, Marissa Talamantes, Katharine D. Andrews and Helene L. Andrews-Polymenis
Department of Microbial Pathogenesis and Immunology, College of Medicine, Texas A&M University, Bryan, Texas, United States of America

Michael McClelland
Department of Microbiology and Molecular Genetics, University of California Irvine, Irvine, California,
United States of America

Giuseppe Bozza, Silvana Savino, Beatrice Aricó, Rino Rappuoli and Mariagrazia Pizza1 Paolo Montanari1, Barbara Benucci1, Marco Biancucci1¤b, Vincenzo Nardi-Dei1, Elena Caproni1, Riccardo Barrile1¤c,
Novartis Vaccines, Siena, Italy

Mirco Capitani, Benedetta Picciani and Michele Sallese
Unit of Genomic Approaches to Membrane Traffic, Fondazione Mario Negri Sud, S. Maria Imbaro (CH), Italy

Marcello Merola
Novartis Vaccines, Siena, Italy
Department of Biology, University of Naples "Federico II", Naples, Italy

Huijuan Hua, Fuguo Xing, Jonathan Nimal Selvaraj, Yan Wang, Yueju Zhao, Lu Zhou, Xiao Liu and Yang Liu
Institute of Agro-Products Processing Science and Technology, Chinese Academy of Agricultural Sciences/Key Laboratory of Agro-Products Processing, Ministry of Agriculture, Beijing, P. R. China

Raquel BranquinhoLuísa V. Peixe
REQUIMTE, Laboratório de Microbiologia, Departamento de Ciências Biológicas, Faculdade de Farmácia, Universidade do Porto, Porto, Portugal

Hugo Osório
IPATIMUP, Instituto de Patologia e Imunologia Molecular da Universidade do Porto, Porto, Portugal
Faculdade de Medicina, Universidade do Porto, Porto, Portugal

Jing Geng and Nelly Henry
Laboratoire Jean Perrin (CNRS FRE 3231), UPMC, Paris, France

Clara Sousa
REQUIMTE, Laboratório de Microbiologia, Departamento de Ciências Biológicas, Faculdade de Farmácia, Universidade do Porto, Porto, Portugal
CEB, Centro de Engenharia Biológica, Universidade do Minho, Braga, Portugal

João Lopes
REQUIMTE, Laboratório de Química Aplicada, Departamento de Ciências Químicas, Faculdade de Farmácia, Universidade do Porto, Porto, Portugal
CBQF, Centro de Biotecnologia e Química Fina, Escola Superior de Biotecnologia, Universidade Católica Portuguesa, Porto, Portugal

Manuela E. Pintado
Departmento de Tecnologia Farmacêutica, Faculdade Farmácia, Universidade de Lisboa, Lisboa, Portugal
REQUIMTE, Laboratório de Microbiologia, Departamento de Ciências Biológicas, Faculdade de Farmácia, Universidade do Porto, Porto, Portugal

Christophe Beloin and Jean-Marc Ghigo
Institut Pasteur, Unitéde Génétique des Biofilms, De´partement de Microbiologie, Paris, France

Miroslava Petrovova, Lukas Dvoracek and Irena Licha
Department of Genetics and Microbiology, Faculty of Science, Charles University, Prague, Czech Republic

Jan Tkadlec and Eliska Streitova
Department of Genetics and Microbiology, Faculty of Science, Charles University, Prague, Czech Republic
Department of Medical Microbiology, 2nd Faculty of Medicine, Charles University, Prague, Czech Republic

Takaaki B. Narita
Graduate School of Science and Technology, Sophia University, Tokyo, Japan
Research Fellow of Japan Society for the Promotion of Science (DC2), Tokyo, Japan

Zhi-hui Chen and Pauline Schaap
College of Life Sciences, University of Dundee, Dundee, United Kingdom

Tamao Saito
Department of Materials and Life Sciences, Sophia University, Tokyo, Japan

Luísa Côrte, Filipa Valente, Mónica Serrano, Cláudio M. Gomes and Adriano O. Henriques
Instituto de Tecnologia Química e Biológica, Universidade Nova de Lisboa, Oeiras, Portugal

Charles P. Moran Jr
Department of Microbiology and Immunology, Emory University School of Medicine, Atlanta, Georgia, United States of America

Anara A. Kamaeva
Department of Microbiology, Orenburg State University, Orenburg, Russia
RSE «Republican Collection of Microorganisms», Astana, Republic of Kazakhstan

Alexey S. Vasilchenko
Department of Microbiology, Orenburg State University, Orenburg, Russia
Laboratory of Disbiosis, Institute of Cellular and Intracellular Symbiosis, RAS, Orenburg, Russia

Dmitry G. Deryabin
Department of Microbiology, Orenburg State University, Orenburg, Russia

Index